T0259813

Lecture Notes in Artificial Intelligence 11511

Subseries of Lecture Notes in Computer Science

More information about this series at http://www.springer.com/series/1244

Diana Cristea · Florence Le Ber ·
Baris Sertkaya (Eds.)

Formal Concept Analysis

15th International Conference, ICFCA 2019 Frankfurt, Germany, June 25–28, 2019 Proceedings

Editors
Diana Cristea
Babeş-Bolyai University
Cluj-Napoca, Romania

Florence Le Ber
University of Strasbourg, ENGEES
Strasbourg, France

Baris Sertkaya
Frankfurt University of Applied Sciences
Frankfurt, Germany

ISSN 0302-9743 ISSN 1611-3349 (electronic)
Lecture Notes in Artificial Intelligence
ISBN 978-3-030-21461-6 ISBN 978-3-030-21462-3 (eBook)
https://doi.org/10.1007/978-3-030-21462-3

LNCS Sublibrary: SL7 – Artificial Intelligence

This Springer imprint is published by the registered company Springer Nature Switzerland AG
The registered company address is: Gewerbestrasse 11, 6330 Cham, Switzerland

Preface

This volume features the papers accepted for the 15th International Conference on Formal Concept Analysis (ICFCA 2019), held during June 25–28, 2019, at Frankfurt University of Applied Sciences, Germany. Formal concept analysis (FCA) is a mathematical field rooted in lattice and order theory which, although being of a theoretical nature, has proved to be of interest to various applied fields such as knowledge discovery and data mining, database theory, data visualization, and many others.

The goal of the International Conference on Formal Concept Analysis is to offer researchers from different backgrounds the possibility to present and discuss their research related to FCA. Since the first ICFCA conference in 2003 in Darmstadt, Germany, it has been held annually in different countries in Europe, Africa, America, and Australia. Since 2015, ICFCA has been held once every two years, alternatively with the Conference on Concept Lattices and Their Applications (CLA).

The field of FCA originated in the 1980s in Darmstadt as a subfield of mathematical order theory, with prior developments in other research groups. Its original motivation was to consider complete lattices as lattices of concepts, drawing motivation from philosophy and mathematics alike. FCA has since then developed into a wide research area with applications far beyond its original motivation, for example, in logic, data mining, learning, and psychology.

There were 36 submissions by authors from 14 different countries. Each paper was reviewed by three members of the Program Committee (in a few cases, four). Fifteen high-quality papers were chosen for publication in this volume, amounting to an acceptance rate of 42%. Eight other works in progress were considered valuable for presentation during the conference, five included as short papers in this volume and three in a supplementary volume.

The papers included in this volume are divided into different sections: The ones in the "Theory" section discuss various theoretical aspects of FCA. In the section "Methods and Applications" you will find papers that present applications of FCA in several different application areas. Finally, in the section "Enhanced FCA" you will find papers that present new trends in FCA, for instance, FCA-based methods for dealing with graphs or relational data.

In addition to the regular presentations, we were delighted to have invited talks by the following four renowned researchers, whose papers you will also find in this volume:

- "Too Much Information: Can AI Cope With Modern Knowledge Graphs?" by Markus Krötzsch (Germany)
- "Concepts in Application Context" by Steffen Staab (Germany)
- "Learning Implications from Data and from Queries" by Sergei Obiedkov (Russia)
- "Elements About Exploratory, Knowledge-Based, Hybrid, and Explainable Knowledge Discovery" by Amedeo Napoli and Miguel Couceiro (France)

Our deepest gratitude goes to all the authors of submitted papers. Choosing ICFCA 2019 as a forum to publish their research was key to the success of the conference. Besides the submitted papers, the high quality of this volume would not have been possible without the strong commitment of the authors, the Program Committee and Editorial Board members, and the external reviewers. Working with the efficient and capable team of local organizers was a constant pleasure. We are deeply indebted to all of them for making this conference a successful forum on FCA.

Last, but not least, we are very grateful to Springer for showing their support of the International Conference on Formal Concept Analysis, as well to the organizations that sponsored this event, namely, Frankfurt University of Applied Sciences and the Deutsche Forschungsgemeinschaft (DFG, German Research Foundation - Project number 419189247). Finally, we would like to emphasize the great help of EasyChair for making the technical duties easier.

June 2019

Diana Cristea
Florence Le Ber
Barış Sertkaya

Organization

Executive Committee

Conference Organizing Committee

Conference Chair

Barış Sertkaya	Frankfurt University of Applied Sciences, Germany

Local Organization

Martine Robert	Frankfurt University of Applied Sciences, Germany

Program and Conference Proceedings

Program Chairs

Diana Cristea	Babeş-Bolyai University, Cluj-Napoca, Romania
Florence Le Ber	Université de Strasbourg, ENGEES, France

Editorial Board

Jaume Baixeries	Polytechnic University of Catalonia, Spain
Karell Bertet	L3I, Université de La Rochelle, France
Peggy Cellier	IRISA, INSA Rennes, France
Florent Domenach	Akita International University, Japan
Sebastien Ferré	Université de Rennes 1, France
Bernhard Ganter	Technische Universität Dresden, Germany
Cynthia-Vera Glodeanu	Technische Universität Dresden, Germany
Robert Godin	Université du Québec à Montréal, Canada
Dmitry Ignatov	Higher School of Economics, Moscow, Russia
Mehdi Kaytoue	Université de Lyon, France
Sergei O. Kuznetsov	Higher School of Economics, Russia
Leonard Kwuida	Bern University of Applied Sciences, Switzerland
Rokia Missaoui	Université du Québec en Outaouais, Canada
Amedeo Napoli	LORIA, Nancy, France
Sergei Obiedkov	Higher School of Economics, Russia
Manuel Ojeda-Aciego	Universidad de Málaga, Spain
Uta Priss	Ostfalia University of Applied Sciences, Germany
Sebastian Rudolph	Technische Universität Dresden, Germany
Christian Sacarea	Babes-Bolyai University, Cluj-Napoca, Romania
Stefan E. Schmidt	Technische Universität Dresden, Germany
Barış Sertkaya	Frankfurt University of Applied Sciences, Germany
Gerd Stumme	Kassel Universität, Germany
Petko Valtchev	Université du Québec à Montréal, Canada
Karl Erich Wolff	University of Applied Sciences, Germany

Honorary Member

Vincent Duquenne	ECP6-CNRS, Université Paris 6, France

Program Committee

Simon Andrews	University of Sheffield, UK
Agnès Braud	ICube, Université de Strasbourg, France
Aleksey Buzmakov	Higher School of Economics, Moscow, Russia
Claudio Carpineto	Fondazione Ugo Bordoni, Italy
Victor Codocedo	Universidad Técnica Federico Santa María, Chile
Jean Diatta	Université de la Réunion, France
Christophe Demko	L3I lab, Université de La Rochelle, France
Xavier Dolques	Université de Strasbourg, ENGEES, France
Alain Gély	Université Paul Verlaine, France
Tom Hanika	Kassel Universität, Germany
Marianne Huchard	LIRMM, Université Montpellier, France
Michal Krupka	Palacký University, Czech Republic
Marzena Kryszkiewicz	Warsaw University of Technology, Poland
Wilfried Lex	Universität Clausthal, Germany
Jesús Medina	Universidad de Cádiz, Spain
Engelbert Mephu Nguifo	LIMOS, Université Clermont Auvergne, France
Lhouari Nourine	Université Clermont Auvergne, France
Jan Outrata	Palacký University, Czech Republic
Jean-Marc Petit	LIRIS, INSA Lyon, France
Pascal Poncelet	LIRMM, Université de Montpellier, France
Sandor Radeleczki	University of Miskolc, Hungary
Henry Soldano	Université Paris 13, France
Laszlo Szathmary	University of Debrecen, Hungary
Andreja Tepavčević	University of Novi Sad, Serbia
Jean-François Viaud	L3I lab, Université de La Rochelle, France

Additional Reviewers

Alexandre Bazin	Université Clermont Auvergne, France
Inma P. Cabrera	Universidad de Málaga, Spain
Pablo Cordero	Universidad de Málaga, Spain
Maria Eugenia Cornejo Piñero	Universidad de Cádiz, Spain
Oscar Defrain	Université Clermont Auvergne, France
Maximilian Felde	Kassel Universität, Germany
Dhouha Grissa	University of Copenhagen, Denmark
Eloisa Ramírez Poussa	Universidad de Cádiz, Spain
Christophe Rey	Université Clermont Auvergne, France
Maximilian Stubbemann	Kassel Universität, Germany
Lauraine Tiogning	University of Yaounde 1, Cameroon
Norbert Tsopze	University of Yaounde 1, Cameroon
Simon Vilmin	Université Clermont Auvergne, France

Sponsoring Institutions

Frankfurt University of Applied Sciences
Deutsche Forschungsgemeinschaft (DFG, German Research Foundation)

Contents

Invited Papers

Theory

Methods and Applications

Invited Papers

Elements About Exploratory, Knowledge-Based, Hybrid, and Explainable Knowledge Discovery

Miguel Couceiro(✉) and Amedeo Napoli

Université de Lorraine, CNRS, Inria, LORIA, 54000 Nancy, France
{miguel.couceiro,amedeo.napoli}@inria.fr

Abstract. Knowledge Discovery in Databases (KDD) and especially pattern mining can be interpreted along several dimensions, namely data, knowledge, problem-solving and interactivity. These dimensions are not disconnected and have a direct impact on the quality, applicability, and efficiency of KDD. Accordingly, we discuss some objectives of KDD based on these dimensions, namely exploration, knowledge orientation, hybridization, and explanation. The data space and the pattern space can be explored in several ways, depending on specific evaluation functions and heuristics, possibly related to domain knowledge. Furthermore, numerical data are complex and supervised numerical machine learning methods are usually the best candidates for efficiently mining such data. However, the work and output of numerical methods are most of the time hard to understand, while symbolic methods are usually more intelligible. This calls for hybridization, combining numerical and symbolic mining methods to improve the applicability and interpretability of KDD. Moreover, suitable explanations about the operating models and possible subsequent decisions should complete KDD, and this is far from being the case at the moment. For illustrating these dimensions and objectives, we analyze a concrete case about the mining of biological data, where we characterize these dimensions and their connections. We also discuss dimensions and objectives in the framework of Formal Concept Analysis and we draw some perspectives for future research.

1 Introduction

Knowledge discovery in databases (KDD) consists in processing possibly large volumes of data in order to discover patterns that can be significant and reusable. It is usually based on three main steps: data preparation, data mining, and interpretation of the extracted patterns (Fig. 1) [48,56]. KDD is interactive and iterative, controlled by an analyst who is a specialist of the domain and is in charge of selecting data and patterns, setting thresholds (frequency, confidence), replaying the process at each step whenever needed, depending on the interpretation of the selected patterns.

In the following, we consider four main dimensions within KDD which are based on data, knowledge, problem-solving, and interactivity.

D. Cristea et al. (Eds.): ICFCA 2019, LNAI 11511, pp. 3–16, 2019.
https://doi.org/10.1007/978-3-030-21462-3_1

- The data dimension: knowledge discovery is data-oriented by nature. This dimension is related to the input of knowledge discovery and involves data preparation, e.g. feature selection, dimensionality reduction, and data transformation. The exploration of the data space and of pattern space are main operations. Moreover, the diversity and quality of data have an influence on the whole KDD process.
- The knowledge dimension [34]: data are related to a particular domain. Hence knowledge discovery is knowledge-oriented and depends on domain knowledge that can be expressed, e.g., as constraints, relations, and preferences. The knowledge dimension is also attached to the control of KDD, possibly involving "meta-mining" [11]. Moreover, the output of KDD, i.e. the discovered patterns, may be represented as actionable knowledge units.
- The problem-solving dimension [9]: knowledge discovery is intended to solve various tasks for human or software agents and may be guided by the task at hand. The problem-solving dimension is dependent on iteration and search strategies. It can be data-directed, pattern-directed or goal-directed, and it can rely on declarative or procedural approaches.
- Interactivity [10]: knowledge discovery is interactive as the analyst may integrate constraints and preferences for guiding the data exploration, especially for minimizing the exploration of the data and pattern spaces. Interaction plays also a role in the evaluation of the quality of the patterns and in the activation of the different replay loops.

These four dimensions are interconnected and correspondences between them can be made explicit, especially in the framework of pattern mining [1]. Such correspondences support the following objectives of KDD:

- KDD is exploratory: the data dimension is related to an interactive exploration of the data and pattern spaces. We should be able to identify "seeds" or "prototypes" for guiding the pattern and data space exploration. Moreover, this exploration should be consistent w.r.t. domain knowledge and associated constraints. Threshold issues w.r.t. analyst queries can be addressed thanks to a skyline analysis within the pattern space or by integrating preferences.
- KDD is knowledge-based: domain knowledge is related to control, i.e. meta-mining, constraints and preferences, explanations, and production of knowledge, materializing the links between knowledge discovery and knowledge engineering. We should be able to define environments within which knowledge discovery and knowledge engineering can be combined in an efficient and operational way.
- KDD is hybrid: and may rely on the combination of numerical and symbolic data mining algorithms for solving problems. In addition, supervised and unsupervised learning methods can interact and be combined as well.
- KDD is expected to be explainable: the output of knowledge discovery may be of different types, e.g. rules and classes or concepts, which can be reused for solving problems and decision making. In this way, elements supporting a decision –and especially an algorithmic decision– should be available [27,44,55].

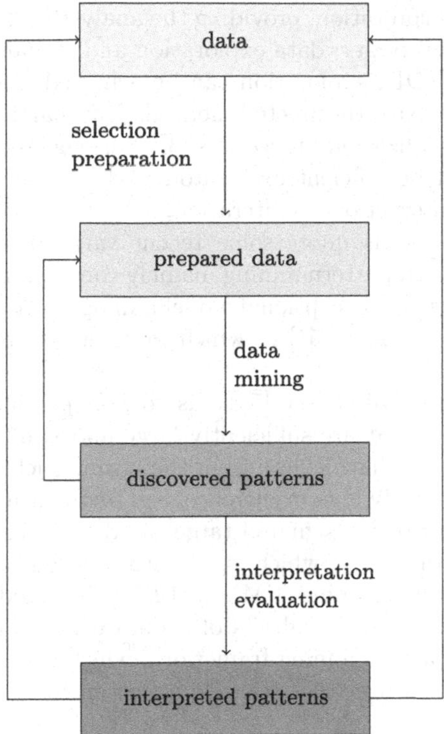

Fig. 1. The KDD loop.

In the following sections, we survey in more detail these different objectives and discuss their importance and their materialization. In the last section of this paper, we propose a synthesis that illustrates how Formal Concept Analysis (FCA [23]) can, more or less, fulfill these objectives. The paper terminates with a large bibliography which illustrates the content of this paper and some relations existing between pattern mining and FCA.

2 KDD Should Be Exploratory

In the KDD loop, exploration is related to data mining where the data space is searched and the patterns are mined. Exploration is also related to the notions of interaction and iteration, involving "replay", which is of main importance within knowledge discovery. This is discussed under different names in the literature, e.g. "exploratory data mining" and the loop "Mine, Interact, Learn, and Repeat" in [8,52], "interactive data mining" in [32], "declarative approaches in data mining" [9], and "exploratory knowledge discovery" in [3] (this list is certainly not exhaustive).

All these approaches are based on interaction and go back to the ideas underlying "exploratory data analysis" (EDA [50]). The goal of EDA is to improve data analysis and result interpretation, providing the analyst with suitable techniques based on computational power, data exploration and visualization methods.

In the context of KDD, exploration can be achieved in various ways, using either data-directed or pattern-directed methods [13], particular interestingness measures [12], and visualization procedures [3]. Nonetheless, the knowledge discovery process should be efficient and automated as much as possible, while keeping facilities for interaction and iteration.

In the same way, let us quote some recent variations about the current exploratory approaches in pattern mining, namely constraint-based pattern mining, subgroup discovery, and exceptional model mining. Constraint-based pattern mining is based on skylines in [51] in which preferences are expressed w.r.t. a dominance relation.

The goal of subgroup discovery [5,41] is to find particular descriptions of subsets of a population that are sufficiently large and statistically unusual, i.e. subsets of the population that deviate from the norm. Such deviations are measured in terms of a relatively high occurrence, e.g. frequent itemset mining, or an unusual distribution for one designated target attribute. The latter is the common use of subgroup discovery, which is, in turn, related to exceptional model mining. Exceptional Model Mining (EMM [6,19,39]) is aimed at capturing a general notion of interestingness in subsets of a dataset. EMM can be considered as a supervised local pattern mining framework, where several target attributes are selected, and a model over these targets is chosen to be the target concept.

3 KDD Should Be Knowledge-Based

There are many links between knowledge discovery and knowledge engineering. At each step of KDD, domain knowledge can be used to guide and to complete the process given in Fig. 1. Actually, a fourth step can be considered in the KDD process, where selected patterns are represented as "actionable knowledge units" (see Fig. 2). This involves a knowledge representation (KR) formalism and, subsequently, actionable units can be reused –by human or software agents– in knowledge graphs or knowledge systems for problem-solving.

Recently, efforts have been made to automate several tasks in KDD. Indeed, this is one main objective of meta-learning to design principles that can make algorithms adaptive to the characteristics of the data [11]. In [31,43], authors consider meta-learning as the application of machine learning techniques to meta-data describing past learning experiences for adapting the learning process and improve the performance of the current model [31,43] (a kind of "analogy"). Meta-learning is often related to the problem of dynamically solving or adjusting learning constraints. Most of the time, if there are no restrictions on the space of hypotheses to be explored by a learning algorithm and no preference criteria for comparing candidate hypotheses, then no inductive method can do better on average than random guessing. Authors in [31,43] make a distinction between two main types of "biases", the representational bias restricts the hypothesis space whereas the preference-bias gives priority to certain hypotheses over others in this space. The most widely addressed meta-learning tasks are algorithm

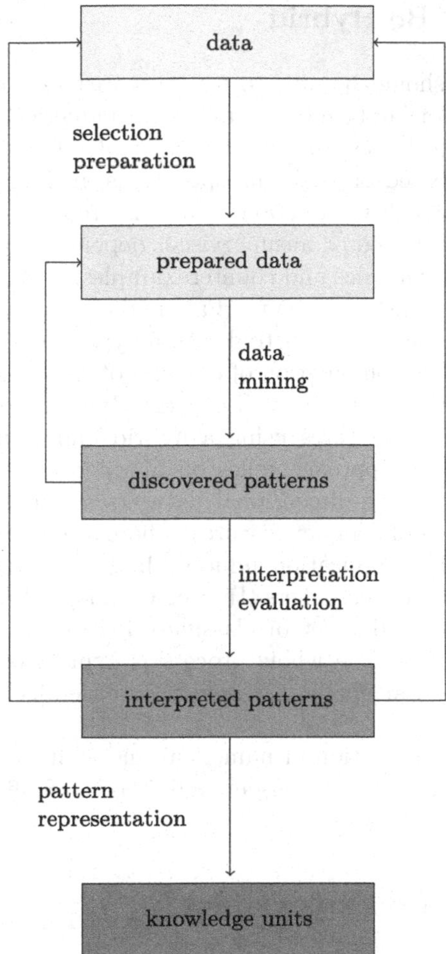

Fig. 2. The augmented KDD loop.

selection and model selection. Algorithm selection is the choice of the appropriate algorithm for a given task, while model selection is the choice of a specific parameter settings that produces a good performance for a given algorithm on a given task.

In the framework of KDD, meta-mining is not only meta-learning, since the KDD loop involves data preparation and pattern interpretation. Thus, the performance of the process and the quality of the discovered patterns and their usage are not only depending on the mining operation but also on data preparation –data selection, data cleaning, and feature selection–, pattern interpretation and possible pattern representation.

4 KDD Should Be Hybrid

Knowledge discovery should be able to work on various datasets with various characteristics, e.g. data can be either symbolic or numerical, or even more complex structured data such as sequences, trees, graphs, texts, linked data... The data management is based on different operations, including for example feature selection, dimensionality reduction, and noise reduction. Also, the mining approaches can be supervised or unsupervised, depending on the task at hand, and the availability of examples and counterexamples. As in every task, there is no universal approach which may be used alone to tackle and to solve all problems. Hence, a reasonable strategy is to design a hybrid process based on several tactics whose output is a combination of outputs of the involved procedures.

Accordingly, in [28], we aimed at discovering in metabolomic data a small set of relevant predictive features using a hybrid and exploratory knowledge discovery approach. This approach relies on adapted classification techniques which should deal with high-dimensional datasets, composed of small sets of individual and large sets of complex features. There are many possible classifiers that can be used and their application induces a bias on the results, calling for the simultaneous use of several classifiers. Hence, we adopted a kind of "ensemble approach" and we designed a set of classifiers instead of using a single one, to reach complementarity. The whole process is exploratory and hybrid as it combines numerical and symbolic classifiers. More details are given farther in Sect. 6.

To conclude, the combination of numerical and symbolic data mining algorithms remains rather rare, even if ensemble methods [18,46] are available without being always satisfactory.

5 KDD Should Be Explainable

Many recent progress in Machine Learning (ML) are mostly due to the success of Deep Learning methods in recognition tasks. However, Deep Learning and other numerical ML approaches are based on complex models, whose outputs and proposed decisions, as accurate as they are, cannot be easily explained to the layman [30]. Indeed, it is interesting to study hybrid ML approaches that combine complex numerical models with explainable symbolic models, in order to make ML methods more "interpretable". The objective is to attach what we could call "integrity constraints" –or kinds of "pre" and "post-conditions" to be fulfilled– to build and then deliver understandable explanations on the work of numerical ML models.

The objective of supervised ML is to learn how to perform a task (e.g. recognition) based on a sufficient number of training examples. The ML algorithm builds a model from the training examples which is then used on test examples. A model can be either symbolic, e.g. a set of rules or a hierarchy of concepts, or numerical, e.g. a set of weights associated with a structure as in neural networks. In practice, numerical models often prove to be more flexible and better suited to capture the complexity of some tasks such as recognition. However, symbolic ML models are more often used in pattern mining and in domains where the learning model should be understandable by human experts, or be related to domain ontologies for knowledge representation and reasoning purposes.

Moreover, ML models approaches based on numerical models are more and more used for complex tasks such as decision making with a strong impact for human users, e.g. make a decision for a student orientation at university. In the latter case it can be very difficult to provide the necessary explanations justifying the decision using these numerical ML models, and especially models based on Deep Learning. Thus there is an emerging research trend whose goal is to provide interpretation and explanations about the decision of numerical ML algorithms such as Deep Learning.

Here, we are interested in understanding the different ways of providing explanations and facilitating interpretation of the outputs of numerical ML models [44]. There are several attempts to build explainable and trustable ML models. As mentioned above, one is to combine symbolic and numerical approaches and to build interactions between both approaches, as in [29] which is based on a combination of numerical learning methods and Formal Concept Analysis [23], or in [47] which is based on a combination of first-order logic and neural network learning.

There are also other initiatives, as in [31,43], on the understandability of a mining process in terms of core components (modules), underlying assumptions, cost functions and optimization strategies being used. One subsequent idea is to understand how Deep Learning models can be decomposed w.r.t. such modules and then to integrate adapted explanation modules.

These are some other possible directions for analyzing a numerical ML model and provide plausible explanations about its output, as for example "neural-symbolic learning and reasoning" combinations [17,49] that should be carefully examined and adapted.

6 An Application in the Mining of Metabolomic Data

In [28], we presented a case study about the mining of metabolomic data using a combination of symbolic and numerical mining methods. Given a dataset composed of individuals described by features, the objective of the experiment is to discover subsets of features that can be *discriminant* and *predictive*. The discrimination power allows to build classes of individuals, where classes include similar individuals and separate dissimilar individuals at the best. The prediction power allows to determine the potential class membership of individuals, e.g. people who will develop the disease under study.

The classification process is split into two main procedures, one being supervised and the other unsupervised (see Fig. 3). The supervised classification procedure is based on the design of NC numerical classifiers (including Random

Forest and SVM in the present case) which are completed by preprocessing and postprocessing operations. This first procedure is applied to a bidimensional dataset composed of individuals and features. The output of this first procedure provides a set of ranked features RF for each of the NC classifiers.

Then the second procedure is based on a unsupervised mining operation applied to RF, the set of ranked features, for discovering the most frequent features, given a threshold set by the analyst. This second procedure is applied to a bidimensional dataset composed this time of $|RF|$ features and NC numerical classifiers. The output of this second procedure provides a set FF of frequent features. These frequent features are the best candidates for becoming best discriminant features. Actually, there is a change of the representation space between the two classification procedures.

Following the classification process and the selection of most discriminant features, the latter are tested for evaluating their predictive capabilities using a ROC analysis [22].

For summarizing, this knowledge discovery strategy relies on two main steps: (i) a concurrent use of multiple classifiers producing a stable set of discriminant features, (ii) a classification of features based on FCA through a change of the problem space representation, where a small set of most relevant features is retained. In this classification process, we can distinguish the following elements:

- KDD is exploratory: two search strategies are run, the first on a data space $individuals \times features$ and the second on a data space $features \times classifiers$. The combination of these two exploration operations produces a set of most discriminant features which are then tested for prediction.
- KDD is knowledge-based: the analyst is asked to control the process and the design of classifiers, to adjust the value of thresholds for dimensionality reduction and feature selection. The analyst plays also a similar role in the prediction analysis.
- KDD is hybrid and involves numerical classifiers as well as more symbolic pattern mining methods. The latter are used for analyzing top-k ranked features w.r.t. discrimination and prediction.
- KDD is explainable thanks to visualization. In particular, the pattern mining procedure involves FCA and the design of a concept lattice where concepts represent classes of top-k ranked features. The distribution of the features within the lattice can be analyzed by the analyst for suggesting further test for prediction analysis.

This experiment shows how and why the different dimensions and objectives of KDD are contributing to deliver a more complete and qualitative analysis of metabolomic data.

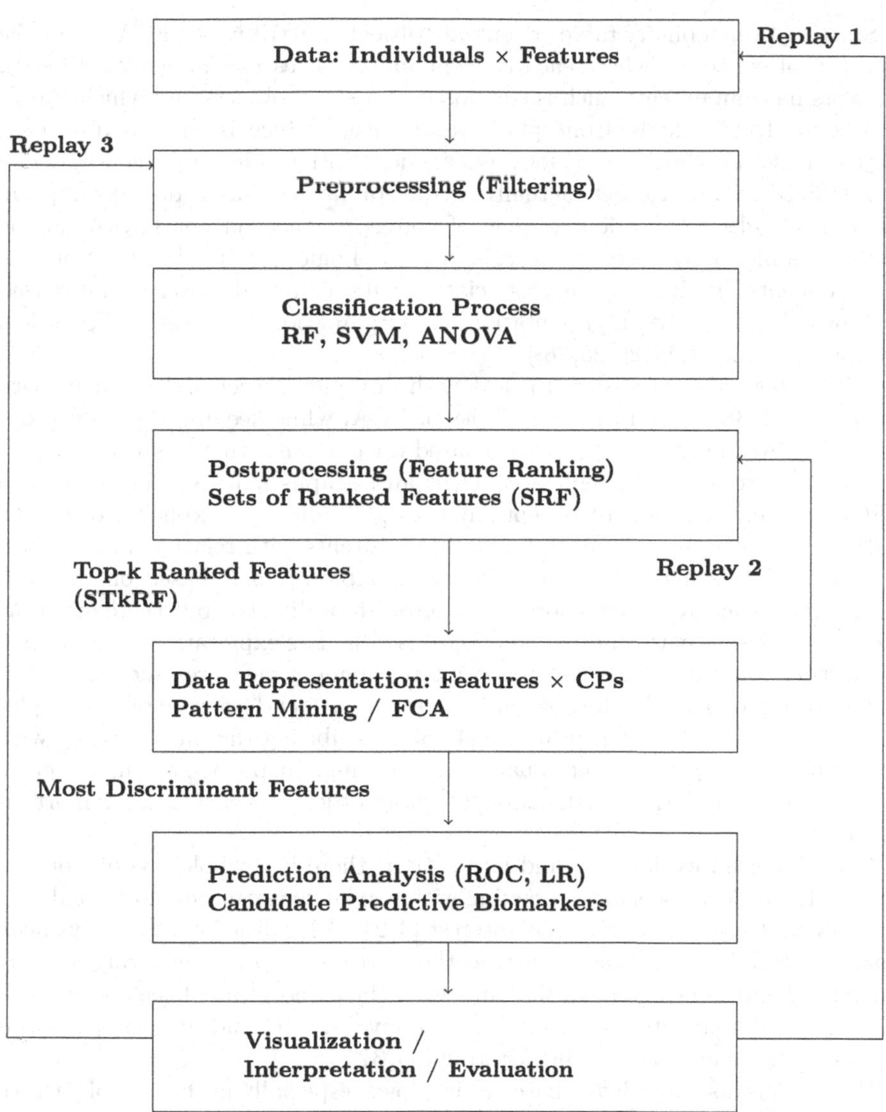

Fig. 3. A hybrid and exploratory approach to metabolomic data mining, with a classification process based on two procedures applied to two different bidimensional datasets. In addition, the "Replay" arrows show that the whole process is interactive and iterative.

7 Discussion: What About FCA?

We have introduced dimensions on which KDD is based, and objectives associated with these dimensions that may characterize KDD. Below we discuss how FCA could be considered w.r.t. these dimensions and objectives.

Starting from a binary table or context "object × attributes", FCA allows the discovery of concepts, where each concept materializes two interrelated views. Concepts have an extent which is composed of a set of objects and which stands for a class of individuals. Concepts have an intent which is composed of a set of attributes and which stands for a (class) description. Moreover, concepts can be organized within a poset, actually a concept lattice, based on subsumption relation. Thanks to the double view of concepts, the concept lattice can be related to a hierarchy of concepts in Description Logics [4]. In addition, from the set of concepts, implications and association rules can be discovered and reused for knowledge discovery, representation and explanation purposes (see practical examples, e.g., in [7,14,20,25,26]).

Now, when it is necessary to deal with complex structured data, pattern structures [24,38] extend the capabilities of FCA, while keeping the good properties of FCA. Many applications are based on pattern structures where object descriptions are intervals, sequences, trees and graphs [36]. Another extension of FCA, namely Relational Concept Analysis [45], allows to explicitly deal with relations between objects and to build object intents with relational attributes.

Hence FCA is naturally exploratory and exploration is based on the concept lattice structure. The exploration is often data-directed but there are some attempts to design pattern-directed processes [13]. The exploration is also aimed at selecting concepts of interest w.r.t. interestingness measures such as stability for example [12,40]. More recently, there is a trend of research on exploration directed by the MDL principle [42,53], exhibiting the links existing with subgroup discovery and exceptional model mining. In particular the selection of initial seeds and the construction of good object coverings are important questions.

Continuing on exploration and interaction, there is a whole line of work on the use of graphical tools for interacting with the lattice structure, for visualizing concepts, and selecting concepts of interest [3,21]. Visualization has always been a main concern for FCA practitioners as the lattice structure can be rather easily understood and interpreted. In the same way, the exploration of the concept lattice allows different types of information retrieval [15,16] and related operations such as recommendation and biclustering [35,37].

But FCA is also knowledge-based as it allows, especially in the case of pattern structures and RCA, to take into account domain knowledge under different forms. For example there is a number of studies on the use of attribute hierarchies within a context (this was initiated in [14], actualized in [24], and then in [3]). Furthermore, there are also other attempts for discovering definitions in the web of data that can be reused as concept definitions in ontologies and knowledge graphs [2].

Furthermore, FCA shows many links with knowledge engineering, while, until now, there does not exist any type of meta-level in FCA. Such a meta-level could take several forms, for example introducing a kind of meta-concepts for providing knowledge about the management of the concepts, or meta-rules for generating or controlling implications and association rules. Still on knowledge construction, we should mention attribute exploration [25] which enables the completion of a context and the exploration of rule sets and concept sets among others.

However, FCA is not hybrid and does not offer any explanation facility strictly speaking. There are some experiments already mentioned showing that FCA can be combined with numerical machine learning methods for performing knowledge discovery tasks. In [29] the concept lattice is also used in an interactive way for concept selection and in a certain sense for providing plausible explanations. Other attempts on hybridization can be found in [33,54].

To conclude, let us mention some research topics that have not received too much attention within FCA, and that we think should deserve more attention in the future, namely the construction of a meta-level, the hybridization with numerical methods, and the production of explicit explanations. This is a vast program, and a series of articles will hopefully emerge in a near future to tackle these open subjects.

References

1. Aggarwal, C.C., Han, J. (eds.): Frequent Pattern Mining. Springer, Cham (2014). https://doi.org/10.1007/978-3-319-07821-2
2. Alam, M., Buzmakov, A., Codocedo, V., Napoli, A.: Mining definitions from RDF annotations using formal concept analysis. In: Yang, Q., Wooldridge, M. (eds.) Proceedings of IJCAI, pp. 823–829. AAAI Press (2015)
3. Alam, M., Buzmakov, A., Napoli, A.: Exploratory knowledge discovery over web of data. Discret. Appl. Math. **249**, 2–17 (2018)
4. Baader, F., Calvanese, D., McGuinness, D., Nardi, D., Patel-Schneider, P. (eds.): The Description Logic Handbook. Cambridge University Press, Cambridge (2003)
5. Belfodil, A., Belfodil, A., Kaytoue, M.: Anytime subgroup discovery in numerical domains with guarantees. In: Berlingerio, M., Bonchi, F., Gärtner, T., Hurley, N., Ifrim, G. (eds.) ECML PKDD 2018. LNCS (LNAI), vol. 11052, pp. 500–516. Springer, Cham (2019). https://doi.org/10.1007/978-3-030-10928-8_30
6. Bendimerad, A.A., Plantevit, M., Robardet, C.: Mining exceptional closed patterns in attributed graphs. Knowl. Inf. Syst. **56**(1), 1–25 (2018)
7. Bertet, K., Demko, C., Viaud, J.-F., Guérin, C.: Lattices, closures systems and implication bases: a survey of structural aspects and algorithms. Theor. Comput. Sci. **743**, 93–109 (2018)
8. De Bie, T.: Subjective interestingness in exploratory data mining. In: Tucker, A., Höppner, F., Siebes, A., Swift, S. (eds.) IDA 2013. LNCS, vol. 8207, pp. 19–31. Springer, Heidelberg (2013). https://doi.org/10.1007/978-3-642-41398-8_3
9. Blockeel, H.: Data mining: from procedural to declarative approaches. New Gener. Comput. **33**(2), 115–135 (2015)
10. Brachman, R.J., Anand, T.: The process of knowledge discovery in databases. In: Fayyad, U.M., Piatetsky-Shapiro, G., Smyth, P., Uthurusamy, R. (eds.) Advances in Knowledge Discovery and Data Mining, pp. 37–57. AAAI Press/MIT Press (1996)
11. Brazdil, P., Giraud-Carrier, C.G., Soares, C., Vilalta, R.: Metalearning - Applications to Data Mining. Cognitive Technologies. Springer, Heidelberg (2009). https://doi.org/10.1007/978-3-540-73263-1

12. Buzmakov, A., Kuznetsov, S.O., Napoli, A.: Scalable estimates of concept stability. In: Glodeanu, C.V., Kaytoue, M., Sacarea, C. (eds.) ICFCA 2014. LNCS (LNAI), vol. 8478, pp. 157–172. Springer, Cham (2014). https://doi.org/10.1007/978-3-319-07248-7_12
13. Buzmakov, A., Kuznetsov, S.O., Napoli, A.: Fast generation of best interval patterns for nonmonotonic constraints. In: Appice, A., Rodrigues, P.P., Santos Costa, V., Gama, J., Jorge, A., Soares, C. (eds.) ECML PKDD 2015. LNCS (LNAI), vol. 9285, pp. 157–172. Springer, Cham (2015). https://doi.org/10.1007/978-3-319-23525-7_10
14. Carpineto, C., Romano, G.: Concept Data Analysis: Theory and Applications. Wiley, Chichester (2004)
15. Codocedo, V., Lykourentzou, I., Napoli, A.: A semantic approach to concept lattice-based information retrieval. Ann. Math. Artif. Intell. **72**, 169–195 (2014)
16. Codocedo, V., Napoli, A.: Formal concept analysis and information retrieval – a survey. In: Baixeries, J., Sacarea, C., Ojeda-Aciego, M. (eds.) ICFCA 2015. LNCS (LNAI), vol. 9113, pp. 61–77. Springer, Cham (2015). https://doi.org/10.1007/978-3-319-19545-2_4
17. d'Avila Garcez, A.S., et al.: Neural-symbolic learning and reasoning: contributions and challenges. In: AAAI Spring Symposium (2015)
18. Dietterich, T.G.: Ensemble methods in machine learning. In: Kittler, J., Roli, F. (eds.) MCS 2000. LNCS, vol. 1857, pp. 1–15. Springer, Heidelberg (2000). https://doi.org/10.1007/3-540-45014-9_1
19. Duivesteijn, W., Feelders, A., Knobbe, A.J.: Exceptional Model Mining - supervised descriptive local pattern mining with complex target concepts. Data Min. Knowl. Discov. **30**(1), 47–98 (2016)
20. Duquenne, V.: Latticial structures in data analysis. Theor. Comput. Sci. **217**, 407–436 (1999)
21. Eklund, P., Villerd, J.: A survey of hybrid representations of concept lattices in conceptual knowledge processing. In: Kwuida, L., Sertkaya, B. (eds.) ICFCA 2010. LNCS (LNAI), vol. 5986, pp. 296–311. Springer, Heidelberg (2010). https://doi.org/10.1007/978-3-642-11928-6_21
22. Fawcett, T.: An introduction to ROC analysis. Pattern Recognit. Lett. **27**(8), 861–874 (2006)
23. Ganter, B., Wille, R.: Formal Concept Analysis - Mathematical Foundations. Springer, Heidelberg (1999). https://doi.org/10.1007/978-3-642-59830-2
24. Ganter, B., Kuznetsov, S.O.: Pattern structures and their projections. In: Delugach, H.S., Stumme, G. (eds.) ICCS-ConceptStruct 2001. LNCS (LNAI), vol. 2120, pp. 129–142. Springer, Heidelberg (2001). https://doi.org/10.1007/3-540-44583-8_10
25. Ganter, B., Obiedkov, S.A.: Conceptual Exploration. Springer, Heidelberg (2016). https://doi.org/10.1007/978-3-662-49291-8
26. Ganter, B., Stumme, G., Wille, R. (eds.): Formal Concept Analysis. LNCS (LNAI), vol. 3626. Springer, Heidelberg (2005). https://doi.org/10.1007/978-3-540-31881-1
27. Grgic-Hlaca, N., Zafar, M.B., Gummadi, K.P., Weller, A.: Beyond distributive fairness in algorithmic decision making: feature selection for procedurally fair learning. In: McIlraith, S.A., Weinberger, K.Q. (eds.) Proceedings of AAAI 2018, pp. 51–60. AAAI Press (2018)
28. Grissa, D., Comte, B., Pétéra, M., Pujos-Guillot, E., Napoli, A.: A hybrid and exploratory approach to knowledge discovery in metabolomic data. Discrete Applied Mathematics (2019, to be published)

29. Grissa, D., Comte, B., Pujos-Guillot, E., Napoli, A.: A hybrid knowledge discovery approach for mining predictive biomarkers in metabolomic data. In: Frasconi, P., Landwehr, N., Manco, G., Vreeken, J. (eds.) ECML PKDD 2016. LNCS (LNAI), vol. 9851, pp. 572–587. Springer, Cham (2016). https://doi.org/10.1007/978-3-319-46128-1_36
30. Guidotti, R., Monreale, A., Ruggieri, S., Turini, F., Gianotti, F., Pedreschi, D.: A survey of methods for explaining black box models. ACM Comput. Surv. **51**(5) (2018)
31. Hilario, M., Nguyen, P., Do, H., Woznica, A., Kalousis, A.: Ontology-based meta-mining of knowledge discovery workflows. In: Jankowski, N., Duch, W., Grabczewski, K. (eds.) Meta-Learning in Computational Intelligence, vol. 358, pp. 273–315. Springer, Heidelberg (2011). https://doi.org/10.1007/978-3-642-20980-2_9
32. Holzinger, A., Dehmer, M., Jurisica, I.: Knowledge discovery and interactive data mining in bioinformatics - state-of-the-art, future challenges and research directions. BMC Bioinform. **15**(S-6), I1 (2014)
33. Hristoskova, A., Boeva, V., Tsiporkova, E.: An integrative clustering approach combining particle swarm optimization and formal concept analysis. In: Böhm, C., Khuri, S., Lhotská, L., Renda, M.E. (eds.) ITBAM 2012. LNCS, vol. 7451, pp. 84–98. Springer, Heidelberg (2012). https://doi.org/10.1007/978-3-642-32395-9_7
34. Janowicz, K., van Harmelen, F., Hendler, J.A., Hitzler, P.: Why the data train needs semantic rails. AI Mag. **36**(1), 5–14 (2015)
35. Kaytoue, M., Codocedo, V., Baixeries, J., Napoli, A.: Three interrelated FCA methods for mining biclusters of similar values on columns. In: Bertet, K., Rudolph, S. (eds.) Proceedings of CLA. CEUR Workshop Proceedings, vol. 1252, pp. 243–254 (2014)
36. Kaytoue, M., Codocedo, V., Buzmakov, A., Baixeries, J., Kuznetsov, S.O., Napoli, A.: Pattern structures and concept lattices for data mining and knowledge processing. In: Bifet, A., et al. (eds.) ECML PKDD 2015. LNCS (LNAI), vol. 9286, pp. 227–231. Springer, Cham (2015). https://doi.org/10.1007/978-3-319-23461-8_19
37. Kaytoue, M., Kuznetsov, S.O., Macko, J., Napoli, A.: Biclustering meets triadic concept analysis. Ann. Math. Artif. Intell. **70**(1–2), 55–79 (2014)
38. Kaytoue, M., Kuznetsov, S.O., Napoli, A., Duplessis, S.: Mining gene expression data with pattern structures in formal concept analysis. Inf. Sci. **181**(10), 1989–2001 (2011)
39. Kaytoue, M., Plantevit, M., Zimmermann, A., Bendimerad, A.A., Robardet, C.: Exceptional contextual subgraph mining. Mach. Learn. **106**(8), 1171–1211 (2017)
40. Kuznetsov, S.O., Makhalova, T.P.: On interestingness measures of formal concepts. Inf. Sci. **442–443**, 202–219 (2018)
41. Lavrac, N., Kavsek, B., Flach, P.A., Todorovski, L.: Subgroup discovery with CN2-SD. J. Mach. Learn. Res. **5**, 153–188 (2004)
42. Makhalova, T.P., Kuznetsov, S.O., Napoli, A.: A first study on what MDL can do for FCA. In: Ignatov, D.I., Nourine, L. (eds.) Proceedings of CLA, CEUR Workshop Proceedings, vol. 2123, pp. 25–36 (2018)
43. Nguyen, P., Hilario, M., Kalousis, A.: Using meta-mining to support data mining workflow planning and optimization. J. Artif. Intell. Res. (JAIR) **51**, 605–644 (2014)
44. Ribeiro, M.T., Singh, S., Guestrin, C.: "Why Should I Trust You?": explaining the predictions of any classifier. In: Krishnapuram, B., Shah, M., Smola, A.J., Aggarwal, C.C., Shen, D., Rastogi, R. (eds.) Proceedings of SIGKDD, pp. 1135–1144. ACM (2016)

45. Rouane-Hacene, M., Huchard, M., Napoli, A., Valtchev, P.: Relational concept analysis: mining concept lattices from multi-relational data. Ann. Math. Artif. Intell. **67**(1), 81–108 (2013)
46. Sagi, O., Rokach, L.: Ensemble learning: a survey. Wiley Interdiscip. Rev. Data Min. Knowl. Discov. **8**(4) (2018)
47. Sourek, G., Aschenbrenner, V., Zelezný, F., Schockaert, S., Kuzelka, O.: Lifted relational neural networks: efficient learning of latent relational structures. J. Artif. Intell. Res. **62**, 140–151 (2018)
48. Tan, P.-N., Steinbach, M., Karpatne, A., Kumar, V.: Introduction to Data Mining, 2nd edn. Pearson, New York (2018)
49. Tran, S.N., d'Avila Garcez, A.S.: Deep logic networks: inserting and extracting knowledge from deep belief networks. IEEE Trans. Neural Netw. Learn. Syst. **29**(2), 246–258 (2018)
50. Tukey, J.W.: Exploratory Data Analysis. Addison-Wesley Publishing Company, Reading (1977)
51. Ugarte, W., et al.: Skypattern mining: from pattern condensed representations to dynamic constraint satisfaction problems. Artif. Intell. **244**, 48–69 (2017)
52. Leeuwen, M.: Interactive data exploration using pattern mining. In: Holzinger, A., Jurisica, I. (eds.) Interactive Knowledge Discovery and Data Mining in Biomedical Informatics. LNCS, vol. 8401, pp. 169–182. Springer, Heidelberg (2014). https://doi.org/10.1007/978-3-662-43968-5_9
53. Vreeken, J., Tatti, N.: Interesting patterns. In: Aggarwal and Han [1], pp. 105–134
54. Yoneda, Y., Sugiyama, M., Washio, T.: Learning graph representation via formal concept analysis. CoRR, abs/1812.03395 (2018)
55. Zafar, M.B., Valera, I., Gomez-Rodriguez, M., Gummadi, K.P., Weller, A.: From parity to preference-based notions of fairness in classification. In: Guyon, I., et al. (eds.) Proceedings of NIPS, pp. 228–238 (2017)
56. Zaki, M.J., Meira Jr., W.: Data Mining and Analysis: Fundamental Concepts and Algorithms. Cambridge University Press, New York (2014)

Too Much Information: Can AI Cope with Modern Knowledge Graphs?

Markus Krötzsch(✉)

TU Dresden, Dresden, Germany
markus.kroetzsch@tu-dresden.de

Abstract. Knowledge graphs play an important role in artificial intelligence (AI) applications – especially in personal assistants, question answering, and semantic search – and public knowledge bases like Wikidata are widely used in industry and research. However, modern AI includes many different techniques, including machine learning, data mining, natural language processing, which are often not able to use knowledge graphs in their full size and complexity. Feature engineering, sampling, and simplification are needed, and commonly achieved with custom preprocessing code. In this position paper, we argue that a more principled integrated approach to this task is possible using declarative methods from knowledge representation and reasoning. In particular, we suggest that modern rule-based systems are a promising platform for computing customised views on knowledge graphs, and for integrating the results of other AI methods back into the overall knowledge model.

1 Introduction

The modern-day rise of artificial intelligence (AI) in research and applications has also sparked a renewed interest in graph-based knowledge representation. So-called *knowledge graphs* (KGs) have become essential resources for intelligent assistants such as Apple's Siri and Amazon's Alexa [48], for question answering features of modern search engines such as Google[1] and Microsoft Bing, and for the new expert systems in the style of IBM's Watson [21]. As with many recent AI breakthroughs, this has partly been a matter of scale: modern *knowledge graphs* are orders of magnitude larger than knowledge bases conceived in early AI. Public KGs such as Wikidata [52], YAGO2 [29] and Bio2RDF [5] comprise hundreds of millions of statements; KGs in parts of the web search industry are significantly larger still (due to, e.g., crawled schema.org annotation data and other resources, such as maps).

The free availability of large knowledge graphs is also creating new opportunities in research. KGs are used to improve the quality of approaches to natural language processing, machine learning, and data mining [43]. Combining techniques from several fields also leads to advances in knowledge representation, e.g., for improved association rule mining [28]. Machine learning meanwhile can

[1] The company's *Google Knowledge Graph* is the origin of the term.

D. Cristea et al. (Eds.): ICFCA 2019, LNAI 11511, pp. 17–31, 2019.
https://doi.org/10.1007/978-3-030-21462-3_2

help KG construction and maintenance, e.g., by employing methods of *knowledge graph embedding* to predict missing information or to detect errors [11,43,54].

Nevertheless, the promise of perfect synergy between knowledge graphs and AI often remains unfulfilled. Most KGs available today are the work of human authors, who have contributed either directly or by creating heavily structured online content (HTML tables, Wikipedia templates, schema.org annotations, etc.) from which KGs could be extracted.[2] When Google decided to migrate its former KG project *Freebase* to Wikidata, it largely relied on human labour [49]. Even closely related AI techniques, such as ontology-based knowledge representation, are used in only a few KG projects.

Why don't we see much more use of neighbouring AI methods for building, maintaining, and using KGs? The answer might be quite simple: the "knowledge graphs" addressed in much of the current research are different from the *real* KGs used in applications. Real KGs are predominantly based on annotated directed graph models, where directed, labelled edges are augmented with additional context information (validity time, data source, trustworthiness, etc.). Related data models are used in Wikidata (Sect. 3), Yago2 [29], and in the widely used *property graph* data model, where graph edges are conceived as complex data objects with additional key-value pairs [46]. Real KGs are also very large.

In contrast, research in other fields of AI often uses small, simplified excerpts of KGs for benchmarking. For example, link prediction through KG embeddings is still frequently evaluated with the *FB15K* dataset, which contains about 600,000 edges for 15,000 Freebase entities [11]. When Google announced the discontinuation of Freebase in 2014, the KG had almost 45M entities and 2.5 billion edges, and it was using annotations, e.g., for time (based on so-called *Compound Value Types*). In fact, KG embeddings often conceive graphs as sets of "triplets" – a simplified form of RDF triples – and n-ary relations, compound structures, quantities, and annotations are poorly supported [43].

These limitations are not specific to KG embeddings, which can actually cope with "triplet" graphs of significant size. Other knowledge-based data mining techniques are also challenged by the scale of modern KGs. For example, Formal Concept Analysis (FCA) requires a projection of graphs to Boolean propositions ("attributes") [22], and studies are typically working with hundreds of attributes at most (see, e.g., [25]). Similar limitations in size and complexity apply to other rule mining approaches [28]. The field of network analysis offers many highly scalable algorithms, e.g., for computing centrality measures such as PageRank, but they need KGs to be preprocessed to obtain directed (and often unlabelled) graphs [38]. Which part of the KG is used to obtain the network has a major impact on the meaning and quality of the results. Ontological knowledge representation and reasoning also has been scaled to hundreds of millions of statements, especially in recent rule engines [7,51], but does not cope well with exceptions and errors, and lacks a general approach for handling annotations [33,35].

[2] Crawling and extracting such content is a difficult task, and a worthy research area in itself, yet the main work of knowledge gathering remains that of the human author.

Simply put: no single AI approach is able to handle current KGs in their full size and complexity. Indeed, although each approach might be further improved, one should not expect future AI to be based on a single principle. New ideas are needed for reconciling different AI formalisms. IBM Watson has pioneered one way of combining the strengths of formerly incompatible methods in question answering [21], but this integration is limited to selecting the best answer at the output level. To fully exploit KGs in AI, we also need principled ways of defining *input* data that can be used by other techniques (including the task of feature engineering in machine learning). In this paper, we conceive this as the task of defining suitable *views* on KG data, and we propose the use of advanced recursive query mechanisms to solve it in a declarative way. Modern rule engines can compute some useful views already, but mostly lack the features needed for integrating other methods. Moreover, for full integration, the outputs of such methods must be reconciled with the KG contents.

In this position paper, we motivate this new approach, discuss the relevant state of the art, and derive open issues for future research needed to realise this idea. The text includes parts of an earlier publication of KGs in ontological modelling [33].

2 What is a Knowledge Graph?

The knowledge graphs in modern applications are characterised by several properties that together distinguish them from more traditional knowledge management paradigms:

(1) **Normalisation:** Information is decomposed into small units of information, interpreted as edges of some form of graph.
(2) **Connectivity:** Knowledge is represented by the relationships between these units.
(3) **Context:** Data is enriched with contextual information to record aspects such as temporal validity, provenance, trustworthiness, or other side conditions and details.

While many databases are normalised in some way (1), the focus on connections (2) is what distinguishes the graph-based view. This is apparent not so much from the data structures – graphs can be stored in many formats –, but from the use of query languages. Graph query languages such as SPARQL [45] or Cypher [44] natively support reachability queries and graph search features that are not supported in other paradigms.[3]

Contextual information (3) is introduced for a variety of reasons. It may simply be convenient for storing additional details that would otherwise be hard to represent in normalisation, or it may be used for capturing meta-information

[3] SQL supports recursive views that resemble the expressivity of linear Datalog, but the standard forbids the use of duplicate elimination (DISTINCT) in their construction, making them quite useless for breadth-first search on graphs that may contain cycles.

such as provenance, trust, and temporal validity. Indeed, knowledge graphs are often used in data integration – graphs are well suited for capturing heterogeneous data sources and their relationships –, and it is natural to retain basic information regarding, e.g., the source and trustworthiness of a particular piece of data.

While we can only speculate about the shape and content of Google's original knowledge graph, we can find the above characteristics in major graph database formats:

- *Property Graph.* The popular graph model that is used in many applications is based on a directed multi-graph with attribute-value annotations that are used to store contextual information on each edge and node [46]. Popular property graph query languages, such as Cypher, support graph search [44].
- *RDF.* The W3C graph data standard encodes directed graphs with different types of edges. Support for storing contextual information has been added to RDF 1.1 by enabling references to named graphs [17]. The SPARQL query language for RDF supports regular path queries and named graphs [26].

Likewise, individual knowledge graphs exhibit these characteristics, for example:

- *Yago* and *Yago2* are prominent knowledge graphs extracted from Wikipedia [29]. The authors extended RDF (at a time before RDF 1.1) with quadruples to capture important contextual information related to time and space.
- *Bio2RDF* is a graph-based data integration effort in the life sciences [5]. It uses an n-ary object model to capture complex and contextual information in plain RDF graphs, i.e., it introduces new identifiers for tuples extracted from relational DBs.
- *Wikidata*, the knowledge graph of Wikipedia, is natively using a graph-like data model that supports expressive annotations for capturing context [52]. Details are discussed in the next section.

3 Wikidata, the Knowledge Graph of Wikipedia

To gain a better understanding of the problem, it is instructive to take a closer look at a modern KG. We choose Wikidata, the knowledge graph of the Wikimedia Foundation, since it is freely accessible and has become a major resource in many applications. Wikidata is a sister project of Wikipedia that aims to gather and manage factual data used in Wikipedia or any other Wikimedia project [52]. Launched in October 2012, the project has quickly grown to become one of the largest and most active in terms of editing. As of March 2019, Wikidata is one of the largest public collections of general knowledge, consisting of more than 680 million statements about more than 55 million entities. Wikidata is prominently used by intelligent agents such as Alexa and Siri [48], but also in many research activities, e.g., in the life sciences [12], in social science [53], and in Formal Concept Analysis [24,25].

The main content of Wikidata are its *statements*, which describe and interrelate the entities. A Wikidata statement can be viewed as an edge in a directed

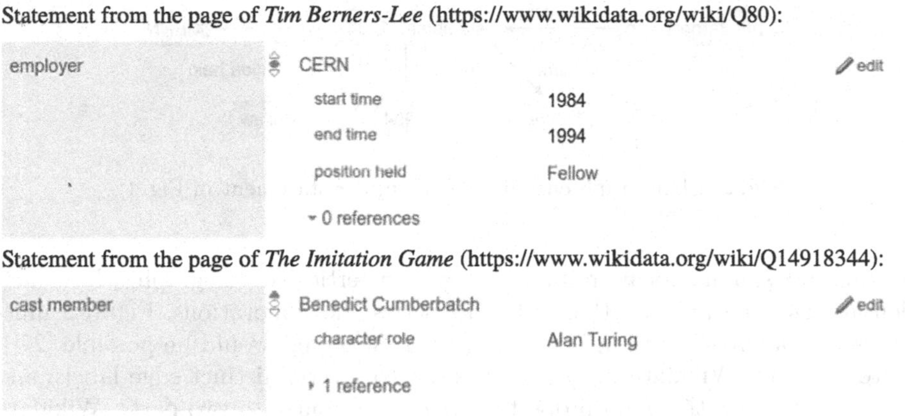

Statement from the page of *Tim Berners-Lee* (https://www.wikidata.org/wiki/Q80):

employer | CERN | edit
start time 1984
end time 1994
position held Fellow
▾ 0 references

Statement from the page of *The Imitation Game* (https://www.wikidata.org/wiki/Q14918344):

cast member | Benedict Cumberbatch | edit
character role Alan Turing
▸ 1 reference

Fig. 1. Two statements from Wikidata

graph that is further annotated by attribute-value pairs and provenance information. For example, Fig. 1 shows two statements as seen by users of wikidata.org. In both cases, the main part of the statement can be read as a directed edge: Berners-Lee's employer has been CERN, and the film *The Imitation Game* has cast member Benedict Cumberbatch. In addition, both statements include contextual information in the form of additional property-value pairs that refer to the statement (rather than to the subject of the statement). As one can see, this additional information includes classical "meta-data" such as validity time and references (collapsed in the figure), but also other details that are more similar to the modelling of n-ary relations.

Statements consist of *properties* (e.g., *employer*, *start date*, *character role*) that are given specific *values*, which may in turn be Wikidata *items* (e.g., *Alan Turing*, *CERN*) or values of specific datatypes (e.g., *1984*, which denotes a date whose precision is limited to the year). Notably, the properties and values used in the main part of the graph are the same as those used to add contextual information, and this can be exploited in queries (e.g., one can ask for actors who have played computer scientists). Moreover, Wikidata allows users to create new properties, and to make statements about them. From a knowledge representation viewpoint, Wikidata properties are therefore a special type of individual rather than a binary predicate.

Some further details of Wikidata's graph model are briefly noted: (1) the same directed relationship may occur with several different annotations, e.g., in the case of Elizabeth Taylor who was married to Richard Burton several times; (2) the same property can be assigned more than one value in the context of some statement, e.g., this is used when several people win an award together (*together with* is the Wikidata property used to annotate the *award received* statement); and (3) the order of statements and statement annotations is not relevant.

Wikidata's rich graph model can of course be encoded in simpler graph structures, which mainly requires some way of encoding statement annotations.

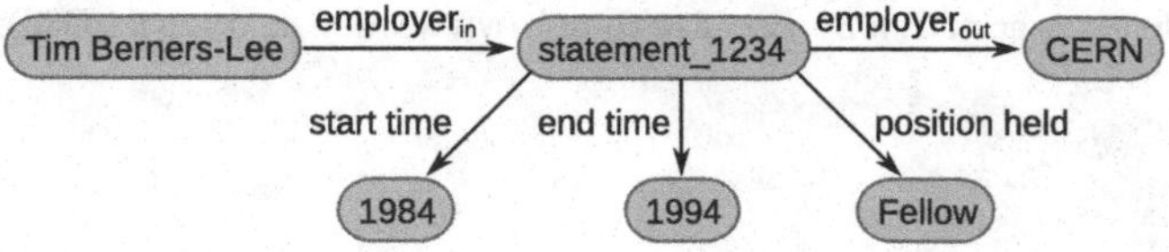

Fig. 2. Plain graph encoding of the upper statement in Fig. 1

To this end, statements are represented by own vertices in the graph, whose incidental edges can indicate their related entities and annotations. Figure 2 illustrates this approach for one encoding (other encodings would be possible [27]). Note that each Wikidata property gives rise to several distinct edge labels, and in particular that the main property *employer* is split into two parts. Wikidata uses a similar encoding to export its contents to RDF. The actual encoding is slightly more complex, e.g., since some data values such as geographic coordinates also need to be encoded using several edges, and since the export already includes several alternative encodings as well as further data [39]. As of March 2019, the RDF export of Wikidata comprises 7.28 billion triples with more than 64,000 RDF properties.[4] It is natural that a more normalised representation like RDF requires significantly more edges to capture the contents of a KG. The magnitude of this increase underlines the challenge that AI methods based on simpler graph models are facing.

The RDF encoding of Wikidata can be downloaded in the form of file exports and queried through a public SPARQL query service, which is based on the current version of this continuously changing KG. Details are described by Malyshev et al. [39]. The public SPARQL service is used as an API in many applications, and serves in the order of 100 million queries per month. Malyshev et al. also published 575 million anonymised queries and provide an initial analysis of their form and content.

4 Ontology-Based Views on Knowledge Graphs

The sheer size of Wikidata creates difficulties for some approaches in data mining and machine learning, but the bigger challenge lies in the complexity of the data, in particular due to the use of complex data types and annotations. Interpreting statements might require background knowledge on Wikidata's modelling conventions or at least a certain amount of common sense.

Example 1. Wikidata specifies many historic facts, and among the most common annotations are those for temporal validity. For example, *London* is said to be located in five administrative regions, including the *Kingdom of Wessex* and two distinct entities called *England*. The only current value is *Greater London*, from which a chain of (temporally current) *located in* statements leads to present-day England. A challenge in this inference is that the times specified

[4] https://grafana.wikimedia.org/dashboard/db/wikidata-query-service.

for each statement along the chain are different, although the intervals overlap. Population numbers and mayors are also commonly temporalised, whereas other transient statements do not specify a temporal context, e.g., London is stated (without any annotations) to be the *capital of* ten entities, including some that no longer exist. Indeed, many entities also specify times for inception (creation) and dissolution (abolishment), which should be considered when interpreting statements.

Notably, the background knowledge needed to correctly interpret such statements might not be encoded in Wikidata at all. Applying data mining or machine learning techniques to the unfiltered data therefore is likely to produce unintended results, even if the scalability issues can be overcome. This suggests that preprocessing is needed to obtain not only smaller but also more regular, hence *predictable*, slices of the data.

Instead of relying on ad hoc preprocessing, we propose to model the required background knowledge explicitly using declarative knowledge representation languages. Indeed, the "slices of the data" that we seek correspond to *views* (under the meaning common in databases). Logical formalisms have been proposed for specifying such views declaratively, comprising approaches such as *Datalog* queries [2], *inclusion dependencies* [37], *tuple-generating dependencies* [13], and *ontology-based data access* (OBDA) [14,32]. It is worth to consider these approaches for use on KGs.

However, the application of formal logic to KGs is not straightforward either, since the latter may not have a native expression in predicate logic. Statements as in Fig. 1 could be represented as n-ary relations, but the arity would be different for each statement depending on how many annotations are given. Indeed, Wikidata does not impose a limit on the number of annotations, and some statements use a large number of them. For a faithful relational representation with predicates of bounded arity, we therefore need to use an encoding like the one in Fig. 2.

With this encoding, any modelling language that is compatible with predicate logic can be used on knowledge graphs, including description logics (DLs) – today's most common ontology languages and the basis of OBDA.

Example 2. Hanika et al. apply Formal Concept Analysis to Wikidata by defining attributes based on the incidence of Wikidata properties to certain objects [25]. Using relation names as in Fig. 2, we can express, e.g., that every entity that is the source of a $\mathsf{mother}_{\mathsf{in}}$ relation should be in a unary predicate Mother (used as an attribute in FCA). In first-order logic, this is expressed as $\forall x.(\exists y.\mathsf{mother}_{\mathsf{in}}(x,y)) \leftrightarrow \mathsf{Mother}(x)$. An equivalent DL sentence is $\exists\mathsf{mother}_{\mathsf{in}}.\top \equiv \mathsf{Mother}$. Similar uses of DL for defining FCA attributes have been considered previously [47].

Note that the formalisation in Example 2 actually defines a two-way relationship between the view and the KG, such that one could, e.g., interpret association rules computed on this view as logical sentences from which new information can be derived about the KG [10]. The logic-based approach thus is more than just a declarative way of preprocessing KGs: it also is in principle invertible, supporting the interpretation of results obtained on a view with respect to the original data. But Example 2 is still too simple. Unary view predicates will not always be enough (e.g., we need "triplets" in KG embeddings), and significantly more complex view definitions would be needed to address use cases as in Example 1.

Unfortunately, if we try to specify more general views, we encounter severe expressivity limits. DLs, for example, offer very little expressivity for deriving binary relations (which could be interpreted as triplets), and are generally too weak to express even simple relationships on rich KG models [33]. The root cause of this restriction is the close connection of DLs to guarded logics, which syntactically ensure tree-like model structures by requiring *guard atoms* that bind certain variables. Our decomposition of statements into several smaller atoms makes it very hard to find guards, and other kinds of guarded logics are therefore similarly restricted on such KG encodings [35].

A more suitable logical formalism might therefore be Datalog, the simple recursive query language that syntactically corresponds to function-free first-order Horn-logic without existential quantifiers. Indeed, Datalog rules can express arbitrary positive relational patterns (conjunctive queries) in their premise, requiring no guard. Their weakness is that they are restricted to the given set of domain elements, i.e., they cannot infer the existence of new entities. Example 2 shows a case where this power is useful (for the "←" part of the equivalence). Datalog can be generalised to support this, leading to so-called *existential rules* or *tuple-generating dependencies*, but then other restrictions must be imposed to retain decidability.

Part of this difficulty arises from the loss of structure that our relational encoding has incurred, since we can no longer distinguish the (non-local, possibly incomplete) connection structure of the KG from the (local, complete) annotations of individual edges. Marx et al. therefore proposed to explicitly include annotations into relational calculus, obtaining what they call *attributed logics* [41]. For example, the upper statement in Fig. 1 could be written in attributed logic as an atom

$$\mathsf{employer}(\mathsf{TimBL}, \mathsf{CERN})@\{\mathsf{start\_time} : 1984, \mathsf{end\_time} : 1994, \mathsf{position} : \mathsf{Fellow}\}.$$

In this syntax, *attributes* (e.g., position) and values (e.g., 1984) are treated like logical terms, and in particular can be variables. In addition, attributed logics allow for quantification over (finite) sets of attributes.

In its full generality, this yields a logic with an undecidable entailment problem, and Marx et al. propose a restricted rule-based language *MARPL* that uses specialised primitives for expressing properties of annotation sets.

Example 3. The following MARPL rule defines a CFellow to be somebody employed by CERN as a fellow, and it copies the respective start and end time (if given):

$$\mathsf{employer}(x, \mathsf{CERN})@Z \wedge \lfloor \mathsf{pos} : \mathsf{Fellow} \rfloor(Z) \to \mathsf{CFellow}@\{\mathsf{start} : Z.\mathsf{start}, \mathsf{end} : Z.\mathsf{end}\}$$

All variables are implicitly quantified universally. Z is bound to the annotation set of the employer fact, and required to contain pos : Fellow and possibly other attribute-value pairs (denoted by the bracket that is opened to the top). The conclusion of the rule is the required fact, using a shortcut notation from [34] to copy (zero or more) values of the start and end attributes.

Theorem 1 ([41]). *Conjunctive query answering with respect to MARPL ontologies is* ExpTime*-complete, both in terms of combined and data complexity.*[5]

Combined complexity therefore matches basic Datalog, but the high data complexity might be surprising. It is caused by MARPL's ability of creating an exponential number of different annotation sets, which is already possible if only a fixed set of attributes is used in annotations. It is not expected that reasoning on real KGs leads to such combinatoric explosions in the relevant annotation sets, but it might still be a strength of the proposal that it can also describe view definitions that are not in polynomial time.

5 Rule Engines as Platforms for Artificial Intelligence

New ontology languages such as MARPL might develop into suitable formalism for expressing views and logical relationships over KGs, but they do not provide the required integration with other AI approaches yet. At the time of this writing, there is not even an implemented reasoning engine for MARPL. If we look for existing reasoners that scale to large amounts of data, (existential) rule reasoners stand out [7]. Indeed, these engines have been developed as extensions of the deductive query language *Datalog*, which has a long tradition in databases, where scalability to large datasets is a prime concern. In recent years, there has been significant progress in this area, and many new rule-based systems have been presented [3,4,8,9,23,42,50]. In this section, we therefore discuss what it would take to develop such systems into a suitable basis for reconciling knowledge graphs with other AI methods.

Existential rule systems, such as RDFox [42] or VLog [50], are often based on bottom-up materialisation of inferences, known as the *chase* in databases. This is the most common reasoning approach for Datalog, and it is naturally extended to existential rules. Common variants of the chase include the *standard* (a.k.a. *restricted*) *chase* [20], the *skolem* (a.k.a. *semi-oblivious*) *chase* [40], and the *core chase* [18]. They differ in the approach taken to determine if the application of a rule can be considered redundant in the sense that it will not contribute to producing distinct new entailments.

However, reasoning with existential rules is undecidable in general, and indeed any chase may fail to terminate in some cases – the algorithm is sound but not complete. Termination of the chase is also undecidable, but many sufficient

[5] Data complexity characterises the worst-case asymptotic complexity of the reasoning problem for a fixed logical theory (i.e., MARPL rule set) with respect to the size of the input data (KG).

criteria have been developed for detecting it [16]. Almost all of these criteria ensure that the skolem chase over a set of rules will terminate *universally*, i.e., on all possible sets of input facts. The following result applies in any such case:

Theorem 2 (Marnette, [40]). *Let Σ be a set of existential rules, such that, for every set of input facts $\mathcal{F}$, the skolem chase over $\Sigma \cup \mathcal{F}$ terminates. Then the skolem chase decides fact entailment over $\Sigma \cup \mathcal{F}$ in polynomial time with respect to the size of $\mathcal{F}$.*

In other words, the data complexity of fact entailment over rules with a universally terminating skolem chase is in PTIME. This data complexity is strictly smaller than for MARPL rules (Theorem 1), i.e., we cannot hope to encode MARPL theories in skolem-chase terminating rules. This restriction applies to virtually every known chase termination criterion, including recent approaches that are not specific to the skolem chase but also feature PTIME data complexity [15].

Surprisingly, however, this restriction is not inherent to other versions of the chase. As has recently been observed, there are universally standard-chase terminating rule sets with non-elementary data complexity [36]. Therefore, there does not seem to be any principled reason why MARPL-style expressive power should not be in reach for existing rule engines. Indeed, we conjecture that MARPL can be translated into existential rules with a suitable encoding. The standard chase is implemented in both RDFox [7] and VLog [51], hence this translation might also allow MARPL reasoning in practice.

A MARPL reasoner would be a first step towards realising the vision described in this paper. It would enable users to replace custom data extraction and projection software with a declarative logical description of a KG view. Further expressive features will be useful to truly cover a significant part of current preprocessing implementations:

(1) Negation: The reasoner needs to be able to check for the absence of data in the KG, e.g., to determine which entities have no time of dissolution.
(2) Quantitative data: Basic range comparisons on input data are important for filtering, e.g., to check if a statement is valid at a specific point in time (within the given validity interval).
(3) Aggregates: especially maximum and minimum values can be of importance, e.g., to find the most recent statement of a certain property.

It would often suffice if these features were supported on input data, which is not affected by recursive computation, and therefore does not raise semantic issues related to, e.g., the use of negation in recursive specifications. Nevertheless, view definitions might benefit from the use of recursion in other places, e.g., to compute excerpts of the KG by computing all entities that are reachable through some property (example: select all monarchs and all of their relatives).

The facts entailed by such recursive view definitions can then be used with other AI methods for data mining or machine learning. This could be done by exporting the entailments, using them as input to other approaches, and re-importing the results into the logical knowledge base. Further logical rules could then be used to draw conclusions that relate to the original data model of the KG.

It would be desirable to achieve a tighter integration that avoids the need for this separate export and import step. A practical way of doing this could be to access the other AI methods through built-in predicates that can be used in rules. This leads to hybrid rule-based systems that can "call" external libraries or sub-systems for optimisation and learning tasks. Some recent rule engines have started to explore this approach, examples being LogicBlox [3] and Vadalog [6].

While built-in functions are not uncommon in rule-based systems, the functions needed here would in fact be aggregates in the sense that they would act on sets of facts (representing views) rather than on single tuples. There are also other conceivable ways of achieving such integration, for example using higher-order built-ins as proposed by Eiter et al. for the case of answer set programming [19]. The main requirement is that the built-in function can receive a complete set of facts, e.g., an unlabelled directed graph that was extracted from the KG.

The use of built-in functions allows us to treat methods from machine learning and data analytics as black boxes. This approach is flexible and versatile, but it does not exploit the characteristics of specific built-in functions. The latter may have significant advantages in terms of overall performance, as has been found by recent works that have explored this idea for (non-recursive) database queries [31]. Similar approaches suggest themselves in the context of rule-based systems.

Accessing data mining and machine learning methods through built-in functions should be contrasted with approaches that attempt to increase the expressivity of the logic in order to be able to implement data analysis algorithms in rules. For example, Aberger et al. present an architecture for computing some recursive network analysis algorithms, such as PageRank centrality, using rule-based specifications [1]. While such iterative algorithms are compatible with the recursive reasoning in rule engines, they require a different control strategy, since they can usually not be iterated until a fixed point is reached but only until some sufficient precision has been obtained. Another approach of combining numeric optimisation with rule reasoning was proposed by Kaminski et al. [30]. It has the potential of integrating linear optimisation with rules in a fully declarative way, but other algorithms such as PageRank seem to be out of scope. We believe that built-in functions can be a middle ground between such a fully integrated approach (which must be carefully restricted to remain within the limits of computability) and today's loose combination of several programs that does not follow any underlying principle.

6 Summary and Outlook

Knowledge graphs are characterised by their heterogeneous and multi-faceted information content, which covers a range of topics at different levels of detail. Significant feature engineering and sampling is required to apply AI techniques from data mining or machine learning to such resources successfully. We have argued that rule-based knowledge representation can be used to achieve this preprocessing in a declarative, principled way.

Besides the immediate benefit of an integrated system that replaces the makeshift processing pipelines in many current works, this approach has the potential of paving the way towards an explainable AI. Indeed, the powerful but opaque methods of statistical optimisation and machine learning would be invoked and interpreted through the knowledge-based framework of logical rules, which is specified by humans and therefore more explicit and easier to verify than learned models.

Looking further ahead, we can see an opportunity for expanding this approach towards new system architectures and design principles that lead to an artificial intelligence that is not just powerful and efficient but also reliable, predictable, and safe.

Acknowledgements. This work is partly supported by Deutsche Forschungsgemeinschaft (DFG, German Research Foundation) in project number 389792660 (TRR 248, Center for Perspicuous Systems), CRC 912 (Highly Adaptive Energy-Efficient Computing, HAEC), and Emmy Noether grant KR 4381/1-1.

References

1. Aberger, C.R., Tu, S., Olukotun, K., Ré, C.: EmptyHeaded: a relational engine for graph processing. In: Özcan, F., Koutrika, G., Madden, S. (eds.) Proceedings of the 2016 ACM SIGMOD International Conference on Management of Data, pp. 431–446. ACM (2016)
2. Abiteboul, S., Hull, R., Vianu, V.: Foundations of Databases. Addison Wesley, Reading (1994)
3. Aref, M., et al.: Design and implementation of the LogicBlox system. In: Sellis, T.K., Davidson, S.B., Ives, Z.G. (eds.) Proceedings of the 2015 ACM SIGMOD International Conference on Management of Data, pp. 1371–1382. ACM (2015)
4. Baget, J.-F., Leclère, M., Mugnier, M.-L., Rocher, S., Sipieter, C.: Graal: a toolkit for query answering with existential rules. In: Bassiliades, N., Gottlob, G., Sadri, F., Paschke, A., Roman, D. (eds.) RuleML 2015. LNCS, vol. 9202, pp. 328–344. Springer, Cham (2015). https://doi.org/10.1007/978-3-319-21542-6_21
5. Belleau, F., Nolin, M., Tourigny, N., Rigault, P., Morissette, J.: Bio2RDF: towards a mashup to build bioinformatics knowledge systems. J. Biomed. Inform. **41**(5), 706–716 (2008)
6. Bellomarini, L., Sallinger, E., Gottlob, G.: The Vadalog system: datalog-based reasoning for knowledge graphs. Proc. VLDB Endowment **11**(9), 975–987 (2018)
7. Benedikt, M., et al.: Benchmarking the chase. In: Sallinger, E., den Bussche, J.V., Geerts, F. (eds.) Proceedings of the 36th Symposium on Principles of Database Systems (PODS 2017), pp. 37–52. ACM (2017)
8. Benedikt, M., Leblay, J., Tsamoura, E.: PDQ: proof-driven query answering over web-based data. Proc. VLDB Endowment **7**(13), 1553–1556 (2014)
9. Bonifati, A., Ileana, I., Linardi, M.: Functional dependencies unleashed for scalable data exchange. In: Baumann, P., et al. (eds.) Proceedings of the 28th International Conference on Scientific and Statistical Database Management (SSDBM 2016), pp. 2:1–2:12. ACM (2016)

10. Borchmann, D.: Towards an error-tolerant construction of $\mathcal{EL}^{\perp}$-ontologies from data using formal concept analysis. In: Cellier, P., Distel, F., Ganter, B. (eds.) ICFCA 2013. LNCS (LNAI), vol. 7880, pp. 60–75. Springer, Heidelberg (2013). https://doi.org/10.1007/978-3-642-38317-5_4
11. Bordes, A., Usunier, N., García-Durán, A., Weston, J., Yakhnenko, O.: Translating embeddings for modeling multi-relational data. In: Burges, C.J.C., Bottou, L., Ghahramani, Z., Weinberger, K.Q. (eds.) Proceedings of the 27th Annual Conference on Neural Information Processing Systems (NIPS 2013), pp. 2787–2795 (2013)
12. Burgstaller-Muehlbacher, S., et al.: Wikidata as a semantic framework for the Gene Wiki initiative. Database 2016, baw015 (2016)
13. Calì, A., Gottlob, G., Lukasiewicz, T.: A general datalog-based framework for tractable query answering over ontologies. In: Paredaens, J., Su, J. (eds.) Proceedings of the 28th Symposium on Principles of Database Systems (PODS 2009), pp. 77–86. ACM (2009)
14. Calvanese, D., Giacomo, G.D., Lembo, D., Lenzerini, M., Rosati, R.: Tractable reasoning and efficient query answering in description logics: the DL-Lite family. J. Autom. Reasoning **39**(3), 385–429 (2007)
15. Carral, D., Dragoste, I., Krötzsch, M.: Restricted chase (non)termination for existential rules with disjunctions. In: Sierra, C. (ed.) Proceedings of the 26th International Joint Conference on Artificial Intelligence (IJCAI 2017), pp. 922–928 (2017). ijcai.org
16. Cuenca Grau, B., et al.: Acyclicity notions for existential rules and their application to query answering in ontologies. J. Artif. Intell. Res. **47**, 741–808 (2013)
17. Cyganiak, R., Wood, D., Lanthaler, M. (eds.): RDF 1.1 Concepts and Abstract Syntax. W3C Recommendation, 25 February 2014. http://www.w3.org/TR/rdf11-concepts/
18. Deutsch, A., Nash, A., Remmel, J.B.: The chase revisited. In: Lenzerini, M., Lembo, D. (eds.) Proceedings of the 27th Symposium on Principles of Database Systems (PODS 2008), pp. 149–158. ACM (2008)
19. Eiter, T., Ianni, G., Schindlauer, R., Tompits, H.: A uniform integration of higher-order reasoning and external evaluations in answer-set programming. In: Kaelbling, L., Saffiotti, A. (eds.) Proceeding 19th Internation Joint Conference on Artificial Intelligence (IJCAI 2005), pp. 90–96. Professional Book Center (2005)
20. Fagin, R., Kolaitis, P.G., Miller, R.J., Popa, L.: Data exchange: semantics and query answering. Theoret. Comput. Sci. **336**(1), 89–124 (2005)
21. Ferrucci, D.A., et al.: Building watson: an overview of the DeepQA project. AI Mag. **31**(3), 59–79 (2010)
22. Ganter, B., Wille, R.: Formal Concept Analysis: Mathematical Foundations. Springer, Heidelberg (1997)
23. Geerts, F., Mecca, G., Papotti, P., Santoro, D.: That's all folks! LLUNATIC goes open source. PVLDB **7**(13), 1565–1568 (2014)
24. González, L., Hogan, A.: Modelling dynamics in semantic web knowledge graphs with formal concept analysis. In: Champin, P., Gandon, F.L., Lalmas, M., Ipeirotis, P.G. (eds.) Proceedings of the 2018 World Wide Web Conference (WWW 2018), pp. 1175–1184. ACM (2018)
25. Hanika, T., Marx, M., Stumme, G.: Discovering implicational knowledge in Wikidata. In: Cristea, D., et al. (eds.) ICFCA 2019, LNAI 11511, pp. 315–323. Springer, Cham (2019)
26. Harris, S., Seaborne, A. (eds.): SPARQL 1.1 Query Language. W3C Recommendation, 21 March 2013. http://www.w3.org/TR/sparql11-query/

27. Hernández, D., Hogan, A., Krötzsch, M.: Reifying RDF: what works well with wikidata? In: Liebig, T., Fokoue, A. (eds.) Proceedings of the 11th International Workshop on Scalable Semantic Web Knowledge Base Systems. CEUR Workshop Proceedings, vol. 1457, pp. 32–47. CEUR-WS.org (2015)
28. Ho, V.T., Stepanova, D., Gad-Elrab, M.H., Kharlamov, E., Weikum, G.: Learning rules from incomplete KGs using embeddings. In: van Erp, M., Atre, M., López, V., Srinivas, K., Fortuna, C. (eds.) Posters & Demonstrations, Industry and Blue Sky Ideas Tracks of the 17th International Semantic Web Conference (ISWC 2018). CEUR Workshop Proceedings, vol. 2180. CEUR-WS.org (2018)
29. Hoffart, J., Suchanek, F.M., Berberich, K., Weikum, G.: YAGO2: a spatially and temporally enhanced knowledge base from Wikipedia. J. Artif. Intell. **194**, 28–61 (2013)
30. Kaminski, M., Grau, B.C., Kostylev, E.V., Motik, B., Horrocks, I.: Foundations of declarative data analysis using limit datalog programs. In: Sierra, C. (ed.) Proceedings of the 26th International Joint Conference on Artificial Intelligence (IJCAI 2017), pp. 1123–1130 (2017). ijcai.org
31. Khamis, M.A., Ngo, H.Q., Nguyen, X., Olteanu, D., Schleich, M.: In-database learning with sparse tensors. In: den Bussche, J.V., Arenas, M. (eds.) Proceedings of the 37th Symposium on Principles of Database Systems (PODS 2018), pp. 325–340. ACM (2018)
32. Kontchakov, R., Lutz, C., Toman, D., Wolter, F., Zakharyaschev, M.: The combined approach to ontology-based data access. In: Walsh, T. (ed.) Proceedings 22nd International Joint Conference on Artificial Intelligence (IJCAI 2011). pp. 2656–2661. AAAI Press/IJCAI (2011)
33. Krötzsch, M.: Ontologies for knowledge graphs? In: Artale, A., Glimm, B., Kontchakov, R. (eds.) Proceedings of the 30th International Workshop on Description Logics (DL 2017). CEUR Workshop Proceedings, vol. 1879. CEUR-WS.org (2017)
34. Krötzsch, M., Marx, M., Ozaki, A., Thost, V.: Attributed description logics: Reasoning on knowledge graphs. In: Lang, J. (ed.) Proceedings of the 27th International Joint Conference on Artificial Intelligence (IJCAI 2018), pp. 5309–5313 (2018). https://doi.org/10.24963/ijcai.2018/743
35. Krötzsch, M., Thost, V.: Ontologies for knowledge graphs: breaking the rules. In: Groth, P., et al. (eds.) ISWC 2016. LNCS, vol. 9981, pp. 376–392. Springer, Cham (2016). https://doi.org/10.1007/978-3-319-46523-4_23
36. Krötzsch, M., Marx, M., Rudolph, S.: The power of the terminating chase. In: Barceló, P., Calautti, M. (eds.) Proceedings of the 22nd International Conference on Database Theory (ICDT 2019). LIPIcs, vol. 127, pp. 3:1–3:17. Schloss Dagstuhl - Leibniz-Zentrum fuer Informatik (2019)
37. Lenzerini, M.: Data integration: a theoretical perspective. In: Popa, L. (ed.) Proceedings of the 21st Symposium on Principles of Database Systems (PODS 2002), pp. 233–246. ACM (2002)
38. Leskovec, J., Rajaraman, A., Ullman, J.D.: Mining of Massive Datasets, 2nd edn. Cambridge University Press, Cambridge (2014)
39. Malyshev, S., Krötzsch, M., González, L., Gonsior, J., Bielefeldt, A.: Getting the most out of wikidata: semantic technology usage in Wikipedia's knowledge graph. In: Vrandečić, D., et al. (eds.) ISWC 2018. LNCS, vol. 11137, pp. 376–394. Springer, Cham (2018). https://doi.org/10.1007/978-3-030-00668-6_23
40. Marnette, B.: Generalized schema-mappings: from termination to tractability. In: Paredaens, J., Su, J. (eds.) Proceedings of the 28th Symposium on Principles of Database Systems (PODS 2009), pp. 13–22. ACM (2009)

41. Marx, M., Krötzsch, M., Thost, V.: Logic on MARS: Ontologies for generalised property graphs. In: Sierra, C. (ed.) Proceedings of the 26th International Joint Conference on Artificial Intelligence (IJCAI 2017), pp. 1188–1194 (2017)
42. Nenov, Y., Piro, R., Motik, B., Horrocks, I., Wu, Z., Banerjee, J.: RDFox: a highly-scalable RDF store. In: Arenas, M., et al. (eds.) ISWC 2015. LNCS, vol. 9367, pp. 3–20. Springer, Cham (2015). https://doi.org/10.1007/978-3-319-25010-6_1
43. Nickel, M., Murphy, K., Tresp, V., Gabrilovich, E.: A review of relational machine learning for knowledge graphs. Proc. IEEE **104**(1), 11–33 (2016)
44. openCypher community: Cypher Query Language Reference, Version 9 (2019). http://www.opencypher.org/resources
45. Prud'hommeaux, E., Seaborne, A. (eds.): SPARQL Query Language for RDF. W3C Recommendation, 15 January 2008. http://www.w3.org/TR/rdf-sparql-query/
46. Rodriguez, M.A., Neubauer, P.: Constructions from dots and lines. Bull. Am. Soc. Inf. Sci. Technol. **36**(6), 35–41 (2010)
47. Rudolph, S.: Exploring relational structures via $\mathcal{FLE}$. In: Wolff, K.E., Pfeiffer, H.D., Delugach, H.S. (eds.) ICCS-ConceptStruct 2004. LNCS (LNAI), vol. 3127, pp. 196–212. Springer, Heidelberg (2004). https://doi.org/10.1007/978-3-540-27769-9_13
48. Simonite, T.: Inside the Alexa-friendly world of Wikidata. WIRED Magazine 27.03 (2019). https://www.wired.com/story/inside-the-alexa-friendly-world-of-wikidata/. Accessed 16 Mar 2019
49. Tanon, T.P., Vrandecic, D., Schaffert, S., Steiner, T., Pintscher, L.: From freebase to wikidata: the great migration. In: Bourdeau, J., Hendler, J., Nkambou, R., Horrocks, I., Zhao, B.Y. (eds.) Proceedings of the 25th International Conference on World Wide Web (WWW 2016), pp. 1419–1428. ACM (2016)
50. Urbani, J., Jacobs, C., Krötzsch, M.: Column-oriented Datalog materialization for large knowledge graphs. In: Schuurmans, D., Wellman, M.P. (eds.) Proceedings of the 30th AAAI Conference on Artificial Intelligence (AAAI 2016), pp. 258–264. AAAI Press (2016)
51. Urbani, J., Krötzsch, M., Jacobs, C., Dragoste, I., Carral, D.: Efficient model construction for horn logic with VLog. In: Galmiche, D., Schulz, S., Sebastiani, R. (eds.) IJCAR 2018. LNCS (LNAI), vol. 10900, pp. 680–688. Springer, Cham (2018). https://doi.org/10.1007/978-3-319-94205-6_44
52. Vrandečić, D., Krötzsch, M.: Wikidata: a free collaborative knowledgebase. Commun. ACM **57**(10), 78–85 (2014)
53. Wagner, C., Graells-Garrido, E., Garcia, D., Menczer, F.: Women through the glass ceiling: gender asymmetries in Wikipedia. EPJ Data Sci. **5**(1), 5 (2016)
54. Wang, Z., Zhang, J., Feng, J., Chen, Z.: Knowledge graph embedding by translating on hyperplanes. In: Brodley, C.E., Stone, P. (eds.) Proceedings of the 28th AAAI Conference on Artificial Intelligence (AAAI 2014), pp. 1112–1119. AAAI Press (2014)

Learning Implications from Data and from Queries

Sergei Obiedkov(✉)

Faculty of Computer Science, National Research University
Higher School of Economics, Moscow, Russia
sergei.obj@gmail.com

Abstract. In this paper, we consider computational problems related to finding implications in an explicitly given formal context or via queries to an oracle. We are concerned with two types of problems: enumerating implications (or association rules) and finding a single implication satisfying certain conditions. We present complexity results for some of these problems and leave others open. The paper is not meant as a comprehensive survey, but rather as a subjective selection of interesting problems.

Keywords: Formal concept analysis · Learning with queries · Attribute exploration · Implications · Association rules

1 Introduction

Implications are an important tool in formal concept analysis, closely related to Horn formulas in logic and association rules in data mining. The aim of this paper is to present a number of computational problems related to implications that either have been solved or, probably, deserve to be solved.

After introducing necessary definitions in Sect. 2, we recall, in Sect. 3, the algorithm for abduction using a formal context from [21], where it is presented in terms of Horn formulas and their characteristic models. The problem here is to find an explanation why particular object contains a particular attribute in form of a valid implication with a minimal premise and this attribute in conclusion. This problem is solvable in polynomial time, but, if we introduce a requirement on the minimal support of implications that could serve as explanations or relax the validity requirement replacing it by a minimal threshold for the confidence, the problem, as we show, becomes NP-hard.

We then move to the problem of learning a logically complete set of implications valid in a formal context. In Sect. 4, we describe two settings for learning: in one the formal context is given explicitly, while in the other it is available only through queries of a certain kind posed to an oracle. The latter is Angluin's framework of exact learning with queries [2], which is known in formal concept analysis in the form of attribute exploration [16]. Implications can be learnt

Supported by the Russian Science Foundation (grant 17-11-01294).

D. Cristea et al. (Eds.): ICFCA 2019, LNAI 11511, pp. 32–44, 2019.
https://doi.org/10.1007/978-3-030-21462-3_3

in total polynomial time when so-called membership and equivalence queries are available [3], but whether the same efficiency can be achieved when learning implications directly from a formal context is a major open problem [22].

In Sect. 5, we recall the notion of Horn approximation and its stronger version from [12] and then consider the existence of probabilistic algorithms for learning approximations in different settings. In some cases, the existence of such algorithms has been established; we state the other cases as open problems.

2 Basic Definitions

We start by recalling some definitions from formal concept analysis (FCA) [17]. Given a *(formal) context* $\mathbb{K} = (G, M, I)$, where G is called a set of *objects*, M is called a set of *attributes*, and the binary relation $I \subseteq G \times M$ specifies which objects have which attributes, the *derivation operators* $(\cdot)^I$ are defined for $A \subseteq G$ and $B \subseteq M$ as follows:

$$A^I = \{m \in M \mid \forall g \in A\colon (g, m) \in I\}$$
$$B^I = \{g \in G \mid \forall m \in B\colon (g, m) \in I\}$$

A^I is the set of attributes shared by objects of A, and B^I is the set of objects having all attributes of B. Often, $(\cdot)'$ is used instead of $(\cdot)^I$. The double application of $(\cdot)'$ defines two closure operators; sets A'' and B'' are said to be *closed*. Closed attribute sets are called *concept intents* of $\mathbb{K}$. The set of all concept intents of $\mathbb{K}$ is denoted by $\operatorname{Int}\mathbb{K}$. For $g \in G$, we call $\{g\}'$ the *object intent* of g.

An *implication* is an expression $A \to B$, where $A, B \subseteq M$ are attribute subsets. We say that $X \subseteq M$ is a *model* of $A \to B$ if $A \nsubseteq X$ or $B \subseteq X$ (notation: $X \models A \to B$); it is a model of an implication set $\mathcal{L}$ if it is a model of every implication in $\mathcal{L}$ (notation: $X \models \mathcal{L}$). In this case, we also say that X is *closed* under $\mathcal{L}$. We denote by $\operatorname{Mod}\mathcal{L}$ the set of all models of an implication set $\mathcal{L}$. It is well known that $\operatorname{Mod}\mathcal{L}$ is closed under intersection. The unique inclusion-minimal model $Y \in \operatorname{Mod}\mathcal{L}$ such that $X \subseteq Y$ is called the *closure of* X *under* $\mathcal{L}$ and is denoted by $\mathcal{L}(X)$.

An implication $A \to B$ *is valid* in the context $\mathbb{K}$ (notation: $\mathbb{K} \models A \to B$) if $A' \subseteq B'$, i.e., every object of the context with all attributes from A also has all attributes from B or, in other words, the set of object intents of $\mathbb{K}$ is a subset of $\operatorname{Mod}\{A \to B\}$.

If $A' = \varnothing$, then $(G, M, I) \models A \to M$. We use special notation for such implications: $A \to \bot$.

We will sometimes omit curly brackets around one-element sets and write, e.g., g' instead of $\{g\}'$ and $A \to m$ instead of $A \to \{m\}$.

Given an implication set, it may be desirable to find an equivalent implication set, i.e., one with the same set of models, containing as few implications as possible. The *Duquenne–Guigues* or *canonical basis* of an implication set is minimal in this sense [18]. The *canonical basis* of a formal context is the canonical basis of the set of its valid implications. Other implication bases have been considered in the literature [1,10,27].

The *support* of an attribute subset $A \subseteq M$ in a formal context $\mathbb{K}$ is defined as $|A'|$. The *support* of an implication $A \to B$ in $\mathbb{K}$, denoted by support$(A \to B, \mathbb{K})$, is $|(A \cup B)'|$. An attribute subset or an implication is called *frequent* if its support is equal to or greater than some specified `minsupp` threshold.

Frequent implications, also known as *exact association rules*, are important in data mining, since they are considered as those strongly supported by data. Such implications may be interesting even if they have counterexamples in data provided that they still exhibit sufficiently high *confidence*, which is defined as follows:

$$\text{confidence}(A \to B, \mathbb{K}) = \frac{|(A \cup B)'|}{|A'|}$$

for $A' \neq \varnothing$. If $A' = \varnothing$, then confidence$(A \to B, \mathbb{K}) = 1$. We usually talk about an *association rule* $A \to B$ when we do not assume that it is a valid implication of our formal context, that is, when its confidence may be below 1. Concise representations for sets of association rules are considered in [23].

We should also mention the close connection between implications and Horn formulas, since many of the results on which we rely in this paper have been obtained for Horn formulas. A propositional *Horn formula* is a conjunction of propositional *Horn clauses*, each such clause being a disjunction of literals at most one of which is without negation. An implication $A \to B$ corresponds to a *Horn formula*:

$$\bigwedge_{b \in B} (b \vee \bigvee_{a \in A} \neg a).$$

An implication set $\mathcal{L}$ corresponds to the conjunction of Horn formulas corresponding to individual implications from $\mathcal{L}$, which is itself a Horn formula.

3 Abduction and Classification Based on Implications and Association Rules

Suppose that, having a context $\mathbb{K} = (G, M, I)$, we would like to suggest an explanation for why some object $g \in G$ has some attribute $m \in \{g\}'$. One possible approach is described in Problem 1.

Problem 1. Given a formal context $\mathbb{K} = (G, M, I)$, a set $B \subseteq M$, and an attribute $m \in B'' \setminus B$, find, if it exists, a minimal (with respect to subset inclusion) *explanation* $A \subseteq B$ such that $\mathbb{K} \models A \to m$ and $A' \neq \varnothing$.

This is a reformulation in terms of formal concept analysis of *abduction* as it is defined in [21]. A similar problem occurs when implications are used to classify newly observed objects with respect to some target attribute m [24].

A simple polynomial-time algorithm solving Problem 1 is described in [21]: search for an object $g \in G$ such that $\mathbb{K} \models B \cap \{g\}' \to m$ and, if found, remove attributes from $B \cap \{g\}'$ to obtain an inclusion-minimal set A such that $A \to \{m\}$ is still valid in $\mathbb{K}$. Thus we have

Theorem 1. *Problem 1 is solvable in polynomial time.*

Can we obtain a similar result if we insist on explanations based on frequent implications or if we are flexible enough to accept explanations based on association rules with relatively high confidence? As we show next, the answer is most likely negative in both cases.

Problem 2. Given a formal context $\mathbb{K} = (G, M, I)$, an attribute $m \in M$, and a $\texttt{minsupp} \in \mathbb{N}$ threshold, decide if there is a subset $A \subseteq M \setminus \{m\}$ such that $\text{support}(A \to m, \mathbb{K}) \geq \texttt{minsupp}$ and $\text{confidence}(A \to m, \mathbb{K}) = 1$.

Note that, in this problem, we do not put any restrictions on which attributes to include in the explanation and we do not require it to be minimal. Nevertheless, the problem becomes hard because of the support requirement.

Theorem 2. *Problem 2 is* NP*-complete.*

Proof. Given a set $A \subseteq M \setminus \{m\}$, it can be checked in polynomial time if it has the properties stated in Problem 2. Therefore, the problem is in NP.

To prove NP-hardness, we describe a reduction from the Vertex Cover problem, which asks whether a graph (V, E) contains a set $V_1 \subseteq V$ of vertices of size at most k such that every edge from E is incident to some $v \in V_1$. For a given graph (V, E), we define the corresponding formal context $\mathbb{K} = (V \cup E, V \cup \{m\}, I)$, where $m \notin V$ and

$$\begin{aligned} I = &\{(v_1, v_2) \mid v_1 \in V, v_2 \in V \cup \{m\}, v_1 \neq v_2\} \\ &\cup \{(e, v) \mid e \in E, v \in V, v \notin e\}. \end{aligned}$$

That is, the context $\mathbb{K}$ contains a single object for every vertex and every edge of graph (V, E); the attributes of $\mathbb{K}$ are vertices of the graph and one additional attribute m. An object corresponding to a vertex v has all attributes except v; an object corresponding to edge (u, v) has all attributes except u, v, and m. We claim that (V, E) has a vertex cover of size at most k if and only if $\text{support}(A \to m, \mathbb{K}) \geq |V| - k$ and $\text{confidence}(A \to m, \mathbb{K}) = 1$ for some $A \subseteq V$.

First, note that $|A'| \geq |V| - |A|$ for every $A \subseteq V$. This is so, because $v \in A'$ if and only if $v \notin A$.

Now suppose that A is a vertex cover of (V, E) and $|A| = k$. Then, for every (u, v) in E, we have $u \in A$ or $v \in A$. This means that, for every $e \in E$, there is $v \in A$ such that object e does not have attribute v in $\mathbb{K}$. Hence, when considering A as an attribute set, $A' \subseteq V$ and $\text{confidence}(A \to m, \mathbb{K}) = 1$, since vIm for every $v \in V$. From $|A| = k$, we have $\text{support}(A \to m, \mathbb{K}) = |A'| \geq |V| - k$.

Conversely, let $A \subseteq M \setminus \{m\}$ satisfy the required conditions for support and confidence. Since $\text{confidence}(A \to m, \mathbb{K}) = 1$, there is no $e \in E$ with $A \subseteq \{e\}'$. This can happen only if every edge is incident to at least some $v \in A$, that is, when A is a vertex cover. From $|A'| = \text{support}(A \to m, \mathbb{K}) \geq |V| - k$, we obtain $|A| \leq k$. Therefore, A is a vertex cover of size at most k.

Thus, the reduction is correct. Clearly, it can be carried out in polynomial time. □

For explanations based on association rules, the problem is hard even for constant confidence thresholds.

Problem 3. Given a formal context $\mathbb{K} = (G, M, I)$ and an attribute $m \in M$, decide if there is a subset $A \subseteq M\backslash\{m\}$ such that confidence$(A \to \{m\}, \mathbb{K}) \geq 1/\alpha$ for some natural constant $\alpha > 1$.

Theorem 3. *Problem 3 is NP-complete.*

Proof. The problem is obviously in NP. To prove NP-hardness, we reduce Problem 2 to this problem. In Problem 2, we are given a context $\mathbb{K} = (G, M, I)$, an attribute $m \in M$, and a $\texttt{minsupp}$ threshold. If $\texttt{minsupp} > |G|$ or $\texttt{minsupp} = 0$, the reduction is trivial. Assume that $0 < \texttt{minsupp} \leq |G|$.

We build a context $\mathbb{K}_c = (H, M, J)$, where $H = G \cup G_1 \cup G_2$, $J = I \cup I_1 \cup I_2$, $I_i \subseteq G_i \times M$ for $i \in \{1, 2\}$, and

G_1 contains $\alpha|G|$ copies of each object $g \in G \setminus \{m\}^I$; for each such copy h of object g, we have $(h, n) \in I_1$ if and only if $(g, n) \in I$;
$G_2 = \{h_1, \ldots, h_{(\alpha-1)\texttt{minsupp}}\}$ and $I_2 = G_2 \times (M \setminus \{m\})$.

Note that $G_2 \neq \varnothing$, since $\alpha > 1$ and $\texttt{minsupp} > 0$. It is clear that this reduction can be carried out in polynomial time. To prove its correctness, we show that support$(A \to \{m\}, \mathbb{K}) \geq \texttt{minsupp}$ and confidence$(A \to \{m\}, \mathbb{K}) = 1$ for some $A \subseteq M \setminus \{m\}$ if an only if confidence$(A \to \{m\}, \mathbb{K}_c) \geq 1/\alpha$.

Suppose that, for some $A \subseteq M \setminus \{m\}$, support$(A \to m, \mathbb{K}) \geq \texttt{minsupp}$ and confidence$(A \to m, \mathbb{K}) = 1$. Then $A \subseteq \{g\}^I$ for no object g in $G \setminus \{m\}^I$ and, consequently, $A \subseteq \{h\}^J$ for no object $h \in G_1$ in $\mathbb{K}_c$. Hence,

$$\text{confidence}(A \to m, \mathbb{K}_c) = \frac{|A^I|}{|A^I| + |G_2|} \geq \frac{\texttt{minsupp}}{\texttt{minsupp} + (\alpha - 1)\texttt{minsupp}} = 1/\alpha.$$

Conversely, suppose that confidence$(A \to m, \mathbb{K}_c) \geq 1/\alpha$ for some subset $A \subseteq M\backslash\{m\}$. In this case, $A \subseteq \{g\}^I$ for no $g \in G\backslash\{m\}^I$, since otherwise A would have been contained in the intents of $\alpha|G|$ copies of g in G_1, as well as in all object intents of G_2, and we would have had confidence$(A \to m, \mathbb{K}_c) < |A^I|/(\alpha|G|) \leq 1/\alpha$. Therefore, confidence$(A \to m, \mathbb{K}) = 1$, support$(A \to m, \mathbb{K}) = |A^I|$, and confidence$(A \to m, \mathbb{K}_c) = |A^I|/(|A^I| + |G_2|)$. If support$(A \to m, \mathbb{K}) < \texttt{minsupp}$, then

$$\text{confidence}(A \to m, \mathbb{K}_c) < \frac{\texttt{minsupp}}{\texttt{minsupp} + (\alpha - 1)\texttt{minsupp}} = 1/\alpha.$$

Thus, support$(A \to m, \mathbb{K}) \geq \texttt{minsupp}$ and the reduction is correct. □

An interesting question, which we leave open, is whether approximation algorithms are possible for the optimization versions of Problems 2 and 3:

Problem 4. Given a formal context $\mathbb{K} = (G, M, I)$ and an attribute $m \in M$, find the maximal value of support$(A \to m, \mathbb{K})$ for any implication $A \to m$ valid in $\mathbb{K}$, where $A \subseteq M \setminus \{m\}$.

Problem 5. Given a formal context $\mathbb{K} = (G, M, I)$ and an attribute $m \in M$, find the maximal value of confidence$(A \to \{m\}, \mathbb{K})$ for any $A \subseteq M \setminus \{m\}$.

4 Exact Learning of Implications

In this section, we consider learning a logically complete implication set of a formal context when the context is available either explicitly or through queries of certain types. Since the size of the canonical basis of a formal context can be exponential in the size of the context [21], we can only hope for an algorithm that runs in total polynomial time (that is, time polynomial in the combined size of input and output) or, more ambitiously, for an algorithm enumerating implications of the canonical basis with a polynomial delay (which, roughly speaking, means spending time polynomial in the input size between outputting two successive implications).

Problem 6. Given a formal context $\mathbb{K} = (G, M, I)$, compute its canonical basis.

The best-known algorithm in formal concept analysis for solving this problem is Next Closure [15]. However, this algorithm takes exponential time in the worst case, since it enumerates all closed subsets of M as a side product. Designing an algorithm for Problem 6 that runs in total polynomial time is a major open question related to other important problems such as enumerating minimal transversals in a hypergraph [14,22]. There have been a number of negative results for this problem. For example, unless P= NP, it is not possible to enumerate the implications of the canonical basis with a polynomial delay either in the lectic order of their premises [13] or in the reverse order [9].

In the framework of exact learning with queries, rather than learning from a training dataset, the learning algorithm has access to a teacher (or oracle), which it can address with certain predefined types of queries [2]. We consider three types of queries defined below.

Definition 1. *A* membership query *for* $\hat{\mathcal{S}} \subseteq 2^M$ *is an attribute subset* $T \subseteq M$*; the answer to the membership query is "yes" if* $T \in \hat{\mathcal{S}}$ *and "no" otherwise.*

Definition 2. *An* equivalence query *for* $\hat{\mathcal{S}} \subseteq 2^M$ *is* $\mathcal{S} \subseteq 2^M$*; the answer to the equivalence query is "yes" if* $\mathcal{S} = \hat{\mathcal{S}}$ *and a* counterexample $T \in \mathcal{S} \triangle \hat{\mathcal{S}}$*, where* $\triangle$ *is symmetric difference, otherwise. A counterexample* T *is* negative *if* $T \in \mathcal{S} \setminus \hat{\mathcal{S}}$ *and* positive *if* $T \in \hat{\mathcal{S}} \setminus \mathcal{S}$*.*

Here the set $\mathcal{S}$ is not necessarily specified explicitly. In the algorithms discussed below, $\mathcal{S}$ is specified by an implication set $\mathcal{L}$ such that $\mathcal{S} = \operatorname{Mod} \mathcal{L}$.

Definition 3. *An* implication query *for* $\hat{\mathcal{S}} \subseteq 2^M$ *is an implication* $A \to B$ *over* M*; the answer to the implication query is "yes" if* $T \models A \to B$ *for all* $T \in \hat{\mathcal{S}}$ *and a* positive counterexample $T \in \hat{\mathcal{S}}$ *such that* $T \not\models A \to B$ *otherwise.*

Equivalence and implication queries are called *restricted* if they return a simple "no" instead of providing a counterexample. We will also talk about an *oracle for a formal context* $\mathbb{K}$ meaning an oracle capable of answering queries for $\operatorname{Int} \mathbb{K}$.

Problem 7. Given an oracle capable of answering membership and equivalence queries for $\mathrm{Mod}\,\hat{\mathcal{L}}$, where $\hat{\mathcal{L}}$ is an implication set over M, compute the canonical basis of $\hat{\mathcal{L}}$.

An algorithm for learning Horn formulas with equivalence and membership queries is described in [3], where it is proved that it requires time polynomial in the number of variables, n, and the number of clauses, m, of the target Horn formula; $O(mn)$ equivalence queries and $O(m^2n)$ membership queries are made in the process. The algorithm starts with the empty hypothesis, compatible with every possible assignment and proceeds until a positive answer is obtained from an equivalence query. If a negative example X is received instead, the algorithm uses membership queries to find an implication $A \to B$ in the current hypothesis $\mathcal{H}$ such that $A \nsubseteq X$ and $A \cap X$ is not a model of the target Horn formula. If such an implication is found, the implication $A \to B$ is replaced by $A \cap X \to B$, which ensures that X is no longer a model of $\mathcal{H}$; otherwise, the implication $X \to \bot$ is added to $\mathcal{H}$. When a positive counterexample X is obtained from an equivalence query, every implication $A \to B$ of which X is not a model is replaced by $A \to B \cap X$ (with some twists if $B = \bot$). We refer the reader to [3] for further details.

In [6], it is shown that this algorithm always produces the canonical basis of the implication set corresponding to the target Horn sentence no matter what examples are received from the equivalence queries. Thus, we have

Theorem 4. *Problem 7 is solvable in total polynomial time.*

Implication queries form the basis of the attribute exploration method from formal concept analysis [16]. The concrete algorithm implementing the method is a modification of the Next Closure algorithm, which, as said above, takes exponential time in the worst case; it may also need an exponential number of queries, since it needs at least one query for every characteristic model of the canonical basis (which is a model that cannot be represented as an intersection of other models), and the number of characteristic models can be exponential in both the number of attributes and the size of the canonical basis [16,21].

However, a membership query can be simulated by a polynomial number of implication queries (see Algorithm 1); this is even possible with restricted implication queries (see Algorithm 2) [7,12]. This gives us

Theorem 5. *Problem 7 is solvable in total polynomial time even if the oracle is capable of answering implication instead of membership queries.*

5 Horn Approximations

We now consider learning approximations of implication sets in settings where no polynomial-time exact algorithms are known. We use two notions of approximation from [12].

Algorithm 1. ISMEMBER(A, *is_valid*$(\cdot)$)

Input: A set $A \subseteq M$ and an implication oracle *is_valid*$(\cdot)$ for some formal context $\mathbb{K} = (G, M, I)$.
Output: **true** if $A = A''$ in $\mathbb{K}$ and **false** otherwise.

```
if is_valid(A → ⊥) then
    return false
B := M
while A ≠ B do
    if is_valid(A → B) returns a counterexample C then
        B := B ∩ C
    else
        return false
return true
```

Algorithm 2. ISMEMBER2(A, *is_valid*$(\cdot)$)

Input: A set $A \subseteq M$ and an implication oracle *is_valid*$(\cdot)$ for some formal context $\mathbb{K} = (G, M, I)$.
Output: **true** if $A = A''$ in $\mathbb{K}$ and **false** otherwise.

```
if is_valid(A → ⊥) then
    return false
for all a ∈ M \ A do
    if is_valid(A → {a}) then
        return false
return true
```

Definition 4. *Let $\mathbb{K} = (G, M, I)$ be a formal context and $\mathcal{L}$ be a set of implications over M. We call $\mathcal{L}$ an* ε-Horn approximation of $\mathbb{K}$ *if*

$$\frac{|\operatorname{Int}\mathbb{K} \bigtriangleup \operatorname{Mod}\mathcal{L}|}{2^{|M|}} \leq \varepsilon.$$

We call $\mathcal{L}$ a ε-strong Horn approximation of $\mathbb{K}$ *if*

$$\frac{|\{A \subseteq M \mid A'' \neq \mathcal{L}(A)\}|}{2^{|M|}} \leq \varepsilon.$$

An ε-strong Horn approximation of $\mathbb{K}$ is always its ε-Horn approximation, but the reverse is not true.

5.1 Learning Implications with Membership Queries

It is not possible to exactly learn an implication set with a polynomial number of queries if only membership queries are available. Indeed, if the target implication set is $\{\varnothing \to M\}$, where M is the attribute set, any membership query for $X \subsetneq M$ will be answered negatively—and such a query has to be asked for every X in order to exclude the implication set $\{\varnothing \to X\}$ from the hypothesis space. Therefore, if only membership queries are available, we may hope only to compute an approximation of the target implication set.

An exact-learning polynomial-time algorithm with equivalence queries can be transformed into a probably approximately correct (PAC) algorithm without equivalence queries [2]. The idea is to use a random sampling strategy to search for a counterexample instead of relying on equivalence queries to provide such counterexamples. More precisely, it is assumed that there is a *sampling oracle* $ex(\cdot)$ that draws an element from the universal set according to some distribution and indicates whether it is an instance of the concept being learnt. When learning an implication set $\hat{\mathcal{L}}$ over M, such an oracle, when called, would return a subset of M with information whether it is closed under $\hat{\mathcal{L}}$.

Consider the following problem of computing an ε-Horn approximation of $\mathbb{K}$:

Problem 8. Given a membership oracle for a formal context $\mathbb{K}$ and parameters $0 < \varepsilon \leq 1$ and $0 < \delta \leq 1$, compute, with probability at least $1 - \delta$, an ε-Horn approximation of $\mathbb{K}$.

To solve Problem 8, we simulate a call to $ex(\cdot)$ by generating a subset $A \subseteq M$ uniformly at random and calling the membership oracle for $\mathbb{K}$ with A as input so as to learn whether A is closed in $\mathbb{K}$. Combined with the HORN1 algorithm from [3] implied in Theorem 4, this results in

Theorem 6. *There is a randomized algorithm that solves Problem 8 in time polynomial in $|M|$, $|\hat{\mathcal{L}}|$, $1/\varepsilon$, and $1/\delta$, where $\hat{\mathcal{L}}$ is the canonical basis of $\mathbb{K}$.*

The reason the strategy described above works is that, at any point of execution of HORN1, either the implication set $\mathcal{L}$ computed by this point is already an ε-Horn approximation of $\mathbb{K}$, or

$$\frac{|\operatorname{Int}\mathbb{K} \bigtriangleup \operatorname{Mod}\mathcal{L}|}{2^{|M|}} > \varepsilon$$

and, therefore, the probability of generating a required counterexample is bounded below by ε and it can be appropriately amplified by repeated trials (see details in [2] or [12]).

Can the algorithm from Theorem 6 be modified to compute ε-strong Horn approximations? If, at some point,

$$\frac{|\{A \subseteq M \mid A'' \neq \mathcal{L}(A)\}|}{2^{|M|}} > \varepsilon,$$

then, by generating X uniformly at random, we obtain X such that $X'' \neq \mathcal{L}(X)$ with probability greater than ε. The problem is that this X is not necessarily a counterexample in the sense required by the algorithm, because it may happen that it is closed neither in $\mathbb{K}$ nor under $\mathcal{L}$. We conjecture that, for the ε-strong Horn approximation, a result similar to Theorem 6 is not possible, but, for now, leave it as an open problem:

Problem 9. Given a membership oracle for a formal context $\mathbb{K}$ and parameters $0 < \varepsilon \leq 1$ and $0 < \delta \leq 1$, compute, with probability at least $1 - \delta$, an ε-strong Horn approximation of $\mathbb{K}$.

5.2 Learning Horn Approximations with Implication Queries

Since a membership query can be simulated by a polynomial number of implication queries (see Algorithms 1 and 2), Theorem 6 immediately gives us a polynomial-time randomized algorithm for computing an ε-Horn approximation of an implication set with (restricted) implication queries.

This algorithm can be easily transformed into an algorithm computing ε-strong Horn approximations [12]. Here the difference from learning with membership queries is that, when we generate X such that $X'' \neq \mathcal{L}(X)$, we can easily obtain from it a counterexample of the type required by an equivalence query. To do this, we compute both X'' and $\mathcal{L}(X)$. We can compute X'', for instance, by querying the oracle to verify the implications of the form $X \to m$, since $X'' = \{m \in M \mid \mathbb{K} \models X \to m\}$. If $X'' \subsetneq \mathcal{L}(X)$, then X'' is a positive a counterexample; otherwise, if $X'' \neq \mathcal{L}(X)$, then $\mathcal{L}(X)$ is a negative counterexample. A different method for producing a counterexample from a randomly generated X is described in [12]. Unlike the method just described, that method is not applicable when only restricted queries are available.

Thus we have:

Problem 10. Given a (restricted) implication oracle for a context $\mathbb{K} = (G, M, I)$ and parameters $0 < \varepsilon \leq 1$ and $0 < \delta \leq 1$, compute, with probability at least $1 - \delta$, an ε- or ε-strong Horn approximation of $\mathbb{K}$.

Theorem 7. *There is a randomized algorithm that solves Problem 10 in time polynomial in* $|M|$, $|\hat{\mathcal{L}}|$, $1/\varepsilon$, *and* $1/\delta$, *where* $\hat{\mathcal{L}}$ *is the canonical basis of* $\mathbb{K}$.

5.3 Learning Horn Approximations from Data

Problem 11. Given a formal context $\mathbb{K} = (G, M, I)$, $0 < \varepsilon \leq 1$, and $0 < \delta \leq 1$, compute, with probability at least $1 - \delta$, an ε- or ε-strong Horn approximation of $\mathbb{K}$.

When we learn implications of an explicitly given formal context, membership queries can be answered in polynomial time by consulting the context: a membership query for A with respect to the context (G, M, I) receives a positive answer if and only if $A'' = A$, which can be checked in time $O(|G||M|)$. Therefore, from Theorem 6, we obtain a randomized algorithm for learning an ε-Horn approximation of a formal context, which was firstly presented (in a different setting) in [21].

Of course, when the context is available, implication queries can also be answered in polynomial time, which, together with Theorem 7, ensures that we can learn ε-strong Horn approximations as well.

Theorem 8. *There is a randomized algorithm that solves Problem 11 in time polynomial in* G, $|\hat{\mathcal{L}}|$, $1/\varepsilon$, *and* $1/\delta$, *where* $\hat{\mathcal{L}}$ *is the canonical basis of* $\mathbb{K}$.

Frequent Implications. It would be interesting to see if the randomized approach described above can be modified to approximately compute frequent implications. We state it as an open problem after appropriately generalizing our notions of approximation.

Definition 5. *Let* $\mathbb{K} = (G, M, I)$ *be a formal context,* $\mathcal{L}$ *be a set of implications over* M*, and* $\mathfrak{C} \subseteq \mathfrak{P}(M)$ *be a collection of "interesting" subsets of* M*. We call* $\mathcal{L}$ *a* conditional ε-Horn approximation of $\mathbb{K}$ with respect to $\mathfrak{C}$ *if*

$$\frac{|(\operatorname{Int}\mathbb{K} \bigtriangleup \operatorname{Mod}\mathcal{L}) \cap \mathfrak{C}|}{|\mathfrak{C}|} \leq \varepsilon.$$

We call $\mathcal{L}$ *a* conditional ε-strong Horn approximation of $\mathbb{K}$ with respect to $\mathfrak{C}$ *if*

$$\frac{|\{A \in \mathfrak{C} \mid A'' \neq \mathcal{L}(A)\}|}{|\mathfrak{C}|} \leq \varepsilon.$$

Problem 12 (Approximate learning of frequent implications from data). Given a formal context $\mathbb{K}$, a `minsupp` threshold, $0 < \varepsilon \leq 1$, and $0 < \delta \leq 1$, compute, with probability at least $1 - \delta$, a conditional ε- or ε-strong Horn approximation of $\mathbb{K}$ with respect to $\{F \subseteq M \mid |F'| \geq \texttt{minsupp}\}$.

6 Conclusion

We have discussed several computational problems related to learning implications and association rules from data and via queries. Some of these problems admit theoretically efficient algorithms, which may still benefit from experimental assessment. An empirical evaluation of the approximation quality of the randomized algorithm for computing the canonical basis implied in Theorem 8 is carried out in [11], but a proper efficiency comparison with deterministic algorithms is still missing.

Regarding algorithms learning implications from queries, other things worth considering include allowing for errors in responses to queries (cf. [4,5]). It would be interesting to adapt the query learning model to learn from multiple oracles with complementary or conflicting views and, based on that, design a framework for collaborative attribute exploration [19,20,26].

We also plan to apply probably approximately correct attribute exploration to complete ontologies based on description logics, as a follow-up to [8], where the classic attribute exploration is used for the same purpose.

References

1. Adaricheva, K., Nation, J.: Discovery of the D-basis in binary tables based on hypergraph dualization. Theor. Comput. Sci. **658**, 307–315 (2017)
2. Angluin, D.: Queries and concept learning. Mach. Learn. **2**(4), 319–342 (1988)
3. Angluin, D., Frazier, M., Pitt, L.: Learning conjunctions of Horn clauses. Mach. Learn. **9**(2–3), 147–164 (1992)

4. Angluin, D., Kriis, M., Sloan, R.H., Turán, G.: Malicious omissions and errors in answers to membership queries. Mach. Learn. **28**(2), 211–255 (1997)
5. Angluin, D., Slonim, D.K.: Randomly fallible teachers: learning monotone DNF with an incomplete membership oracle. Mach. Learn. **14**(1), 7–26 (1994)
6. Arias, M., Balcázar, J.L.: Construction and learnability of canonical Horn formulas. Mach. Learn. **85**(3), 273–297 (2011)
7. Arias, M., Balcázar, J.L., Tîrnăucă, C.: Learning definite Horn formulas from closure queries. Theor. Comput. Sci. **658**(Part B), 346–356 (2017)
8. Baader, F., Ganter, B., Sertkaya, B., Sattler, U.: Completing description logic knowledge bases using formal concept analysis. In: Veloso, M.M. (ed.) Proceedings IJCAI 2007, pp. 230–235. AAAI Press (2007)
9. Babin, M.A., Kuznetsov, S.O.: Computing premises of a minimal cover of functional dependencies is intractable. Discrete Appl. Math. **161**(6), 742–749 (2013)
10. Bertet, K., Monjardet, B.: The multiple facets of the canonical direct unit implicational basis. Theor. Comput. Sci. **411**(22), 2155–2166 (2010)
11. Borchmann, D., Hanika, T., Obiedkov, S.: On the usability of probably approximately correct implication bases. In: Bertet, K., Borchmann, D., Cellier, P., Ferré, S. (eds.) ICFCA 2017. LNCS (LNAI), vol. 10308, pp. 72–88. Springer, Cham (2017). https://doi.org/10.1007/978-3-319-59271-8_5
12. Borchmann, D., Hanika, T., Obiedkov, S.: Probably approximately correct learning of Horn envelopes from queries. Discrete Appl. Math. (2019, in press)
13. Distel, F.: Hardness of enumerating pseudo-intents in the lectic order. In: Kwuida and Sertkaya [25], pp. 124–137
14. Distel, F., Sertkaya, B.: On the complexity of enumerating pseudo-intents. Discrete Appl. Math. **159**(6), 450–466 (2011)
15. Ganter, B.: Two basic algorithms in concept analysis. In: Kwuida and Sertkaya [25], pp. 312–340
16. Ganter, B., Obiedkov, S.: Conceptual Exploration. Springer, Heidelberg (2016). https://doi.org/10.1007/978-3-662-49291-8
17. Ganter, B., Wille, R.: Formal Concept Analysis: Mathematical Foundations. Springer, Heidelberg (1999). https://doi.org/10.1007/978-3-642-59830-2
18. Guigues, J.L., Duquenne, V.: Famille minimale d'implications informatives résultant d'un tableau de données binaires. Mathématiques et Sciences Humaines **24**(95), 5–18 (1986)
19. Hanika, T., Zumbrägel, J.: Towards collaborative conceptual exploration. In: Chapman, P., Endres, D., Pernelle, N. (eds.) ICCS 2018. LNCS (LNAI), vol. 10872, pp. 120–134. Springer, Cham (2018). https://doi.org/10.1007/978-3-319-91379-7_10
20. Jäschke, R., Rudolph, S.: Attribute exploration on the web. Preprint (2013). www.qucosa.de
21. Kautz, H., Kearns, M., Selman, B.: Horn approximations of empirical data. Artif. Intell. **74**(1), 129–145 (1995)
22. Khardon, R.: Translating between Horn representations and their characteristic models. J. Artif. Intell. Res. (JAIR) **3**, 349–372 (1995)
23. Kryszkiewicz, M.: Concise representations of association rules. In: Hand, D.J., Adams, N.M., Bolton, R.J. (eds.) Pattern Detection and Discovery. LNCS (LNAI), vol. 2447, pp. 92–109. Springer, Heidelberg (2002). https://doi.org/10.1007/3-540-45728-3_8. ISBN 978-3-540-45728-2
24. Kuznetsov, S.O.: Fitting pattern structures to knowledge discovery in big data. In: Cellier, P., Distel, F., Ganter, B. (eds.) ICFCA 2013. LNCS (LNAI), vol. 7880, pp. 254–266. Springer, Heidelberg (2013). https://doi.org/10.1007/978-3-642-38317-5_17

25. Kwuida, L., Sertkaya, B. (eds.): ICFCA 2010. LNCS (LNAI), vol. 5986. Springer, Heidelberg (2010). https://doi.org/10.1007/978-3-642-11928-6
26. Obiedkov, S., Romashkin, N.: Collaborative conceptual exploration as a tool for crowdsourcing domain ontologies. In: Proceedings of Russian and South African Workshop on Knowledge Discovery Techniques Based on Formal Concept Analysis, CEUR Workshop Proceedings, vol. 1552, pp. 58–70 (2015)
27. Wild, M.: The joy of implications, aka pure Horn formulas: mainly a survey. Theoretical Computer Science **658**, 264–292 (2017)

Concepts in Application Context

Steffen Staab[1,2](✉)

[1] Institute for Web Science and Technologies, Universität Koblenz-Landau, Koblenz, Germany
staab@uni-koblenz.de
[2] WAIS Research Group, University of Southampton, Southampton, UK
s.r.staab@soton.ac.uk
https://west.uni-koblenz.de
https://wais.ecs.soton.ac.uk

Abstract. Formal concept analysis (FCA) derives a hierarchy of concepts in a *formal context* that relates objects with attributes. This approach is very well aligned with the traditions of Frege, Saussure and Peirce, which relate a signifier (e.g. a word/an attribute) to a mental concept evoked by this word and meant to refer to a specific object in the real world. However, in the practice of natural languages as well as artificial languages (e.g. programming languages), the *application context* often constitutes a latent variable that influences the interpretation of a signifier. We present some of our current work that analyzes the usage of words in natural language in varying application contexts as well as the usage of variables in programming languages in varying application contexts in order to provide conceptual constraints on these signifiers.

Keywords: FCA · Semantics · Programming · Word embeddings

1 Introduction

In this talk we want to bridge between and explore several research communities that consider the conceptual analysis of various kinds of objects: First, the natural language processing research community which has seen a rapid growth and enormous successes derived from better conceptual understanding of words. Second, the programming language research community which has always used conceptual analysis, including formal concept analysis, in the past, but has not fully exploited its potential in the past decade.

In both research areas we observe that there is a need for conceptual analysis that varies and adapts according to the *context of use* of signifiers/attributes. We try to derive such need from the fundamental consideration of how we use words in communication situations, that arise both in the use of natural language and the use of programming languages.

It is germane to this undertaking that we cannot and do not want to claim completeness of our exploration and our knowledge of the above mentioned fields.

D. Cristea et al. (Eds.): ICFCA 2019, LNAI 11511, pp. 45–52, 2019.
https://doi.org/10.1007/978-3-030-21462-3_4

However, we hope to stimulate a fruitful discussion for extending methods of FCA and furthering its uptake in various fields.

2 Concept Analysis as a Foundational Means in Communication

The core model of Formal Concept Analysis [15] represents objects and attributes in a *formal context*:

Definition 1 (Formal Context). *A formal context* $\mathbb{K} := (G, M, I)$ *is a triple comprised of a set of objects* G*, a set of attributes* M *and an indicidence relation* $I \subseteq G \times M$ *encoding that "g has attribute m" iff* gIm.

It computes the formal concepts of this context by:

Definition 2 (Formal Concept). *For* $A \subseteq G$ *and* $B \subseteq M$*, define* $A' := \{m \in M | \forall g \in A : gIm\}$, $B' := \{g \in G | \forall m \in B : gIm\}$*. Then a pair* (A, B) *with* $A' = B$ *and* $B' = A$ *is a formal concept.* A *is the extent,* B *is the intent of the concept.*

We may bridge between formal concept analysis and the usage of words, or more generally *signifiers*, in communication. Following the tradition of Frege, Saussure or Pierce, we may illustrate the usage of signifiers in Fig. 1.[1]

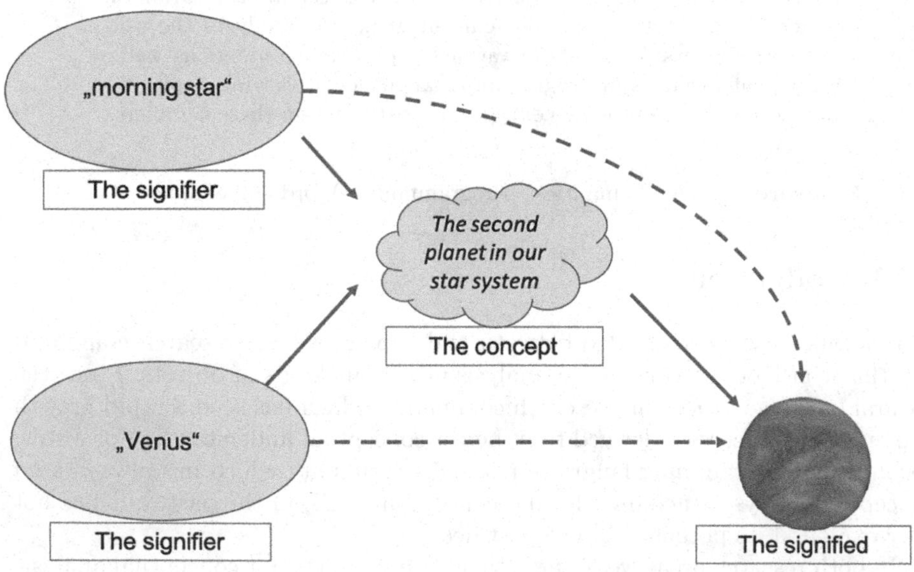

Fig. 1. A signifier evokes a concept in the mind of a recipient of communication allowing her to identify the signified object.

[1] NASA - NSSDC Photo Gallery Version: ftp://nssdcftp.gsfc.nasa.gov/photo_gallery/hi-res/planetary/venus/pvo_uv_790226.tiff, Public Domain, https://commons.wikimedia.org/w/index.php?curid=10914.

Indeed, different signifiers such as "Venus" or "morning star" may be used to evoke the same mental concept in the receiver of a message, which allows him to dereference, i.e. to point, to the actual object, which might be the corresponding planet in our solar system. We may understand formal concept analysis as a means that simulates mental concept formation. Investigating the *object*, which is referred to as planet Venus, we may come up with its attributes, such as its orbit around the sun, but also its naming attributes "Venus" and "morning star". Based on these attributes and the set of objects concerned, which is a singleton in this specific example, we may form the concept in our mind and use the naming attributes, "Venus" or "morning star", to point to the intended object.

3 Concepts in Natural Language Processing Using Word Embeddings

3.1 Global Representations of Signifiers in Concept Space

Methods of natural language processing usually come with very limited knowledge of objects in the world, but rather with massive text corpora based on which the usage of signifiers may be analysed. In this vein, we have analysed the usage of signifiers in their *textual contexts* using formal concept analysis in order to determine concepts and conceptual relationships [3]. Being based on standard formal concept analysis our approach mapped each signifier into a binary vector representation indicating the words that would co-occur (or not) with the signifier in a given corpus.

Likewise approaches had pursued such representation of signifiers in word space [7] much earlier. However, they could not preserve structural information (e.g. *isa*-relationships) and they suffered from the very high dimensionality of these representations, as the length of vectors representing signifiers was given by the number of words that occurred in a text corpus (i.e. in the order of 10^4 to 10^6).

Word2Vec [9] has been a turning point in such research, efficiently and effectively mapping these vectors into a lower dimensional space with vector lengths significantly below 1000. Even some structural information is partially preserved by the offset between different representations of signifiers in this *concept space*. Thus, the vector offset between concept referred to by signifiers such as "king" and "queen" resembles the vector offset between "man" and "woman".

3.2 Context-Dependent Representations of Signifiers in Concept Space

Figure 2[2] illustrates the problem with all of these approaches. Depending on their textual contexts signifiers like "morning star" and "Venus" may alternatively

[2] Sketch of mace from https://commons.wikimedia.org/wiki/File:Boeheim_Morgenstern_01.jpg#/media/File:Boeheim_Morgenstern_01.jpg, Public domain.

Table 1. Various textual contexts of "morning star" evoking varying concepts of planet Venus (rows 1, 2) and mace (3, 4).

1	Morning star, most commonly used as a name for the planet Venus when it appears in the east before sunrise
2	The Egyptians knew the morning star as Tioumoutiri and the evening star as Ouaiti
3	A morning star is any of several medieval club-like weapons consisting of a shaft with an attached ball adorned with one or more spikes
4	The Morning star is normally considered to be a one handed weapon but it was also a polearm weapon

refer to weapons (cf. Table 1, rows 3 and 4) or a Greek goddess, rather than to the planet Venus (cf. Table 1, rows 1 and 2).

More recent and most successful approaches such as ELMo [10] and BERT [4] also analyse the words that co-occur with signifiers, however they map each individual occurrence of a signifier into an individual vector in concept space. A signifier like "Venus" is no longer represented by a single vector in concept space, but rather by samples from a probability distribution that reflects in which word contexts it has been observed. We posit that the analysis of such a probability distribution lends itself also to an abstraction that discretizes thousands (or more) observations of a signifier into a more manageable set of concept representations [11], though further experiments are still to be performed.

4 Concepts in Programming Languages Using Queries

Formal concept analysis has been successfully used in several stages of the software engineering lifecycle. For example, Hesse and Tilley [5] describe the use of formal concept analysis in the early stages of the software development process. Core is the elucidation and formation of important programming concepts, such as classes and components:

> *"FCA allows a 'crossing of perspectives'—between the functional view represented by the use cases and the data view implied by the 'things' occurring there."*

Just like natural language signifiers appear in a variety of contexts, the same may apply to signifiers, terms and expressions, in programming languages. An often-used software architecture relies on one or several databases that facilitate the integration of data and serve a multitude of applications (cf. Fig. 3).

Each particular data object is put into a variety of application contexts. Each application context assumes its own set of concepts. These are often formed by queries.

Consider the following SPARQL query [1] which selects all researchers ?X who are members of some research group that is part of one of the institutes of the department of computer science, "uniko:informatik".

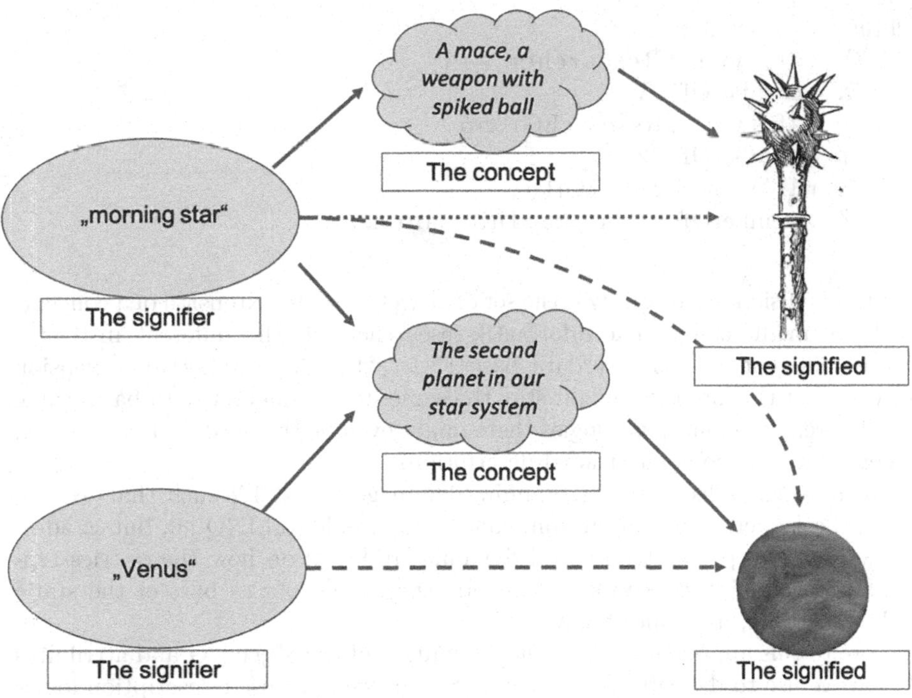

Fig. 2. A "morning star" may allude to different concepts and objects.

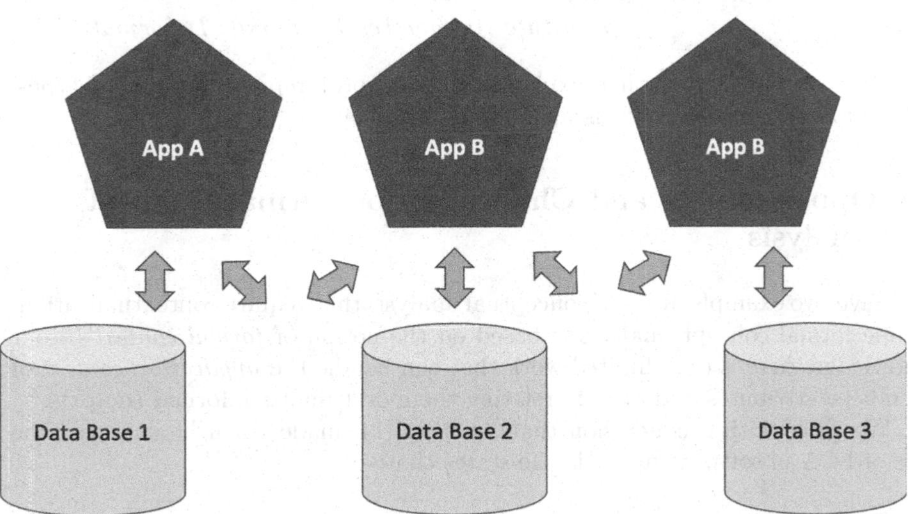

Fig. 3. Integration data bases serve a multitude of applications, which put objects in a variety of contexts.

```
SELECT ?X
WHERE {
    ?X rdf:type :Researcher.
    ?X :memberOf ?Y.
    ?Y rdf:type :ResearchGroup.
    ?Y :memberOf ?Z.
    ?Z rdf:type :Institute.
    ?Z :memberOf    uniko:Informatik.
}
```

The extension of this query is the set of objects, i.e. the extension of a concept, which we might call "uniko:Informatik-researcher". If the different institutes operate different databases ("data base 1" to "data base 3"), their extension will vary, but the intension might stay the same. If the different data bases come with different schemata/ontologies there might even be the need to describe such concepts using various intensional descriptions.

We have extended the programming language Scala [12] such that we can embed these queries into programm code, comparable to LINQ [8], but in addition we derive types statically and dynamically based on how the queries represent programming concepts and include these concepts as part of the static (and, sometimes, dynamic) analysis [6].

Core to this approach is the *ad hoc* definition of the objects G and attributes M that we use to describe the concepts. So far, we have used description logics [2] to do this type inference. Hence, the above query would receive the type:

$$\begin{aligned}&Researcher \sqcap \exists memberOf.\\&\quad\exists(ResearchGroup \sqcap \exists.memberOf.\\&\qquad(Institute \sqcap \exists memberOf.\{uniko : Informatik\}))\end{aligned}$$

However, we are further exploring closed-world representations of types, where formal concept analysis might come into play.

5 Opportunities and Challenges for Formal Concept Analysis

We gave two example areas of conceptual analysis that require contextualization. While formal concept analysis is based on the notion of *formal context* to our knowledge there is only limited work that can handle the *application context* of signifiers/attributes and objects relating them over multiple formal contexts.[3]

We agree with a conclusion that Snelting [14] made when he surveyed the use of FCA in software analysis. He states that:

[3] See [13] for a survey on the integration of description logics and formal concept analysis. It includes the representation of multiple formal contexts to represent concepts in non-hierarchical relationships.

"Future work in lattice theory must show whether the structure theory of concept lattices can be extended in such a way that typical "local" situations occuring in software analysis can be handled."

We see this as a tremendous opportunity in software engineering, but also in natural language processing. In the latter case, black box systems building on ELMo or BERT may excel when it comes to tasks such as question answering, but when the system is to be made transparent there is a need to elucidate and discretize structures to a level of generalization understandable for the human user—and FCA may come in handy for this purpose.

References

1. Sparql 1.1 query language. Technical report, W3C Recommendation, 21 March 2013. http://www.w3.org/TR/sparql11-query/
2. Baader, F., Horrocks, I., Lutz, C., Sattler, U.: Introduction to Description Logic. Cambridge University Press, Cambridge (2017)
3. Cimiano, P., Hotho, A., Staab, S.: Learning concept hierarchies from text corpora using formal concept analysis. J. Artif. Intell. Res. **24**, 305–339 (2005). https://doi.org/10.1613/jair.1648
4. Devlin, J., Chang, M., Lee, K., Toutanova, K.: BERT: pre-training of deep bidirectional transformers for language understanding. CoRR abs/1810.04805 (2018). http://arxiv.org/abs/1810.04805
5. Hesse, W., Tilley, T.: Formal concept analysis used for software analysis and modelling. In: Ganter, B., Stumme, G., Wille, R. (eds.) Formal Concept Analysis. LNCS (LNAI), vol. 3626, pp. 288–303. Springer, Heidelberg (2005). https://doi.org/10.1007/11528784_15
6. Leinberger, M., Lämmel, R., Staab, S.: The essence of functional programming on semantic data. In: Yang, H. (ed.) ESOP 2017. LNCS, vol. 10201, pp. 750–776. Springer, Heidelberg (2017). https://doi.org/10.1007/978-3-662-54434-1_28
7. Lin, D., Pantel, P.: Induction of semantic classes from natural language text. In: Proceedings of the Seventh ACM SIGKDD International Conference on Knowledge Discovery and Data Mining, San Francisco, CA, USA, 26–29 August 2001, pp. 317–322 (2001). http://portal.acm.org/citation.cfm?id=502512.502558
8. Meijer, E., Beckman, B., Bierman, G.M.: LINQ: reconciling object, relations and XML in the .net framework. In: Proceedings of the ACM SIGMOD International Conference on Management of Data, Chicago, Illinois, USA, 27–29 June 2006, p. 706 (2006). https://doi.org/10.1145/1142473.1142552
9. Mikolov, T., Sutskever, I., Chen, K., Corrado, G.S., Dean, J.: Distributed representations of words and phrases and their compositionality. In: 27th Annual Conference on Neural Information Processing Systems, Lake Tahoe, Nevada, United States, 5–8 December 2013, pp. 3111–3119 (2013). http://papers.nips.cc/paper/5021-distributed-representations-of-words-and-phrases-and-their-compositionality
10. Peters, M.E., et al.: Deep contextualized word representations. In: Proceedings of the 2018 Conference of the North American Chapter of the Association for Computational Linguistics: Human Language Technologies, NAACL-HLT 2018, New Orleans, Louisiana, USA, 1–6 June 2018, vol. 1 (Long Papers), pp. 2227–2237 (2018). https://aclanthology.info/papers/N18-1202/n18-1202

11. Schmelzeisen, L., Staab, S.: Learning taxonomies of concepts and not words using contextualized word representations: a position paper. CoRR abs/1902.02169 (2019). http://arxiv.org/abs/1902.02169
12. Seifer, P., Leinberger, M., Lämmel, R., Staab, S.: Semantic query integration with reason. Art Sci. Eng. Program. **3**(3) (2019). https://doi.org/10.22152/programming-journal.org/2019/3/13
13. Sertkaya, B.: A survey on how description logic ontologies benefit from FCA. In: Proceedings of the 7th International Conference on Concept Lattices and Their Applications, Sevilla, Spain, 19–21 October 2010, pp. 2–21 (2010). http://ceur-ws.org/Vol-672/paper2.pdf
14. Snelting, G.: Concept lattices in software analysis. In: Ganter, B., Stumme, G., Wille, R. (eds.) Formal Concept Analysis. LNCS (LNAI), vol. 3626, pp. 272–287. Springer, Heidelberg (2005). https://doi.org/10.1007/11528784_14
15. Wille, R.: Restructuring lattice theory: an approach based on hierarchies of concepts. In: Rival, I. (ed.) Ordered Sets, vol. 83, pp. 445–470. Springer, Dordrecht (1982). https://doi.org/10.1007/978-94-009-7798-3_15

Theory

Direct and Binary Direct Bases for One-Set Updates of a Closure System

Kira Adaricheva[1(✉)] and Taylor Ninesling[2]

[1] Department of Mathematics, Hofstra University, Hempstead, NY 11549, USA
kira.adaricheva@hofstra.edu
[2] Hofstra University, Hempstead, NY 11549, USA
tninesling1@pride.hofstra.edu

Abstract. We introduce a concept of a *binary-direct* implicational basis and show that the shortest binary-direct basis exists and it is known as the *D*-basis introduced in Adaricheva, Nation, Rand [4]. Using this concept we approach the algorithmic solution to the Singleton Horn Extension problem, as well as the one set removal problem, when the closure system is given by the canonical direct or binary-direct basis. In this problem, a new closed set is added to or removed from the closure system forcing the re-write of a given basis. Our goal is to obtain the same type of implicational basis for the new closure system as was given for original closure system and to make the basis update an optimal process.

Keywords: Closure system · Horn-to-Horn belief revision · Singleton Horn Extension problem · Direct basis · Canonical direct basis · The *D*-basis · Ordered direct basis

1 Introduction

The dynamic update of evolving knowledge bases and ontologies is a routine procedure in the realm of Artificial Intelligence. These applications require tractable representations, such as Horn logic or various versions of descriptive logic. The interest in Horn logic is easily explained by the fact that the reasoning in Horn logic is effective, while the reasoning in general propositional logic is intractable.

If some knowledge base is represented by a (definite) Horn formula Σ in variables $X = \{x_1, \ldots, x_n\}$, then the set of its models $\mathcal{F}_\Sigma$ forms a lower subsemilattice in 2^X, which is often referred to as a *closure system* on X, or a *Moore family* on X. Alternately, one can associate with Σ a *closure operator* φ on X, so that models from $\mathcal{F}_\Sigma$ are exactly the closed sets of φ. Also, Σ can be interpreted as a set of *implications* defining the closure operator φ. The general connections between Horn formulas (in propositional and first order logic), closure operators and their models were surveyed recently in [1].

The knowledge base requires an update if some of the models expire or the new models need to be incorporated into the existing base. In the current work

D. Cristea et al. (Eds.): ICFCA 2019, LNAI 11511, pp. 55–72, 2019.
https://doi.org/10.1007/978-3-030-21462-3_5

we tackle the problem of re-writing the Horn formula Σ, when a new model A has to be added or existing model A has to be removed from the family $\mathcal{F}_\Sigma$.

To avoid misconception, we note that adding set A may result in adding more than just single set to $\mathcal{F}_\Sigma$. Some proper subsets of A may be added as well, which are intersections of A with members of $\mathcal{F}_\Sigma$. This is due to the requirement that the updated family of models must be described by a Horn formula as well, see the classical result in [13].

In the case of the removal of A, some more sets have to be removed, if A is the intersection of other sets in $\mathcal{F}_\Sigma$. In this paper, we discuss the case when set A is meet-irreducible in $\mathcal{F}_\Sigma$, thus, only A may be removed.

If the closure operator is encoded through the reduced formal context, the update of the closure system corresponds to adding or removing a row of the table.

In case of addition of a new set to $\mathcal{F}_\Sigma$, the algorithmic solution for the update of the basis was given in the framework of relational databases in [12], and improvement of the algorithm was suggested in [20]. The latter publication is the conference proceedings version of the longer and more detailed publication [19]. Note that in this algorithm the one-set update was considered as one step of iterative process of generating a canonical direct basis of a closure system. The problem was also addressed in the general framework of closure systems, including the FCA framework in [16] and iterative algorithms in [18], and the framework of Horn-to-Horn belief revision in [5]. In the latter paper, the problem was called the Singleton Horn Extension (SHE) Problem.

In our work we considered two special cases of the SHE problem: when formula Σ is given by the *canonical direct basis* of implications defining closure operator φ, and when it is given by its refined form, the *D-basis*. We will assume that one needs an algorithmic solution that provides at the output an updated formula $\Sigma^*(A)$ that is canonical direct, or, respectively, the D-basis of the extended closure system.

These two cases will be addressed in Sects. 3 and 5. Note that one step of iteration process in [19] also deals with the case of canonical direct basis. Our approach uses a new data structure associated with the basis that allows us to improve the performance time.

In Sect. 4 we introduce the concept of the binary-direct basis of a closure system and show that the D-basis is the shortest binary-direct basis among all the binary-direct implicational bases for the closure system. This allows to extend approach used in the update of the canonical direct basis for the new family of bases, including the D-basis.

In Sect. 6 we present the algorithm of removal one single set from the closure system, assuming that the set is meet-irreducible in the system. The canonical direct basis update is then reduced to the well-known problem of the hypergraph dualization, for which the algorithmic solutions are numerous, see [10,11,14].

The last section is devoted to the results of algorithmic implementations and testing on various closure systems.

2 Preparing Implicational Bases for Updates

Prior to the update of the basis, we must do some preparations with regard to the new closed set being added to the family. We assume that the existing basis to update corresponds to a standard closure system (sometimes referred to as a $T\frac{1}{2}$ system [21]). We choose to operate on the basis representing the equivalent reduced closure system, which is a convenient intermediate form as mentioned in [2]. Note that the basis for the reduced closure system will be a superset of the basis for the standard system, but they are equivalent. An algorithm for retrieving a standard system from a reduced system is describe in [2]. Since, we are converting from the standard form to the reduced form, we reverse the process which is described as follows.

Say we have a closure system $\langle X, \varphi\rangle$ with corresponding standard system $\langle S, \varphi_S\rangle$. Note that $S \subset X$, and φ_S is the restriction of operator φ to S. If we have $a \in X$ but $a \notin S$, there is some $B \subset S$ equivalent to a. This means that the φ-closures of these two sets in $\langle X, \varphi\rangle$ are equal, or in terms of implications, $B \to a$ and $a \to b, \forall b \in B$. If $|B| = 1$, we call this a binary equivalence. In the standard system, we remove as many elements from the base set as possible, given all equivalences. However, only the binary equivalences are used to remove elements when deriving the reduced system. So, for each non-binary equivalence $a \leftrightarrow B$, we take $\Sigma' = (\Sigma \cup (B \to a) \cup \{b \to a : b \in B\})^{tr}$ where tr simply includes all transitive implications $a \to c$ where some $a \to b$ and $b \to c$ are already in the basis. Once we take the expand the basis in this way for each binary equivalence, we obtain the basis corresponding to the desired reduced system.

When a new set is to be added to the basis, some binary equivalences may be broken, and we will need to update the base set and the basis. Say we have some binary equivalence $x \leftrightarrow y$ and want to add new closed set A. If $x, y \in A$ or $x, y \notin A$, the equivalence remains. However, say $x \notin A, y \in A$. Then we must add to the basis implications, $x \to y$ and $y \to x$. Note that y no longer implies x, so $x \to y$ defines the new part of the partial order while we need $y \to x$ so the update process can add in weaker implications that still hold. Additionally, if x was already in our base set, we must add in a copy of each implication containing x, replacing x with y. After we add these implications, we must again take the transitive closure of the basis. After this is completed for each broken equivalence, we can perform the update procedure.

Our aim for the update is to produce the basis for the updated closure system in its reduced form. However, the basis of the standard closure system is often our desired output. In that case, we can simply apply the aforementioned algorithm from [2] to our reduced form. In terms of the basis itself, this process is exactly the process of maximal set projection described after Lemma 13.5 in [12] where we project the closed sets into the base set for the standard system.

3 Update of the Canonical Direct Basis of Implications

In [5], the SHE problem was addressed in the case when the formula Σ describing the knowledge base is assumed to be a conjunction of *prime implicates* of the

Horn belief set. Translating this into the language of closure systems, one would call Σ *the canonical direct basis*, a type of implicational basis surveyed in [7].

Recall that a formula/implicational set $\Sigma = \{C \to d : C \cup \{d\} \subseteq X\}$ is called *direct* for closure operator φ on X, if, for any $Y \subseteq X$,

$$\varphi(Y) = Y \cup \{d : (C \to d) \in \Sigma, C \subseteq Y\}.$$

In other words, the closure of any subset Y can be computed by checking the bodies (left sides) of implications of Σ with respect to set Y and expanding Y by their consequents when possible. Each implication of Σ is attended only once during the process. Recall that the computation of the closure of Y is generally performed through multiple iteration of Σ and expansion of Y, see the theoretical background in [19], or through the algorithm known as Forward Chaining [9] or LinClosure [12]. The canonical direct basis is the smallest implicational set contained in all direct bases defining the same closure operator φ on X, see [7].

3.1 Body-Building Formula

The algorithmic solution for the SHE problem in [5] was given in the form of *body-building* formula $\Sigma(A)$, which was produced given a set of implications/formula Σ that forms the canonical direct basis of a closure system, and a new set A that needs to be added to the closure system $\mathcal{F}_\Sigma$. We will denote the extended closure system $\mathcal{F}_\Sigma(A) = \mathcal{F}_\Sigma \cup \{F \cap A : F \in \mathcal{F}_\Sigma\}$.

To describe the body-building formula, consider splitting of Σ into two subsets: implications $\Sigma_t(A)$ which are true on A, and implications $\Sigma_f(A)$ which fail on A.

If $\sigma = (C \to d) \in \Sigma_f(A)$, then implication fails on set A, i.e., $C \subseteq A$ and $d \notin A$. Denote $\sigma(A) = \{C \cup x \to d : x \in X \setminus (A \cup d)\}$. Note that $\sigma(A) = \emptyset$, if $X \setminus (A \cup d) = \emptyset$.

Then the body-building formula can be given as

$$\Sigma(A) = \Sigma_t(A) \cup \bigcup_{\sigma \in \Sigma_f(A)} \sigma(A)$$

In other words, the new formula preserves all implications that are true on A and replaces every implication that fails on A by a subset $\sigma(A)$ of new formulas. Each of the new formulas extend the body of a failing implication by a single element not in A and distinct from a consequent.

The formula came up as a consequence to earlier work [15], where the body-building formula was provided to a special extension of the closure system, namely, to the one corresponding to the *saturation operator* φ^* associated with given operator φ. The necessary background for the saturation operator can be found in [6].

In our current work we analyze further the solution for the one-set extension of a closure system. The first observations are collected in the following theorem. We note that item (3) was mentioned in [19] without proof, and we include the proof for completeness of the exposition.

Theorem 1. *Let $\mathcal{F}_\Sigma$ be a closure system with basis Σ and let $A \subseteq X$ be a set not in $\mathcal{F}_\Sigma$. Consider extended closure system $\mathcal{F}_\Sigma(A)$ and body-building formula $\Sigma(A)$.*

(1) Closure system $\mathcal{F}_\Sigma(A)$ comprises the sets that satisfy $\Sigma(A)$.
(2) Any set $P \not\subseteq A$ that satisfies $\Sigma(A)$ is in $\mathcal{F}_\Sigma(A)$.
(3) If Σ is direct, then $\mathcal{F}_\Sigma(A)$ is defined by basis $\Sigma(A)$, moreover, $\Sigma(A)$ is direct.

Proof. The proofs of (1) and (2) are straightforward.

(3) Let φ be a closure operator on X corresponding to $\mathcal{F}_\Sigma$. Define

$$\varphi^*(Y) = \begin{cases} \varphi(Y) & Y \not\subseteq A \\ \varphi(Y) \cap A & Y \subseteq A \end{cases}$$

Then φ^* defines a coarsest closure system that includes all φ-closed sets and set A. Thus, φ^* is a closure operator for closure system $\mathcal{F}_\Sigma(A)$.

Define the expected direct closure operator for $\mathcal{F}_\Sigma(A)$,

$$\pi(Y) = Y \cup \{b : (Z \to b) \in \Sigma(A),\ Z \subseteq Y\}$$

First, note that any element of $\pi(Y)$ is either an element of Y, or it is the consequent of some $Z \to p \in \Sigma(A)$ corresponding to some $C \to b \in \Sigma$ where $C \subseteq Z \subseteq Y$. Thus, $\pi(Y) \subseteq \varphi(Y)$. If $Y \subseteq A$, and we have some $p \in \pi(Y)$, then $p \in Y \subseteq A$ or there is some $Z \to p \in \Sigma(A)$. However, p must be an element of A, otherwise $Z \not\subseteq A$ by the body-building process. So, $\pi(Y) \subseteq \varphi^*(Y)$.

Let $Y \not\subseteq A$. Then, $\varphi^*(Y) = \varphi(Y) = Y \cup \{b : (Z \to b) \in \Sigma,\ Z \subseteq Y\}$ because Σ is direct. If $b \in A$, or $Z \not\subseteq A$, then $Z \to b$ is not removed and $b \in \pi(Y)$. Assume $Z \subseteq A$ and $b \notin A$. Then that implication is removed, but we add $Z \cup \{s\} \to b$ to $\Sigma(A)$ for $s \notin A \cup \{b\}$. In this case there exists $s \notin A \cup \{b\}$ such that $s \in Y \setminus A$. Thus, for this particular s, $Z \cup \{s\} \subseteq Y$ and once again $b \in \pi(Y)$. So, $\varphi^*(Y) = \pi(Y)$ for $Y \not\subseteq A$.

Now let $Y \subseteq A$.

$$\begin{aligned} \varphi^*(Y) &= \varphi(Y) \cap A \\ &= (Y \cup \{b : (Z \to b) \in \Sigma,\ Z \subseteq Y\}) \cap A \\ &= (Y \cap A) \cup (\{b : (Z \to b) \in \Sigma,\ Z \subseteq Y\} \cap A) \\ &= Y \cup (\{b : (Z \to b) \in \Sigma,\ Z \subseteq Y\} \cap A) \end{aligned}$$

Consider the set $P = \varphi^*(Y) \setminus Y$. Since $\varphi^*(Y) = Y \cup P$, it suffices to show that $P \subseteq \pi(Y)$. Let $p \in P$. Then $p \in A$, and there is some $Z \to p \in \Sigma$. However, since $p \in A$, $Z \to p \in \Sigma(A)$ and thus $p \in \pi(Y)$. So, $P \subseteq \pi(Y)$, and thus $\varphi^*(Y) \subseteq \pi(Y)$. Since for all $Y \subseteq X$, $\varphi^*(Y) = \pi(Y)$, $\Sigma(A)$ is direct.

It turns out that without the assumption about the directness of Σ, the body-building formula $\Sigma(A)$ may lack implications to define the updated closure system.

Example 1. Take base set $X = \{z_1, z_2, z_3, d, u\}$ with the basis $\Sigma = \{z_1z_2 \rightarrow d, z_3d \rightarrow u\}$ and closure system $\mathcal{F}$. Apparently, this basis is not direct, since it misses the resolution implication $z_1z_2z_3 \rightarrow u$.

Consider new set $A = \{z_1, z_2, z_3, u\}$ and consider its subset $Z = \{z_1, z_2, z_3\}$. Since the closure of Z in original system is X, Z is not an intersection of A with any set from $\mathcal{F}$, it should not be added when A is added.

On the other hand, the implicational set $\Sigma(A) = \{z_3d \rightarrow u\}$ holds on Z. Therefore, $\Sigma(A)$ allows more closed sets than $\mathcal{F}(A)$.

In general, one needs to add to $\Sigma(A)$ implications $C \rightarrow d$ that follow from Σ and such that $C \cup d \subseteq A$. Say, in this example, one needs additional implication $z_1z_2z_3 \rightarrow u$.

If Σ is the *canonical* direct basis, the formula $\Sigma(A)$ may not be the *canonical* direct basis of the updated closure system $\mathcal{F}(A)$.

Example 2. Indeed, consider $X = \{a, b, c, d, e\}$ and $\Sigma = \{e \rightarrow d, ad \rightarrow e, bc \rightarrow d, abc \rightarrow e\}$. If the new set $A = \{a, b, c\}$, then the body-building formula would require to replace $abc \rightarrow e$ by $abcd \rightarrow e$, but stronger implication $ad \rightarrow e$ is already in Σ. Similarly, one would need to replace $bc \rightarrow d$ by $bce \rightarrow d$, and Σ has stronger implication $e \rightarrow d$. Therefore, $\Sigma(A)$ is not canonical direct basis.

3.2 Modified Body-Building Formula

The last example highlights an approach to algorithmic solution to SHE that allows to update the canonical direct basis without the need to reduce implications in the body-building formula. For this we consider a modification of body-building procedure.

For the basis Σ and $d \in X$ we will call $\Sigma_d = \{C \rightarrow d\} \subseteq \Sigma$ a d-sector of Σ. With each $C \rightarrow d$ in Σ_d we will store a list $E_C = \{e_1, \dots e_n\} \subseteq X$ such that $\{e_i\} = E_i \setminus C$ for some $E_i \in \Sigma_d$. Since every implication in Σ_d has d as its consequent, we simply store the pair (C, E_C) for each implication in the sector. For Example 2, the sectors would be $\Sigma_d = \{(bc, \{e\}), (e, \emptyset)\}$, and $\Sigma_e = \{(abc, \{d\}), (ad, \emptyset)\}$.

Consider the modification to the body-building formula. Given basis Σ and new set A, let $\sigma = (C \rightarrow d)$ be an implication from Σ, i.e., $(C, E_C) \in \Sigma_d$, and suppose that σ fails on A. Define $\sigma^*(A) = \{C \cup x \rightarrow d : x \in X \setminus (A \cup d \cup E_C)\}$. Then modified body-building formula is

$$\Sigma^*(A) = \Sigma_t(A) \cup \bigcup_{\sigma \in \Sigma_f(A)} \sigma^*(A)$$

Theorem 2. *If Σ is a canonical direct basis of closure system $\mathcal{F}$, and $\mathcal{F}$ is being extended by new set A, then $\Sigma^*(A)$ is the canonical direct basis of $\mathcal{F}(A)$.*

Proof. We want to identify implications in direct basis $\Sigma(A)$ which should be deleted to make it canonical direct.

If $C \to d$ and $G \to d$ are two implications in Σ, then $C \not\subseteq G$ and $G \not\subseteq C$, because Σ is canonical direct. If both these implications are in $\Sigma_f(A)$, then $C, G \subseteq A$, and for any $x, y \in X \setminus (A \cup d)$ one has $C \cup x \not\subseteq G \cup y$ and $G \cup y \not\subseteq C \cup x$.

If both implications are in $\Sigma_t(A)$, then they are also in $\Sigma(A)$ without modification.

Thus, the only possibility that one body part is a subset of the other in $\Sigma(A)$ is when $C \to d$ is in $\Sigma_f(A)$, and $G \to d$ is in $\Sigma_t(A)$. Thus, we would have $G \subseteq C \cup x$, for some $x \in X \setminus (A \cup d)$. Given that $G \not\subseteq C$, it implies $\{x\} = G \setminus C$, therefore, $x \in E_C$. Thus, we do not need to add implication $C \cup x \to d$ whenever $x \in E_C$.

4 Binary-Direct Basis of a Closure System and the *D*-Basis

This part of the work is devoted to the algorithmic solution for the case when Σ is the D-basis for the closure operator φ and updated formula $\Sigma^*(A)$ is expected to be the D-basis of the expanded closure system.

The D-basis was introduced in [4] as a refined and shorter version of the canonical direct basis: the former is a subset of the latter, while the D-basis still possessing the form of the directness property, known as *ordered direct* [4]. The closure of any subset Y can be computed attending the implications of the D-basis Σ only once, when it is done in the specific order.

The part of the basis containing implications $x \to y$, i.e. implications with only two variables from X, is called *binary*, and it plays a special role in the computation of the closures.

We will assume that the binary part Σ^b of basis Σ is *transitive*, i.e., if $a \to b$ and $b \to c$ are in Σ^b, then $a \to c$ is also in Σ^b.

We will use notation $Y_\downarrow = \{c \in X : (y \to c) \in \Sigma^b \text{ for some } y \in Y\}$.

Using this notation, the D-basis describes a partial order $\langle X, \ll\rangle$ such that, for subsets $Y, Z \subseteq X$, $Z \ll Y$ if $Z \subseteq Y_\downarrow$. In this case, we say that Z *refines* Y.

The following statement describes the relation between the canonical direct and the D-basis of a closure system.

Proposition 1. *[4] The D-basis of a closure system can be obtained from the canonical direct basis Σ_{cd} by removing every implication $C \to d$ for which there exists $D \to d$ in Σ_{cd} with $D \subseteq C_\downarrow$.*

Note that D cannot be simply an expansion of C, because both implications are in the canonical direct basis. Therefore, $|C \setminus D| \geq 1$.

It is well-known that the direct bases of (X, φ) are characterized by the property:

$$\varphi(Y) = Y \cup \{d : (C \to d) \in \Sigma, C \subseteq Y\}.$$

In order to characterize the D-basis we introduce the following definition.

Definition 1. Basis Σ for a closure system (X, φ) is called b-direct (a shortcut for ‘binary-direct’), if for every $Y \subseteq X$, $\varphi(Y) = Y_\downarrow \cup \{d : (C \to d) \in \Sigma, C \subseteq Y_\downarrow\}$

Proposition 2. *Any direct basis is b-direct, and every b-direct basis is ordered direct but the inverse statements are not true.*

We only mention why the inverse statement are not true. As for the first statement, the computation of the closure for a b-direct basis is performed in two stages: first, the binary implications are applied, and then the computation can be done on the expanded set as for the direct basis. In particular, the D-basis which is b-direct is not direct.

For the second statement, we notice that closure systems without cycles may have ordered basis shorter than the D-basis, namely, the E-basis [4]. According to Theorem 4 below, the D-basis is the shortest b-direct basis, therefore, the E-basis is ordered direct but not b-direct.

Thus, the property of being b-direct is stronger than ordered directness. After the binary implications are applied, the order of the remaining implications does not matter, like in a direct basis. This is in contrast with the E-basis, for example, where the order of non-binary implications is rather specific.

The following statement is proved similar to Theorem 14 in [7].

Theorem 3. *Basis Σ is b-direct iff Σ^b is transitive and for any $A \to b$ and $C \cup b \to d$ in Σ there exists $G \subseteq (A \cup C)_\downarrow$ such that $G \to d$ is also in Σ.*

The following statement generalizes the description of the canonical direct basis in [7].

Theorem 4. *Let (X, φ) be any closure system.*

(1) There exists a smallest b-direct basis, i.e. Σ_{bd} such that it is contained in any b-direct basis for a given closure system.
(2) Σ_{bd} satisfies property: for any two distinct $Z \to d$, $Y \to d$ in Σ_{bd}, $Z \not\subseteq Y_\downarrow$.
(3) Basis Σ_{bd} is the D-basis.

The proof is done by observing that the Σ_{bd} basis can be obtained from the canonical direct as described in Proposition 1, therefore it satisfies property in (2), and removing any implication from it will bring to a failure of the property of the b-direct basis.

5 Algorithm of the D-Basis Update in SHE Problem

Now we describe an effective algorithm for the solution of SHE problem, when the implicational basis Σ is the D-basis of the associated closure operator φ.

First, we follow the steps of 2 to prepare the basis for the update and make sure it is in reduced form. Our reasons for working with the reduced system are twofold. First, the D-basis describes a partial order $\langle X, \ll \rangle$. A definition for the relation $\ll$ is that for $Y, Z \subseteq X$, $Y \ll Z$ iff $\forall y \in Y, \exists z \in Z\ y \in \varphi(z)$. If the system is not reduced, then $\ll$ does not satisfy the property of anti-symmetry and is thus not a partial order. Second, the preparation of the basis is easier when only binary equivalences are present. An optimization for the preparation procedure is evident when we have broken equivalence $x \leftrightarrow y$ and we add in all $C \cup y \to d$ such that $C \cup x \to d \in \Sigma$. When $x \to y$ holds, $C \cup y \to d \ll C \cup x \to d$, so we replace $C \cup x \to d$ by $C \cup y \to d$ instead of simply adding them in as in the general case. Now, we perform the actual update procedure.

Let A be a new set that needs to be added to $\mathcal{F}_\Sigma$.

We will denote $\geq$ the partial order imposed on X by the binary part of the basis Σ^b: $y \geq x$ iff $y \to x$ is in Σ^b. We denote $\Sigma^b_f = \{(a \to c) \in \Sigma^b : a \in A, c \notin A\}$ the set of binary implications failing on A and $\Sigma^b_t(A) = \Sigma^b \setminus \Sigma^b_f(A)$ the set of all binary implications that hold on A.

We will denote by $\geq_A$ a partial order imposed on X by $\Sigma^b_t(A) = \Sigma^b(A)$, i.e., the binary part of updated basis $\Sigma(A)$. Apparently, $y \geq_A x$ implies $y \geq x$. Moreover, the inverse holds when $x \in A$ or $x, y \in X \setminus A$. Similar to $Y_\downarrow$ we use $Y_{\downarrow A} = \{c : y \geq_A c \text{ for some } y \in Y\}$. We say that implication $Z \to d$ $\ll_A$-refines $Y \to d$, or $Z \ll_A Y$, if $Z \subseteq Y_{\downarrow A}$.

We define the set of *target* elements $T(A) = \{c \in X \setminus A : (a \to c) \in \Sigma^b_f\}$. So, target elements are heads of binary implications that fail on set A. Whenever target element c is in the body of some implication $C \to d$ in Σ^{nb}, there may be another implication with element from A replacing c which refines $C \to d$ in the updated basis $\Sigma(A)$. Thus, we need a process to add such implications to the basis in case they are part of $\Sigma(A)$.

The update of Σ proceeds in several stages.

(I) For each $x \in T(A)$ define $A_x = \{a \in A : a \geq x \text{ and } a \text{ is minimal in } A \text{ with this property}\}$. Note that $a \ngeq_A x$, for all $a \in A_x$ and that $A_x \subseteq A$. Elements from A_x are replacements of target element x, if it appears in the body of any implication in Σ^{nb}. Also note that, for any $a \in A$ and $x \in X \setminus A$ such that $a \geq x$, there exists $a' \in A_x$ with $a \geq_A a' \geq x$.

(II) We will call this part of the procedure *A-Lift*, indicating that some new implications Σ^L will be added to Σ that replace elements x from the bodies of existing implications by elements in A_x. The A-lift adds implications which may be needed in the body-building phase but have refinements in Σ.

More precisely, if $C \to d$ is a non-binary implication, and C has elements from $T(A)$, then we want to add implications $C' \to d$, when at least one element $x \in C \cap T(A)$ is replaced by some element a from A_x.

We could use notation $\binom{a}{x} C'' \to d$ for one instance of A-Lift, which records new implication together with element $x \in C$ which is lifted to $a \in A_x$, so that $C = x \cup C''$ and $C' = a \cup C''$.

Note that $|C'| \leq |C|$ and that $C \ll C'$ in the old Σ^b, but it is no longer true in $\Sigma^b_t(A)$. Also note that in the case where $C'' \cap a_{\downarrow A} \neq \emptyset$, we can add a stronger implication than $C'' \cup a$. For example, if we have $C'' = D \cup b$ for $b \in a_{\downarrow A}$, then $C'' \cup a = D \cup b \cup a$, so $D \cup a \ll_A C'' \cup a$. For the A-lift, we then want to lift $C' \to d$ to new implications $\binom{a}{x} (C'' \setminus a_{\downarrow A}) \to d$.

Example 3. Consider $X = \{x, y, d, a, a'\}$ and $\Sigma = \{a \to x, a' \to y, xy \to d\}$. Note that Σ_{cd} for the closure system defined by Σ will also have $ay \to d$, $xa' \to d$ and $aa' \to d$, but implication $xy \to d$ refines all three.

If $A = \{a, a'\}$, then binary implications $a \to x, a' \to y$ do not hold on A, and the set of targets is $T(A) = \{x, y\}$.

We have $A_x = \{a\}, A_y = \{a'\}$, and implications $\Sigma^L = \{\binom{a}{x} y \to d, x \binom{a'}{y} \to d, \binom{a}{x}\binom{a'}{y} \to d\}$ are obtained by A-Lift from $xy \to d$.

Note that implication $aa' \to d$ in Σ^L, does not hold on A, thus, one needs to modify it further on the body-building stage of the algorithm.

(III) Any $(Y \to d) \in \Sigma^L$ may have a $\ll_A$-refinement in set of implications $\Sigma^{nb} \cup \Sigma^L$.

The following observation shows how to identify implications in Σ^L that may have a $\ll_A$-refinement, and thus, can be removed. The refinement may be an original implication from Σ, of from the A-Lift of an original implication.

Proposition 3. *Suppose* $(X \to d) \in \Sigma^{nb} \cup \Sigma^L$ *and* $(\binom{a_y}{y} Y \to d) \in \Sigma^L$. *If* $X \not\ll Y \cup y$ *and* $X \ll_A Y \cup a_y$, *then* $a_{y_{\downarrow A}} \setminus y_\downarrow$ *has an element from* X.

Example 4. Let us modify Example 3 by adding element $a_y \in X$, removing implication $a' \to y$ and adding implications $a_y \to y$, $a_y \to a'$ and $xa' \to d$. Let $A = \{a, a', a_y\}$.

Then $(x \binom{a_y}{y} \to d) \in \Sigma^L$ and it can be refined by $xa' \to d$. We have $a_{y_{\downarrow A}} \setminus y_\downarrow = \{a_y, a'\}$, which is in the body of implication $xa' \to d$.

This allows to select implications $X \to d$ which might be $\ll_A$-refinements of implication $(\binom{a_y}{y} Y \to d) \in \Sigma^L$: compute set $a_{y_{\downarrow A}} \setminus y_\downarrow$ and check its intersection with X. By the definition, $a_{y_{\downarrow A}} \setminus y_\downarrow \subseteq A$, only implications $X \to d$ such that $X \cap A \neq \emptyset$ should be checked as possible refinements of implications in Σ^L.

At this stage we remove implications from Σ^L, if they can be $\ll_A$-refined within $\Sigma_t^b(A) \cup \Sigma^{nb} \cup \Sigma^L$. We continue using notation Σ^L for A-lift implications that remain.

(IV) In this stage of the algorithm the body-building technique is applied to implications of $\Sigma \cup \Sigma^L$.

Implications from $\Sigma \cup \Sigma^L$ that hold on A will be included in $\Sigma(A)$ without change.

Now consider $A' \to d$ in $\Sigma \cup \Sigma^L$ that fails on A.

For $A' \subseteq X$, we will use the notation $A'_{\uparrow A} = \{x \in X : x \geq_A a$ for some $a \in A'\}$.

For any body-building by elements in $X \setminus (A \cup d_{\uparrow A})$ we need to choose $\geq_A$-minimal elements in poset $\langle X \setminus (A \cup d_{\uparrow A}), \geq_A\rangle$. Indeed, if $x_1 \geq_A x_2$ for $x_1, x_2 \in X \setminus (A \cup d_{\uparrow A})$, then $A' \cup x_2 \ll_A A' \cup x_1$.

When it happens that minimal element x_m in $\langle X \setminus (A \cup d_{\uparrow_A}), \geq_A \rangle$ also satisfies $x_m \geq_A a$ for some $a \in A'$, then apparently implication $A' \cup x_m \to d$ can be $\ll_A$-refined to $(A' \setminus a) \cup x_m \to d$. Therefore, extension by any element $x_m \in A'_{\uparrow_A}$ should be modified to *replacement* of a by x_m.

Note that this cannot happen for A-Lift $\begin{pmatrix} a \\ x \end{pmatrix} A'' \to d$. In this case $x_m \geq_A x$, thus, cannot be minimal in $\langle X \setminus (A \cup d_{\uparrow_A}), \geq_A \rangle$.

(V) Body-building may generate implications that can be removed due to $\ll_A$-refinements. Also note that some implications added by the body-building process may $\ll_A$-refine implications of Σ^L.

Proposition 4. *Suppose x_m is minimal in $\langle X \setminus (A \cup d_{\uparrow_A}), \geq_A \rangle$. If $(X' \to d) \in \Sigma \cup \Sigma^L$ is a $\ll_A$-refinement for $A' \cup x_m \to d$, then $x_m \in X'$ or $x_m \geq_A a$ for some $a \in X' \cap A$.*

Indeed, by assumption, $X' \not\ll_A A'$. Then $X' \ll_A A' \cup x_m$ means that $x_m \geq_A x$ for some $x \in X'$. The only element $x \in X \setminus (A \cup d_{\uparrow_A})$ with this property is x_m. Since X' cannot have elements in $d_{\uparrow_A}$, we have $x_m \in X$ or $x_m \geq_A a$ for some $a \in X' \cap A$.

Thus, body-building process identifies implications that could be potentially refined.

Note that no body-building replacement/extension is a refinement of the other. Indeed, consider $A' \cup x' \to d$ and $A'' \cup x'' \to d$ extended or replace with minimal elements $x', x'' \in X \setminus (A \cup d_{\uparrow_A})$, and suppose $(A' \cup a') \ll_A (A'' \cup a'')$. If $x' = x''$, then $A' \ll_A A''$, but we assumed that all refinements were applied at stage (III). Since $x'' \not\geq_A x'$, then $a'' \geq_A x'$ for some $a'' \in A''$, a contradiction.

After applying stages (I)–(V) one obtains implicational set $\Sigma^*(A)$.

Example 5. Return again to Example 3. Recall that A-lift of implication $xy \to d$ comprises three implications: $ay \to d$, $xa' \to d$ and $aa' \to d$. While the first two belong to $\Sigma^*(A)$, the last one fails on A, therefore, it needs body-building update on stage (IV).

These are $aa'y \to d$ and $aa'x \to d$, but both can be refined on stage (V) by $ay \to d$, $xa' \to d \in \Sigma^L$, respectively. Here, according to Proposition 4, minimal elements x, y used for extension are in the bodies of other implications in Σ^L.

Additionally, the body-building will be used for implications $a \to x$, $a' \to y$, giving $ay \to x$, $ad \to x$, $a'x \to y$ and $a'd \to y$.

Theorem 5. *$\Sigma^*(A)$ is the D-basis of modified closure system $\mathcal{F}(A)$.*

Proof. We assume that Σ is the D-basis of closure system $\mathcal{F}$ associated with closure operator φ, thus,

$$\varphi(Y) = Y_\downarrow \cup \{d : (C \to d) \in \Sigma, C \subseteq Y_\downarrow\}.$$

Let φ_A be the closure operator associated with $\mathcal{F}(A)$, i.e., $\varphi_A(Y) = \varphi(Y)$, when $Y \not\subseteq A$, and $\varphi_A(Y) = \varphi(Y) \cap A$ for $Y \subseteq A$.

We want to show that $\Sigma(A)$ defined through the algorithm is b-direct basis for φ_A, i.e.

$$(*) \qquad \varphi_A(Y) = Y_{\downarrow_A} \cup \{d : (C \to d) \in \Sigma(A), C \subseteq Y_{\downarrow_A}\}.$$

Indeed, this would show that $\Sigma(A)$ is the b-direct basis for φ_A; then stages (III) and (V) of the algorithm are intended for the refinement, so their goal is to make the basis $\Sigma(A)$ to satisfy property (2) of Theorem 4. As the result, $\Sigma(A)$ would be the shortest b-direct basis, the D-basis by Theorem 4.

First, we observe that any $\sigma \in \Sigma^L$ holds on $\mathcal{F}$. Similarly, any body-building implication of stage (IV) holds on $\mathcal{F}$. Moreover, $\geq_A \subseteq \geq$, therefore, $Y_{\downarrow_A} \cup \{d : (C \to d) \in \Sigma(A), C \subseteq Y_{\downarrow_A}\} \subseteq \varphi(Y)$.

If $Y \not\subseteq A$, in order to confirm $(*)$ we would need to show $\varphi(Y) \subseteq Y_{\downarrow_A} \cup \{d : (C \to d) \in \Sigma(A), C \subseteq Y_{\downarrow_A}\}$. So assume that $y_0 \in Y \setminus A$.

First, show that $Y_\downarrow \subseteq Y_{\downarrow_A} \cup \{d : (C \to d) \in \Sigma(A), C \subseteq Y_{\downarrow_A}\}$.

Assume that $z \in Y_\downarrow \setminus Y_{\downarrow_A}$. This means $y \geq z$ for some $y \in Y \cap A$ and $z \notin A$. On body-building stage (IV) we would add implication $yy_m \to z$, for some minimal element y_m in $\langle X \setminus A, \geq_A\rangle$ such that $y_0 \geq_A y_m$. Then $yy_m \subseteq Y_{\downarrow_A}$ and $yy_m \to z$ or its refinement in $\Sigma(A)$. Therefore, $z \in Y_{\downarrow_A} \cup \{d : (C \to d) \in \Sigma(A), C \subseteq Y_{\downarrow_A}\}$.

Now consider $C \to d$ in Σ, where $C \subseteq Y_\downarrow$. There are two cases to consider.

(1) $C \not\subseteq Y_{\downarrow_A}$, because for some element $c \in C \cap (X \setminus A)$ we have $y \geq c$ for $y \in Y \cap A$, thus, $y \not\geq_A c$.

Take element $c_m \in A$ which is $\geq$-minimal element in A such that $y \geq c_m \geq c$.

Then replacing all such $c \in C$ by c_m will be an A-lift implication $C' \to d$ added on stage (II). Note that $C' \subseteq Y_{\downarrow_A}$. If $C' \not\subseteq A$, then $C' \to d$ or its refinement is in $\Sigma(A)$. If $C' \subseteq A$, then $C' \cup y_m \to d$ is in $\Sigma(A)$, for some minimal element y_m in $\langle X \setminus A, \geq_A\rangle$ such that $y_0 \geq_A y_m$. In either case, $d \in \{d : (C \to d) \in \Sigma(A), C \subseteq Y_{\downarrow_A}\}$.

(2) $C \subseteq Y_{\downarrow_A}$, but $C \subseteq A$, therefore, $C \to d$ is not in $\Sigma(A)$. In this case $C \cup y_m \to d$ or its $\geq_A$-refinement is in $\Sigma(A)$, for some minimal element y_m in $\langle X \setminus A, \geq_A\rangle$ such that $y_0 \geq_A y_m$. Apparently, $C \cup y_m \subseteq Y_{\downarrow_A}$.

Now consider $Y \subseteq A$. Then $Y_{\downarrow_A} \subseteq A$ and for any $C \to d$ in $\Sigma(A)$ we must have $d \in A$. Therefore, $Y_{\downarrow_A} \cup \{d : (C \to d) \in \Sigma(A), C \subseteq Y_{\downarrow_A}\} \subseteq \varphi(Y) \cap A$, and we need only to show the inverse inclusion.

Apparently, $Y_\downarrow \cap A \subseteq Y_{\downarrow_A}$. So we take $d \in A \cap \{d : (C \to d) \in \Sigma, C \subseteq Y_\downarrow\}$. For any $c \in C \cap (X \setminus A)$ we can find $c_m \geq_A c$, a minimal element in $\langle A, \geq_A\rangle$ such that $y \geq_A c_m \geq c$, for some $y \in A$. Replacing all such elements c by c_m makes an A-lift $C' \to d$ of implication $C \to d$, which is added on stage (II). Apparently, $C' \subseteq Y_{\downarrow_A}$, therefore, $d \in \{d : (C \to d) \in \Sigma(A), C \subseteq Y_{\downarrow_A}\}$ as desired.

6 Removal of a Closed Set

In this section we consider the case when closure system $\mathcal{F}_\Sigma$ on set X defined by the set of implications Σ is modified by the removal of one closed set $A \in \mathcal{F}_\Sigma$.

The remaining family $\mathcal{F}_\Sigma \setminus \{A\}$ will be again a closure system if and only if A is a meet-irreducible in $\mathcal{F}$: $A = B \cap C$, for some $B, C \in \mathcal{F}$, implies $B = A$ or $C = A$.

In general, one would need to remove more than just single set A, but this can be achieved through the iteration process, where one removes a meet-irreducible set on each step of iteration process. Apparently, there are various paths leading to removal of set A.

In the case when the closure system is defined by a context with set of objects $\mathcal{O}$ and set of attributes $\mathcal{A}$, the reduced context with the same closure system on the set of objects $\mathcal{O}^* \subseteq \mathcal{O}$ will contain only rows corresponding the meet-irreducible elements of the closure system. Thus, removal of any row in the *reduced* context corresponds to the removal of meet-irreducible element of closure system.

These considerations prompt to consider the case of meet-irreducible $A \in \mathcal{F}$, which will be our assumption.

We also assume that Σ is the canonical direct basis of closure system $\mathcal{F}_\Sigma$. Our goal is to find the canonical direct basis Σ^* for $\mathcal{F}^* = \mathcal{F} \setminus \{A\}$.

Lemma 1. *Let ϕ, ϕ^* be closure operators corresponding to closure systems $\mathcal{F}, \mathcal{F} \setminus \{A\}$, respectively, where A is some meet-irreducible element of $\mathcal{F}$. Then $\phi^*(Y) = \phi(Y)$, for any $Y \not\subseteq A$.*

It follows that the closure changes only for subsets of A, thus, we should expect new implications $Z \to x$ with $Z \subseteq A$. To describe premises and consequents of these new implications, we introduce further notations.

Consider (unique) upper cover A^* of A in $\mathcal{F}$, and let $A^* \setminus A = \{d_1, \ldots, d_k\}$.

Recall that a pair $S = \langle B, \mathcal{B} \rangle$, where B is a set and $\mathcal{B} \subseteq 2^B$ is a family of subsets of B is called a *hypergraph.* Subset $U \subseteq B$ is called *a transversal* of hypergraph $S = \langle B, \mathcal{B} \rangle$, if $U \cap V \neq \emptyset$, for every $V \in \mathcal{B}$.

Define hypergraph $H = \langle A, \mathcal{T} \rangle$ as follows: $C \subseteq A \in \mathcal{T}$ iff $C = A \setminus \phi(A')$, for some $A' \subseteq A$. In other words, $\mathcal{T}$ is the collection of complements (in A) of ϕ-closed subsets of A.

Let $Y_1, \ldots, Y_t$ be minimal (with respect to $\subseteq$ relation) transversals of hypergraph $H = \langle A, \mathcal{T} \rangle$.

Lemma 2. *If $Y \to d$ is any implication that holds on $\mathcal{F}^* = \mathcal{F} \setminus \{A\}$ and fails on A, then $Y_j \subseteq Y$, for some $j \leq t$, and $d = d_i$, for some $i \leq k$.*

Proof. It is clear that for any implication $Y \to d$ failing on A we have $Y \subseteq A$ and $d \notin A$. Since it holds on A^*, we must have $d \in A^*$, i.e. $d = d_i$ for some $i \leq k$.

Apparently, $Y \to d$ holds on set $Z \in \mathcal{F}^*$, if $d \in Z$. So we consider $Z \in \mathcal{F}^*$ such that $d \notin Z$. Then $Y \to d$ holds on Z, when $Y \not\subseteq Z$, which also implies $Y \not\subseteq Z \cap A$. Thus, we only need to consider the case of closed sets $Z \subseteq A$. Then $Y \not\subseteq Z$ iff $Y \cap A \setminus Z \neq \emptyset$. Since it is true for any $Z \in \mathcal{F}^*$, $Z \subseteq A$, we conclude that Y is a transversal of hypergraph $G = \langle A, \mathcal{T} \rangle$. It follows that Y contains some minimal transversal Y_j, $j \leq t$.

Lemma 3. *Let* $\Sigma_1 = \{Y_j \to d_i : j \leq t, i \leq k\}$. *Then* $\Sigma \cup \Sigma_1$ *is a direct basis of* $\mathcal{F}^*$.

The proof of Lemma is given in the full version of the paper.

Note that $\Sigma \cup \Sigma_1$ is not necessarily the *canonical* direct basis of $\mathcal{F}^*$.

Example 6. Consider closure system on $X = \{m_1, m_2, x, d\}$ defined by set of implications $\Sigma = \{d \to x, m_1m_2x \to d\}$. Set $A = \{m_1, m_2\}$ is closed and meet-irreducible, its upper cover is X. If we want to remove it, then the hypergraph is $H = \langle A, \{\{m_1\}, \{m_2\}\}\rangle$, thus, it has a unique minimal transversal $\{m_1, m_2\}$. Then we need to add implications $m_1m_2 \to x$ and $m_1m_2 \to d$. It follows that original implication $m_1m_2x \to d$ can be removed.

7 Algorithmic Solutions and Testing

We will present the results of code implementations of two algorithms discussed in Sects. 3 and 5 and their testing on some bench-mark examples developed in earlier code implementations for the D-basis outputs. In [17], the algorithm produces the D-basis, when the input is some set of implications defining the closure operator. In [3], the closure operator is encoded in the context, and the extraction of the D-basis is done by reduction to known solutions of the hypergraph dualization problem.

We will present the bounds of algorithmic complexity and compare them with the actual time distributions based on parameters such as the sizes of the input and output.

7.1 Canonical Direct Update Implementation

When we consider the complexity of the update for the Canonical Direct basis, we compare the modified algorithm to a naive implementation of the original body building formula described in [5]. When we apply the body building formula, the first step is to remove and replace the implications which are invalid, given the new closed set we wish to add. Going through the basis once, each implication is either kept or removed. If an implication is removed, we add implications where the premise is extended by elements of the base set. If the basis contains n implications and the base set has size x, then the application of this formula has complexity $O(n \cdot x)$.

Once this pass is completed, the resulting basis is a direct basis for the desired closure system. However, there may be extra implications which must be eliminated. Let the number of implications in the updated basis be $m \leq n \cdot x$. To remove extra implications, each implication is compared to each other implication in the basis, and if there is a stronger implication with the same consequent, the weaker implication is removed. Since this compares each implication to each other, this step of the process has complexity $O(m^2)$. Once the basis is reduced to be minimal with respect to this condition, we have produced the Canonical Direct basis for the updated closure system.

With the application of the modified body building formula, we follow the same steps for the first part, but the use of *d-sectors* allows us to restrict the addition of extra implications. So, the produced basis is the Canonical Direct basis, produced with $O(n \cdot x)$ complexity and has m implications. The added complexity to this comes in building the enriched data structure for the new basis. We keep track of singleton skewed differences between implication premises among sectors. So in each sector, each implication is compared to each other and if the set difference between the premises is a singleton, we store it in the data structure. This must be done for each affected sector. We must update d-sectors where d is not an element of the new closed set to add. In the worst case, we have all m implications in a single sector and they must all be prepared, so our complexity for this step is also $O(m^2)$.

While the two algorithms have the same worst case complexity, the use of d-sectors provides practical time benefits. The worst case is that all implications are in a single sector, it is common for implications to be split among several or many sectors.

To test the practical difference between the two algorithms, the algorithms were implemented in Scala and run on random examples. For each of the 1000 examples, a random binary table between the size of 10×10 and 15×15 is generated. Then, the Canonical Direct basis of the table is generated and a random subset of the base set is chosen as the new closed set for the update. Then the update is performed with each of the algorithms and compared (Table 1).

We also compared the algorithm to an implementation of Algorithm 3 in Wild [19,20], which also mentions earlier implementation in Mannila and Räihä [12] for computation of a direct basis. While the goal of this algorithm is to generate an entire basis from a given Moore family, the algorithm does so iteratively with a subroutine which performs the same action as our algorithm. The algorithm is similar to the original body building formula, the main difference being that the algorithm in [19] is implemented for the Canonical Direct basis of full implications as opposed to the Canonical Direct unit basis.

Table 1. Average update times by number of broken implications

Broken	Naive update (ms)	Wild update (ms)	Modified update (ms)
10	91.1	23.9	18.9
20	190.8	45.7	30.4
40	194.6	51.8	28.3
Overall	198.5	54.2	27.7

Comparatively Wild's algorithm performs much better than the original body building formula, likely due to the smaller number of implications in his condensed basis. Overall, our modified body building formula outperformed Wild's algorithm, especially in cases where the number of broken was large. The modified body building formula completed the update faster than the naive update in all 1000 examples. However, Wild's update algorithm, while slower on average, outperformed the modified formula in 133 cases. Of these cases, 110 occurred when at most 20 implications were broken. The number of broken implications in these examples had an average of 30 and ranged from 0 to 140.

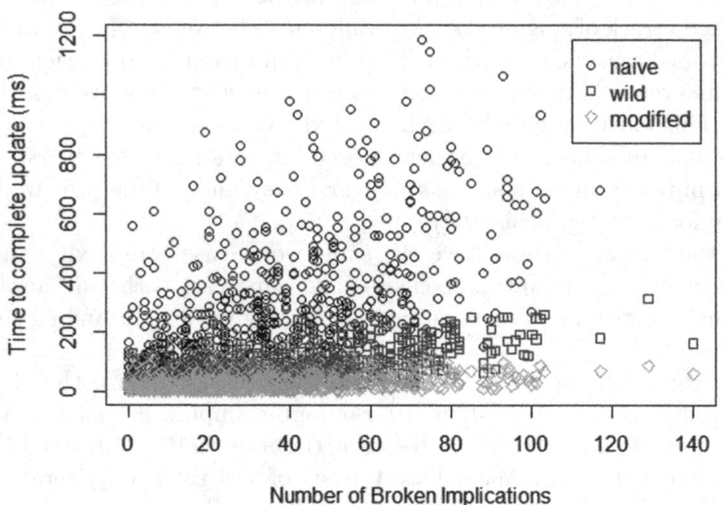

As we see in the plot of the example data, the data generally follows 3 bands, one for each algorithm with the modified algorithm generally having the lowest update time.

7.2 *D*-Basis Update Implementation

We have also tested implementation of the D-basis algorithm update described in Sect. 5 of this paper. The following 10-by-22 benchmark binary matrix generates a closure system on the set X of its 22 columns, with Moore family given by X as well as intersections of all sets represented by rows of the matrix. As usual, an implication $Y \to b$, $Y \subseteq X, b \in X$, holds in the matrix, if for any row r, the entry of r in column b is equal to 1 as long as the entries in all $y \in Y$ equal to 1.

$$
\begin{array}{cccccccccccccccccccccc}
1&0&0&0&0&1&0&0&0&1&1&0&1&1&0&1&1&1&0&1&1&1\\
1&1&1&0&0&1&0&0&1&1&1&0&0&0&1&1&0&1&0&1&0&1\\
1&1&1&1&0&1&1&1&1&1&1&1&1&0&0&0&0&0&0&0&0&1\\
1&1&0&1&0&1&1&1&0&0&0&0&1&1&1&0&0&0&1&1&1&1\\
0&0&1&1&1&1&1&1&1&1&1&1&1&1&1&1&1&1&1&0&0&0\\
0&0&0&1&1&1&1&1&0&0&1&1&0&0&1&1&0&0&0&1&1&0\\
1&1&1&1&1&1&1&1&0&1&0&1&0&1&0&1&0&1&0&1&0&1\\
1&1&0&1&0&1&1&1&1&1&1&0&1&1&1&1&1&1&0&1&1&1\\
1&1&0&0&0&1&1&1&0&0&0&0&0&0&0&1&1&1&1&0&0&0\\
0&0&0&0&1&1&1&1&0&0&0&0&1&1&1&1&0&0&0&0&1&1
\end{array}
$$

We applied the algorithm of Sect. 5 adding one row at a time and updating the D-basis.

The column 6 has all ones, which implies any implication $x \to 6$ holds, for any $x \in X$. Also, columns 7 and 8 are identical, therefore, the following implications hold: $7 \to 8$ and $8 \to 7$. The rest of the D-basis has 720 unit implications.

We were able to obtain the canonical direct basis from Sect. 3 adding one row at a time, then use the algorithm of [17] to obtain the D-basis from the canonical direct. The canonical direct basis has 1534 unit implications.

Alternately, we verified the D-basis obtaining by algorithm of [3].

We tested the update of the given matrix adding an additional row, in particular, with indicators for the set of columns $\{6, 7, 8, 17\}$. Across 5 runs, the D-basis updated in average 148 ms while the canonical direct took 171 ms. The testing was done on a laptop with 4 cores, 16 GB RAM, running at around 2.7 GHz clock speed.

Acknowledgements. The first results of the paper were presented on the poster session of ICFCA-2017 in Rennes, France, and both authors' participation in the conference was supported by the research fund of Hofstra University. We thank Sergey Obiedkov for pointing to the important publication of Marcel Wild [19], and we thank Justin Cabot-Miller from Hofstra University for his support in producing valuable test cases in the implementation phase.

References

1. Adaricheva, K., Nation, J.B.: Bases of closure systems. In: Grätzer, G., Wehrung, F. (eds.) Lattice Theory: Special Topics and Applications, vol. 2. Springer, Basel (2016). https://doi.org/10.1007/978-3-319-44236-5_6
2. Adaricheva, K., Nation, J.B.: Lattices of algebraic subsets and implicational classes. In: Grätzer, G., Wehrung, F. (eds.) Lattice Theory: Special Topics and Applications, vol. 2. Springer, Basel (2016). https://doi.org/10.1007/978-3-319-44236-5_4
3. Adaricheva, K., Nation, J.B.: Discovery of the D-basis in binary tables based on hypergraph dualization. Theoret. Comput. Sci. **658**, 307–315 (2017)
4. Adaricheva, K., Nation, J.B., Rand, R.: Ordered direct implicational basis of a finite closure system. Disc. Appl. Math. **161**, 707–723 (2013)
5. Adaricheva, K., Sloan, R., Szörenyi, B., Turan, G.: Horn belief contraction: remainders, envelopes and complexity. In: Proceedings of the KR 2012, pp. 107–115 (2012)
6. Caspard, N., Monjardet, B.: The lattices of closure systems, closure operators, and implicational systems on a finite set: a survey. Disc. Appl. Math. **127**, 241–269 (2003)
7. Bertet, K., Monjardet, B.: The multiple facets of the canonical direct unit implicational basis. Theoret. Comput. Sci. **411**, 2155–2166 (2010)
8. Boros, E., Elbassioni, K., Gurvich, V., Khachiyan, L.: Generating dual-bounded hypergraphs. Optim. Meth. Softw. **17**, 749–781 (2002)
9. Dowling, W., Gallier, J.H.: Linear-time algorithms for testing the satisfiability of propositional Horn formulae. J. Logic Program. **3**, 267–284 (1984)
10. Fredman, M., Khachiyan, L.: On the complexity of dualization of monotone disjunctive normal forms. J. Algorithms **21**, 618–628 (1996)

11. Khachiyan, L., Boros, E., Elbassioni, K., Gurvich, V.: An efficient implementation of a quasi-polynomial algorithmfor generating hypergraph transversals and its application in joint generation. Disc. Appl. Math. **154**, 2350–2372 (2006)
12. Mannila, H., Räihä, K.J.: The Design of Relational Databases. Addison-Wesley, Reading (1992)
13. McKinsey, J.: The decision problem for some classes of sentences without quantifiers. J. Symbolic Logic **8**, 61–76 (1943)
14. Murakami, K., Uno, T.: Efficient algorithms for dualizing large scale hypergraphs. Disc. Appl. Math. **170**, 83–94 (2014)
15. Langlois, M., Sloan, R., Szörenyi, B., Turan, G.: Horn complements: towards Horn-to-Horn belief revision. In: Proceedings of the AAAI 2008, pp. 466–471 (2008)
16. Rudolph, S.: Succintness and tractability of closure operator representations. Theoret. Comput. Sci. **658**, 327–345 (2017)
17. Rodríguez-Lorenzo, E., Adaricheva, K., Cordero, P., Enciso, M., Mora, A.: Formation of the D-basis from implicational systems using simplification logic. Int. J. Gen. Syst. **46**(5), 547–568 (2017)
18. Valtchev, P., Missaoui, R.: Building concept (Galois) lattices from parts: generalizing the incremental methods. In: Delugach, H.S., Stumme, G. (eds.) ICCS-ConceptStruct 2001. LNCS (LNAI), vol. 2120, pp. 290–303. Springer, Heidelberg (2001). https://doi.org/10.1007/3-540-44583-8_21
19. Wild, M.: Computations with finite closure systems and implications, Preprint N 1708. Technische Hochschule Darmstadt, pp. 1–22 (1994)
20. Wild, M.: Computations with finite closure systems and implications. In: Du, D.-Z., Li, M. (eds.) COCOON 1995. LNCS, vol. 959, pp. 111–120. Springer, Heidelberg (1995). https://doi.org/10.1007/BFb0030825
21. Wild, M.: The joy of implications. Theoret. Comput. Sci. **658**, 264–292 (2017)

Reduction and Introducers in d-contexts

Alexandre Bazin[1] and Giacomo Kahn[2(✉)]

[1] Université de Lorraine, CNRS, Inria, LORIA, 54000 Nancy, France
contact@alexandrebazin.com

[2] Université d'Orléans, INSA Centre Val de Loire, LIFO, Orléans, France
giacomo@kahn.science

Abstract. Concept lattices are well-known conceptual structures that organise interesting patterns – the concepts – extracted from data. In some applications, the size of the lattice can be a problem, as it is often too large to be efficiently computed and too complex to be browsed. In others, redundant information produces noise that makes understanding the data difficult. In classical FCA, those two problems can be attenuated by, respectively, computing a substructure of the lattice – such as the AOC-poset – and reducing the context. These solutions have not been studied in d-dimensional contexts for $d > 3$. In this paper, we generalise the notions of AOC-poset and reduction to d-lattices, the structures that are obtained from multidimensional data in the same way that concept lattices are obtained from binary relations.

1 Introduction

Formal concept analysis (FCA) is a mathematical framework introduced in the 1980s that allows for the application of lattice theory to data analysis. While it is now widely used, its main drawback, scalability, remains. Many ways to improve it have been proposed over the years. Some focus on reducing the size of the dataset by grouping or removing attributes or objects while others prefer to consider only a substructure of the lattice.

The end result of two of those approaches – respectively reduced contexts and AOC-posets – are of interest in our work. The first is the state of the dataset in which no element can be considered *redundant*. The second is the substructure of the lattice composed of its elements that best describe each individual object or attribute. Both have been extensively studied in traditional, bidimensional FCA. However, it is not the case in Polyadic Concept Analysis (PCA), the multidimensional generalization of FCA. The notion of reducibility of polyadic contexts have only proposed in the 3-dimensional case [1] and, to the best of our knowledge, no work on introducer multidimensional concepts exist.

PCA gives rise to an even greater number of patterns and so scalability is even more of an issue. Additionally, we believe that introducer concepts can be of use to better analyse d-dimensional data as each element is associated to multiple concepts instead of one. For these reasons, in this work, we define and study both the reducibility and the introducer d-ordered set of d-contexts.

D. Cristea et al. (Eds.): ICFCA 2019, LNAI 11511, pp. 73–88, 2019.
https://doi.org/10.1007/978-3-030-21462-3_6

The article is divided as follows: in the next section, Sect. 2, we give the necessary definitions and preliminaries in order to ensure a smooth reading of the paper. Section 3 is focused on dataset reduction while Sect. 4 introduces introducers. We conclude by highlighting some problems that may – or may not – be of interest in the future.

2 Formal and Polyadic Concept Analysis

2.1 Formal Concept Analysis

Formal Concept Analysis (FCA) is a mathematical framework that revolves around *formal contexts* as a condensed representation of lattices. This framework allows one to inject the powerful mathematical machinery of lattice theory into data analysis. FCA has been introduced in the 1980's by a research team led by Rudolph Wille in Darmstadt. It is based on previous work by Garrett Birkhoff on lattice theory [2] and by Marc Barbut and Bernard Monjardet [3].

In this section, we give the basic definitions of FCA. For an informative book, the reader can refer to [4]. From now on, we will freely alternate between notations ab and $\{ab\}$ to denote the set $\{a, b\}$.

Definition 1. *A* (formal) context *is a triple* $(\mathcal{O}, \mathcal{A}, \mathcal{R})$ *where* $\mathcal{O}$ *and* $\mathcal{A}$ *are finite sets and* $\mathcal{R} \subseteq \mathcal{O} \times \mathcal{A}$ *is a relation between them. We call* $\mathcal{O}$ *the set of (formal) objects and* $\mathcal{A}$ *the set of (formal) attributes.*

A formal context can be naturally represented as a cross table, as shown in Fig. 1. A pair $(o, a) \in \mathcal{R}$ corresponds to a cross in cell (o, a) of the cross table. Such a pair is read "object o has attribute a". Since many datasets can be represented as binary relations such as the one in Fig. 1, FCA finds natural applications in data analysis.

	a_1	a_2	a_3	a_4	a_5
o_1	×		×		×
o_2		×		×	×
o_3	×	×	×		
o_4			×	×	
o_5	×	×			×
o_6	×		×	×	
o_7	×	×		×	×

Fig. 1. An example context with $\mathcal{O} = \{o_1, o_2, o_3, o_4, o_5, o_6, o_7\}$ and $\mathcal{A} = \{a_1, a_2, a_3, a_4, a_5\}$. A cross in a cell (o, a) is read "object a has attribute a". A maximal rectangle of crosses is highlighted.

To allow one to efficiently jump from a set of objects to the set of attributes that describes it, and vice versa, two derivation operators are defined. For a set O of objects and a set A of attributes, they are defined as follows:

$$\cdot' : 2^{\mathcal{O}} \mapsto 2^{\mathcal{A}}$$

$$O' = \{a \in \mathcal{A} | \forall o \in O, (o, a) \in \mathcal{R}\}$$

and

$$\cdot' : 2^{\mathcal{A}} \mapsto 2^{\mathcal{O}}$$

$$A' = \{o \in \mathcal{O} | \forall a \in A, (o, a) \in \mathcal{R}\}.$$

The $\cdot'$ derivation operator maps a set of objects (resp. attributes) to the set of attributes (resp. objects) that they share. The composition of the two derivation operators ($\cdot''$) forms a Galois connection. As such, it forms a closure operator (an extensive, increasing and idempotent operator). Depending on which set the composition of operators is applied on, we can have two closure operators: $\cdot'' : 2^{\mathcal{O}} \mapsto 2^{\mathcal{O}}$ or $\cdot'' : 2^{\mathcal{A}} \mapsto 2^{\mathcal{A}}$.

A set X such that $X = X''$ is said to be closed.

Definition 2. *A pair* (O, A) *where* $O \subseteq \mathcal{O}$ *and* $A \subseteq \mathcal{A}$ *are closed,* $A = O'$ *and* $O = A'$ *is called a* concept. O *is called the* extent *of the concept while* A *is called the* intent *of the concept.*

A concept corresponds to a maximal rectangle of crosses in the cross table that represents a context, up to permutation on rows and columns. In Fig. 1, the concept $(o_2o_7, a_2a_4a_5)$ is highlighted. We denote by $\mathcal{L}(\mathcal{C})$ the set of all concepts of a context $\mathcal{C}$.

The concepts of $\mathcal{L}(\mathcal{C})$ can be ordered. Let (O_1, A_1) and (O_2, A_2) be concepts of a context. We say that (O_1, A_1) is a subconcept of (O_2, A_2) (denoted $(O_1, A_1) \leq (O_2, A_2)$) if $O_1 \subseteq O_2$. As the Galois connection that rises from the derivation is antitone, this is equivalent to $A_2 \subseteq A_1$. The concept (O_2, A_2) is a superconcept of (O_1, A_1) if $O_2 \supseteq O_1$ (and then $A_1 \supseteq A_2$).

The set of concepts from a context ordered in this way (by inclusion on the extents) forms a complete lattice called the *concept lattice* of the context. Additionally, every complete lattice is the concept lattice of some context, as stated in the basic theorem of formal concept analysis [4].

Two types of concepts can be emphasised.

Definition 3. *Let* o *be an object of a formal context. Then, the concept* (o'', o') *is called an* object-concept. *It is also called the* introducer *of the object* o.

Definition 4. *Let* a *be an attribute of a formal context. Then, the concept* (a', a'') *is called an* attribute-concept. *It is also called the* introducer *of the attribute* a.

For example, in Fig. 1, the emphasised concept is the concept (o_2'', o_2'), and it introduces o_2: it is the least concept that contains o_2. In [4], such concepts are denoted respectively $\tilde{\gamma}o$ for object-concepts and $\tilde{\mu}a$ for attribute-concepts. A concept can be both an object-concept and an attribute-concept. Concepts that are neither attribute-concepts nor object-concepts are called plain-concepts. Those concepts are Sect. 4's main characters.

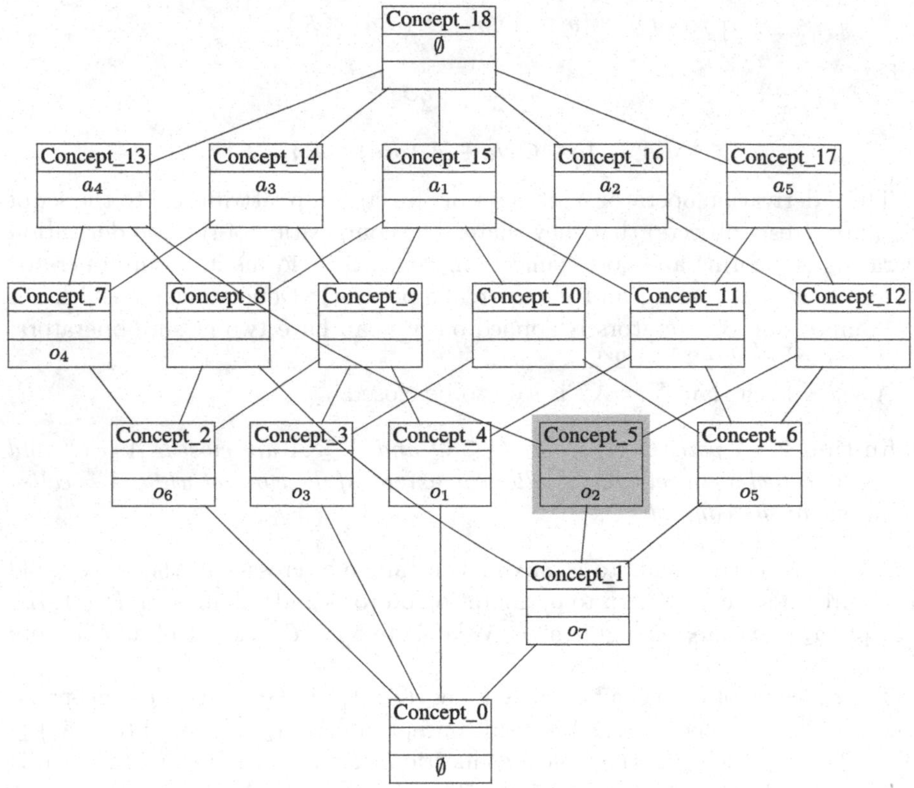

Fig. 2. Concept lattice (with simplified labels) corresponding to the example in Fig. 1.

The simplified representation of a concept lattice is a representation where the labels of the concepts are limited, in order to avoid redundancy. The label for a particular object appears only in the smallest concept that contains it (its introducer). Reversely, the label for a particular attribute appears only in the greatest concept that contains it (its introducer). The other labels are inferred using the inheritance property (the objects going up and attributes going down). Figure 2 shows to the concept lattice from the context of Fig. 1, with simplified labels. One can check that the concepts of Fig. 2 correspond to maximal boxes of incidence in Fig. 1. The highlighted concept $(o_2o_7, a_2a_4a_5)$ corresponds to the highlighted concept of Fig. 1. The concept named `Concept_10` has both labels empty. By applying the inheritance, we can retrieve that `Concept_10` is in fact the concept $(o_3o_5o_7, a_1a_2)$.

This representation allows for a clear reading into the concepts. It allows to see at first glance which concepts are objects-concepts and attribute-concepts (by definition, the introducers have non-empty labels). The plain-concepts are, quite literally, plain.

2.2 Polyadic Concept Analysis

Polyadic Concept Analysis is a natural generalisation of FCA. It has been introduced firstly by Lehmann and Wille [5,6] in the triadic case, and then generalised by Voutsadakis [7].

It deals with d-ary relations between sets instead of binary ones. More formally, a d-context can be defined in the following way.

Definition 5. *A* d-context *is a* $(d+1)$*-tuple* $\mathcal{C} = (\mathcal{S}_1, \ldots, \mathcal{S}_d, \mathcal{R})$ *where the* $\mathcal{S}_i, i \in \{1, \ldots, d\}$*, are sets called the* dimensions *and* $\mathcal{R}$ *is a* d*-ary relation between them.*

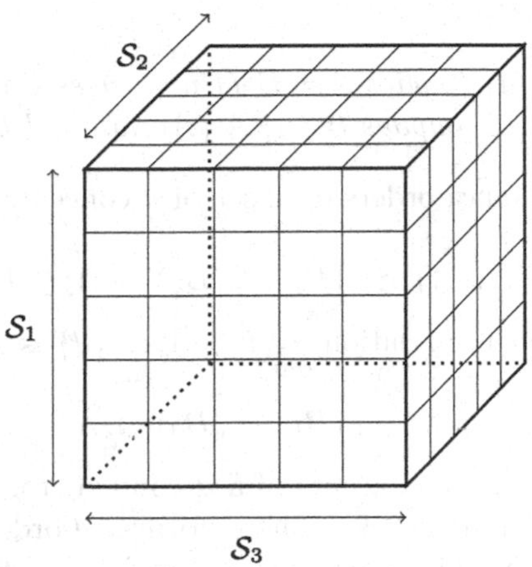

Fig. 3. Visual representation of a 3-context without its crosses.

A d-context can be represented as a $|\mathcal{S}_1| \times \ldots \times |\mathcal{S}_d|$ cross table, as shown in Fig. 3. For technical reasons, most of our examples figures will be drawn in two or three dimensions.

When needed, one can represent a d-context by separating the "slices" of the cross table. For instance, Fig. 4 shows a 3-context which dimensions are *Numbers*, *Latin*, and *Greek*.

	a	b	c	a	b	c	a	b	c
α	×	×		×			×		
β	×			×			×	×	
γ	×			×		×			×
		1			2			3	

Fig. 4. A 3-context $\mathcal{C} = (Numbers, Greek, Latin, \mathcal{R})$ where $Numbers = \{1, 2, 3\}$, $Greek = \{\alpha, \beta, \gamma\}$ and $Latin = \{a, b, c\}$. The relation $\mathcal{R}$ is the set of crosses, that is $\{(1, \alpha, a), (1, \alpha, b), (1, \beta, a), (1, \gamma, a), (2, \alpha, a), (2, \beta, a), (2, \gamma, a), (2, \gamma, c), (3, \alpha, a), (3, \beta, a), (3, \beta, b), (3, \gamma, c)\}$.

In the same way as in the 2-dimensional case, d-dimensional maximal boxes of incidence have an important role.

Definition 6. *A* d-concept *of* $\mathcal{C} = (\mathcal{S}_1, \dots, \mathcal{S}_d, \mathcal{R})$ *is a* d*-tuple* $(X_1, \dots, X_d)$ *such that* $\prod_{i\in\{1,\dots,d\}} X_i \subseteq \mathcal{R}$ *and, for all* $i \in \{1, \dots, d\}$*, there is no* $k \in \mathcal{S}_i \setminus X_i$ *such that* $\{k\} \times \prod_{j\in\{1,\dots,d\}\setminus\{i\}} X_j \subseteq R$*.*

When the dimensionality is clear from the context, we will simply call d-concepts concepts.

Definition 7. *Let* $\lesssim_i$*,* $i \in \{1, \dots, d\}$*, be quasi-orders on a set* P*. The equivalence relation* $\sim_i$ *is defined as* $\lesssim_i \cap \gtrsim_i$*. Then,* $\mathcal{P} = (P, \lesssim_1, \dots, \lesssim_d)$ *is a* d*-ordered set if, for* A *and* B *in* P*:*

1. $A \sim_i B$, $\forall i \in \{1, \dots, d\}$ *implies* $A = B$ *(Uniqueness Condition)*
2. $A \lesssim_{i_1} B, \dots, A \lesssim_{i_{d-1}} B$ *implies* $B \lesssim_{i_d} A$ *(Antiordinal Dependency)*

Let us now define d quasi-orders on a set of d-concepts of a d-context $\mathcal{C}$:

$$(A_1, \dots, A_d) \lesssim_i (B_1, \dots, B_d) \Leftrightarrow A_i \subseteq B_i$$

The resulting equivalence relation $\sim_i, i \in \{1, \dots, d\}$ is then:

$$(A_1, \dots, A_d) \sim_i (B_1, \dots, B_d) \Leftrightarrow A_i = B_i$$

We can see that the set of concepts of a d-context, together with the quasi-orders and equivalence relations defined here, forms a d-ordered set. Additionally, the existence of some particular joins makes this d-ordered set a d-lattice. For a more detailed definition, we refer the reader to Voutsadakis' seminal paper on Polyadic Concept Analysis [7].

We can see that concept lattices are in fact 2-ordered sets that satisfy both the uniqueness condition and the antiordinal dependency.

In order to fully understand d-lattices, let us illustrate the definition with a small digression about graphical representation. In 2 dimensions – i.e. concept lattices – the two orders (on the extent and on the intent) are dual and only one is usually mentioned (the set of concepts ordered by inclusion on the extent, for example). Thus, their representation is possible with Hasse diagrams. From dimension 3 and up, the representation of d-ordered sets is harder. For example, in dimension 3, as of the time of writing, no good (in the sense that it allows to represent any 3-ordered set) graphical representation exists. However there is still a possible representation in the form of triadic diagrams.

Figure 5 shows an example of a *triadic diagram*, a representation of a 3-ordered set. Let us explain how this diagram should be read. The white circles in the central triangle are the concepts. The lines of the triangle represent the equivalence relation between concepts: the horizontal lines represent $\sim_1$, the north-west to south-east lines represent $\sim_2$ and the south-west to north-east lines represent $\sim_3$. Recall that two concepts are equivalent with respect to $\sim_i$ if they have the same coordinate on dimension i. This coordinate can be read by following the dotted line until the diagrams outside the triangle. In [5], Lehmann and Wille call the external diagrams the *extent diagram*, *intent diagram* and *modi diagram* depending on which dimension they represent. Here, to pursue the permutability of the dimensions further, we simply denote them by *dimension diagram* for dimension 1, 2 or 3.

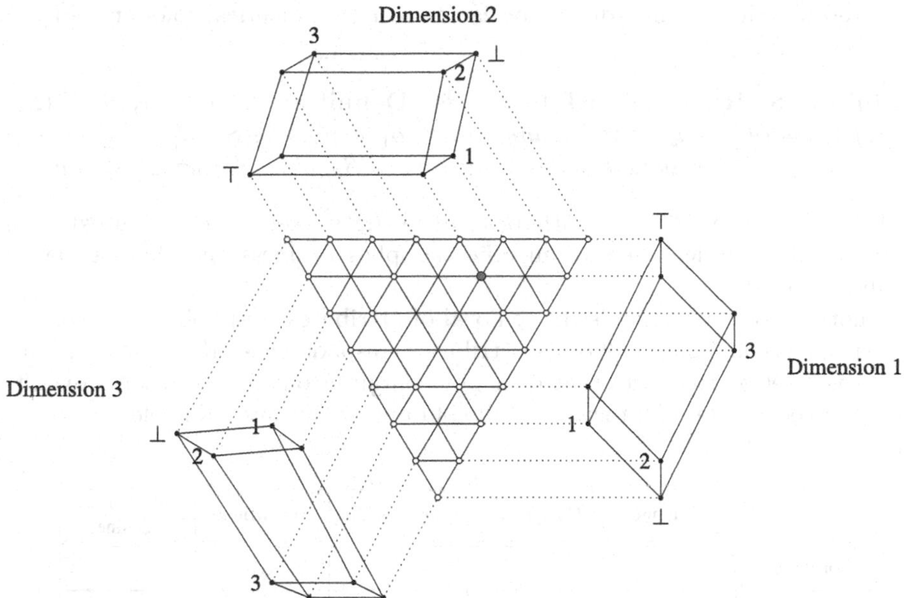

Fig. 5. Representation of a powerset 3-lattice. The red concept is (13, 13, 23). (Color figure online)

If we position ourselves on the red concept of Fig. 5, we can follow the dotted line up to the dimension 1 diagram and read its coordinate on the first dimension. It is $\{13\}$. Following the same logic for dimension 2 and 3, we can see that the red concept is $(13, 12, 23)$. The concept on the south-east of the red concept is on the same equivalence line, with respect to $\sim_2$. We know that its second dimension will be $\{12\}$ too. When we follow the lines to the other dimension diagrams, we see that it is $(23, 12, 13)$.

Here, every dimension diagram is a powerset on 3 elements. It is not always the case, as the three dimension diagrams are not always isomorphic.

We mentioned earlier that this representation does not allow to draw any 3-lattice (with straight lines). In [5], Lehmann and Wille explain this by saying that a violation of the so-called 'Thomsen Conditions [8]' and their ordinal generalisation [9] may appear in triadic contexts.[1]

[1] I was unable to put my hands on those two books in order to learn more about that. If you have a copy, I am most interested.

3 Reduction in *d*-contexts

3.1 Reduction in 2-Contexts

Two concept lattices from different contexts can be isomorphic. Reduction is a way of reaching a canonical context (up to isomorphism), the standard context, for any finite lattice. Reduction can be described as a two steps process. The first step of reduction in 2-dimension is the fusion of identical rows or columns (clarification).

Definition 8 (Reformulated from [4, Definition 23]). *A context $(\mathcal{S}_1, \mathcal{S}_2, \mathcal{R})$ is called* ***clarified*** *if for any objects o_1 and o_2 in S_1, $o_1' = o_2'$ implies that $o_1 = o_2$ and for any attributes a_1 and a_2 in $\mathcal{S}_2$, $a_1' = a_2'$ implies $a_1 = a_2$.*

In [4, Definition 23], the authors use the Fig. 6 example of a context that represents the service offers of an office supplies business and the associated clarified context.

Another possible action is the removal of attributes (resp. objects) that can be written as combination of other attributes (resp. objects). If a is an attribute and A is a set of attributes that does not contain a but have the same extent, then a is **reducible**. Full rows and full columns are always reducible.

	Furniture	Computers	Copy-machine	Typewriters	Specialised machines
Consulting	×	×	×	×	×
Planning	×	×			
Installation	×	×	×	×	×
Instruction		×	×	×	×
Training		×			
Spare parts	×	×	×	×	×
Repairs	×	×	×	×	×
Service contracts		×	×	×	

	Furniture	Computers	Copy-machine and Typewriters	Specialised machines
Consulting, Installation, Spare parts, repairs	×	×	×	×
Planning	×	×		
Instruction		×	×	×
Training		×		
Service contracts		×	×	

Fig. 6. Context and clarified context (from [4])

In a lattice, an element x is $\vee$-irreducible if $x = a \vee b$ implies $x = a$ or $x = b$ with a and b two elements of the lattice. An element x is $\wedge$-irreducible if $x = a \wedge b$ implies $x = a$ or $x = b$, with a and b two elements of the lattice.

Definition 9 (Reformulated from [4, Definition 24]). *A context* $(\mathcal{S}_1, \mathcal{S}_2, \mathcal{R})$ *is called* ***row reduced*** *if every object-concept is $\vee$-irreducible and* ***column reduced*** *if every attribute-concept is $\wedge$-irreducible. A context that is both row reduced and column reduced is* ***reduced.***

This yields that for every finite lattice L, there is a unique (up to isomorphism) reduced context such that L is the concept lattice of this context. This context is called the standard context of L. This standard context can be obtained from any finite context by first clarifying the context and then deleting all the objects that can be represented as intersection of other objects and attributes that can be represented as intersection of other attributes.

3.2 Generalisation to d-dimensions

A first generalisation to 3-dimensional contexts was made in [1] by Rudolph, Sacarea and Troanca. Here, we generalise further to the d-dimensional case.

To define reduction in multidimensional contexts, we need to recall some definitions. The following notations are borrowed from [7]. A d-context gives rise to numerous[2] k-contexts, with $k \in \{2, \dots, d\}$. Those k-contexts correspond to partitions $\pi = (\pi_1, \dots, \pi_k)$ of $\{1, \dots, d\}$ into k disjoint subsets. The k-context corresponding to π is then $\mathcal{C}^\pi = (\prod_{i \in \pi_1} \mathcal{S}_i, \dots, \prod_{i \in \pi_k} \mathcal{S}_i, \mathcal{R}^\pi)$, where $(s^{(1)}, \dots, s^{(k)}) \in \mathcal{R}^\pi$ if and only if $(s_1, \dots, s_d) \in R$ with $s_i \in s^{(j)} \Leftrightarrow i \in \pi_j$. The contexts $\mathcal{C}^\pi$ are essentially the context $\mathcal{C}$ flattened by merging dimensions with the Cartesian product.

In the following, we consider only binary partitions with a singleton on one side and all the other dimensions on the other. Let $\pi = (i, \{1, \dots, d\} \setminus i)$ be such a partition of $\{1, \dots, d\}$. Then, from the context $\mathcal{C} = (\mathcal{S}_1, \dots, \mathcal{S}_d, \mathcal{R})$ we obtain the dyadic context $\mathcal{C}^\pi = (\mathcal{S}_i, \prod_{j \in \{1,\dots,d\} \setminus \{i\}} \mathcal{S}_j, \mathcal{R}^\pi)$ where $(a, b) \in \mathcal{R}^\pi$ if and only if $a \in \mathcal{S}_i$ and $b = b_1 \times \dots \times b_{d-1} \in \prod_{j \in \{1,\dots,d\} \setminus i} \mathcal{S}_j$ and $(a, b_1, \dots, b_{d-1}) \in \mathcal{R}$.

We refer the reader to Figs. 4 and 7 for a graphical representation of this transformation. Such a binary partition gives rise to the dyadic derivation operators $X \mapsto X^{(\pi)}$ on the 2-context $\mathcal{C}^\pi$.

	(1,a)	(1,b)	(1,c)	(2,a)	(2,b)	(2,c)	(3,a)	(3,b)	(3,c)
α	×	×		×			×		
β	×			×			×	×	
γ	×			×		×			×

Fig. 7. Let $\mathcal{C}$ be our context from Fig. 4. Let $\pi = (\{Greek\}, \{Number, Latin\})$. This figure represents the 2-context $\mathcal{C}^\pi$.

[2] Stirling number of the second kind, or number of ways of arranging d dimensions into k slots.

Let x be an element of a dimension i of a d-context $\mathcal{C}$. Then we denote by C_x the $(d-1)$-context $C_x = (\mathcal{S}_1, \ldots, \mathcal{S}_{i-1}, \mathcal{S}_{i+1}, \ldots, \mathcal{S}_d, \mathcal{R}_x)$ with $R_x = \{(x_1, \ldots, x_{i-1}, x_{i+1}, \ldots, x_d) \mid (x_1, \ldots, x_{i-1}, x, x_{i+1}, \ldots, x_d) \in \mathcal{R}\}$.

We first define clarified d-contexts. Just as in the 2-dimensional case, our definition is equivalent to the fusion of identical $(d-1)$-dimensional layers.

Definition 10. *A d-context $(\mathcal{S}_1, \ldots, \mathcal{S}_d, \mathcal{R})$ is called* ***clarified*** *if, for all i in $\{1, \ldots, d\}$ and any x_1, x_2 in $\mathcal{S}_i$, $x_1^{(\pi)} = x_2^{(\pi)}$ implies $x_1 = x_2$, with $\pi = (i, \{1, \ldots, d\} \setminus i)$.*

As with the 2-dimensional case, we provide an example in Fig. 8.

	a	b	c	a	b	c	a	b	c
α	×	×		×	×				
β			×	×	×	×			×
γ	×	×					×	×	×
		1			2			3	

	$(\alpha,1)$	$(\alpha,2)$	$(\alpha,3)$	$(\beta,1)$	$(\beta,2)$	$(\beta,3)$	$(\gamma,1)$	$(\gamma,2)$	$(\gamma,3)$
a	×	×			×		×		×
b	×	×			×		×		×
c				×	×	×			×

	a and b	c	a and b	c	a and b	c
α	×		×			
β		×	×	×		×
γ	×				×	×
	1		2		3	

Fig. 8. A 3-context $\mathcal{C} = (Numbers, Greek, Latin, \mathcal{R})$ (above) and the 2-context $\mathcal{C}^{\pi}$ with $\pi = (Latin, \{Greek, Numbers\})$. We can see that $a^{(\pi)} = b^{(\pi)} = \{(\alpha,1), (\alpha,2), (\beta,2), (\gamma,1), (\gamma,3)\}$, which means that a and b can be aggregated into a new attribute "a and b". Below, the corresponding clarified 3-context.

Definition 11. *A clarified d-context $\mathcal{C} = (\mathcal{S}_1, \ldots, \mathcal{S}_d, \mathcal{R})$ is called i-reduced if every object-concept from $\mathcal{C}^{(\pi)}$, with $\pi = (i, \{1, \ldots, d\} \setminus \{i\})$, is $\vee$-irreducible. A d-context is* ***reduced*** *if it is i-reduced for all i in $\{1, \ldots, d\}$.*

The most important property of reduction in the two dimensional setting is that the lattice structure obtained from the concepts of a given context is isomorphic to the one of its reduced context. The following proposition proves that the same holds for our definition of reduction in d dimensions.

Proposition 1. *Let $\mathcal{C} = (\mathcal{S}_1, \ldots, \mathcal{S}_d, \mathcal{R})$ be a d context. Let π be a binary partition $(i, \{1, \ldots, d\} \setminus \{i\})$. Let y_i be an element of $\mathcal{S}_i$ and Y_i be a subset of $\mathcal{S}_i$ such that y_i is not in Y_i and $y_i^{(\pi)} = Y_i^{(\pi)}$. Then,*

$$\mathcal{L}(\mathcal{C}) \cong \mathcal{L}\left(\left(\mathcal{S}_1, \ldots, \mathcal{S}_i \setminus \{y_i\}, \ldots, \mathcal{S}_d, \mathcal{R} \cap \left(\mathcal{S}_i \setminus \{y_i\} \times \prod_{j \in \{1,\ldots,d\} \setminus \{i\}} \mathcal{S}_j\right)\right)\right).$$

Proof. Without loss of generality, let us assume that y_i is an element of $\mathcal{S}_1$. We have to ensure that if $(X_1, \ldots, X_d)$ is a concept of $\mathcal{C}$, then $(X_1 \setminus \{y_i\}, \ldots, X_d)$ is a concept in the reduced context. In $\mathcal{C}$, $(X_1 \setminus \{y_i\}, \ldots, X_d)$ is a d-dimensional box full of crosses. We have to show that removing y_i from X_1 does not allow it to be extended on any other dimension. As $(X_1, \ldots, X_d)$ is a d-concept, its components X_j, $j \neq i$ form a $(d-1)$-concept in the intersection of all the layers induced by the elements of X_i. As $\mathcal{C}$ is not reduced because of y_i, $\mathcal{C}_{y_i}$ is the intersection of at least two layers $\mathcal{C}_a$ and $\mathcal{C}_b$. Obviously, a and b are elements of X_1. If $(X_2, \ldots, X_d)$ is not a $(d-1)$-concept in the intersection of the layers $\mathcal{C}_x$, $x \in X_1 \setminus \{y_i\}$, then it can be extended using crosses that both $\mathcal{C}_a$ and $\mathcal{C}_b$ share. This means that $\mathcal{C}_{y_i}$ should have them too, preventing $(\mathcal{S}_1, \ldots, \mathcal{S}_d, \mathcal{R})$ from being a concept in the first place. This ensures that $(X_1 \setminus \{y_i\}, \ldots, X_d)$ is indeed a concept.

This proposition states that removing a reducible element from a d-context does not change the structure of the underlying d-lattice. If we keep track of the reduced and clarified elements during the process, it is possible not to lose information (by creating aggregate attributes or objects for example). It is still a deletion from the dataset. In the next section we speak about introducer concepts in a multidimensional setting.

4 Introducer Concepts

4.1 Introducer d-concepts

Reduction induces a loss of information in a dataset since reducible elements are erased. Lots of applications cannot afford this loss of information and have to use other ways of reducing the complexity. Another structure, smaller than the concept lattice, has been introduced by Godin and Mili [10] in 1993. This structure consists in the restriction of the lattice to the set of introducer concepts. In the general case, the properties that make a concept lattice a lattice are lost and we are left with a simple poset. Since dyadic FCA deals with objects and attributes, such a poset is also called an Attribute-Object-Concept poset, or AOC-poset for short. In this section, we introduce the introducer concepts in a d-dimensional setting.

Due to the unicity of the dyadic closure (one component of a concept leaves only one choice for the other), each element of a dimension has only one introducer. This bounds the size of an AOC-poset by the number of objects plus the number of attributes of a context, when a concept lattice can have an exponential number of objects or attributes. As we will see in the following, this property is lost when we go multidimensional.

Definition 12. *Let i be a dimension called the height while all other dimensions are called the width. Let x be an element of dimension i. The concepts with maximal width such that x is in the height are the introducer concepts of x.*

	a	b	c	a	b	c	a	b	c	a	b	c
α	×	×		×			×			×	×	
β	×			×			×	×			×	×
γ	×			×		×		×	×		×	×
		1			2			3			4	

Fig. 9. This 3-context shall serve as an example of our definitions.

Let us consider the 3-context from Fig. 9 as an example. The set of introducers of element a, denoted I_a is $I_a = \{(12, \alpha\beta\gamma, a), (123, \alpha\beta, a), (1234, \alpha, a)\}$. For the element 3, we have $I_3 = \{(123, \alpha\beta, a), (3, \beta, ab), (34, \beta\gamma, b), (34, \gamma, bc)\}$.

We denote by $\mathcal{I}(\mathcal{S}_i)$ the union of the introducer concepts of all the elements of a dimension i and by $\mathcal{I}(\mathcal{C})$ the set of all the introducer concepts of a context $\mathcal{C}$.

Proposition 2. *$(\mathcal{I}(\mathcal{C}), \lesssim_1, \dots, \lesssim_d)$ is a d-ordered set.*

Proof. Let $A = (A_1, \dots, A_d)$ and $B = (B_1, \dots, B_d)$ be elements of $\mathcal{I}(\mathcal{C})$. We recall that $A_i \subseteq B_i \Leftrightarrow A \lesssim_i B$ and that $A_i = B_i \Leftrightarrow A \sim_i B$. Without loss of generality, A is an introducer for an element of dimension i and B for an element of dimension j. If, for all k between 1 and d, $A \sim_k B$, then for all k, $A_k = B_k$ and $A = B$ (Uniqueness Condition).

Let us suppose that A and B are such that $A \lesssim_i B$, $\forall i \in \{1, \dots, d\} \setminus \{k\}$. Then, necessarily, $B_k \subseteq A_k$ or A would not be a d-concept. Hence, $B \lesssim_k A$ (Antiordinal Dependency).

From now on we can mirror the terminology of the 2-dimensional case, where we have complete lattices and attribute-objects-concepts partially ordered set, and use complete d-lattices and introducers d-ordered sets.

The following proposition links introducer d-concepts with the $(d-1)$-concepts that arise on a layer of a d-context.

Proposition 3. *Let x be an element of $\mathcal{S}_i$. If $(X_1, \dots, X_{i-1}, X_{i+1}, \dots, X_d)$ is a $(d-1)$-concept of $\mathcal{C}_x$, then there exists some X_i such that $(X_1, \dots, \{x\} \cup X_i, \dots, X_d)$ is an introducer of x. If $(X_1, \dots, \{x\} \cup X_i, \dots, X_d)$ is an introducer of x, then there exists a $(d-1)$-concept $(X_1, \dots, X_{i-1}, X_{i+1}, \dots, X_d)$ in $\mathcal{C}_x$.*

Proof. We suppose, without loss of generality, that x is in $\mathcal{S}_1$. The $(d-1)$-concepts of $\mathcal{C}_x$ are of the form $(X_2, \dots, X_d)$. If $(x, X_2, \dots, X_d)$ is a d-concept of $\mathcal{C}$, then it is maximal in width and has x in its height, so it is an introducer of x.

If $(x, X_2, \ldots, X_d)$ is not a d-concept of $\mathcal{C}$, it means that it can be augmented only on the first dimension (since $(X_2, \ldots, X_d)$ is maximal in $\mathcal{C}_x$). Thus, there exists a d-concept $(\{x\} \cup X_1, X_2, \ldots, X_d)$ that is maximal in width and has x in its height and is, as such, an introducer for x.

Suppose that there is a $X = (X_1, \ldots, X_d)$ that is an introducer of x but that is not obtained from a $(d-1)$-concept of $\mathcal{C}_x$ by extending X_1. It means that $(X_2, \ldots, X_d)$ is not maximal in $\mathcal{C}_x$ (or else it would be a $(d-1)$-concept). Then there exists a d-concept $Y = (Y_1, Y_2, \ldots, Y_d)$ with $x \in Y_1 \subseteq X_1$ and $X_i \subseteq Y_i$ for all i between 1 and d. This is in contradiction with the fact that X is an introducer of x.

Proposition 3 states that every $(d-1)$-concept of a layer $\mathcal{C}_x$ maps to an introducer of x in $\mathcal{C}$ and that every introducer of x is the image of a $(d-1)$-concept of $\mathcal{C}_x$. This proposition results in a naive algorithm to compute the set of introducer concepts of a d-context. It is sufficient to compute the $(d-1)$-concepts of the $(d-1)$-contexts obtained by fixing an element of a dimension.

Algorithm 1 computes the introducers for each element of a dimension i. For a given element $x \in \mathcal{S}_i$, we compute $\mathcal{T}(\mathcal{C}_x)$. Then, for each $(d-1)$-concept $X \in \mathcal{T}(\mathcal{C}_x)$, we build the set X_i needed to extend X into a d-concept. An element y is added to X_i when $y \times \prod_{j \neq i} X_j \subseteq \mathcal{R}$, that is if there exists a $(d-1)$-dimensional box full of crosses (but not necessarily maximal) in $\mathcal{R}$, at level y. The final set X_i always contains at least x. To compute the set introducer concepts for a d-context, one needs to call Algorithm 1 on each dimension of a d-context. In some applications, it may be useful to compute the introducer concepts with respect to a given quasi-order $\lesssim_i$.

Algorithm 1. INTRODUCERDIM($\mathcal{C}, i$)

Input: $\mathcal{C}$ a d-context, $i \in \{1, \ldots, d\}$ a dimension
Output: $\mathcal{I}(\mathcal{S}_i)$ the set of introducer concepts of elements of dimension i

$I \leftarrow \emptyset$
foreach $x \in \mathcal{S}_i$ **do**
    $C \leftarrow \emptyset$
    foreach $X = (X_1, \ldots, X_{i-1}, X_{i+1}, \ldots, X_d) \in \mathcal{T}(\mathcal{C}^x)$ **do**
        $X_i \leftarrow \emptyset$
        foreach $y \in \mathcal{S}_i$ **do**
            if $\prod_{j \neq i} X_j \times y \subseteq \mathcal{R}$ **then**
                $X_i \leftarrow X_i \cup y$
        $C \leftarrow C \cup (X_1, \ldots, X_i, \ldots, X_d)$
    $I \leftarrow I \cup C$
return I

4.2 Combinatorial Intuition: Powerset d-lattices

Unlike the 2-dimensional case, where introducers are unique and the size of the AOC-poset is thus bounded by $|\mathcal{S}_1| + |\mathcal{S}_2|$, in the general case, it is bounded by $\mathbb{K}_{d-1} \times \sum_{i\in\{1,\ldots,d\}} |\mathcal{S}_i|$, with $\mathbb{K}_d$ the maximal number of d-concepts in a d-context. Since a 1-context has only one 1-concept, this bound is reached in the 2-dimensional case.

Let us consider a powerset 3-lattice $\mathfrak{T}(5)$ on a ground set of size 5. It is well known[3] that $\mathfrak{T}(5)$ has $3^5 = 243$ concepts. However, the size of the introducer set of the powerset trilattice $\mathfrak{T}(5)$ is 30. Indeed, by Proposition 3 we know that there exists a mapping between the 2-concepts of each layer induced by fixing an element of a dimension and the introducers. As, by definition, every layer of the context inducing a powerset trilattice has two 2-concepts, the number of introducer concepts in a powerset 3-lattice on a ground set of five elements is then bounded by $3 \times 5 \times 2$. This number is reached as the unique "hole" in each layer intersects all the concepts of the other layers. A more formal proof is given for a more general proposition below (Proposition 4).

In fact, for any powerset 3-lattice on a ground set of size n, the corresponding introducer d-ordered set has $3 \times 2 \times n$ elements. Figure 10 shows a 4-adic contranominal scale on three elements. Figure 11 shows the introducers of the powerset 3-lattice on a ground set of 3 elements.

		a	b	c	a	b	c	a	b	c
A	α		×	×	×	×	×	×	×	×
	β	×	×	×	×	×	×	×	×	×
	γ	×	×	×	×	×	×	×	×	×
B	α	×	×	×	×	×	×	×	×	×
	β	×	×	×	×		×	×	×	×
	γ	×	×	×	×	×	×	×	×	×
C	α	×	×	×	×	×	×	×	×	×
	β	×	×	×	×	×	×	×	×	×
	γ	×	×	×	×	×	×	×	×	
			1			2			3	

Fig. 10. This is a 4-adic contranominal scale where the empty cells have been framed. Every layer induced by fixing an element (for example A) has three 3-concepts (in $\mathcal{C}^A$ we have $(123, \alpha\beta\gamma, bc)$, $(123, \beta\gamma, abc)$ and $(23, \alpha\beta\gamma, abc)$).

Proposition 4. *Let d be an integer. A powerset d-lattice $\mathfrak{T}^d(n)$ on a ground set of n elements has d^n elements. Its corresponding introducer d-ordered set has $d \times (d-1) \times n$ elements.*

Proof. Let $\mathcal{C}$ be a d-dimensional contranominal scale, that gives rise to $\mathfrak{T}^d(n)$. Let x be an element of a dimension. The $(d-1)$-context $\mathcal{C}_x$ has only one hole

[3] Not that well known, but it is said in this paper [11].

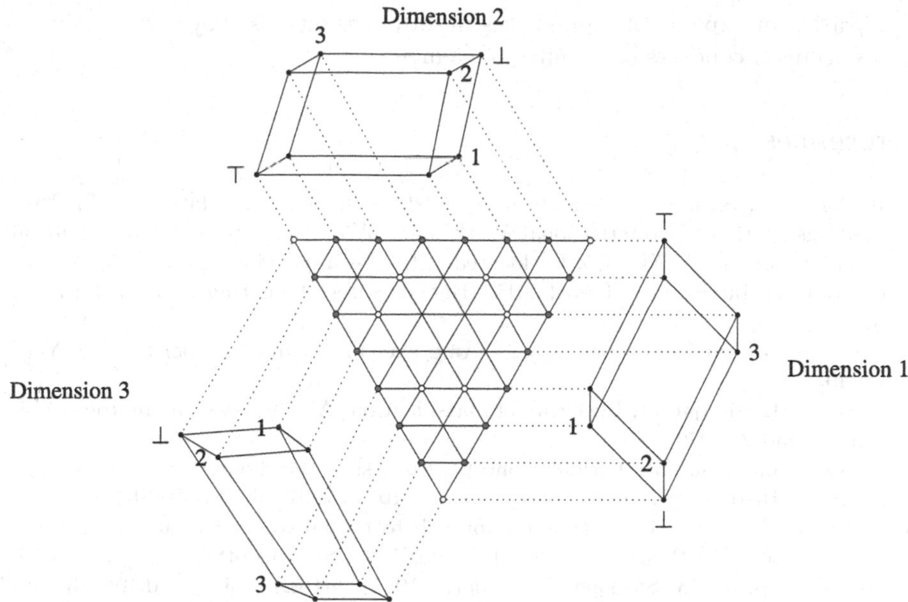

Fig. 11. This represents a powerset 3-lattice. Its introducer concepts are filled in red. (Color figure online)

(by definition of a contranominal scale). This implies that each $(d-1)$-layer induced by fixing an element of a dimension has $d-1$ concepts. By Proposition 3, we know that there exists a mapping between the $(d-1)$-concepts of the layers and the introducer concepts of the context. Since we have d dimensions and n layers by dimension, the number of introducer concepts is bounded by $d \times (d-1) \times n$.

Moreover, let $X = (X_1, \ldots, X_d)$ be an introducer concept for element x of dimension i. Then $X_i = x$. Indeed, by definition of a contranominal scale, there will be a 'hole' per layer of the context $\mathcal{C}$ that will be in the width of X.

This ensures that the $d \times (d-1) \times n$ introducer concepts that arise from Proposition 3 are distinct and that this number is reached in the case of powerset d-lattices.

5 Open Problems

With respect to reduction, it would be interesting to investigate the relation of irreducible elements to the Dedekind-MacNeille completion of d-ordered sets presented by Voutsadakis in [12].

Some interesting questions remain open regarding the number of introducer concepts compared to the number of concepts, both experimentally and theoretically. Furthermore, it would be interesting to use introducer concepts in visualisation or explainable approaches in d-dimensions, as they give insight on the less general concepts containing an element.

References

1. Rudolph, S., Sacarea, C., Troanca, D.: Reduction in triadic data sets. In: Proceedings of the 4th International Workshop "What Can FCA do for Artificial Intelligence?" FCA4AI 2015, Co-located with the International Joint Conference on Artificial Intelligence (IJCAI 2015), Buenos Aires, Argentina, 25 July 2015, pp. 55–62 (2015)
2. Birkhoff, G.: Lattice Theory, vol. 25. American Mathematical Society, New York (1940)
3. Barbut, M., Monjardet, B.: Ordre et classification. Algebre et Combinatoire, Volumes 1 and 2 (1970)
4. Ganter, B., Wille, R.: Formal Concept Analysis - Mathematical Foundations. Springer, Heidelberg (1999). https://doi.org/10.1007/978-3-642-59830-2
5. Lehmann, F., Wille, R.: A triadic approach to formal concept analysis. In: Ellis, G., Levinson, R., Rich, W., Sowa, J.F. (eds.) ICCS-ConceptStruct 1995. LNCS, vol. 954, pp. 32–43. Springer, Heidelberg (1995). https://doi.org/10.1007/3-540-60161-9_27
6. Wille, R.: The basic theorem of triadic concept analysis. Order **12**(2), 149–158 (1995)
7. Voutsadakis, G.: Polyadic concept analysis. Order **19**(3), 295–304 (2002)
8. Krantz, D.H., Suppes, P., Luce, R.D.: Foundations of Measurement. Academic Press, New York (1971)
9. Wille, U.: Geometric representation of ordinal contexts. Ph.D. thesis, University Gießen (1995)
10. Godin, R., Mili, H.: Building and maintaining analysis-level class hierarchies using Galois lattices. In: 8th Conference on Object-Oriented Programming Systems, Languages, and Applications (OOPSLA), pp. 394–410 (1993)
11. Biedermann, K.: Powerset trilattices. In: Mugnier, M.-L., Chein, M. (eds.) ICCS-ConceptStruct 1998. LNCS, vol. 1453, pp. 209–221. Springer, Heidelberg (1998). https://doi.org/10.1007/BFb0054916
12. Voutsadakis, G.: Dedekind-Macneille completion of n-ordered sets. Order **24**(1), 15–29 (2007)

Dualization in Lattices Given by Implicational Bases

Oscar Defrain(✉) and Lhouari Nourine

LIMOS, Université Clermont Auvergne, Aubière, France
{oscar.defrain,lhouari.nourine}@uca.fr

Abstract. It was recently proved that the dualization in lattices given by implicational bases is impossible in output-polynomial time unless P = NP. In this paper, we show that this result holds even when the premises in the implicational base are of size at most two. In the case of premises of size one—when the lattice is distributive—we show that the dualization is possible in output quasi-polynomial time whenever the graph of implications is of bounded maximum induced matching. Lattices that share this property include distributive lattices coded by the ideals of an interval order.

Keywords: Lattice dualization · Transversals enumeration · Implicational base · Distributive lattices · Interval order

1 Introduction

The *dualization of a monotone Boolean function* is ubiquitous in many areas including database theory, logic, artificial intelligence and pattern mining [8,9, 11,17,20]. When defined on *Boolean lattices*, the problem is equivalent to the enumeration of the *minimal transversals* of a hypergraph, arguably one of the most studied open problems in algorithmic enumeration by now [10]. In this case, the best known algorithm is due to Fredman and Khachiyan and runs in output quasi-polynomial time [14]. An enumeration algorithm is said to be running in *output-polynomial* time if its running time is bounded by a polynomial in the sizes of both the input and the output [6,18]. When generalized to any lattice, it was recently proved by Babin and Kuznetsov in [1,2] that the dualization is impossible in output-polynomial time unless P = NP. This result holds under two different settings, when the lattice is given by an *implicational base*, or by the ordered set of its *irreducible elements* (by its *context* in FCA terminology). In the first case, the proof is based on a result of Kavvadias et al. on the intractability of enumerating the maximal models of a Horn expression [19]. The constructed implicational base, however, has an implication with a premise of unbounded size, and the tractability status of the dualization remained open in the case of implications with premises of bounded size. In this paper, we address this problem with the following result.

This work has been supported by the ANR project GraphEn ANR-15-CE40-0009.

D. Cristea et al. (Eds.): ICFCA 2019, LNAI 11511, pp. 89–98, 2019.
https://doi.org/10.1007/978-3-030-21462-3_7

Theorem 1. *The dualization in lattices given by implicational bases is impossible in output-polynomial time unless* $\mathsf{P} = \mathsf{NP}$*, even when the premises in the implicational base are of size at most two.*

In the case of premises of size one, the problem is still open. The best known algorithm is due to Babin and Kuznetsov and runs in output sub-exponential time [2]. We show that it can be solved in output quasi-polynomial time whenever the *graph of implications* is of bounded maximum induced matching; see Theorem 4. Our approach is similar to the one of [21] as we show that the problem can be reduced to the dualization in Boolean lattices, which allows us to use the algorithm of Fredman and Khachiyan for the dualization of monotone Boolean functions. Lattices that share this property include distributive lattices coded by the ideals of an *interval order*.

The rest of the paper is organized as follows. In Section 2 we introduce necessary concepts and definitions. Theorems 1 and 4 are respectively proved in Sects. 3 and 4. We conclude with future research directions in Sect. 5.

2 Preliminaries

A *partial order* on a set X (or *poset*) is a binary relation $\leq$ on X which is reflexive, anti-symmetric and transitive, denoted by $P = (X, \leq)$. Two elements x and y of P are said to be *comparable* if $x \leq y$ or $y \leq x$, otherwise they are said to be *incomparable*. The *comparability graph* of a poset $P = (X, \leq)$ is the graph $G(P)$ defined on vertex set X and where two vertices x and y are adjacent if they are comparable. We note $x < y$ if $x \leq y$ and $x \neq y$. If an element u of P is such that both $x \leq u$ and $y \leq u$ then u is called *upper bound* of x and y; it is called *least upper bound* of x and y if moreover $u \leq v$ for every upper bound v of x and y. Note that two elements of a poset may or may not have a least upper bound. The least upper bound (also known as *supremum* or *join*) of x and y, if it exists, is denoted by $x \vee y$. Dually, an element u that is such that both $u \leq x$ and $u \leq y$ is called *lower bound* of x and y; it is called *greatest lower bound* of x and y if moreover $v \leq u$ for every lower bound v of x and y. The greatest lower bound (also known as *infimum* or *meet*) of x and y, if it exists, is denoted by $x \wedge y$. A subset of a poset in which every two elements are comparable is called a *chain*. A subset of a poset in which no two distinct elements are comparable is called an *antichain*. A poset is an *interval order* if it corresponds to an ordered collection of intervals on the real line such that $[x_1, x_2] < [x_3, x_4]$ if and only if $x_2 < x_3$. The 2+2 poset is the union of two disjoint 2-elements chains. It is well known that interval orders are 2+2-free, that is, they do not induce the 2+2 poset as a suborder [5,13]. A set $I \subseteq X$ is called *ideal* of P if $x \in I$ and $y \leq x$ imply $y \in I$. If $x \in I$ and $x \leq y$ imply $y \in I$, then I is called *filter* of P. Note that the complementary of an ideal is a filter, and vice versa. For every $x \in P$ we associate the *principal ideal of* x (or simply *ideal of* x), denoted by $\downarrow x$, and defined by $\downarrow x = \{y \in X \mid y \leq x\}$. The *principal filter of* $x \in X$ is the dual $\uparrow x = \{y \in X \mid x \leq y\}$. The set of all subsets of X is denoted by 2^X. The set of all ideals of P is denoted by $\mathcal{I}(P)$. Clearly $\mathcal{I}(P) \subseteq 2^X$. If S is a subset of X, we respectively denote by $\downarrow S$ and $\uparrow S$ the sets defined by $\downarrow S = \bigcup_{x \in S} \downarrow x$ and $\uparrow S = \bigcup_{x \in S} \uparrow x$, and denote by $\text{Min}(S)$ and $\text{Max}(S)$ the sets of minimal and maximal elements of S with respect to $\leq$ in P. The following notion is central in this paper.

Definition 1. *Let $P = (X, \leq)$ be a poset and $\mathcal{B}^+$, $\mathcal{B}^-$ be two antichains of P. We say that $\mathcal{B}^+$ and $\mathcal{B}^-$ are dual in P if $\downarrow \mathcal{B}^+ \cup \uparrow \mathcal{B}^- = X$ and $\downarrow \mathcal{B}^+ \cap \uparrow \mathcal{B}^- = \emptyset$.*

Note that the problem of deciding whether two antichains $\mathcal{B}^+$ and $\mathcal{B}^-$ of a poset P are dual can be solved in polynomial time in the size of P: first compute the ideal of $\mathcal{B}^+$, remove it from P, and then compute the minimal elements among the remaining ones to check whether you obtain $\mathcal{B}^-$. Clearly, each of these three steps require to iterate only once through the comparabilities of P. The task becomes difficult when the poset is not fully given, but only an implicit coding—of possibly logarithmic size in the size of P—is given: this is usually the case when considering dualization problems in lattices.

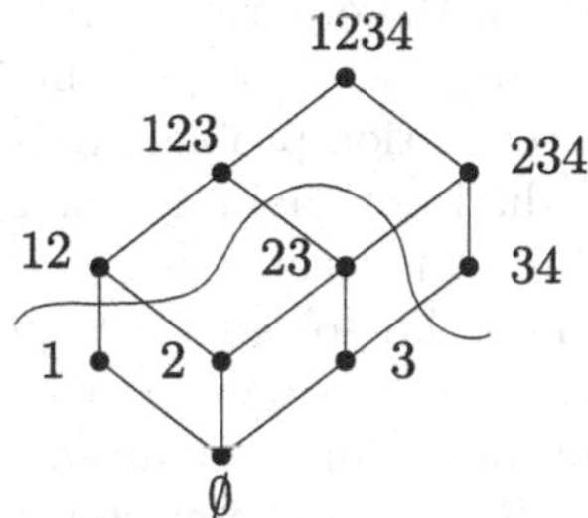

Fig. 1. The lattice $\mathcal{L}(\Sigma)$ of closed sets of the implicational base $\Sigma = \{13 \rightarrow 2,\ 4 \rightarrow 3\}$ on ground set $X = \{1, 2, 3, 4\}$, and the border (curved line) formed by the two dual antichains $\mathcal{B}^+ = \{\{1\}, \{2, 3\}\}$ and $\mathcal{B}^- = \{\{1, 2\}, \{3, 4\}\}$ of $\mathcal{L}(\Sigma)$. For better readability, closed sets are denoted without braces in the lattice, i.e., 123 stands for $\{1, 2, 3\}$.

A *lattice* is a poset in which every two elements have an infimum and a supremum; see [4,7,16]. It is called *distributive* if for any three elements x, y, z of the lattice,

$$x \wedge (y \vee z) = (x \wedge y) \vee (x \wedge z).$$

An implicational base Σ on ground set X, denoted by $\langle X, \Sigma \rangle$, is a set of implications of the form $A \rightarrow B$ where $A \subseteq X$ and $B \subseteq X$; see [22]. In this paper we only consider Σ in its equivalent form where $|B| = 1$ for every implication, and denote by $A \rightarrow b$ such implications. Then we call *premise* of $A \rightarrow b$ the set A, and *conclusion* of $A \rightarrow b$ the element b. The *graph of implications* of Σ is the graph $G(\Sigma)$ defined on vertex set X and where two vertices x and y are adjacent if there exists $A \rightarrow b \in \Sigma$ such that $x \in A$ and $y = b$. A set S is *closed* in Σ if for every implication $A \rightarrow b$ of Σ, either $b \in S$ or $A \not\subseteq S$. To Σ we associate the *closure operator* ϕ which maps every subset S of X to the smallest closed set of Σ containing S, and that we denote by $\phi(S)$. Then, we note Σ^ϕ the set of all closed sets of Σ. It is well known that every lattice can be represented as the set of all closed sets of an implicational base, ordered by inclusion. To Σ we associate $\mathcal{L}(\Sigma) = (\Sigma^\phi, \subseteq)$ such a lattice. An example of a lattice of closed sets of an implicational base is given in Fig. 1. If Σ is empty, then $\mathcal{L}(\Sigma) = (2^X, \subseteq)$ and the lattice is Boolean. If Σ only has premises of size one, then the lattice is distributive [7,16] and it is in fact a characterization [3]. Note that in general $\mathcal{L}(\Sigma)$ may be of exponential size in the size of Σ: it is in particular the case when the implicational base is empty.

In this paper, we are concerned with the following decision problem and its generation versions.

Dualization in Lattices Given by Implicational Bases (DUAL)
Input: An implicational base $\langle X, \Sigma \rangle$ and two antichains $\mathcal{B}^+, \mathcal{B}^-$ of $\mathcal{L}(\Sigma)$.
Question: Are $\mathcal{B}^+$ and $\mathcal{B}^-$ dual in $\mathcal{L}(\Sigma)$?

A positive instance of DUAL is given in Fig. 1. In its generation version, this problem calls for enumerating every inclusion-wise minimal closed set of Σ that is not a subset of any B in $\mathcal{B}^+$, or the other way around (that is, every maximal closed set of Σ that is not a superset of any B in $\mathcal{B}^-$). In the following, we denote by DUALIZATION the generation problem in its first version, that is, the problem of enumerating the dual antichain $\mathcal{B}^-$ of $\mathcal{B}^+$ in $\mathcal{L}(\Sigma)$, given $\langle X, \Sigma \rangle$ and $\mathcal{B}^+$ as input. Recently in [2], it was shown that DUAL is coNP-complete, hence that DUALIZATION cannot be solved in output-polynomial time unless P = NP. When the implicational base is empty—when the lattice is Boolean—the problem admits an output quasi-polynomial time algorithm running in time $N^{o(\log N)}$ where $N = |\mathcal{B}^+| + |\mathcal{B}^-|$, and the existence of an output-polynomial time algorithm solving the problem is now open for more than 35 years [8, 10,14]. In the case of premises of size one—when the lattice is distributive—the best known algorithm runs in output sub-exponential time $2^{O(n^{0,67} \log^3 N)}$ where $N = |\mathcal{B}^+| + |\mathcal{B}^-|$ and where n is the size of the ground set on which Σ is defined [2]. Output quasi-polynomial time algorithms are known for subclasses of distributive lattices, including products of chains [11,12]. For general distributive lattices, the existence of an output quasi-polynomial time algorithm is open.

In the following, we focus on the case where the premises in the implicational base are of size at most two.

3 Dualization in Lattices Given by Implicational Bases

We show that it is coNP-complete to decide whether two antichains of a lattice given by an implicational base are dual, even when the premises in the implicational base are of size at most two. The reduction is based on the one of Kavvadias et al. in [19], except that we manage to hide the implication of unbounded size in one of the two antichains.

Theorem 2. *The problem* DUAL *is coNP-complete for implicational bases with premises of size at most two.*

Proof. Membership in coNP follows from the fact that checking whether $\downarrow \mathcal{B}^+ \cap \uparrow \mathcal{B}^- \neq \emptyset$, or whether there exists some set $F \subseteq X$ that is closed in Σ, and that is such that both $F \notin \downarrow \mathcal{B}^+$ and $F \notin \uparrow \mathcal{B}^-$, can be done in polynomial time in the sizes of $\langle X, \Sigma \rangle$, $\mathcal{B}^+$ and $\mathcal{B}^-$.

We show completeness by reducing One-in-Three 3Sat, restricted to positive literals, to the complement of Dual. This restricted case of One-in-Three 3Sat remains NP-complete [15,19]. Consider a n-variables, m-clauses instance of One-in-Three 3Sat

$$\phi(x_1, \ldots, x_n) = \bigwedge_{j=1}^{m} C_j = \bigwedge_{j=1}^{m} (c_{j,1} \vee c_{j,2} \vee c_{j,3})$$

where $x_1, \ldots, x_n$ and $C_1, \ldots, C_m$ are respectively the variables and the clauses of ϕ, and where every variable appears in at least one clause ($c_{j,i}$ denotes the variable that appears in clause j at position i). We construct an instance of Dual as follows. Let $X = \{x_1, \ldots, x_n, y_1, \ldots, y_m, z\}$ be the ground set made of one element x per variable of ϕ, one element y per clause of ϕ, and an additional special element z. Let Σ be the implicational base defined by

$$\Sigma = \left\{ \begin{array}{rcll|c} c_{j,1}c_{j,2} & \rightarrow & z & (1) & \\ c_{j,1}c_{j,3} & \rightarrow & z & (2) & \\ c_{j,2}c_{j,3} & \rightarrow & z & (3) & \\ zc_{j,1} & \rightarrow & y_j & (4) & j \in [m] \\ zc_{j,2} & \rightarrow & y_j & (5) & \\ zc_{j,3} & \rightarrow & y_j & (6) & \\ y_j & \rightarrow & z & (7) & \end{array} \right\},$$

and let

$$\mathcal{B}^+ = \{B_j = X \setminus \{y_j, c_{j,1}, c_{j,2}, c_{j,3}\} \mid j \in [m]\},$$
$$\mathcal{B}^- = \{F = \{y_1, \ldots, y_m, z\}\}.$$

Clearly, Σ, $\mathcal{B}^+$ and $\mathcal{B}^-$ are constructed in polynomial time in the size of ϕ. Moreover, every $B_j \in \mathcal{B}^+$ is closed in Σ (observe that no literal in $\{c_{1,j}, c_{2,j}, c_{3,j}\}$ is the conclusion of an implication of Σ, and that y_j cannot be implied without any literal of $\{c_{1,j}, c_{2,j}, c_{3,j}\}$). As B_j is the only set of $\mathcal{B}^+$ containing y_j for every $j \in [m]$, no two sets in $\mathcal{B}^+$ are inclusion-wise comparable. Hence $\mathcal{B}^+$ is an antichain of $\mathcal{L}(\Sigma)$. Also, $\mathcal{B}^-$ is an antichain of $\mathcal{L}(\Sigma)$ as it is a singleton and its element F is closed in Σ. Are $\mathcal{B}^+$ and $\mathcal{B}^-$ dual in $\mathcal{L}(\Sigma)$? We show that the answer is no if and only if there is a one-in-three truth assignment of ϕ.

We prove the first implication. Let us suppose that $\mathcal{B}^+$ and $\mathcal{B}^-$ are not dual in $\mathcal{L}(\Sigma)$. Since $\downarrow \mathcal{B}^+ \cap \uparrow \mathcal{B}^- = \emptyset$, there must be some closed set $F' \subseteq X$ such that both $F' \notin \downarrow \mathcal{B}^+$ and $F' \notin \uparrow \mathcal{B}^-$. In the following, we consider an inclusion-wise minimal such set F'. Since $F \setminus \{z\}$ is not closed in Σ, and $F \setminus \{y_j\} \subseteq B_j$ for every $j \in [m]$, we deduce that $F' \nsubseteq F$. Then $F' \cap \{x_1, \ldots, x_n\} \neq \emptyset$. We show that $z \notin F'$. Let $x \in F' \cap \{x_1, \ldots, x_n\}$ and suppose that $z \in F'$. Then by implications (4) to (6), $y_j \in F'$ for all $j \in [m]$ such that $x \in C_j$. Hence for every clause C_j

containing x, we have $|F' \cap \{y_j, c_{j,1}, c_{j,2}, c_{j,3}\}| \geq 2$. Thus $F' \setminus \{x\} \not\subseteq B_j$ for any $j \in [m]$. Since $F' \setminus \{x\}$ is closed, this contradicts the fact that F' is minimal such that $F' \notin \downarrow \mathcal{B}^+$. So F' does not contain z. Clearly $F' \cap \{y_1, \dots, y_m\} = \emptyset$, or otherwise by implication (7), F' would contain z. Also $|F' \cap C_j| = 1$ for every $j \in [m]$, or otherwise by implications (1) to (3), F' would contain z. Hence F' is a one-in-three truth assignment of ϕ.

We prove the other implication. Let T be a one-in-three truth assignment of ϕ. As $T \subseteq \{x_1, \dots, x_n\}$ and $|T \cap C_j| = 1$ for all $j \in [m]$, T is closed in Σ. Furthermore $T \not\subseteq B_j$ for any $B_j \in \mathcal{B}^+$. Since at last $T \not\supseteq F$, we obtain that both $T \notin \downarrow \mathcal{B}^+$ and $T \notin \uparrow \mathcal{B}^-$, and conclude that $\mathcal{B}^+$ and $\mathcal{B}^-$ are not dual in $\mathcal{L}(\Sigma)$. □

As a consequence, there is no algorithm solving Dualization in output-polynomial time unless $\mathsf{P} = \mathsf{NP}$, even when the premises in the implicational base are of size at most two. This proves Theorem 1.

4 Distributive Lattices and Graphs of Implications of Bounded Maximum Induced Matching

In this section, we restrict ourselves to the case where the implicational base only has premises of size one, which characterizes distributive lattices [3,7]. It is well known that in that case, Σ can be considered acyclic and defines a poset P on the same ground set, where $x \leq y$ if and only if $y \to x$. Then $G(P) = G(\Sigma)$, $\phi = \downarrow$, and $\mathcal{L}(\Sigma) = \mathcal{L}(P) = (\mathcal{I}(P), \subseteq)$, where $\downarrow$ denotes the ideal in P. In the following, we find it more convenient to place ourselves in this context where Σ is given as a poset, as in [2]. We show that if $G(P)$ is of bounded maximum induced matching, then the dualization can be solved in output quasi-polynomial time using the algorithm of Fredman and Khachiyan [14] for the dualization of monotone Boolean functions.

An *induced matching* in a graph G is a set of edges M such that no two edges in M share a vertex, or are joined by an edge in G. Comparability graphs of antichains, chains, and of the union of k incomparable chains respectively have a maximum induced matching of size zero, one, and k. Posets with a comparability graph of bounded maximum induced matching include interval orders as they are 2+2-free; see Sect. 2 and [5,13].

In what follows, let $(P = (X, \leq), \mathcal{B}^+)$ be an instance of Dualization where the implicational base coding the lattice is given as a poset. Observe that $\mathcal{B}^+$ is given as antichain of $\mathcal{L}(P)$, and *not* of P, which is a crucial point. In fact, $\mathcal{B}^+$ and its dual antichain $\mathcal{B}^-$ are families of ideals of P. In the following, when using the notations $\leq$, $\mathrm{Max}(\cdot)$ and $\downarrow$ we will refer to P and not to the lattice. We denote by $\mathcal{H}$ the *complementary hypergraph* of $\mathcal{B}^+$ defined by $\mathcal{H} = \{X \setminus B \mid B \in \mathcal{B}^+\}$, and by $\mathcal{E}_x$ the set of *incident edges* of some element $x \in X$, defined by $\mathcal{E}_x = \{E \in \mathcal{H} \mid x \in E\}$. We call *vertex* an element of X, and *hyperedge* (or *edge*) a set of $\mathcal{H}$. A *transversal* of $\mathcal{H}$ is a set of vertices $T \subseteq X$ that intersects every edge $E \in \mathcal{H}$. It is called *minimal* if it does not contain any transversal as a proper subset. The set of minimal transversals of $\mathcal{H}$ is denoted by $Tr(\mathcal{H})$.

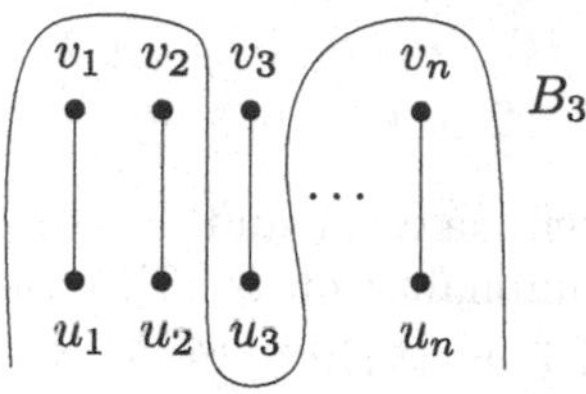

Fig. 2. A copy of n incomparable two-elements chains, the 2+2+...+2 poset. By taking $\mathcal{B}^+ = \{X \setminus \{u_i, v_i\} \mid i \in [n]\}$, we get $\mathcal{H} = \{\{u_i, v_i\} \mid i \in [n]\}$, $\mathcal{B}^- = \{\{u_1, \dots, u_n\}\}$ and $Tr(\mathcal{H}) = \{\{z_1, \dots, z_n\} \mid (z_1, \dots, z_n) \in \{u_1, v_1\} \times \cdots \times \{u_n, v_n\}\}$.

Lemma 1. *Every edge of $\mathcal{H}$ is a filter of P.*

Proof. As $\mathcal{B}^+ \subseteq \mathcal{I}(P)$, every $B \in \mathcal{B}^+$ is an ideal of P, and $X \setminus B$ a filter of P. □

Lemma 2. *If T is a transversal of $\mathcal{H}$ then $Max(T)$ is. Hence, every minimal transversal of $\mathcal{H}$ is an antichain of P.*

Proof. Let T be a transversal of $\mathcal{H}$ and x, y be two elements of T such that $x \leq y$. By Lemma 1, every edge of $\mathcal{H}$ is a filter of P. Hence $\mathcal{E}_x \subseteq \mathcal{E}_y$ and $T \setminus \{x\}$ is still a transversal of $\mathcal{H}$. We deduce that $\text{Max}(T)$ is a transversal of $\mathcal{H}$, hence that every minimal transversal of $\mathcal{H}$ is an antichain of P. □

Lemma 3. *For every $I \in \mathcal{B}^-$ there exists some $T \in Tr(\mathcal{H})$ such that $I = \downarrow T$ and $T = Max(I)$.*

Proof. Let $I \in \mathcal{B}^-$. Since $I \not\subseteq B$ for any $B \in \mathcal{B}^+$, I is a transversal of $\mathcal{H}$. By Lemma 2, $T = \text{Max}(I)$ is a transversal of $\mathcal{H}$. Since I is a minimal ideal such that $I \not\subseteq B$ for any $B \in \mathcal{B}^+$, $I \setminus \{x\}$ is not a transversal of $\mathcal{H}$ for any $x \in \text{Max}(I)$. Hence T is minimal, and $I = \downarrow T$ and $T = \text{Max}(I)$. □

As a consequence, one can enumerate $\mathcal{B}^-$ from $Tr(\mathcal{H})$ by checking for every $T \in Tr(\mathcal{H})$ whether its ideal belongs to $\mathcal{B}^-$, and discarding the solution if not. Clearly, testing such a condition can be done in polynomial time in the sizes of P and $\mathcal{B}^+$ by checking whether $I \setminus \{x\} \not\subseteq B$ for any two $x \in \text{Max}(I)$ and $B \in \mathcal{B}^+$. Hence, enumerating $\mathcal{B}^-$ can be done in total time

$$N^{o(\log N)} + |Tr(\mathcal{H})| \cdot \text{poly}(|P| + |\mathcal{B}^+|)$$

where $N = |\mathcal{H}| + |Tr(\mathcal{H})|$, by constructing $\mathcal{H}$ in time $\text{poly}(|P| + |\mathcal{B}^+|)$, using the algorithm of [14] for the enumeration of $Tr(\mathcal{H})$ in time $N^{o(\log N)}$, and discarding at most $|Tr(\mathcal{H})|$ solutions with a cost of $\text{poly}(|P| + |\mathcal{B}^+|)$ per solution. In the following, we refer to this algorithm as Algorithm $\mathcal{A}$. The limitations of this procedure is that the size of $Tr(\mathcal{H})$ may be exponential in the size of $\mathcal{B}^-$, and that the described algorithm may run in a time which is exponential in the sizes of P, $\mathcal{B}^+$ and $\mathcal{B}^-$. An example of one such instance is given in Fig. 2. However, we will show that this is not the case whenever the comparability graph of P is of bounded maximum induced matching.

Lemma 4. *Let $T_1, T_2 \in Tr(\mathcal{H})$ such that $T_1 \subsetneq\downarrow T_2$. Then for all $y \in T_2$ there exists some $x \in T_1$ such that $x \leq y$ and such that $x \not\leq z$ for any $z \in T_2 \setminus \{y\}$.*

Proof. First recall that by Lemma 2, T_1 and T_2 are antichains of P. They may intersect. Let $y \in T_2$. By minimality of T_2, $T_2 \setminus \{y\}$ is not a transversal. By Lemmas 1 and 2, neither is $\downarrow (T_2 \setminus \{y\})$ or else $\mathrm{Max}(\downarrow (T_2 \setminus \{y\})) = T_2 \setminus \{y\}$ is a transversal, which is absurd. Hence, $T_1 \not\subseteq\downarrow (T_2 \setminus \{y\})$. Let $x \in T_1 \setminus \downarrow (T_2 \setminus \{y\})$. Either $x = y$, and the lemma holds (as T_1 and T_2 are antichains), or $x \neq y$ and we conclude that $x \leq y$ and that $x \not\leq z$ for any $z \in T_2 \setminus \{y\}$. This situation is depicted in Fig. 3. □

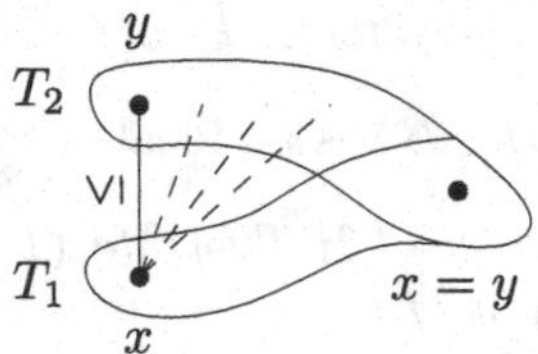

Fig. 3. The situation of Lemma 4.

Theorem 3. *The size of $Tr(\mathcal{H})$ is bounded by $n^k \cdot |\mathcal{B}^-|$ where n is the number of elements of P, and k is the size of a maximum induced matching of $G(P)$.*

Proof. Let k be the size of a maximum induced matching of $G(P)$, and let $I \in \mathcal{B}^-$. By Lemma 3, $T_0 = \mathrm{Max}(I)$ belongs to $Tr(\mathcal{H})$. Let $T_1, \ldots, T_p \in Tr(\mathcal{H})$ be the p sets such that $T_0 \neq T_i$, $T_i \neq T_j$ and $T_0 \subseteq\downarrow T_i$ for all $i, j \in [p]$. Clearly, no T_i, $i \in [p]$ is a solution of $\mathcal{B}^-$ as its corresponding ideal contains T_0 as a subset. Let us consider $T = T_i$ for one such i, and the two sets $A = T \setminus T_0$ and $B = T_0 \setminus T$. By Lemma 4, for all $y \in A$ there exists $x \in B$ such that $x \leq y$ and $x \not\leq z$ for any $z \in A \setminus \{y\}$. As a consequence, A is of size at most k or else we get $k+1$ incomparable 2-elements chains induced by every such x and y, hence an induced matching of size $k+1$ in $G(P)$. Since every such T is uniquely characterized by its intersection with A (as from A there is a unique way of covering T_0 by selecting elements of $T_0 \setminus \downarrow A$), we conclude that

$$p \leq \sum_{i=1}^{k} \binom{n}{i}$$

hence that $|Tr(\mathcal{H})| \leq n^k \cdot |\mathcal{B}^-|$ where n is the number of elements in P. □

As a consequence, the sizes of $\mathcal{H}$ and $Tr(\mathcal{H})$ are bounded by a polynomial in $|P|+|\mathcal{B}^+|+|\mathcal{B}^-|$ whenever the comparability graph of P is of bounded maximum induced matching. This shows that, under such a condition, it is still reasonable to test each of the minimal transversals generated by Algorithm $\mathcal{A}$ even though some might not lead to a solution of $\mathcal{B}^-$. We conclude with the next theorem.

Theorem 4. *There is an algorithm that, for every $k \in \mathbb{N}$, given a poset P such that $G(P)$ has no induced matching of size k, and an antichain $\mathcal{B}^+$ of the lattice $\mathcal{L}(P) = (\mathcal{I}(P), \subseteq)$, enumerates the dual antichain $\mathcal{B}^-$ of $\mathcal{B}^+$ in $\mathcal{L}(P)$ in quasi-polynomial time $N^{o(\log N)}$ where $N = |P| + |\mathcal{B}^+| + |\mathcal{B}^-|$.*

Proof. Let $k \in \mathbb{N}$ and P be a poset such that $G(P)$ has no induced matching of size k. Let $\mathcal{B}^+$ be an antichain of $\mathcal{L}(P)$, and $\mathcal{H} = \{X \setminus B \mid B \in \mathcal{B}^+\}$ be the complementary hypergraph of $\mathcal{B}^+$. By Theorem 3, $|Tr(\mathcal{H})| \leq |P|^k \cdot |\mathcal{B}^-|$. Since $|\mathcal{H}| \leq |\mathcal{B}^+|$, $|\mathcal{H}| + |Tr(\mathcal{H})| \leq (|P| + |\mathcal{B}^+| + |\mathcal{B}^-|)^c$ where $\mathcal{B}^-$ is the dual antichain of $\mathcal{B}^+$ in $\mathcal{L}(P)$, and where c is a constant that depends in k. Then, the running time of Algorithm $\mathcal{A}$ on instance $(P, \mathcal{B}^+)$ is bounded by

$$M^{o(\log M)} + |Tr(\mathcal{H})| \cdot \text{poly}(|P| + |\mathcal{B}^+|)$$

where $M = |\mathcal{H}| + |Tr(\mathcal{H})|$, hence by

$$N^{c \cdot o(\log N^c)} + \text{poly}(N) = N^{o(\log N)}$$

where $N = |P| + |\mathcal{B}^+| + |\mathcal{B}^-|$. □

5 Conclusion

We proved that the dualization in lattices given by implicational bases is impossible in output-polynomial time unless $\mathsf{P} = \mathsf{NP}$, even when the premises in the implicational base are of size at most two. In the case of premises of size one—when the lattice is distributive—we showed that the problem admits an output quasi-polynomial time algorithm whenever the graph of implications is of bounded maximum induced matching, using the algorithm of Fredman and Khachiyan for the dualization of monotone Boolean functions. Lattices that share this property include distributive lattices coded by the ideals of an interval order.

We leave open the question whether the dualization in distributive lattices can be solved in output quasi-polynomial time in general. Subclasses of interest include distributive lattices coded by the ideals of a poset of bounded dimension. Superclasses of interest include lattices coded by acyclic[1] implicational bases, a subclass of convex geometries: it is easily observed that the constructed implicational base of Theorem 1 is not acyclic, hence that the theorem does not hold for such a class.

For future research, we would be interested in knowing whether the notions of Sect. 4 can be generalized to implicational bases with premises of arbitrary or constant size.

[1] Here the graph of implications is considered directed, where there is an arc (ab) in $G(\Sigma)$ if there exists $A \to b \in \Sigma$, $a \in A$, and the implicational base is said to be acyclic if its graph of implications is; see [22].

References

1. Babin, M.A., Kuznetsov, S.O.: Enumerating minimal hypotheses and dualizing monotone boolean functions on lattices. In: Valtchev, P., Jäschke, R. (eds.) ICFCA 2011. LNCS (LNAI), vol. 6628, pp. 42–48. Springer, Heidelberg (2011). https://doi.org/10.1007/978-3-642-20514-9_5
2. Babin, M.A., Kuznetsov, S.O.: Dualization in lattices given by ordered sets of irreducibles. Theor. Comput. Sci. **658**, 316–326 (2017)
3. Birkhoff, G.: Rings of sets. Duke Math. J. **3**(3), 443–454 (1937)
4. Birkhoff, G.: Lattice Theory, vol. 25. American Mathematical Society, New York (1940)
5. Bogart, K.P.: An obvious proof of Fishburn's interval order theorem. Discrete Mathe. **118**(1–3), 239–242 (1993)
6. Creignou, N., Kröll, M., Pichler, R., Skritek, S., Vollmer, H.: A complexity theory for hard enumeration problems. Discrete Appl. Math. (2019). https://doi.org/10.1016/j.dam.2019.02.025
7. Davey, B.A., Priestley, H.A.: Introduction to Lattices and Order. Cambridge University Press, Cambridge (2002)
8. Eiter, T., Gottlob, G.: Identifying the minimal transversals of a hypergraph and related problems. SIAM J. Comput. **24**(6), 1278–1304 (1995)
9. Eiter, T., Gottlob, G., Makino, K.: New results on monotone dualization and generating hypergraph transversals. SIAM J. Comput. **32**(2), 514–537 (2003)
10. Eiter, T., Makino, K., Gottlob, G.: Computational aspects of monotone dualization: a brief survey. Discrete Appl. Math. **156**(11), 2035–2049 (2008)
11. Elbassioni, K.M.: An algorithm for dualization in products of lattices and its applications. In: Möhring, R., Raman, R. (eds.) ESA 2002. LNCS, vol. 2461, pp. 424–435. Springer, Heidelberg (2002). https://doi.org/10.1007/3-540-45749-6_39
12. Elbassioni, K.M.: Algorithms for dualization over products of partially ordered sets. SIAM J. Discrete Math. **23**(1), 487–510 (2009)
13. Fishburn, P.C.: Intransitive indifference with unequal indifference intervals. J. Math. Psychol. **7**(1), 144–149 (1970)
14. Fredman, M.L., Khachiyan, L.: On the complexity of dualization of monotone disjunctive normal forms. J. Algorithms **21**(3), 618–628 (1996)
15. Garey, M.R., Johnson, D.S.: Computers and Intractability, vol. 29. W. H. Freeman, New York (2002)
16. Grätzer, G.: Lattice Theory: Foundation. Springer, Basel (2011). https://doi.org/10.1007/978-3-0348-0018-1
17. Gunopulos, D., Mannila, H., Khardon, R., Toivonen, H.: Data mining, hypergraph transversals, and machine learning. In: PODS, pp. 209–216. ACM (1997)
18. Johnson, D.S., Yannakakis, M., Papadimitriou, C.H.: On generating all maximal independent sets. Inf. Process. Lett. **27**(3), 119–123 (1988)
19. Kavvadias, D.J., Sideri, M., Stavropoulos, E.C.: Generating all maximal models of a boolean expression. Inf. Process. Lett. **74**(3–4), 157–162 (2000)
20. Nourine, L., Petit, J.M.: Extending set-based dualization: application to pattern mining. In: Proceedings of the 20th European Conference on Artificial Intelligence, pp. 630–635. IOS Press (2012)
21. Nourine, L., Petit, J.M.: Dualization on partially ordered sets: preliminary results. In: Kotzinos, D., Choong, Y.W., Spyratos, N., Tanaka, Y. (eds.) ISIP 2014. CCIS, vol. 497, pp. 23–34. Springer, Cham (2016). https://doi.org/10.1007/978-3-319-38901-1_2
22. Wild, M.: The joy of implications, aka pure horn formulas: mainly a survey. Theor. Comput. Sci. **658**, 264–292 (2017)

"*Properties of Finite Lattices*" by S. Reeg and W. Weiß, Revisited

In Memoriam Peter Burmeister (1941–2019)

Bernhard Ganter(✉)

Institut für Algebra, Technische Universität Dresden, Dresden, Germany
bernhard.ganter@tu-dresden.de

Abstract. We review an attribute exploration form 1990, which was never published, although the results are impressive. We suggest a method for making implication lists better readable and demonstrate its effect on the canonical basis obtained from that exploration by Reeg and Weiß.

Attribute Exploration is one of the useful techniques that emerged from Formal Concept Analysis [1]. It had been implemented immediately after it was developed in 1984, and in 1986 became part of Peter Burmeister's popular `ConImp` software [2].

It was Rudolf Wille who initiated several substantial application instances of this method. One of these, a comprehensive classification of properties of finite lattices, was elaborated by two of Wille's students, Stefan Reeg and Wolfgang Weiß. They had learnt about the topic in one of Wille's seminars, jointly worked through the problems and eventually submitted their findings as a master thesis ("Diplomarbeit") [3] in 1990. The authors chose 51 lattice properties from the literature and determined the logical dependencies between these properties (almost) completely. Their work results in an irredundant collection of 100 implications and a formal context of 84 separating examples for the 51 properties. The concept lattice has 1768 elements.

These results never occurred in print, although they are quite interesting. A manuscript by Reeg, Skorsky, Wille, and Weiß (cited in [4]) apparently remained unpublished. One of the reasons is that the work is rather technical. To give an impression, we cite one of the 100 implications proved by Reeg and Weiß:

(RW31) {lower balanced, Kurosh-Ore property, dual Kurosh-Ore property, complemented, sectionally semicomplemented, semicomplemented, almost complemented, dually semicomplemented, dually almost complemented, atomistic, finite, width:3} $\to$ {sectionally complemented}

This example is of medium size. Their list contains shorter implications, but considerably longer ones as well. This comes from using the *canonical basis* as a key ingredient of Attribute Exploration. Is has a tendency of producing long implications.

D. Cristea et al. (Eds.): ICFCA 2019, LNAI 11511, pp. 99–109, 2019.
https://doi.org/10.1007/978-3-030-21462-3_8

Since the time when this investigation was performed, Attribute Exploration was further developed. Together with Sergei Obiedkov the present author published a state-of-the art book [5] in 2016, including up-to-date versions of the algorithms. On this basis we have now started reconsidering old explorations, hoping that the advanced methods can ease the work and smoothen the results. An obstacle is that sufficiently sophisticated software for modern Attribute Exploration is not yet available. A 2020 version of Burmeister's `ConImp` is missing! Nevertheless we have tried, with some ad-hoc coding, to make the results of Reeg and Weiß better accessible. Our results are below.

1 The Lattice Properties Under Investigation

Reeg and Weiß selected most of the lattice properties for their investigation from three standard lattice theory books: Birkhoff's *"Lattice Theory"* [6], the *"Algebraic Theory of Lattices"* by Crawley and Dilworth [7], and Grätzer's *"General Lattice Theory"* [8], and from an earlier investigation by Wille [9]. Moreover they asked leading lattice theorists for additional suggestions and eventually came to a collection of 51 properties, shown in Fig. 1. About the *c-condition*[1] we learnt from A. Day, who also named it *semi-convex.* Property No. 47 in that list is "finite". One might wonder why it is listed at all, since the word *"finite"* is already part of the title. But Reeg and Weiß keep more than they promise: they classify lattices of *finite length*, not only finite ones.

A glance at Fig. 1 immediately suggests some very elementary simplifications. We see that by far not all of the listed properties are not equivalent: two properties which are not separated by a vertical or horizontal line are equivalent for lattices of finite length. For example, the two properties "*Boolean*" and "*uniquely complemented*" are equivalent, not for lattices in general, but for lattices of finite length. This is non-trivial, and requires a proof or a citation. Fortunately it is a corollary to Theorem 12 in [6]. Thus we may accept the pair of implications

$$\textit{Boolean} \leftrightarrow \textit{uniquely complemented}$$

for lattices of finite length as true and use them as *background implications.* For the other equivalences we can do the same, see Fig. 2 for a list. Attribute Exploration with background knowledge is well understood (see [5]), and the case of implicational background knowledge is particularly easy. Implications can be added at any time to the knowledge base without terminating the exploration procedure.

Figure 1 also reminds us that the exploration domain of finite lattices has a natural automorphism: duality. Again, this is well understood. The algorithms for Attribute Exploration under automorphisms (or, more generally, under *c-endomorphisms*) are discussed in the above-mentioned book [5]. And again, in our case it is particularly easy. It simply amounts to the following: For every valid implication, the *dual implication*, in which all properties are replaced by

[1] $x \wedge y = x \wedge z$ and $x \vee z = y \vee z$ together imply $x \leq z$.

	Property			Dual Property
1	Boolean	[6]		
29	uniquely complemented	[6]		
2	Stone	[6]	3	dually Stone
4	Browerian	[6]		
5	distributive	[6]		
6	join-distributive	[1]	7	meet-distributive
8	SD-meet	[1]	9	SD-join
10	0-distributive	[8]	11	1-distributive
14	pseudocomplemented	[8]	15	dually pseudocomplemented
12	c-condition	[1]		
13	dual c-condition			
16	modular	[6]		
46	metric	[6]		
17	locally modular	[7]	18	dually locally modular
19	M-symmetric	[8]	20	dually M-symmetric
21	upper semimodular	[6]	22	lower semimodular
23	balanced	[4]		
24	upper balanced	[4]	25	lower balanced
26	graded	[6]		
27	Kurosh-Ore property	[7]	28	dual Kurosh-Ore property
30	relatively complemented	[6]		
31	relatively semicomplemented	[9]		
32	dually relatively semicomplemented	[9]		
33	sectionally complemented	[8]	34	dually sectionally complemented
35	complemented	[6]		
36	sectionally semicomplemented	[9]	37	dually sectionally semicomplemented
44	atomistic	[6]	45	coatomistic
38	semicomplemented	[9]	40	dually semicomplemented
39	almost complemented	[9]	41	dually almost complemented
42	geometric	[6]	43	dually geometric
47	finite			
48	self-dual	[6]		
49	width:2	[6]		
50	width:3	[6]		
51	breadth:2	[6]		

Fig. 1. The lattice properties investigated by Reeg and Weiß, and where the definitions can be found.

their duals, must be valid as well. We therefore list only one specimen of each dual pair of implications, but increase the implication counter by two, except for self-dual implications. And if an implication is not valid, then the dual of any counterexample refutes the dual implication.

There are a few more potential background implications to be read off from Fig. 1. Property 48, "*self-dual*", yields, together with any other property, the dual of that property as, for example, in

$$upper\ semimodular,\ self\text{-}dual \longrightarrow lower\ semimodular.$$

And we may express the fact that the two conditions "*breadth:2*" and "*breadth:3*" are mutually exclusive, as

$$\textit{breadth:2, breadth:3} \longrightarrow \bot \ .$$

Instead of working with such an *indefinite implication*, one may use the implication that these two properties together imply all other ones. A list of such implications is given in Fig. 3.

2 A Better Readable List of Implications

The canonical basis is, as we have seen, sometimes unpleasant to read. For practical work we therefore suggest a modified list[2] that is better "human readable". But we emphasize that this is not meant to introduce a new mathematical definition. For theory considerations, the canonical basis remains the one to use.

Any list $\mathcal{L}$ of implications over some set M induces a closure system. A set $X \subseteq M$ is $\mathcal{L}$-*closed* iff it respects all implications in $\mathcal{L}$, i.e., iff for all $P \to Q \in \mathcal{L}$ it holds that

$$P \subseteq X \quad \text{implies} \quad Q \subseteq X.$$

We denote the corresponding closure operator by $\mathcal{L}(\cdot)$. Then $\mathcal{L}(X)$ is the smallest $\mathcal{L}$-closed set containing X.

Each closure operator $\mathcal{L}(\cdot)$ induces a quasi-order $\leq_{\mathcal{L}}$ on its carrier M by

$$x \leq_{\mathcal{L}} y :\Longleftrightarrow y \in \mathcal{L}(\{x\}).$$

Elements $x, y \in M$ are called $\mathcal{L}$-*equivalent* iff they occur in exactly the same $\mathcal{L}$-closed sets. Formally,

$$x \equiv_{\mathcal{L}} y :\Longleftrightarrow (x \leq_{\mathcal{L}} y \text{ and } y \leq_{\mathcal{L}} x) \qquad (\Longleftrightarrow \mathcal{L}(\{x\}) = \mathcal{L}(\{y\})).$$

It is straightforward that factorizing M by this equivalence relation turns $\leq_{\mathcal{L}}$ into an order on the factor set. The closure $\mathcal{L}(\emptyset)$ of the empty set, if non-empty, is the largest class.

In order to make an implication list better readable, we perform a series of modifications, without changing the associated closure system (for finite M). Logically, the resulting list of implications is equivalent to the initial one.

1. Pick a *representative* element from each $\equiv_{\mathcal{L}}$-equivalence class, except for $\mathcal{L}(\emptyset)$.
2. In all implications, remove the elements of $\mathcal{L}(\emptyset)$, and replace each other element by the representative in its equivalence class.
3. Make each premise disjoint from its conclusion, i.e., replace

$$P \to C \quad \text{by} \quad P \to C \setminus P$$

whenever $P \cap C \neq \emptyset$.

[2] I recently learnt from A. Hotho and G. Stumme that my construction is very similar to that of the *canonical cover* for functional dependencies [10].

4. Whenever $A \to B$ is an implication in the list and p is a property contained in A or in B, then remove all proper consequences of p from that set: if $A \to B$ is in the list and $X = A$ or $X = B$, then replace X by
$$X \setminus \bigcup_{p \in X} (\mathcal{L}(\{p\}) \setminus \{p\}).$$
5. For each $\equiv_{\mathcal{L}}$-equivalence class $\{p_1, p_2, \ldots, p_n\}$ other than $\mathcal{L}(\emptyset)$, having more than one element, add implications $p_1 \to p_2$, $p_2 \to p_3, \ldots, p_n \to p_1$ that cycle through the equivalence class.
6. If $\mathcal{L}(\emptyset) \neq \emptyset$, add the implication $\emptyset \to \mathcal{L}(\emptyset)$ to the list.
7. Merge implications with the same premise:
$$\text{replace } P \to C_1 \text{ and } P \to C_2 \text{ by } P \to C_1 \cup C_2.$$
8. Remove redundant implications.

The outcome is an irredundant implication list which has the same closed sets as $\mathcal{L}$ and satisfies the following:

1. For each non-representative property $q \notin \mathcal{L}(\emptyset)$ there is exactly one implication, and that is of the form $\{q\} \to \{s\}$, where $q \equiv_{\mathcal{L}} s$. The closure of q contains a unique representative element, the one equivalent to q.
2. For each representative property p there is a unique implication of the form $p \to C$. C contains one non-representative element, and that is equivalent to p, provided there is such an element. Otherwise, C contains no non-representative elements.[3]
3. All other implications contain only representative elements. If $X = A$ or $X = B$ for such an implication $A \to B$ in the list, then X does not contain two distinct elements $p \neq q$ such that $q \in \mathcal{L}(p)$.

For a *proof* that the procedure has the desired effect, first note that the steps (3), (7), and (8) do not change the expressivity of an implication list, and that steps (1), (2), (5), and (6) together correspond to the *clarification* process for formal contexts. The crucial step is (4), where each premise and each conclusion s shortened to the set of its minimal elements wrt. the $\leq_{\mathcal{L}}$-quasiorder. This is applied to implications consisting of representative elements only. Restricted to the set of representatives, $\leq_{\mathcal{L}}$ is an order, and $\mathcal{L}(\{p\}) \setminus \{p\}$ contains only elements which are strictly larger than p. Elements which in X are minimal wrt. $\leq_{\mathcal{L}}$ therefore are never removed. All other elements are implied by the minimal elements below them.

It is straightforward to integrate this procedure directly into the exploration algorithm. For simplifying single implications, all we need to know are the closures of the at most one-element sets of properties. So one possible strategy is to ask for these bits of information first and then use them both as background

[3] Note that in our example we have sometimes split such implications in two, one of which is contained in Fig. 2, the other in Fig. 5.

implications and for simplifying the implications which the exploration suggests. An even easier way is to generate all questions as usual and to use only those unit-premise implications for simplification which were already discovered. Recall that each question asked by the standard Attribute Exploration algorithm is of the form

$$P \to P'',$$

where P is the smallest pseudo-intent the closure of which is not yet known. "Smallest" refers to the so-called *lectic order*, which is an order-extension of the $\subseteq$-order. Therefore $P \to P''$ can only be asked when the closures of all pseudo-intents properly contained in P are known. Singleton sets are either in the closure of the empty set (in which case the empty set is a pseudo-intent), are closed or are pseudo-intents. Therefore the already acquired knowledge always suffices for simplifying P, but not necessarily for Q.

3 Application to the List by Reeg and Weiß

As an instance of this procedure we transform the outcome of the exploration by Reeg and Weiß into our "readable" form. Starting from the canonical basis $\mathcal{L}$ of their collection of counter-examples, we derive the list of implications displayed in Figs. 2, 3, 4 and 5. Only one of each dual pair of implications is listed, but the implication counter is advanced twice if necessary.

1. {Boolean} $\leftrightarrow$ {uniquely complemented}
3. {distributive} $\leftrightarrow$ {Browerian}
5. {pseudocomplemented} $\leftrightarrow$ {0-distributive} (+ *dual*)
9. {c-condition} $\leftrightarrow$ {dual c-condition}
10. {modular} $\leftrightarrow$ {metric}
12. {upper semimodular} $\leftrightarrow$ {M-symmetric} (+ *dual*)
16. {relatively complemented} $\to$ {relatively semicomplemented}
17. {relatively semicomplemented} $\to$ {dually relatively semicomplemented}
18. {dually relatively semicomplemented} $\to$ {relatively complemented}
19. {atomistic} $\leftrightarrow$ {sectionally semicomplemented} (+ *dual*)
23. {semicomplemented} $\leftrightarrow$ {almost complemented} (+ *dual*)

Fig. 2. Implications connecting equivalent properties.

Figure 2 shows the implications of the form $\{m\} \to \{n\}$, where $m \equiv_{\mathcal{L}} n$. Double arrows abbreviate pairs of implications. Some of these implications have the same premise as other implications in Fig. 4 (the starred ones). They should be merged according to step (5) of our procedure. We have left them separate for greater clarity, but counted them only once.

Figure 4 lists the implications with premise of size one. Such implications define the $\leq_{\mathcal{L}}$-quasiorder on the properties, and since we restrict to representative properties, they induce an order. Each listed implication is of the form

$$\{p\} \to C,$$

27. {Stone, self-dual} → {dually Stone} (+ *dual*)
29. {SD-meet, self-dual} → {SD-join} (+ *dual*)
31. {pseudocomplemented, self-dual} → {dually pseudocomplemented} (+ *dual*)
33. {upper balanced, self-dual} → {lower balanced} (+ *dual*)
35. {Kurosh-Ore property, self-dual} → {dual Kurosh-Ore property} (+ *dual*)
37. {sectionally complemented, self-dual} → {dually sectionally complemented} (+ *dual*)
39. {semicomplemented, self-dual} → {dually semicomplemented} (+ *dual*)
41. {width:2, width:3} → ⊥.

Fig. 3. Some obvious implications.

1* {Boolean} → {Stone, dually Stone, geometric, dually geometric, self-dual}
42. {Stone} → {distributive} (+ *dual*)
3* {distributive} → {join-distributive, meet-distributive, modular}
44. {join-distributive} → {SD-meet, locally modular} (+ *dual*)
46. {SD-meet} → {c-condition, pseudocomplemented} (+ *dual*)
9* {c-condition} → {finite}
11* {modular} → {locally modular, dually locally modular, balanced}
48. {locally modular} → {upper semimodular, Kurosh-Ore property}(+ *dual*)
13* {upper semimodular} → {upper balanced, graded} (+ *dual*)
49. {balanced} → {upper balanced, lower balanced}
17* {relatively complemented} → {balanced, sectionally complemented, dually sectionally complemented}
20* {atomistic} → {lower balanced, semicomplemented}(+ *dual*)
50. {complemented} → {semicomplemented, dually semicomplemented}
51. {sectionally complemented} → {complemented, atomistic} (+ *dual*)
53. {geometric} → {upper semimodular, dual Kurosh-Ore property, relatively complemented} (+ *dual*)
55. {width:3} → {finite}
56. {width:2} → {SD-meet, SD-join, breadth:2}
57. {breadth:2} → {Kurosh-Ore property, dual Kurosh-Ore property}

Fig. 4. Implications with a singleton premise. A star indicates that an implication with the same premise was already listed in Fig. 2 and that therefore the implication counter is not advanced.

where p is a representative property and C is the set of its representative upper neighbours. The reader may wonder why implication 55 states that *width:3* implies *finite*, while implication (56) with the premise *width:2* does not have *finite* in its conclusion. This is because we list only immediate consequences. Being *finite* follows from *width:2*, but not immediately. By (56), *width:2* implies *SD-meet*, which by (46) implies the *c-condition*, which then, according to (9*) enforces *finite* for lattices of finite length.

For a more detailed understanding we demonstrate the effect of the procedure on a single implication. As an example, we use the implication RW31, mentioned in the introduction. Its conclusion consists of a single property, *sectionally complemented*, which is $\equiv_{\mathcal{L}}$-equivalent to no other property, not in $\mathcal{L}(\emptyset)(=\emptyset)$, and therefore is representative. Thus this conclusion remains unchanged. The premise contains three non-representative properties, *sectionally semicomplemented*, *almost complemented*, and *dually almost complemented*. These are replaced by their respective representatives, *atomistic*, *semicomplemented*, and *dually semicomplemented*, which are already present anyway.

58. {semicomplemented, dually semicomplemented, width:3} → {complemented}
59. {lower balanced, width:3} → {Kurosh-Ore property, dual Kurosh-Ore property} (+ *dual*)
61. {upper balanced, semicomplemented} → {coatomistic} (+ *dual*)
63. {semicomplemented, coatomistic, breadth:2} → {complemented} (+ *dual*)
65. {Kurosh-Ore property, dual Kurosh-Ore property, complemented, atomistic, width:3} → {sectionally complemented} (+ *dual*)
67. {upper balanced, graded, Kurosh-Ore property, dual Kurosh-Ore property, width:3} → {locally modular} (+ *dual*)
69. {upper balanced, lower balanced} → {balanced}
70. {balanced, sectionally complemented, coatomistic, self-dual} → {dually sectionally complemented} (+ *dual*)
72. {balanced, Kurosh-Ore property, dual Kurosh-Ore property, width:3} → {modular}
73. {lower semimodular, dually semicomplemented, atomistic} → {sectionally complemented} (+ *dual*)
75. {lower semimodular, Kurosh-Ore property, sectionally complemented} → {dually geometric} (+ *dual*)
77. {lower semimodular, balanced} → {Kurosh-Ore property} (+ *dual*)
79. {upper semimodular, Kurosh-Ore property} → {locally modular} (+ *dual*)
81. {locally modular, dually sectionally complemented} → {modular, geometric, dually geometric} (+ *dual*)
83. {locally modular, balanced, dual Kurosh-Ore property} → {modular} (+ *dual*)
85. {modular, geometric, dually geometric, breadth:2} → {self-dual}
86. {modular, geometric, dually geometric, width:3} → {self-dual}
87. {c-condition, semicomplemented} → {dually pseudocomplemented} (+ *dual*)
89. {c-condition, lower balanced} → {meet-distributive} (+ *dual*)
91. {pseudocomplemented, semicomplemented} → {dually pseudocomplemented, complemented} (+ *dual*)
93. {pseudocomplemented, dually semicomplemented, width:3} → {SD-meet} (+ *dual*)
95. {pseudocomplemented, dually pseudocomplemented, complemented, atomistic} → {Boolean} (+ *dual*)
97. {complemented, width:2} → {self-dual}
98. {meet-distributive, SD-meet} → {distributive} (+ *dual*)
100. {Boolean, breadth:2} → {width:2}

Fig. 5. The remaining implications of the readable basis.

atomistic implies, according to implication (20*), *lower balanced* and *semicomplemented*, so these properties are deleted in step (4) of the procedure. The same happens to *finite*, because it is a consequence of *width:3*. What is left is exactly implication (65) of Fig. 5.

4 Proofs, Examples, and an Open Problem

We do not prove these implications here. We did actually not even indend to check the proofs given by Reeg and Weiß systematically. Although the simplified implications look easier than the original ones, it is still tiresome to work out a complete proof. Some implications are straightforward, some, like (69), follow immediately from the definitions. Others, e.g. (89), are a bit tricky. One should remember that not only the implications in Fig. 5 require a proof, but all of them. Recall that already the second one,

$$\textit{uniquely complemented} \rightarrow \textit{Boolean},$$

was not obvious.

We also do not present the list of the 84 counterexamples for the non-valid implications. There is simply not sufficient space for that in this publication. Many of the examples are small, but there are larger and even infinite ones, which are however easy to describe. One of the examples is a non-desarguesian projective plane of order 9, a lattice of size 184, constructed over a proper near field.

And then there is the task of verifying that the given counterexamples have the properties as claimed. That requires some other 4000 argumentations, because one has to verify all incidences of a formal context of size 84×51. Most of the properties can be checked by efficient algorithms, but we do not know of any implementation that is convenient to use.

Reeg and Weiß did not solve all problems that occured during the exploration. Their formal context contains one fictitious example (No. 84), refuting an implication which they could not decide. The question is if a lattice which is

> *of finite length*, *atomistic*, *coatomistic*, has the *Kurosh-Ore property* and the *dual Kurosh-Ore property*,

necessarily also is

complemented.

The question remains unanswered even if some or all of the properties

graded, *finite*, *self-dual*

are added to the premise. We are not aware of an answer to this question, even 30 years after it was asked.

5 Better Readable Proofs

A typical application of exploration results is providing proofs or disproofs of implications. A user may ask if an implication

$$A \to B$$

holds in the respective domain or not. This question can be answered by checking if $B \subseteq A''$ is true in the formal context of examples. If $B \not\subseteq A''$, then $A' \not\subseteq B'$, and every element of $A' \setminus B'$ refutes the implication $A \to B$. Here is an example output of such an instance. The question asked is if every lattice of finite length which is both *0-distributive* and *modular* also is *distributive*. The numbers refer to the list of counterexamples given by Reeg and Weiss [3]:

```
Does {0-distributive, modular} imply {distributive}?

No! Counterexample(s): 11, 38, 48.
```

If there are no refuting examples, then the implication $A \to B$ holds and a *proof* may be compiled from the implication list $\mathcal{L}$ which was collected in the exploration process. Each implication in $\mathcal{L}$ is considered to be proven, and when the list is complete, $A \to B$ holds if and only if $B \subseteq \mathcal{L}(A)$. Computing $\mathcal{L}(A)$ is straightforward (see [5] for a details): A set X, initially with $X := A$, is iteratively extended using $\mathcal{L}$. Whenever an implication $U \to V$ is found in $\mathcal{L}$ such that $U \subseteq X$ and $V \nsubseteq X$, then X is replaced by $X \cup V$. If no such implication exists, then $X = \mathcal{L}(A)$. The implications used for generating $\mathcal{L}(A)$ together provide a proof of $A \to \mathcal{L}(A)$, and in case that $B \subseteq \mathcal{L}(A)$ also provide a proof of $A \to B$.

Such proofs often can be simplified. We demonstrate this by the next example, in which we ask if every *modular* lattice of finite length which satisfies the *c-condition* necessarily is *distributive*. The first proof uses the canonical basis. It was simplified by removing redundant implications and by removing implied properties which are not used in the sequel.

```
Does {c-condition, modular} imply {distributive}?

Yes! Here is a proof:

{modular} ==> {dually locally modular, dually M-symmetric,
lower semimodular, upper balanced, lower balanced, graded,
dual Kurosh-Ore property}

{c-condition} ==> {dual c-condition, finite}

{c-condition, dual c-condition, lower balanced, finite} ==>
{meet-distributive, SD-join, 1-distributive}

{c-condition, dual c-condition, upper balanced, finite} ==>
{SD-meet, 0-distributive, pseudocomplemented}

{1-distributive} ==> {dually pseudocomplemented}

{meet-distributive, SD-meet, SD-join, 0-distributive,
1-distributive, c-condition, dual c-condition,
pseudocomplemented, dually pseudocomplemented, dually locally
modular, dually M-symmetric, lower semimodular, lower
balanced, graded, dual Kurosh-Ore property, finite} ==>
{distributive}.
```

The second proof, automatically derived from the "human readable" basis with the same simplifications, is more transparent. The numbers refer to the implications in Figs. 2, 3, 4 and 5.

```
Does {c-condition, modular} imply {distributive}?

Yes! Here is a proof:
```

```
           (11)
{modular}  ==>  {balanced}
            (49)
{balanced}  ==>  {upper balanced, lower balanced}
                                   (89)
{c-condition, lower balanced}  ==>  {meet-distributive}
                                   (90)
{c-condition, upper balanced}  ==>  {join-distributive}
                        (45)
{meet-distributive}  ==>  {SD-join}
                                   (99)
{join-distributive, SD-join}  ==>  {distributive}.
```

6 Conclusion

We suggest a method for making implication lists better readable, and apply it to an unpublished exploration from 1990. The modified, yet logically equivalent implications indeed seem more appealing than the ones in the original canonical basis, at least to a human reader.

At the same time it becomes clearer that the combined results in the work by S. Reeg and W. Weiß are quite substantial. We found no way to sufficiently compress the other parts of their findings (examples and proofs) to fit into this publication. We conclude that Attribute Exploration indeed can systematically collect and store structural knowledge, and can combine simple pieces of information effectively.

But although the basic algorithms are well documented in [5], there is still a lot of work to do in the development of an exploration software to be widely used. Ideally, it should support collaborative explorations, cope with occasional faulty input and also honour each individual author's contributions. Such an implementation is not yet in reach.

References

1. Ganter, B., Wille, R.: Formal Concept Analysis: Mathematical Foundations. Springer, Heidelberg (1998). https://doi.org/10.1007/978-3-642-59830-2
2. Burmeister, P.: ConImp: Ein Programm zur Formalen Begriffsanalyse. In: [11], pp. 25–56
3. Reeg, S., Weiß, W.: Properties of finite lattices. Master's thesis, Technische Universität Darmstadt (1990) (Diplomarbeit)
4. Stern, M.: Semimodular Lattices. Teubner, Stuttgart (1991)
5. Ganter, B., Obiedkov, S.: Conceptual exploration. Springer, Heidelberg (2016). https://doi.org/10.1007/978-3-662-49291-8
6. Birkhoff, G.: Lattice Theory, 3rd edn. Volume 25 of Colloquium Publications. American Mathematical Society, Providence (1967)
7. Crawley, P., Dilworth, R.: Algebraic Theory of Lattices. Prentice-Hall, Englewood Cliffs (1973)
8. Grätzer, G.: General Lattice Theory. Birkhäuser Verlag, Basel (2003)
9. Wille, R.: Halbkomplementäre Verbände. Math. Zeitschrift **94**, 1–31 (1966)
10. Kemper, A., Eickler, A.: Datenbanksysteme: Eine Einführung. Oldenbourg (2009)
11. Stumme, G., Wille, R. (eds.): Begriffliche Wissensverarbeitung – Methoden und Anwendungen. Springer, Heidelberg (2000)

Joining Implications in Formal Contexts and Inductive Learning in a Horn Description Logic

Francesco Kriegel(✉)

Institute of Theoretical Computer Science,
Technische Universität Dresden, Dresden, Germany
francesco.kriegel@tu-dresden.de

Abstract. A joining implication is a restricted form of an implication where it is explicitly specified which attributes may occur in the premise and in the conclusion, respectively. A technique for sound and complete axiomatization of joining implications valid in a given formal context is provided. In particular, a canonical base for the joining implications valid in a given formal context is proposed, which enjoys the property of being of minimal cardinality among all such bases. Background knowledge in form of a set of valid joining implications can be incorporated. Furthermore, an application to inductive learning in a Horn description logic is proposed, that is, a procedure for sound and complete axiomatization of Horn-$\mathcal{M}$ concept inclusions from a given interpretation is developed. A complexity analysis shows that this procedure runs in deterministic exponential time.

Keywords: Inductive learning · Data mining · Axiomatization · Formal Concept Analysis · Joining implication · Horn description logic · Concept inclusion

1 Introduction

Formal Concept Analysis (abbrv. FCA) [10] is subfield of lattice theory that allows to analyze data-sets that can be represented as formal contexts. Put simply, such a formal context binds a set of objects to a set of attributes by specifying which objects have which attributes. There are two major techniques that can be applied in various ways for purposes of data mining, machine learning, knowledge management, knowledge visualization, etc. On the one hand, it is possible to describe the hierarchical structure of such a data-set in form of a formal concept lattice [10]. On the other hand, the theory of implications (dependencies between attributes) valid in a given formal context can be axiomatized in a sound and complete manner by the so-called canonical base [11], which furthermore contains a minimal number of implications w.r.t. the properties of soundness and completeness. So far, some variations of the canonical base have been developed,

D. Cristea et al. (Eds.): ICFCA 2019, LNAI 11511, pp. 110–129, 2019.
https://doi.org/10.1007/978-3-030-21462-3_9

e.g., incorporation of valid background knowledge [29], constraining premises and conclusions in implications by some closure operator [3], and incorporation of arbitrary background knowledge [22], among others. The canonical base in its default form as well as its variations can be be computed by the algorithm *NextClosures* [17,22] in a highly parallel way such the necessary computation time is almost inverse linear proportional to the number of available CPU cores.

Description Logic (abbrv. DL) [2] belongs to the field of knowledge representation and reasoning. DL researchers have developed a large family of logic-based languages, so-called *description logics* (abbrv. DLs). These logics allow their users to explicitly represent knowledge as *ontologies*, which are finite sets of (human- and machine-readable) axioms, and provide them with automated inference services to derive implicit knowledge. The landscape of decidability and computational complexity of common reasoning tasks for various description logics has been explored in large parts: there is always a trade-off between expressibility and reasoning costs. It is therefore not surprising that DLs are nowadays applied in a large variety of domains [2]: agriculture, astronomy, biology, defense, education, energy management, geography, geoscience, medicine, oceanography, and oil and gas. Furthermore, the most notable success of DLs is that these constitute the logical underpinning of the *Web Ontology Language* (abbrv. OWL) [13] in the *Semantic Web*.

Within this document, we propose the new notion of so-called *joining implications* in FCA. More specifically, we assume that there are two distinct sets of attributes: the first one containing the attributes that may occur in premises of implications, while conclusions must only contain attributes from the second set. A canonical base for the joining implications valid in a given formal context is developed and it is proven that it has minimal cardinality among all such bases. Then, an application to inductive learning in a *Horn description logic* [24] is provided. Roughly speaking, such a Horn DL is obtained from some DL by disallowing any disjunctions. Reasoning procedures can then work deterministically, i.e., *reasoning by case* is not required [14]. Hornness is not a new notion: Horn clauses in first-order logic are disjunctions of an arbitrary number of negated atomic formulae and at most one non-negated atomic formula. It is easy to see that such Horn clauses have an implicative character, since $\neg\phi_1 \vee \ldots \vee \neg\phi_n \vee \psi$ is equivalent to $\phi_1 \wedge \ldots \phi_n \rightarrow \psi$. A *logic program* is a set of Horn clauses, and a *Datalog program* is a function-free logic program [7]. All commonly known Horn description logics can be translated into Datalog—more specifically, each Horn-$\mathcal{DL}$ TBox $\mathcal{T}$ can be translated into some Datalog program $\mathcal{D}$ such that, for each simple ABox $\mathcal{A}$, the ontology $\mathcal{T} \cup \mathcal{A}$ is satisfiable if, and only if, the Datalog programm $\mathcal{D} \cup \mathcal{A}$ is satisfiable. For deeper insights please consider [12,15,24]. The most important advantage of Horn fragments is that these often have a significantly lower computational complexity. Using the canonical base of joining implications, we show how the Horn-$\mathcal{M}$ concept inclusions valid in a given interpretation can be axiomatized. This continues a line of research that combines FCA and DL for the sake of inductive learning, cf. [4,5,9,18,19,21,27] just to name a few.

Due to space constraints some technical lemmas and some proofs have been moved to a technical report [20].

2 Joining Implications in Formal Contexts

Throughout this section, assume that $\mathbb{K} := (G, M, I)$ is some *formal context*, that is, G is a set of *objects*, M is a set of *attributes*, and $I \subseteq G \times M$ is an *incidence relation*. If $(g, m) \in I$, then we say that g *has* m. It is well-known that the two mappings $\cdot^I\colon \wp(G) \to \wp(M)$ and $\cdot^I\colon \wp(M) \to \wp(G)$ defined below constitute a *Galois connection*, cf. [10].

$$A^I := \{\, m \in M \mid (g, m) \in I \text{ for each } g \in A \,\} \quad \text{for any } A \subseteq G$$
$$B^I := \{\, g \in G \mid (g, m) \in I \text{ for each } m \in B \,\} \quad \text{for any } B \subseteq M$$

In particular, this means that the following statements hold true for any sets $A, C \subseteq G$ and $B, D \subseteq M$.

1. $A \subseteq B^I$ if, and only if, $B \subseteq A^I$ if, and only if, $A \times B \subseteq I$
2. $A \subseteq A^{II}$
3. $A^I = A^{III}$
4. $A \subseteq C$ implies $C^I \subseteq A^I$
5. $B \subseteq B^{II}$
6. $B^I = B^{III}$
7. $B \subseteq D$ implies $D^I \subseteq B^I$

An *implication* over M is a term $X \to Y$ where $X, Y \subseteq M$. It is valid in $\mathbb{K}$ if $X^I \subseteq Y^I$ is satisfied, i.e., if each object that has all attributes in X also has all attributes in Y, and we shall then write $\mathbb{K} \models X \to Y$. A *model* of $X \to Y$ is a set $U \subseteq M$ such that $X \subseteq U$ implies $Y \subseteq U$, denoted as $U \models X \to Y$. An implication set $\mathcal{L}$ *entails* an implication $X \to Y$ if any model of $\mathcal{L}$, i.e., any set that is a model of all implications in $\mathcal{L}$, is also a model of $X \to Y$, and we denote this by $\mathcal{L} \models X \to Y$.

We are now interested in a restricted form of implications. In particular, we restrict the sets of attributes that may occur in the *premise* X and in the *conclusion* Y, respectively, of every implication $X \to Y$. Thus, let further M_{p} be a set of *premise attributes* and let M_{c} be a set of *conclusion attributes* such that $M_{\mathsf{p}} \cup M_{\mathsf{c}} \subseteq M$ holds true. For each $\mathsf{x} \in \{\mathsf{p}, \mathsf{c}\}$, we define the subcontext $\mathbb{K}_{\mathsf{x}} := (G, M_{\mathsf{x}}, I_{\mathsf{x}})$ where $I_{\mathsf{x}} := I \cap (G \times M_{\mathsf{x}})$. Furthermore, we may also write X^{x} instead of $X^{I_{\mathsf{x}}}$ for subsets $X \subseteq G$ or $X \subseteq M_{\mathsf{x}}$. Please note that then each pair $(\cdot^{\mathsf{x}}, \cdot^{\mathsf{x}})$ is a Galois connection between $(\wp(G), \subseteq)$ and $(\wp(M_{\mathsf{x}}), \subseteq)$, that is, similar statements like above are valid.

Definition 1. *A* joining implication *from M_{p} to M_{c}, or simply* pc-implication, *is an expression $X \to Y$ where $X \subseteq M_{\mathsf{p}}$ and $Y \subseteq M_{\mathsf{c}}$. It is valid in $\mathbb{K}$, written $\mathbb{K} \models X \to Y$, if $X^{\mathsf{p}} \subseteq Y^{\mathsf{c}}$ holds true.*

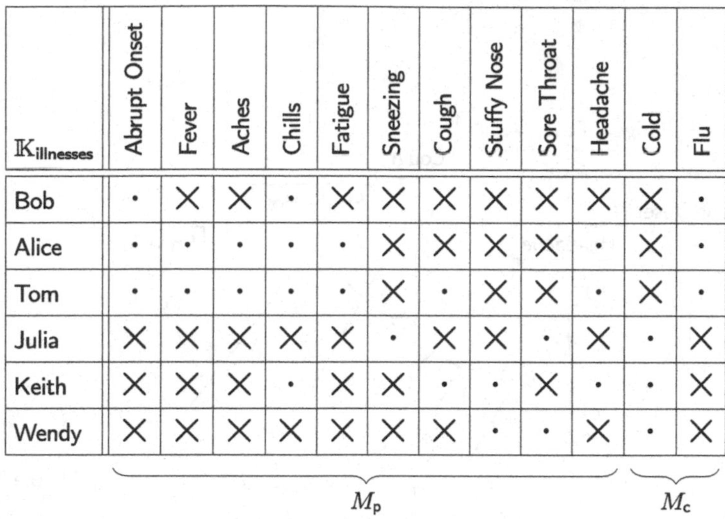

$\mathbb{K}_{\text{illnesses}}$	Abrupt Onset	Fever	Aches	Chills	Fatigue	Sneezing	Cough	Stuffy Nose	Sore Throat	Headache	Cold	Flu
Bob	·	×	×	·	×	×	×	×	×	×	×	·
Alice	·	·	·	·	·	×	×	×	×	·	×	·
Tom	·	·	·	·	·	×	·	×	×	·	×	·
Julia	×	×	×	×	×	·	×	×	·	×	·	×
Keith	×	×	×	·	×	×	·	·	×	·	·	×
Wendy	×	×	×	×	×	×	×	·	·	×	·	×

Fig. 1. The formal context $\mathbb{K}_{\text{illnesses}}$.

Example. Consider the formal context $\mathbb{K}_{\text{illnesses}}$ in Fig. 1. It considers six persons as objects and their symptoms and illnesses as attributes. Furthermore, we regard the symptoms as premise attributes and the illnesses as conclusion attributes. Note that, in general, it is not required that the sets M_{p} and M_{c} form a partition of the attribute set. For other use cases both could overlap, one could be contained in the other, or their union could be a strict subset of the whole attribute set. The concept lattice is displayed in Fig. 2.[1]

The expression {Cold, Cough} $\to$ {Chills} is no pc-implication, since the attribute Cold must not occur in a premise and, likewise, the attribute Chills must not occur in a conclusion. The expression {Sneezing, Cough, Stuffy Nose} $\to$ {Cold} is a well-formed joining implication and it is valid in $\mathbb{K}_{\text{illnesses}}$, since {Sneezing, Cough, Stuffy Nose}^{p} = {Bob, Alice} is a subset of {Cold}^{c} = {Bob, Alice, Tom}. Furthermore, the expression {Abrupt Onset} $\to$ {Cold} is a well-formed joining implication as well, but it is not valid in $\mathbb{K}_{\text{illnesses}}$, as {Abrupt Onset}^{p} = {Julia, Keith, Wendy} is not a subset of {Cold}^{c} = {Bob, Alice, Tom}.

In the following, we shall characterize the set of all joining implications valid in $\mathbb{K}$. Of course, the pc-implication set

$$\mathsf{Imp}_{\mathsf{pc}}(\mathbb{K}) := \{\, X \to X^{\mathsf{pc}} \mid X \subseteq M_{\mathsf{p}} \,\}$$

contains only valid pc-implications and further entails any valid pc-implication, since $\mathbb{K} \models X \to Y$ is equivalent to $Y \subseteq X^{\mathsf{pc}}$ and so $\{X \to X^{\mathsf{pc}}\} \models X \to Y$.

[1] We have not introduced the notion of a concept lattice here, since it is not needed for our purposes; the interested reader is rather referred to [10].

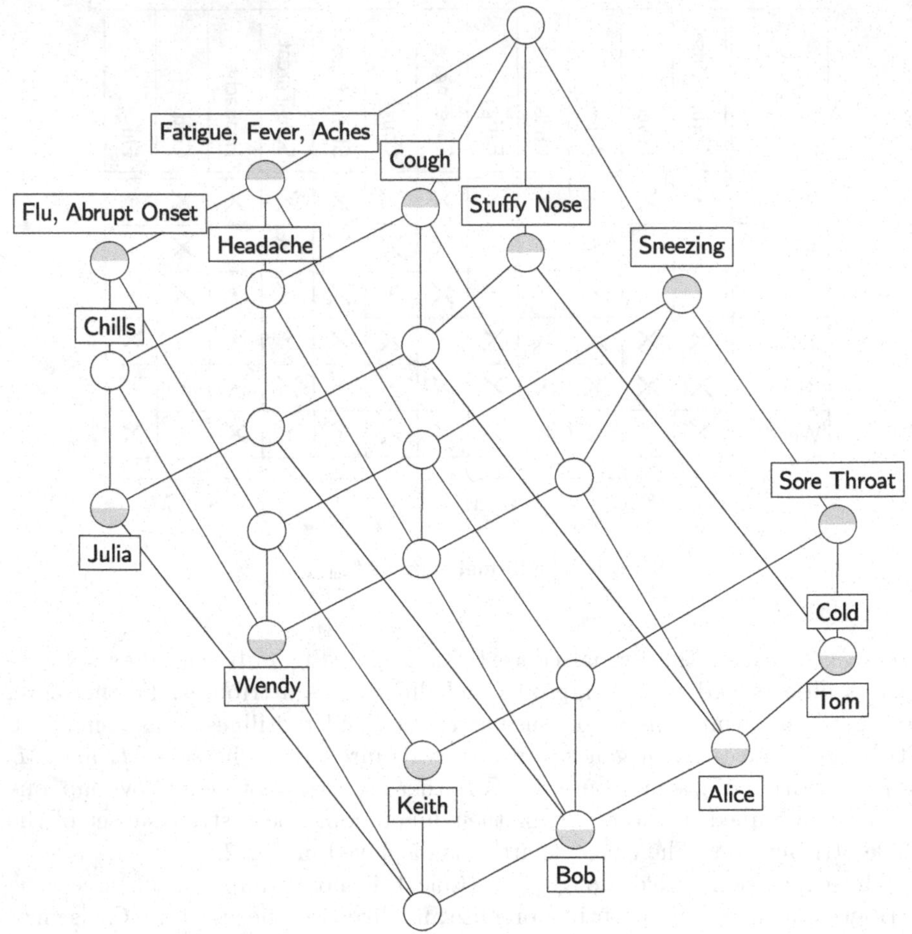

Fig. 2. The concept lattice of $\mathbb{K}_{\text{illnesses}}$.

Remark that a *closure operator* on M is some mapping $\phi\colon \wp(M) \to \wp(M)$ with the following properties for all subsets $X, Y \subseteq M$.

1. $X \subseteq \phi(X)$ *(extensive)*
2. $X \subseteq Y$ implies $\phi(X) \subseteq \phi(Y)$ *(monotonic)*
3. $\phi(\phi(X)) = \phi(X)$ *(idempotent)*

It is easy to verify that, for each Galois connection (f, g), the compositions $f \circ g$ and $g \circ f$ are closure operators. It is well-known that each implication set $\mathcal{L}$ induces a corresponding closure operator $\phi_{\mathcal{L}}$ such that the models of $\mathcal{L}$ are exactly the closures of $\phi_{\mathcal{L}}$, cf. [10,22]: for each $U \subseteq M$, the closure $\phi_{\mathcal{L}}(U)$ is the smallest superset of U such that $X \subseteq \phi_{\mathcal{L}}(U)$ implies $Y \subseteq \phi_{\mathcal{L}}(U)$ for any implication $X \to Y$ in $\mathcal{L}$. In particular, we can readily verify the following.

$$\phi_{\mathcal{L}}(U) = U^{\mathcal{L}} := \bigcup\{\, U^{\mathcal{L},n} \mid n \in \mathbb{N} \,\}$$

$$\text{where} \qquad V^{\mathcal{L},n+1} := (V^{\mathcal{L},1})^{\mathcal{L},n}$$

$$\text{and} \qquad V^{\mathcal{L},1} := V \cup \{\, Y \mid X \to Y \in \mathcal{L} \text{ and } X \subseteq V \,\} \quad \text{for each } V \subseteq M$$

For the above joining implication set, we easily get that $\phi^{\mathsf{pc}}_{\mathbb{K}} := \phi_{\mathsf{Imp}_{\mathsf{pc}}(\mathbb{K})}$ satisfies

$$\phi^{\mathsf{pc}}_{\mathbb{K}}(X) = X \cup (X \cap M_{\mathsf{p}})^{\mathsf{pc}}$$

for any $X \subseteq M$.

An implication $X \to Y$ is *valid* in a closure operator ϕ, written $\phi \models X \to Y$, if $Y \subseteq \phi(X)$ holds true, cf. [17]. Please note that this coincides with the notion of validity in a formal context $\mathbb{K}$ if we consider the closure operator $\phi_{\mathbb{K}} \colon X \mapsto X^{II}$ and, likewise, entailment by an implication set $\mathcal{L}$ is the same as validity in $\phi_{\mathcal{L}}$. Now consider an implication set $\mathcal{L}$. We say that $\mathcal{L}$ is *sound* for ϕ if $\phi \models \mathcal{L}$ holds true, that is, if $\phi \models X \to Y$ is satisfied for each implication $X \to Y \in \mathcal{L}$. Furthermore, $\mathcal{L}$ is *complete* for ϕ if, for any implication $X \to Y$, it holds true that $\phi \models X \to Y$ implies $\mathcal{L} \models X \to Y$.

Definition 2. *An implication set is* join-sound *or* pc-sound *if it is sound for* $\phi^{\mathsf{pc}}_{\mathbb{K}}$*, and it is* join-complete *or* pc-complete *if it is complete for* $\phi^{\mathsf{pc}}_{\mathbb{K}}$*. Fix some* pc-*sound implication set* $\mathcal{S}$*. A* pc-*implication set is called* joining implication base *or* pc-implication base *relative to* $\mathcal{S}$ *if it is* pc-*sound and its union with* $\mathcal{S}$ *is* pc-*complete.*

Obviously, the above $\mathsf{Imp}_{\mathsf{pc}}(\mathbb{K})$ is a joining implication base relative to $\emptyset$.

Further fix some implication set $\mathcal{S}$ as well as a closure operator ϕ such that $\phi \models \mathcal{S}$. Now remark that a *pseudo-closure* of ϕ relative to $\mathcal{S}$ is a set $P \subseteq M$ such that $P \neq \phi(P)$ and $P \models \mathcal{S}$ (i.e., $P = \phi_{\mathcal{S}}(P)$) hold true and $Q \subsetneq P$ implies $\phi(Q) \subseteq P$ for each pseudo-closure Q of ϕ relative to $\mathcal{S}$. We shall denote the set of all pseudo-closures of ϕ relative to $\mathcal{S}$ as $\mathsf{PsClo}(\phi, \mathcal{S})$. Then, the *canonical implication base* of ϕ relative to $\mathcal{S}$ is defined as $\mathsf{Can}(\phi, \mathcal{S}) := \{\, P \to \phi(P) \mid P \in \mathsf{PsClo}(\phi, \mathcal{S}) \,\}$, and it is sound for ϕ and is further complete for ϕ relative to $\mathcal{S}$, i.e., $\phi \models X \to Y$ if, and only if, $\mathsf{Can}(\phi, \mathcal{S}) \cup \mathcal{S} \models X \to Y$ for each implication $X \to Y$, cf. [11,17,29]. It is easy to see that we can replace each implication $P \to \phi(P)$ by $P \to \phi(P) \setminus P$ to get an equivalent implication set.

Our aim for the sequel of this section is to find a *canonical* representation of the valid joining implications of some formal context, i.e., we shall provide a joining implication base that has *minimal* cardinality among all joining implication bases. For this purpose, we consider the canonical implication base of the above closure operator $\phi^{\mathsf{pc}}_{\mathbb{K}}$ and show how we can modify it to get a *canonical joining implication base.* We start with showing that we can rewrite any join-sound and join-complete implication set into a joining implication base in a certain normal form. For the remainder of this section, fix some arbitrary join-sound joining implication set $\mathcal{S}$ that is used as background knowledge.

Lemma 3. *Fix some join-sound implication set $\mathcal{L}$ over M. Further assume that $\mathcal{L} \cup \mathcal{S}$ is join-complete, and define the following set of joining implications.*

$$\mathcal{L}_{\mathsf{pc}} := \{ X \cap M_{\mathsf{p}} \to (X \cap M_{\mathsf{p}})^{\mathsf{pc}} \mid X \to Y \in \mathcal{L} \}$$

Then, $\mathcal{L}_{\mathsf{pc}}$ is a joining implication base relative to $\mathcal{S}$.

Proof. Since $\mathcal{L}_{\mathsf{pc}} \subseteq \mathsf{Imp}_{\mathsf{pc}}(\mathbb{K})$ obviously holds true, we know that $\mathcal{L}_{\mathsf{pc}}$ is join-sound. For join-completeness we show that $\mathcal{L}_{\mathsf{pc}} \cup \mathcal{S} \models \mathsf{Imp}_{\mathsf{pc}}(\mathbb{K})$. Thus, consider some $Z \subseteq M_{\mathsf{p}}$. As $\mathcal{L} \cup \mathcal{S}$ is join-complete, it must hold true that $\mathcal{L} \cup \mathcal{S} \models Z \to Z^{\mathsf{pc}}$, that is, there are implications $X_1 \to Y_1$, ..., $X_n \to Y_n$ in $\mathcal{L} \cup \mathcal{S}$ such that the following statements are satisfied.

$$\begin{aligned}
X_1 &\subseteq Z \\
X_2 &\subseteq Z \cup Y_1 \\
X_3 &\subseteq Z \cup Y_1 \cup Y_2 \\
&\vdots \\
X_n &\subseteq Z \cup Y_1 \cup Y_2 \cup \cdots \cup Y_{n-1} \\
Z^{\mathsf{pc}} &\subseteq Z \cup Y_1 \cup Y_2 \cup \cdots \cup Y_{n-1} \cup Y_n
\end{aligned}$$

Let $L := \{ k \mid k \in \{1, \ldots, n\} \text{ and } X_k \to Y_k \in \mathcal{L} \setminus \mathcal{S} \}$ and $S := \{1, \ldots, n\} \setminus L$. Since $\mathcal{L}$ is join-sound, we have $Y_k \subseteq X_k \cup (X_k \cap M_{\mathsf{p}})^{\mathsf{pc}}$ for each index $k \in L$. Define $X_{n+1} := Z^{\mathsf{pc}}$. An induction on $k \in \{1, \ldots, n+1\}$ shows the following.

$$\begin{aligned}
X_k \subseteq Z &\cup \bigcup \{ Y_i \mid i \in \{1, \ldots, k-1\} \cap S \} \\
&\cup \bigcup \{ (X_i \cap M_{\mathsf{p}})^{\mathsf{pc}} \mid i \in \{1, \ldots, k-1\} \cap L \}
\end{aligned}$$

Of course, $X_k \cap M_{\mathsf{p}} \subseteq X_k$ is satisfied for any index $k \in L$. We conclude that $\{ X_k \to Y_k \mid k \in S \} \cup \{ X_k \cap M_{\mathsf{p}} \to (X_k \cap M_{\mathsf{p}})^{\mathsf{pc}} \mid k \in L \}$ entails $Z \to Z^{\mathsf{pc}}$ and we are done. □

The transformation from Lemma 3 can now immediately be applied to the canonical implication base of the closure operator $\phi_{\mathbb{K}}^{\mathsf{pc}}$ to obtain a joining implication base, which we call *canonical*. This is due to fact that, by definition, $\mathsf{Can}(\phi_{\mathbb{K}}^{\mathsf{pc}})$ is both pc-sound and pc-complete.

Proposition 4. *The following is a joining implication base relative to $\mathcal{S}$ and is called* canonical joining implication base *or* canonical pc-implication base *of $\mathbb{K}$ relative to $\mathcal{S}$.*

$$\mathsf{Can}_{\mathsf{pc}}(\mathbb{K}, \mathcal{S}) := \{ P \cap M_{\mathsf{p}} \to (P \cap M_{\mathsf{p}})^{\mathsf{pc}} \mid P \in \mathsf{PsClo}(\phi_{\mathbb{K}}^{\mathsf{pc}}, \mathcal{S}) \}$$

Proof. Remark that $\phi_{\mathbb{K}}^{\mathsf{pc}}(P) = P \cup (P \cap M_{\mathsf{p}})^{\mathsf{pc}}$ holds true and, consequently, the canonical implication base for $\phi_{\mathbb{K}}^{\mathsf{pc}}$ relative to $\mathcal{S}$ evaluates to

$$\mathsf{Can}(\phi_{\mathbb{K}}^{\mathsf{pc}}, \mathcal{S}) = \{ P \to (P \cap M_{\mathsf{p}})^{\mathsf{pc}} \mid P \in \mathsf{PsClo}(\phi_{\mathbb{K}}^{\mathsf{pc}}, \mathcal{S}) \}.$$

We already know that $\mathsf{Can}(\phi_{\mathbb{K}}^{\mathsf{pc}}, \mathcal{S})$ is join-sound and its union with $\mathcal{S}$ is join-complete. Since $(\mathsf{Can}(\phi_{\mathbb{K}}^{\mathsf{pc}}, \mathcal{S}))_{\mathsf{pc}} = \mathsf{Can}_{\mathsf{pc}}(\mathbb{K}, \mathcal{S})$ holds true, an application of Lemma 3 shows that $\mathsf{Can}_{\mathsf{pc}}(\mathbb{K}, \mathcal{S})$ is indeed a joining implication base relative to $\mathcal{S}$. □

Example. We continue with investigating our exemplary formal context $\mathbb{K}_{\mathsf{illnesses}}$. In order to compute the canonical joining implication base of it (relative to $\emptyset$), we first need to construct the canonical base of the closure operator $\phi_{\mathbb{K}_{\mathsf{illnesses}}}^{\mathsf{pc}}$.[2]

$$\mathsf{Can}(\phi_{\mathbb{K}_{\mathsf{illnesses}}}^{\mathsf{pc}}, \emptyset) = \left\{ \begin{array}{r} \{\mathsf{Headache,\ Sore\ Throat}\} \to \{\mathsf{Cold}\} \\ \{\mathsf{Abrupt\ Onset}\} \to \{\mathsf{Flu}\} \\ \{\mathsf{Sore\ Throat,\ Stuffy\ Nose}\} \to \{\mathsf{Cold}\} \\ \{\mathsf{Flu,\ Sore\ Throat,\ Chills}\} \to \{\mathsf{Cold}\} \\ \{\mathsf{Stuffy\ Nose,\ Sneezing}\} \to \{\mathsf{Cold}\} \\ \{\mathsf{Chills}\} \to \{\mathsf{Flu}\} \\ \{\mathsf{Sore\ Throat,\ Cough}\} \to \{\mathsf{Cold}\} \end{array} \right\}$$

Now applying the transformation from Lemma 3 yields the following set of joining implications, which is the canonical joining implication base. In particular, only the fourth implication is altered.

$$\mathsf{Can}_{\mathsf{pc}}(\mathbb{K}_{\mathsf{illnesses}}, \emptyset) = \left\{ \begin{array}{rl} \{\mathsf{Headache,\ Sore\ Throat}\} \to & \{\mathsf{Cold}\} \\ \{\mathsf{Abrupt\ Onset}\} \to & \{\mathsf{Flu}\} \\ \{\mathsf{Sore\ Throat,\ Stuffy\ Nose}\} \to & \{\mathsf{Cold}\} \\ \{\mathsf{Sore\ Throat,\ Chills}\} \to & \{\mathsf{Flu,\ Cold}\} \\ \{\mathsf{Stuffy\ Nose,\ Sneezing}\} \to & \{\mathsf{Cold}\} \\ \{\mathsf{Chills}\} \to & \{\mathsf{Flu}\} \\ \{\mathsf{Sore\ Throat,\ Cough}\} \to & \{\mathsf{Cold}\} \end{array} \right\}$$

The canonical base of $\mathbb{K}_{\mathsf{illnesses}}$, which coincides with the canonical base of the induced closure operator $\phi_{\mathbb{K}_{\mathsf{illnesses}}}$, is as follows. Note that it is sound and complete for *all* implications valid in $\mathbb{K}_{\mathsf{illnesses}}$, i.e., no constraints on premises and conclusions are imposed.

[2] The result has not been obtained by hand, but instead the implementation of the algorithm *NextClosures* [17] in *ConceptExplorer FX* [16] has been utilized. Thus, no intermediate computation steps are provided.

$$\mathsf{Can}(\mathbb{K}_{\mathsf{illnesses}}, \emptyset) = \left\{ \begin{aligned} \{\mathsf{Fever}\} &\to \{\mathsf{Fatigue,\ Aches}\} \\ \{\mathsf{Sore\ Throat}\} &\to \{\mathsf{Sneezing}\} \\ \{\mathsf{Chills}\} &\to \left\{ \begin{aligned} &\mathsf{Headache,\ Flu,\ Fatigue,\ Cough,} \\ &\mathsf{Fever,\ Aches,\ Abrupt\ Onset} \end{aligned} \right\} \\ \{\mathsf{Cold}\} &\to \{\mathsf{Sore\ Throat,\ Stuffy\ Nose,\ Sneezing}\} \\ \{\mathsf{Headache}\} &\to \{\mathsf{Fatigue,\ Cough,\ Fever,\ Aches}\} \\ \left\{ \begin{aligned} &\mathsf{Headache,\ Flu,\ Fatigue,\ Cough,} \\ &\mathsf{Fever,\ Aches,\ Abrupt\ Onset} \end{aligned} \right\} &\to \{\mathsf{Chills}\} \\ \{\mathsf{Aches}\} &\to \{\mathsf{Fatigue,\ Fever}\} \\ \{\mathsf{Stuffy\ Nose,\ Sneezing}\} &\to \{\mathsf{Sore\ Throat,\ Cold}\} \\ \{\mathsf{Fatigue}\} &\to \{\mathsf{Fever,\ Aches}\} \\ \{\mathsf{Sore\ Throat,\ Sneezing,\ Cough}\} &\to \{\mathsf{Stuffy\ Nose,\ Cold}\} \\ \{\mathsf{Fatigue,\ Stuffy\ Nose,\ Fever,\ Aches}\} &\to \{\mathsf{Headache,\ Cough}\} \\ \{\mathsf{Fatigue,\ Cough,\ Fever,\ Aches}\} &\to \{\mathsf{Headache}\} \\ \{\mathsf{Abrupt\ Onset}\} &\to \{\mathsf{Flu,\ Fatigue,\ Fever,\ Aches}\} \\ \{\mathsf{Flu}\} &\to \{\mathsf{Fatigue,\ Fever,\ Aches,\ Abrupt\ Onset}\} \end{aligned} \right\}$$

If we apply the transformation from Lemma 3 to $\mathsf{Can}(\mathbb{K}_{\mathsf{illnesses}}, \emptyset)$, then we obtain the following set of joining implications. Obviously, it is not complete, since it does not entail the valid joining implication $\{\mathsf{Headache,\ Sore\ Throat}\} \to \{\mathsf{Cold}\}$.

$$\left\{ \begin{aligned} \{\mathsf{Chills}\} &\to \{\mathsf{Flu}\} \\ \{\mathsf{Stuffy\ Nose,\ Sneezing}\} &\to \{\mathsf{Cold}\} \\ \{\mathsf{Sore\ Throat,\ Sneezing,\ Cough}\} &\to \{\mathsf{Cold}\} \\ \{\mathsf{Abrupt\ Onset}\} &\to \{\mathsf{Flu}\} \end{aligned} \right\}$$

We close this section with two further important properties of the canonical joining implication base. On the one hand, we shall show that it has minimal cardinality among all joining implication bases or, more generally, even among all join-sound, join-complete implication bases. On the other hand, we investigate the computational complexity of actually computing the canonical joining implication base.

Proposition 5. *The canonical joining implication base* $\mathsf{Can}_{\mathsf{pc}}(\mathbb{K}, \mathcal{S})$ *has minimal cardinality among all implication sets that are join-sound and have a union with* $\mathcal{S}$ *that is join-complete.*

Proof. Consider some implication set $\mathcal{L}$ such that $\mathcal{L} \cup \mathcal{S}$ is join-sound and join-complete. According to Lemma 3, we can assume that—without loss of generality—$\mathcal{L} \subseteq \mathsf{Imp}_{\mathsf{pc}}(\mathbb{K})$ holds true. In particular, note that $|\mathcal{L}_{\mathsf{pc}}| \leq |\mathcal{L}|$ is always true.

Join-soundness and join-completeness of $\mathcal{L} \cup \mathcal{S}$ yield that $\mathcal{L} \cup \mathcal{S}$ and $\mathsf{Can}(\phi_{\mathbb{K}}^{\mathsf{pc}}, \mathcal{S}) \cup \mathcal{S}$ are equivalent. It is well-known [11,29] that $\mathsf{Can}(\phi_{\mathbb{K}}^{\mathsf{pc}}, \mathcal{S})$ has minimal cardinality among all implication bases for $\phi_{\mathbb{K}}^{\mathsf{pc}}$ relative to $\mathcal{S}$, and so it follows that $|\mathcal{L}| \geq |\mathsf{Can}(\phi_{\mathbb{K}}^{\mathsf{pc}}, \mathcal{S})|$.

Clearly, the choice $\mathcal{L} := \mathsf{Can}_{\mathsf{pc}}(\mathbb{K}, \mathcal{S})$ implies $|\mathsf{Can}_{\mathsf{pc}}(\mathbb{K}, \mathcal{S})| \geq |\mathsf{Can}(\phi_{\mathbb{K}}^{\mathsf{pc}}, \mathcal{S})|$. It is further apparent that $|\mathsf{Can}_{\mathsf{pc}}(\mathbb{K}, \mathcal{S})| \leq |\mathsf{Can}(\phi_{\mathbb{K}}^{\mathsf{pc}}, \mathcal{S})|$ holds true and we infer that, in particular, $\mathsf{Can}_{\mathsf{pc}}(\mathbb{K}, \mathcal{S})$ and $\mathsf{Can}(\phi_{\mathbb{K}}^{\mathsf{pc}}, \mathcal{S})$ must contain the same number of implications. □

The next proposition shows that computing the canonical joining implication base is not more expensive than computing the canonical implication base where no constraints on the premises and conclusions must be satisfied. It uses the fact that canonical implication bases of closure operators can be computed using the algorithm *NextClosures* [17].

Proposition 6. *The canonical joining implication base can be computed in exponential time, and there exist formal contexts for which the canonical joining implication base cannot be encoded in polynomial space.*

Proof. The canonical implication base of the closure operator $\phi_{\mathbb{K}}^{\mathsf{pc}}$ relative to some background implication set $\mathcal{S}$ can be computed in exponential time by means of the algorithm *NextClosures*, cf. [5,17], which is easy to verify. The transformation of $\mathsf{Can}(\phi_{\mathbb{K}}^{\mathsf{pc}}, \mathcal{S})$ into $\mathsf{Can}_{\mathsf{pc}}(\mathbb{K}, \mathcal{S})$ as described in Lemma 3 can be done in polynomial time.

Kuznetsov and Obiedkov have shown in [26, Theorem 4.1] that the number of implications in the canonical implication base $\mathsf{Can}(\mathbb{K})$ of a formal context $\mathbb{K} := (G, M, I)$ can be exponential in $|G| \cdot |M|$. Clearly, if we let $\mathcal{S} := \emptyset$ and set both M_{p} and M_{c} to M, then $\mathsf{Can}(\mathbb{K})$ and $\mathsf{Can}_{\mathsf{pc}}(\mathbb{K}, \mathcal{S})$ coincide. □

We have seen in the running example that the canonical pc-implication base can be used to characterize implications between symptoms and diagnoses/illnesses. A further applications is, for instance, formal contexts encoding observations between attributes satisfied *yesterday* and *today*, i.e., we could construct the canonical base of pc-implications and then use it as a forecast stating which combinations of attributes being satisfied *today* would imply which combinations of attributes being satisfied *tomorrow*. In general, we could think of the premise attributes as *observable attributes* and the conclusion attributes as *goal/decision attributes*. By constructing the canonical pc-implication base from some formal context in which the goal/decision attributes have been manually assessed, we would obtain a set of rules with which we could analyze new data sets for which only the observable attributes are specified.

3 The Description Logic Horn-$\mathcal{M}$

A *Horn description logic* [12,15,24] is some description logic that, basically, does not allow for any usage of disjunction. While Hornness decreases expressiveness, it often also significantly lowers the computational complexity of some common reasoning tasks, e.g., *instance checking* or *query answering*. These are, thus, of importance in practical applications where computation times and costs must not be too high.

In the sequel of this section, we introduce the description logic Horn-$\mathcal{M}$, which is the Horn variant of $\mathcal{M} := \mathcal{ALQ}^{\geq}\mathcal{N}^{\leq}(\mathsf{Self})$ [18]. Restrictions are imposed on concept inclusions only and, generally speaking, premises must always be $\mathcal{EL}^* := \mathcal{EL}^{\bot}(\mathsf{Self})$ concept descriptions while conclusions may be arbitrary $\mathcal{M}^{\leq 1} := \mathcal{ALQ}^{\geq}\mathcal{N}^{\leq 1}(\mathsf{Self})$ concept descriptions, that is, $\mathcal{M}$ concept descriptions except that in unqualified smaller-than restrictions $\exists \leq n.\, r$ only the case $n = 1$ is allowed. More specifically, a Horn-$\mathcal{M}$ *concept inclusion* is an expression $C \sqsubseteq D$ where the *concept descriptions* C and D are built by means of the following grammar. Beforehand, fix some *signature* Σ, which is a disjoint union of a set Σ_{I} of *individual names*, a set Σ_{C} of *concept names*, and a set Σ_{R} of *role names*. In the below grammar, A can be replaced by an arbitrary concept name from Σ_{C} and, likewise, r can be replaced by an arbitrary role name from Σ_{R}.

$$C := \bot \mid \top \mid A \mid C \sqcap C \mid \exists\, r.\, C \mid \exists\, r.\, \mathsf{Self}$$
$$D := \bot \mid \top \mid A \mid \neg A \mid D \sqcap D \mid \exists \geq n.\, r.\, D \mid \exists \leq 1.\, r \mid \forall\, r.\, D \mid \exists\, r.\, \mathsf{Self}$$

As usual, we denote by $\mathcal{DL}(\Sigma)$ the set of all $\mathcal{DL}$ concept descriptions over Σ for each description logic $\mathcal{DL}$. The *role depth* $\mathsf{rd}(E)$ of a concept description E is the maximal number of nestings of restrictions within E.[3] We then further denote by $\mathcal{DL}_d(\Sigma)$ the set of all $\mathcal{DL}$ concept descriptions over Σ with a role depth not greater than d. Note that the above syntactic characterization follows easily from the results in [12,15,24]. A finite set of concept inclusions is called *terminological box* (abbrv. TBox).

As it has already been pointed out in [15], the following properties can be expressed in a sufficiently strong Horn DL, e.g., in Horn-$\mathcal{M}$.

Inclusion of Simple Concepts. $A \sqsubseteq B$ states that each individual being A is also B.

Concept Disjointness. $A \sqcap B \sqsubseteq \bot$ states that there are no individuals that are both A and B.

Domain Restrictions. $\exists\, r.\, \top \sqsubseteq A$ states that each individual having an r-successor must be an A.

Range Restrictions. $\top \sqsubseteq \forall\, r.\, A$ states that each individual being an r-successor must be an A.

Functionality Restrictions. $\top \sqsubseteq \exists \leq 1.\, r$ states that each individual has at most one r-successor.

Participation Constraints. $A \sqsubseteq \exists\, r.\, B$ states that each individual that is an A has an r-successor that is a B.

[3] Formally, the *role depth* is recursively defined as follows: $\mathsf{rd}(\bot) := \mathsf{rd}(\top) := \mathsf{rd}(A) := \mathsf{rd}(\neg A) := 0$, and $\mathsf{rd}(E \sqcap F) := \mathsf{rd}(E) \vee \mathsf{rd}(F)$, and $\mathsf{rd}(\exists\, r.\, E) := \mathsf{rd}(\exists \geq n.\, r.\, E) := \mathsf{rd}(\forall\, r.\, E) := 1 + \mathsf{rd}(E)$, and $\mathsf{rd}(\exists \leq 1.\, r) := \mathsf{rd}(\exists\, r.\, \mathsf{Self}) := 1$.

A *concept assertion* is an expression $a \sqsubseteq E$ where a is an individual name from Σ_{I} and E is some concept description, and further a *role assertion* is an expression $(a,b) \sqsubseteq r$ where $a, b \in \Sigma_{\mathsf{I}}$ and r is some role name from Σ_{R}. A finite set of concept and role assertions is called *assertional box* (abbrv. ABox). The union of a terminological and an assertional box yields an *ontology*. We often call the assertional part of an ontology the *data* and the terminological part of an ontology the *schema*. If a question of the form $\mathcal{O} \models \alpha$? is to be decided, then we also call the axiom α the *query*.

An *interpretation* $\mathcal{I}$ is a pair $(\Delta^{\mathcal{I}}, \cdot^{\mathcal{I}})$ consisting of a non-empty set $\Delta^{\mathcal{I}}$ of *objects*, called *domain*, and an *extension mapping* $\cdot^{\mathcal{I}}$ such that $a^{\mathcal{I}} \in \Delta^{\mathcal{I}}$ for $a \in \Sigma_{\mathsf{I}}$, $A^{\mathcal{I}} \subseteq \Delta^{\mathcal{I}}$ for each $A \in \Sigma_{\mathsf{C}}$, and $r^{\mathcal{I}} \subseteq \Delta^{\mathcal{I}} \times \Delta^{\mathcal{I}}$ for each $r \in \Sigma_{\mathsf{R}}$. The extension mapping is then extended to all concept descriptions in the following recursive manner; the names of these concept descriptions are shown in the right column.

$$
\begin{aligned}
\bot^{\mathcal{I}} &:= \emptyset && \textit{(bottom concept description)}\\
\top^{\mathcal{I}} &:= \Delta^{\mathcal{I}} && \textit{(top concept description)}\\
(\neg A)^{\mathcal{I}} &:= \Delta^{\mathcal{I}} \setminus A^{\mathcal{I}} && \textit{(negated concept name)}\\
(E \sqcap F)^{\mathcal{I}} &:= E^{\mathcal{I}} \cap F^{\mathcal{I}} && \textit{(conjunction)}\\
(\exists\, r.\, E)^{\mathcal{I}} &:= \{\, \delta \mid (\delta, \epsilon) \in r^{\mathcal{I}} \text{ and } \epsilon \in E^{\mathcal{I}} \text{ for some } \epsilon \,\} && \textit{(existential restr.)}\\
(\exists {\geq} n.\, r.\, E)^{\mathcal{I}} &:= \{\, \delta \mid |\{\, \epsilon \mid (\delta, \epsilon) \in r^{\mathcal{I}} \text{ and } \epsilon \in E^{\mathcal{I}} \,\}| \geq n \,\} && \textit{(qualified at-least r.)}\\
(\exists {\leq} 1.\, r)^{\mathcal{I}} &:= \{\, \delta \mid |\{\, \epsilon \mid (\delta, \epsilon) \in r^{\mathcal{I}} \,\}| \leq 1 \,\} && \textit{(local functionality restriction)}\\
(\exists\, r.\, \mathsf{Self})^{\mathcal{I}} &:= \{\, \delta \mid (\delta, \delta) \in r^{\mathcal{I}} \,\} && \textit{(existential self-restriction)}\\
(\forall\, r.\, E)^{\mathcal{I}} &:= \{\, \delta \mid (\delta, \epsilon) \in r^{\mathcal{I}} \text{ implies } \epsilon \in E^{\mathcal{I}} \text{ for each } \epsilon \,\} && \textit{(value restriction)}
\end{aligned}
$$

Now a concept inclusion $C \sqsubseteq D$ is *valid* in $\mathcal{I}$ if $C^{\mathcal{I}} \subseteq D^{\mathcal{I}}$ holds true, written $\mathcal{I} \models C \sqsubseteq D$. A concept assertion $a \sqsubseteq E$ is *valid* in $\mathcal{I}$ if $a^{\mathcal{I}} \in E^{\mathcal{I}}$ is satisfied, and we shall denote this as $\mathcal{I} \models a \sqsubseteq E$. Likewise, a role assertion $(a,b) \sqsubseteq r$ is *valid* in $\mathcal{I}$ if $(a^{\mathcal{I}}, b^{\mathcal{I}}) \in r^{\mathcal{I}}$ holds true, and we symbolize this as $\mathcal{I} \models (a,b) \sqsubseteq r$. If $\mathcal{O}$ is an ontology, then $\mathcal{I}$ is a *model* of $\mathcal{O}$ if $\mathcal{I} \models \alpha$ holds true for each axiom $\alpha \in \mathcal{O}$, and we shall denote this as $\mathcal{I} \models \mathcal{O}$. Furthermore, an ontology $\mathcal{O}_1$ *entails* another ontology $\mathcal{O}_2$, written $\mathcal{O}_1 \models \mathcal{O}_2$, if each model of $\mathcal{O}_1$ is a model of $\mathcal{O}_2$ too. In case $\mathcal{O} \models \{\alpha\}$ for some single axiom α, we shall omit set parenthesis and simply write $\mathcal{O} \models \alpha$. Note that, if $x * y$ is an axiom and $\mathcal{Z}$ is either an interpretation or an ontology, then we sometimes write $x *_{\mathcal{Z}} y$ instead of $\mathcal{Z} \models x * y$.

There are several standard reasoning tasks as follows.

Knowledge Base Consistency. Given an ontology $\mathcal{O}$, is there a model of $\mathcal{O}$?

Concept Satisfiability. Given a concept description E and an ontology $\mathcal{O}$, is there a model of $\mathcal{O}$ in which E has a non-empty extension?

Concept Subsumption. Given two concept descriptions C and D and an ontology $\mathcal{O}$, does $\mathcal{O}$ entail $C \sqsubseteq D$?

Instance Checking. Given an individual a, a concept description E, and an ontology $\mathcal{O}$, does $\mathcal{O}$ entail $a \sqsubseteq E$?

There are two approaches to determining the computational complexity of the above tasks.

Combined Complexity. This is the default. Necessary time and space for solving the reasoning problem is measured as a function in the size of the whole input. For instance, if $a \in_{\mathcal{O}} E$ is to be decided, then time and space requirements are measured as a function of $||a \in E|| + ||\mathcal{O}||$.

Data Complexity. Determining data complexity is more meaningful for practical purposes, as in most cases the size of the stored data easily outgrows the size of the schema and query. In particular, time and space needed for solving the reasoning problem is measured as a function in the size of the ABox only. If, e.g., $a \in_{\mathcal{O}} E$ is to be decided where $\mathcal{O}$ is the union of an ABox $\mathcal{A}$ and some TBox $\mathcal{T}$, then necessary time and space is only measured as a function of $||\mathcal{A}||$.

So far, the computational complexity of reasoning in $\mathcal{M}$ and its sibling Horn-$\mathcal{M}$ has not been determined and, thus, we shall catch up on this here. Since for a large variety of description logics complexity results have been obtained, we can immediately find the following results for $\mathcal{M}^- := \mathcal{ALQ}^{\geq}\mathcal{N}^{\leq}$, the sublogic of $\mathcal{M}$ in which we disallow existential self-restrictions $\exists\, r.\,\mathsf{Self}$. Note that we always consider the case of a *general* TBox, i.e., where no restrictions are imposed on the concept inclusions (except those possibly implied by Hornness).

Concept subsumption in $\mathcal{M}^-$ is **EXP**-*complete (combined complexity).* Since $\mathcal{M}^-$ is a sublogic of $\mathcal{SHIQ}$ and concept subsumption in $\mathcal{SHIQ}$ is in **EXP** [28,30], it follows that concept subsumption in $\mathcal{M}^-$ is in **EXP** as well. Furthermore, $\mathcal{FL}_0$ is a sublogic of $\mathcal{M}^-$ in which concept subsumption is **EXP**-hard [1]. We conclude that concept subsumption in $\mathcal{M}^-$ must be **EXP**-hard too.

Concept subsumption in Horn-$\mathcal{M}^-$ *is* **EXP**-*complete (combined complexity).* Horn-$\mathcal{M}^-$ is a sublogic of Horn-$\mathcal{SHIQ}$ and for the latter concept subsumption is known to be in **EXP** [24]. Thus, concept subsumption in Horn-$\mathcal{M}^-$ is also in **EXP**. Since $\mathcal{ELF}$ is a sublogic of Horn-$\mathcal{M}^-$ in which concept subsumption is **EXP**-hard [1], we infer that the same problem in Horn-$\mathcal{M}^-$ must be **EXP**-hard too.

Instance checking in $\mathcal{M}^-$ is **co NP**-*complete (data complexity).* Instance checking in $\mathcal{SHIQ}$ is in **co NP** (data complexity) [14] and since $\mathcal{M}^-$ is a sublogic of $\mathcal{SHIQ}$, it follows that instance checking in $\mathcal{M}^-$ is also in **co NP** (data complexity). Furthermore, $\mathcal{EL}^{kf}$ is a sublogic of $\mathcal{M}^-$ in which instance checking is **co NP**-hard (data complexity) [23], and this result immediately transfers to $\mathcal{M}^-$.

Instance checking in $\mathsf{Horn}\text{-}\mathcal{M}^-$ *is* **P**-*complete (data complexity).* As instance checking in $\mathsf{Horn}\text{-}\mathcal{SHIQ}$ is in **P** (data complexity) [14] and $\mathsf{Horn}\text{-}\mathcal{M}^-$ is a sublogic of $\mathsf{Horn}\text{-}\mathcal{SHIQ}$, we conclude that the similar problem in $\mathsf{Horn}\text{-}\mathcal{M}^-$ is in **P** (data complexity) as well. Furthermore, $\mathcal{EL}$ is a sublogic of $\mathsf{Horn}\text{-}\mathcal{M}^-$ and instance checking in $\mathcal{EL}$ is **P**-hard (data complexity) [6]. Consequently, instance checking in $\mathsf{Horn}\text{-}\mathcal{M}^-$ is **P**-hard as well.

We see that terminological reasoning in $\mathsf{Horn}\text{-}\mathcal{M}^-$ is not cheaper than in $\mathcal{M}^-$, but that assertional reasoning with knowledge bases containing both a schema (TBox) and data (ABox) is considerably cheaper in $\mathsf{Horn}\text{-}\mathcal{M}^-$ than in $\mathcal{M}^-$ if we only take into account the size of the ABox (data complexity), unless $\mathbf{P} = \mathbf{NP}$. It is obvious that the hardness results transfer from $\mathcal{M}^-$ to $\mathcal{M}$ and accordingly for the Horn variants. Furthermore, since $\mathcal{M}$ and $\mathsf{Horn}\text{-}\mathcal{M}$ can each be seen as a sublogic of $\mu\mathcal{ALCQ}$ in which concept subsumption is **EXP**-complete [8,25], we can infer that concept subsumption in $\mathcal{M}$ as well as in $\mathsf{Horn}\text{-}\mathcal{M}$ is **EXP**-complete (combined complexity) as well. Unfortunately, the author cannot provide sharp upper bounds for the data complexity of instance checking in $\mathcal{M}$ and $\mathsf{Horn}\text{-}\mathcal{M}$. If one takes a closer look on the proofs in [15], one could get the impression that it might suffice to include the case $\pi_y(\exists\, R.\,\mathsf{Self}, X) := R(X, X)$ for the translation of concept descriptions into first-order logic. While the author conjectures that this extended translation allows for obtaining the same complexity results, it is necessary to check whether all later steps in the proof indeed work as before.

Henceforth, it makes sense to use a $\mathsf{Horn}\text{-}\mathcal{M}$ TBox as the schema for *ontology-based data access* (abbrv. OBDA) applications. This motivates the development of a procedure that can learn $\mathsf{Horn}\text{-}\mathcal{M}$ concept inclusions from observations in form of an interpretation.

The next section makes use of the notion of a *model-based most specific concept description*, which we shall define now. Fix some description logic $\mathcal{DL}$, an interpretation $\mathcal{I}$, a subset $X \subseteq \Delta^{\mathcal{I}}$, as well as some role-depth bound $d \in \mathbb{N}$. The *model-based most specific concept description* (abbrv. MMSC) of X in $\mathcal{I}$ is then some $\mathcal{DL}$ concept description E that satisfies the following conditions.

1. $\mathsf{rd}(E) \leq d$
2. $X \subseteq E^{\mathcal{I}}$
3. $X \subseteq F^{\mathcal{I}}$ implies $E \sqsubseteq_\emptyset F$ for each $\mathcal{DL}$ concept description F where $\mathsf{rd}(F) \leq d$.

Since MMSCs are unique up to equivalence, we shall denote these as $X^{\mathcal{I}_d^{\mathcal{DL}}}$. In [18] the author has shown how MMSCs can be computed in the description logic $\mathcal{M}$. For any sublogic of $\mathcal{M}$, the computation method can suitably be adapted by simply ignoring unsupported concept constructors.

It is easy to see that this MMSC mapping $\cdot^{\mathcal{I}_d^{\mathcal{DL}}} \colon \wp(\Delta^{\mathcal{I}}) \to \mathcal{DL}_d(\Sigma)$ is the *adjoint* of the extension mapping $\cdot^{\mathcal{I}} \colon \mathcal{DL}_d(\Sigma) \to \wp(\Delta^{\mathcal{I}})$, that is, the pair of both constitutes a galois connection just like it is the case for the pair of derivation operators $\cdot^I$ induced by a formal context. This implies that the following statements hold true, where X and Y are arbitrary subsets of the domain $\Delta^{\mathcal{I}}$, and E and F are any $\mathcal{DL}$ concept descriptions with a role depth of at most d.

1. $X \subseteq E^{\mathcal{I}}$ if, and only if, $X^{\mathcal{I}_d^{\mathcal{DL}}} \sqsubseteq_\emptyset E$
2. $X \subseteq X^{\mathcal{I}_d^{\mathcal{DL}}\mathcal{I}}$
3. $X^{\mathcal{I}_d^{\mathcal{DL}}} = X^{\mathcal{I}_d^{\mathcal{DL}}\mathcal{I}\mathcal{I}_d^{\mathcal{DL}}}$
4. $X \subseteq Y$ implies $X^{\mathcal{I}_d^{\mathcal{DL}}} \sqsubseteq_\emptyset Y^{\mathcal{I}_d^{\mathcal{DL}}}$
5. $E^{\mathcal{I}\mathcal{I}_d^{\mathcal{DL}}} \sqsubseteq_\emptyset E$
6. $E^{\mathcal{I}} \equiv_\emptyset E^{\mathcal{I}\mathcal{I}_d^{\mathcal{DL}}\mathcal{I}}$
7. $E \sqsubseteq_\emptyset F$ implies $E^{\mathcal{I}} \subseteq F^{\mathcal{I}}$

Compared to the FCA setting, we have replaced intent descriptions using sets of attributes by intent descriptions using $\mathcal{DL}$ concept descriptions.

4 Inductive Learning in Horn-$\mathcal{M}$

Now fix some finitely representable interpretation $\mathcal{I}$ over a signature Σ, and further let $d \in \mathbb{N}$ be a role-depth bound. Similarly to [5,9,18], we define the *induced formal context* $\mathbb{K}_{\mathcal{I},d} := (\Delta^{\mathcal{I}}, M, I)$ where $M := M_{\mathsf{p}} \cup M_{\mathsf{c}}$, the premise attribute set is defined by

$$M_{\mathsf{p}} := \{\bot\} \cup \Sigma_{\mathsf{C}} \cup \{\exists r.\mathsf{Self} \mid r \in \Sigma_{\mathsf{R}}\} \cup \{\exists r.X^{\mathcal{I}_{d-1}^{\mathcal{EL}^*}} \mid r \in \Sigma_{\mathsf{R}} \text{ and } \emptyset \neq X \subseteq \Delta^{\mathcal{I}}\}$$

while the conclusion attribute set is given as

$$\begin{aligned} M_{\mathsf{c}} := &\{\bot\} \cup \{A, \neg A \mid A \in \Sigma_{\mathsf{C}}\} \cup \{\exists r.\mathsf{Self},\ \exists{\leq}1.r,\ \forall r.\bot \mid r \in \Sigma_{\mathsf{R}}\} \\ &\cup \left\{ \mho\, r.X^{\mathcal{I}_{d-1}^{\mathcal{M}^{\leq 1}}} \;\middle|\; \begin{array}{l} \mho \in \{\exists{\geq}n \mid 1 \leq n \leq |\Delta^{\mathcal{I}}|\} \cup \{\forall\}, \\ r \in \Sigma_{\mathsf{R}}, \text{ and } \emptyset \neq X \subseteq \Delta^{\mathcal{I}} \end{array} \right\}, \end{aligned}$$

and $(\delta, C) \in I$ if $\delta \in C^{\mathcal{I}}$. Our interest is to axiomatize the Horn-$\mathcal{M}$ concept inclusions valid in $\mathcal{I}$. Of course, it holds true that $\bigsqcap\mathbf{X} \sqsubseteq \bigsqcap\mathbf{Y}$ is a Horn-$\mathcal{M}$ concept inclusion for each subset $\mathbf{X} \subseteq M_{\mathsf{p}}$ and each subset $\mathbf{Y} \subseteq M_{\mathsf{c}}$. As in [5,9, 18], such a concept inclusion is valid in $\mathcal{I}$ if, and only if, the joining implication $\mathbf{X} \to \mathbf{Y}$ is valid in the induced formal context $\mathbb{K}_{\mathcal{I},d}$. As we are only interested in axiomatizing these concept inclusions that are valid in $\mathcal{I}$ and are no tautologies, we define the following joining implication set that we shall use as background knowledge on the FCA side.

$$\begin{aligned} \mathcal{S} := &\{\{C\} \to \{D\} \mid C \in M_{\mathsf{p}},\ D \in M_{\mathsf{c}}, \text{ and } C \sqsubseteq_\emptyset D\} \\ &\cup \{\{C, \exists r.\mathsf{Self}\} \to \{D\} \mid C \in M_{\mathsf{p}},\ r \in \Sigma_{\mathsf{R}},\ D \in M_{\mathsf{c}}, \text{ and } C \sqcap \exists r.\mathsf{Self} \sqsubseteq_\emptyset D\} \end{aligned}$$

We will see at the end of this section that the model-based most specific concept descriptions $X^{\mathcal{I}_d}$ can have an exponential size w.r.t. $|\Delta^{\mathcal{I}}|$ and d in $\mathcal{M}^{\leq 1}$. Since the problem of deciding subsumption in Horn-$\mathcal{M}$ is **EXP**-complete, we infer that a naïve approach of computing $\mathcal{S}$ needs double exponential time. However, a more sophisticated analysis yields that most concept inclusions cannot be valid. In particular, a concept description from M_{p} only contains concept names and existential (self-)restrictions and, thus, these can never be subsumed (w.r.t. $\emptyset$) by a concept description from M_{c} containing a negated concept name, a local functionality restriction, a qualified at-least restriction where $n > 1$, or a value restriction. Thus, we conclude from the characterization in [18, Section 8] that $\mathcal{S}$ does not contain any implication $\{C\} \to \{D\}$ or $\{C, \exists\, r.\mathsf{Self}\} \to \{D\}$ except for the trivial cases where $C = \bot$, $C = D$, or $D = \exists\, r.\mathsf{Self}$ (only for the second form), and it can hence be computed in single exponential time. Even in the case where the tautological TBox $\mathcal{S}$ is not that simple, e.g., for another description logic where subsumption is also **EXP**-complete and model-based most specific concept descriptions can have exponential sizes, we could also dispense with the expensive computation of $\mathcal{S}$, since the canonical base can then still be computed in single exponential time with the only drawback that it could contain tautologies.

In the remainder of this section, we show how the techniques from Sect. 2 can be applied to axiomatize Horn-$\mathcal{M}$ concept inclusions valid in $\mathcal{I}$. Note that the proofs are suitable adaptations of those for the $\mathcal{EL}^{\bot}$ case [5] or of those for the $\mathcal{M}$ case [18]. We first show that the TBox consisting of the concept inclusions $C \sqsubseteq C^{\mathcal{I}\mathcal{I}_d^{\mathcal{M}^{\leq 1}}}$ for all $\mathcal{EL}^*$ concept descriptions C with a role depth not exceeding d is sound and complete. As a further step, we prove by means of structural induction that also the TBox containing the concept inclusions $\bigsqcap \mathbf{C} \sqsubseteq (\bigsqcap \mathbf{C})^{\mathcal{I}\mathcal{I}_d^{\mathcal{M}^{\leq 1}}}$ where $\mathbf{C}$ is a subset of M_{p} is sound and complete. Furthermore, when computing the MMSC of a conjunction $\bigsqcap \mathbf{C}$ where $\mathbf{C} \subseteq M_{\mathsf{p}}$ we do not have to do this on the DL side, which is expensive, but it suffices to compute the result $\mathbf{C}^{\mathsf{pc}}$ on the FCA side by applying the derivation operators $\cdot^{\mathsf{p}}$ and $\cdot^{\mathsf{c}}$. The conjunction $\bigsqcap \mathbf{C}^{\mathsf{pc}}$ is then (equivalent to) the MMSC in the DL $\mathcal{M}^{\leq 1}$. The proofs for the three aforementioned statements can be found in the technical report [20].

The main result for inductive learning of Horn-$\mathcal{M}$ concept inclusions is as follows. It states that (the premises of) each pc-implication base of the induced context $\mathbb{K}_{\mathcal{I},d}$ give rise to a base of Horn-$\mathcal{M}$ concept inclusions for $\mathcal{I}$.

Proposition 7. *If $\mathcal{L}$ is a joining implication base for $\mathbb{K}_{\mathcal{I},d}$ relative to $\mathcal{S}$, then the following TBox is sound and complete for the* Horn-$\mathcal{M}$ *concept inclusions that are valid in $\mathcal{I}$ and have role depths not exceeding d.*

$$\{\bigsqcap \mathbf{C} \sqsubseteq (\bigsqcap \mathbf{C})^{\mathcal{I}\mathcal{I}_d^{\mathcal{M}^{\leq 1}}} \mid \mathbf{C} \to \mathbf{D} \in \mathcal{L}\}$$

Instantiating the previous proposition with the canonical pc-implication base now yields the following corollary.

Corollary 8. *The following* Horn-$\mathcal{M}$ *TBox, called* canonical Horn-$\mathcal{M}$ concept inclusion base *for $\mathcal{I}$ and d, is sound and complete for the* Horn-$\mathcal{M}$ *concept inclusions that are valid in $\mathcal{I}$ and have role depths at most d.*

$$\mathsf{Can}_{\mathsf{Horn}\text{-}\mathcal{M}}(\mathcal{I}, d) := \{\bigsqcap (P \cap M_{\mathsf{p}}) \sqsubseteq \bigsqcap (P \cap M_{\mathsf{p}})^{\mathsf{pc}} \mid P \in \mathsf{PsClo}(\phi_{\mathcal{I},d}^{\mathsf{Horn}\text{-}\mathcal{M}}, \mathcal{S})\}$$

The closure operator $\phi_{\mathcal{I},d}^{\mathsf{Horn}\text{-}\mathcal{M}} \colon \wp(M) \to \wp(M)$ is defined by $X \mapsto X \cup (X \cap M_{\mathsf{p}})^{\mathsf{pc}}$.

In the sequel of this section, we investigate the computational complexity of computing the canonical Horn-$\mathcal{M}$ concept inclusion base. As it turns out, the complexity is the same as for computing the canonical pc-implication base—both can be obtained in exponential time. Afterwards, we investigate whether we can show that the canonical Horn-$\mathcal{M}$ concept inclusion base has minimal cardinality.

Proposition 9. *The canonical* Horn-$\mathcal{M}$ *concept inclusion base for a finitely representable interpretation $\mathcal{I}$ and role depth bound $d \geq 1$ can be computed in exponential time with respect to d and the cardinality of the domain $\Delta^{\mathcal{I}}$, and further there exist finitely representable interpretations $\mathcal{I}$ for which the canonical* Horn-$\mathcal{M}$ *concept inclusion base cannot be encoded in polynomial space.*

Note that in order to save space for representing the model-based most specific concept descriptions, we could also represent them in the form $X^{\mathcal{I}}\restriction_d$ where $X^{\mathcal{I}}$ is the model-based most specific concept description without any bound on the role depth and $E\restriction_d$ denotes the *unraveling* of some concept description E (formulated in a DL with greatest fixed-point semantics) up to role depth d. In general, these unbounded MMSCs $X^{\mathcal{I}}$ only exist in extensions of the considered DL with greatest fixed-point semantics. The advantage is that then the size of $X^{\mathcal{I}}\restriction_d$ is exponential only in $|\Delta^{\mathcal{I}}|$ but not in d.

The author conjectures that, for each finitely representable interpretation $\mathcal{I}$, the canonical Horn-$\mathcal{M}$ concept inclusion base $\mathsf{Can}_{\mathsf{Horn}\text{-}\mathcal{M}}(\mathcal{I}, d)$ has *minimal cardinality* among all Horn-$\mathcal{M}$ concept inclusion bases for $\mathcal{I}$ and d. However, it is not immediately possible to suitably adapt the minimality proof for the $\mathcal{EL}$ case described in [5,9], since not all notions from $\mathcal{EL}$ are available in more expressive description logics. The crucial point is that we need the validity of the following claim, which resembles [9, Lemma 5.16] or [5, Lemma A.9], respectively, for our case of Horn-$\mathcal{M}$.

Claim. Fix some Horn-$\mathcal{M}$ TBox $\mathcal{T} \cup \{C \sqsubseteq D\}$ in which all occurring concept descriptions have role depths not exceeding d. Further assume that $\mathcal{I}$ is a finitely representable model of $\mathcal{T}$ such that, for each subconcept $\exists\, r. X$ of C, the filler X is (equivalent to) some model-based most specific concept description of $\mathcal{I}$ in the description logic $\mathcal{EL}^*$; more specifically, we assume that $Y \equiv Y^{\mathcal{I}\mathcal{I}^{\mathcal{EL}^*}_{d-1}}$ is satisfied for each $\exists\, r. Y \in \mathsf{Conj}(C)$. If $C \not\sqsubseteq_{\emptyset} D$ and $C \sqsubseteq_{\mathcal{T}} D$, then $C \sqsubseteq_{\emptyset} E$ and $C \not\sqsubseteq_{\emptyset} F$ holds true for some concept inclusion $E \sqsubseteq F$ contained in $\mathcal{T}$.

However, the author has just developed a computation procedure for so-called most specific consequences, cf. [19,21, Definition 3], in a description logic that is more expressive than $\mathcal{EL}^{\bot}$. A proof of the above claim can then be obtained as a by-product. This will be subject of a future publication.

5 Conclusion

In Formal Concept Analysis, a restricted form of implications has been introduced: so-called *joining implications*. From the underlying attribute set M two subsets M_{p} and M_{c} are declared, and then only those implications are considered in which the premises only contain attributes from M_{p} and in which the conclusions only contain attributes from M_{c}. A canonical base for the joining implications valid in some given formal context has been devised, and it has been proven that it has minimal cardinality and can be computed in deterministic exponential time.

The former results have then been applied to the problem of inductive learning in the Horn description logic Horn-$\mathcal{M}$. More specifically, we have proposed a canonical base for the Horn-$\mathcal{M}$ concept inclusions valid in a given interpretation. While the author conjectures that it has minimal cardinality, it has been demonstrated that it can be computed in deterministic exponential time.

Future research could deliver the proof for the claimed minimality of the canonical Horn-$\mathcal{M}$ concept inclusion base, or could investigate means that allow for the integration of existing knowledge to make incremental learning possible. The author believes that both tasks can be tackled as soon as computation procedures for most specific consequences in Horn-$\mathcal{M}$ are available. This will be subject of future publications.

Eventually, the author wants to point out that incremental learning from a sequence of interpretations [19,21, Section 8.4 for the $\mathcal{EL}^{\bot}$ case] is probably more practical than *model exploration* or *ABox exploration* [9], since new observations that could show invalidity of concept inclusions are not requested from the expert at a certain time point, but are rather processed upon availability (*"push instead of pull"*). However, completeness of the eventual result for the considered domain of interest is only achieved if *all* typical individuals occur in the sequence at some time point.

Acknowledgments. The author gratefully thanks Sebastian Rudolph for the very idea of learning in Horn description logics as well as for a helpful discussion on basics of Horn description logics. The author further thanks the reviewers for their constructive remarks.

References

1. Baader, F., Brandt, S., Lutz, C.: Pushing the $\mathcal{EL}$ envelope. In: Kaelbling, L.P., Saffiotti, A. (eds.) IJCAI-05, Proceedings of the Nineteenth International Joint Conference on Artificial Intelligence, Edinburgh, Scotland, UK, July 30–August 5 2005, pp. 364–369. Professional Book Center (2005)
2. Baader, F., Horrocks, I., Lutz, C., Sattler, U.: An Introduction to Description Logic. Cambridge University Press, New York (2017)
3. Belohlávek, R., Vychodil, V.: Closure-based constraints in formal concept analysis. Discrete Appl. Math. **161**(13–14), 1894–1911 (2013)
4. Borchmann, D.: Learning terminological knowledge with high confidence from erroneous data. Doctoral thesis, Technische Universität Dresden, Dresden, Germany (2014)
5. Borchmann, D., Distel, F., Kriegel, F.: Axiomatisation of general concept inclusions from finite interpretations. J. Appl. Non-Class. Logics **26**(1), 1–46 (2016)

6. Calvanese, D., De Giacomo, G., Lembo, D., Lenzerini, M., Rosati, R.: Data complexity of query answering in description logics. In: Doherty, P., Mylopoulos, J., Welty, C.A. (eds.) Proceedings, Tenth International Conference on Principles of Knowledge Representation and Reasoning, Lake District of the United Kingdom, 2–5 June 2006, pp. 260–270. AAAI Press (2006)
7. Dantsin, E., Eiter, T., Gottlob, G., Voronkov, A.: Complexity and expressive power of logic programming. ACM Comput. Surv. **33**(3), 374–425 (2001)
8. De Giacomo, G., Lenzerini, M.: A uniform framework for concept definitions in description logics. J. Artif. Intell. Res. **6**, 87–110 (1997)
9. Distel, F.: Learning description logic knowledge bases from data using methods from formal concept analysis. Doctoral thesis, Technische Universität Dresden, Dresden, Germany (2011)
10. Ganter, B., Wille, R.: Formal Concept Analysis: Mathematical Foundations. Springer, Heidelberg (1999). https://doi.org/10.1007/978-3-642-59830-2
11. Guigues, J.L., Duquenne, V.: Famille minimale d'implications informatives résultant d'un tableau de données binaires. Mathématiques et Sciences Humaines **95**, 5–18 (1986)
12. Hernich, A., Lutz, C., Papacchini, F., Wolter, F.: Horn-Rewritability vs. PTime query evaluation in ontology-mediated querying. In: Lang, J. (ed.) Proceedings of the Twenty-Seventh International Joint Conference on Artificial Intelligence, IJCAI 2018, Stockholm, Sweden, 13–19 July 2018, pp. 1861–1867. ijcai.org (2018)
13. Hitzler, P., Krötzsch, M., Rudolph, S.: Foundations of Semantic Web Technologies. Chapman and Hall/CRC Press, Boca Raton (2010)
14. Hustadt, U., Motik, B., Sattler, U.: Data complexity of reasoning in very expressive description logics. In: Kaelbling, L.P., Saffiotti, A. (eds.) IJCAI-05, Proceedings of the Nineteenth International Joint Conference on Artificial Intelligence, Edinburgh, Scotland, UK, July 30 - August 5 2005, pp. 466–471. Professional Book Center (2005)
15. Hustadt, U., Motik, B., Sattler, U.: Reasoning in description logics by a reduction to disjunctive datalog. J. Autom. Reason. **39**(3), 351–384 (2007)
16. Kriegel, F.: Concept Explorer FX (2010–2019), Software for Formal Concept Analysis with Description Logic Extensions. https://github.com/francesco-kriegel/conexp-fx
17. Kriegel, F.: NextClosures with constraints. In: Huchard, M., Kuznetsov, S. (eds.) Proceedings of the Thirteenth International Conference on Concept Lattices and Their Applications, Moscow, Russia, 18–22 July 2016. CEUR Workshop Proceedings, vol. 1624, pp. 231–243. CEUR-WS.org (2016)
18. Kriegel, F.: Acquisition of terminological knowledge from social networks in description logic. In: Missaoui, R., Kuznetsov, S.O., Obiedkov, S. (eds.) Formal Concept Analysis of Social Networks. LNSN, pp. 97–142. Springer, Cham (2017). https://doi.org/10.1007/978-3-319-64167-6_5
19. Kriegel, F.: Most specific consequences in the description logic $\mathcal{EL}$. LTCS-Report 18–11, Chair of Automata Theory, Institute of Theoretical Computer Science, Technische Universität Dresden, Dresden, Germany (2018, accepted for publication in Discrete Applied Mathematics). https://tu-dresden.de/inf/lat/reports#Kr-LTCS-18-11
20. Kriegel, F.: Joining implications in formal contexts and inductive learning in a horn description logic (Extended Version). LTCS-Report 19–02, Chair of Automata Theory, Institute of Theoretical Computer Science, Technische Universität Dresden, Dresden, Germany (2019). https://tu-dresden.de/inf/lat/reports#Kr-LTCS-19-02

21. Kriegel, F.: Most specific consequences in the description logic $\mathcal{EL}$. Discrete Applied Mathematics (2019). https://doi.org/10.1016/j.dam.2019.01.029
22. Kriegel, F., Borchmann, D.: NextClosures: parallel computation of the canonical base with background knowledge. Int. J. Gen. Syst. **46**(5), 490–510 (2017)
23. Krisnadhi, A., Lutz, C.: Data complexity in the $\mathcal{EL}$ family of description logics. In: Dershowitz, N., Voronkov, A. (eds.) LPAR 2007. LNCS (LNAI), vol. 4790, pp. 333–347. Springer, Heidelberg (2007). https://doi.org/10.1007/978-3-540-75560-9_25
24. Krötzsch, M., Rudolph, S., Hitzler, P.: Complexities of horn description logics. ACM Trans. Comput. Logic **14**(1), 2:1–2:36 (2013)
25. Kupferman, O., Sattler, U., Vardi, M.Y.: The complexity of the graded μ-Calculus. In: Voronkov, A. (ed.) CADE 2002. LNCS (LNAI), vol. 2392, pp. 423–437. Springer, Heidelberg (2002). https://doi.org/10.1007/3-540-45620-1_34
26. Kuznetsov, S.O., Obiedkov, S.A.: Some decision and counting problems of the Duquenne-Guigues basis of implications. Discrete Appl. Math. **156**(11), 1994–2003 (2008)
27. Rudolph, S.: Relational exploration: combining description logics and formal concept analysis for knowledge specification. Doctoral thesis, Technische Universität Dresden, Dresden, Germany (2006)
28. Schild, K.: A correspondence theory for terminological logics: preliminary report. In: Mylopoulos, J., Reiter, R. (eds.) Proceedings of the 12th International Joint Conference on Artificial Intelligence, Sydney, Australia, 24–30 August 1991, pp. 466–471. Morgan Kaufmann (1991)
29. Stumme, G.: Attribute exploration with background implications and exceptions. In: Bock, H.H., Polasek, W. (eds.) Studies in Classification, Data Analysis, and Knowledge Organization, pp. 457–469. Springer, Heidelberg (1996). https://doi.org/10.1007/978-3-642-80098-6_39
30. Tobies, S.: Complexity results and practical algorithms for logics in knowledge representation. Doctoral thesis, Rheinisch-Westfälische Technische Hochschule Aachen, Aachen, Germany (2001)

Lattices of Orders

Christian Meschke(✉)

Dresden, Germany
Christian.Meschke@gmail.com

Abstract. Ordering order relations is a well-established topic of order theory. Traditionally, the concept of order extension plays an important role and leads to fundamental results like Dushnik and Miller's Theorem stating that every order is the intersection of linear extensions [1]. We introduce an alternative but still quite elementary way to order relations. The resulting *lattices of orders* can be viewed as a generalisation of the lattices of permutations from [2] and accordingly maintain a very high degree of symmetry. Furthermore, the resulting lattices of orders form complete sublattices of it's quasiorder counterpart, the *lattices of quasiorders*, which are also introduced. We examine the basic properties of these two classes of lattices and present their contextual representations.

1 Introduction and Basic Notions

The concept of ordering order relations is a well-established field in order theory. Typically it is dealt within the context of dimension theory [1,3], where one is interested in representing a given order relation as the intersection of linear extensions. The linear orders, as is well-known, form the maximal elements with respect to the extension-order. Nevertheless, the extension-order gives rise to lattices of orders. For this purpose one has to restrict oneself to suborders of a fixed (linear) order, like in [4], or has to add an additional top element, like in [5].

In [6] we investigated the possibilities to combine two posets in a way that the order inside of the input posets remains unaffected. Given two disjoint posets $(X, \leq_X)$ and $(Y, \leq_Y)$ we investigated the pairs (A, B) with $A \subseteq X \times Y$ and $B \subseteq Y \times X$ that give rise to an order relation $\leq_X \cup \leq_Y \cup A \cup B$. As it turned out, these so-called *proper mergings* form a complete lattice when ordered in the following way: $(A_1, B_1) \leq (A_2, B_2)$ iff $A_1 \subseteq A_2$ and $B_1 \supseteq B_2$. An example of a simple lattice of proper mergings is given in Fig. 2.

When we wanted to generalise the setting to more than two input posets, we started with the simplest possible example: three singleton posets. The result, which is depicted in Fig. 3, directly led to a promising way of ordering order relations. The aim of this article is to generally introduce the resulting *lattices of orders* and to describe their basic properties. In the finite case the lattices of orders can be seen as a generalisation of the lattices of permutations presented in [2]. Moreover, the lattices of orders are complete sublattices of the lattices of quasiorders, which are also going to be introduced. Accordingly, Fig. 3 does

D. Cristea et al. (Eds.): ICFCA 2019, LNAI 11511, pp. 130–151, 2019.
https://doi.org/10.1007/978-3-030-21462-3_10

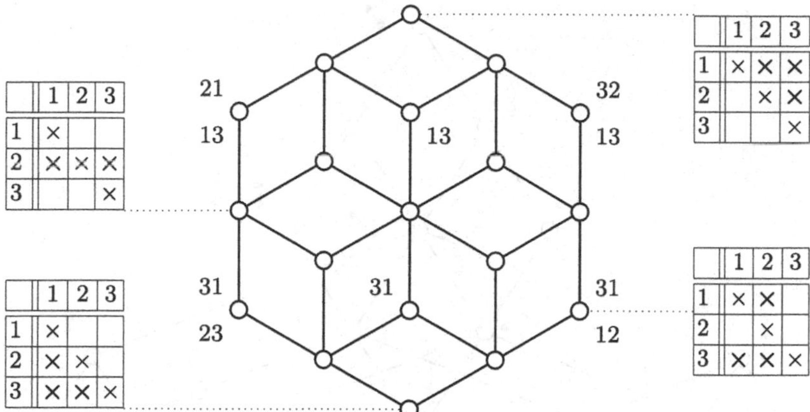

Fig. 1. The lattice of orders on the set $\{1, 2, 3\}$ with the positive pole L given by $1\ L\ 2\ L\ 3$. The labelling is a reduced labelling inspired by the common labelling of concept lattices. For a better readability we leave away brackets and commata and simply write xy when referring to the pair (x, y).

not just show a possible generalisation of the lattice of proper mergings of three singleton posets, but the lattice of orders on a three-element set. The 19 different order relations are each represented by it's line diagram. Another representation of the same lattice is shown in Fig. 1. The labelling in this diagram is inspired by the reduced labelling as it is common in Formal Concept Analysis [7]. It is deduced from the contextual representation of the lattices of orders that we are going to present in the final section of this article.

Preliminary, we need to recall some basic notions. Let R and S be binary relations on a set X. The **relation product** of R and S is denoted by $R \circ S$. Hence, a pair (x, y) belongs to $R \circ S$ iff there is an $a \in X$ with $x\ R\ a\ S\ y$. The **transitive closure** of R is denoted by R°, i.e., $R^\circ := \bigcup_{n \geq 1} R^n$ with $R^1 := R$ and with $R^{n+1} := R \circ R^n$. Furthermore, $R^{-1} := \{(y, x) \mid (x, y) \in R\}$ denotes the **inverse** of R and $\Delta_X := \{(x, x) \mid x \in X\}$ denotes the **diagonal**. Whenever the role of X is obvious we simply write Δ. We put $R_{\neq} := R \setminus \Delta$, $R^{\complement} := X^2 \setminus R$ and define, for $a \in X$,

$$^{R}a := \{x \in X \mid x\ R\ a\} \quad \text{and} \quad a^{R} := \{x \in X \mid a\ R\ x\}.$$

Additionally we need to provide clarity about some basic notions of order theory. We simply say **order** when we mean a partial order relation. In comparison, when referring to a pair (X, P) of a set X equipped with an order relation P on X, we call it, an **ordered set**, or a **poset**, as it is common. A binary relation is called a **quasiorder** if it is transitive and reflexive. Furthermore, the **covering relation** of an order P on X is denoted by $\text{Cover}(P)$, i.e., $(a, b) \in \text{Cover}(P)$ iff $(a, b) \in P_{\neq}$ and for every $x \in X$ with $a\ P\ x\ P\ b$ it follows that $a = x$ or $x = b$. We say that two elements x and y are **neighboured** in P if x is either a lower

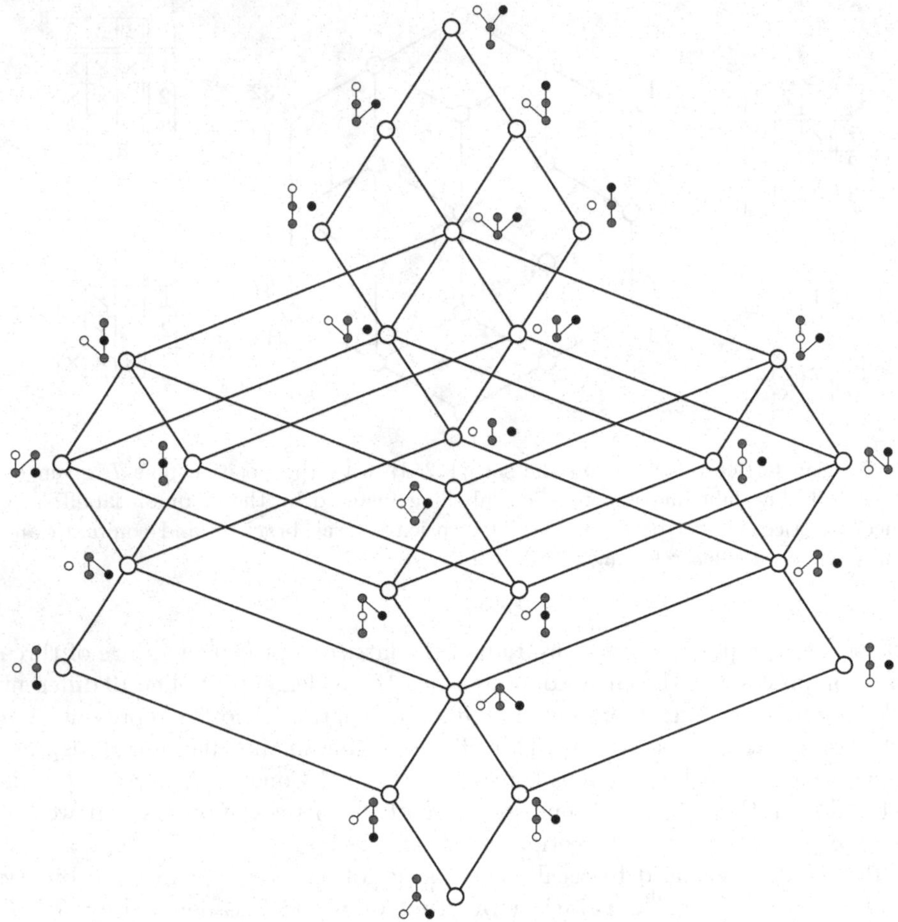

Fig. 2. The lattice of proper mergings [6] of a two-element chain and a two-element antichain ∘ •. (Color figure online)

or an upper cover of y. Moreover, for $(x, y) \in P$ we define the corresponding **interval** in P to be the set

$$[x, y]_P := \{a \in X \mid x\ P\ a\ P\ y\}.$$

We say P fulfils the **finite interval property**, short **(FIP)**, if the interval $[x, y]_P$ is finite for every $(x, y) \in P$. For order relations P and Q with $P \subseteq Q$ we say that P is a **suborder** of Q, or equivalently, that Q is an **extension** of P. The set of all suborder relations of P is denoted by $\mathcal{S}_P$. The **dual** of a quasiorder relation Q is denoted by $Q^d := Q^{-1}$. For additional information about the basic notions of order theory we refer the reader to [8].

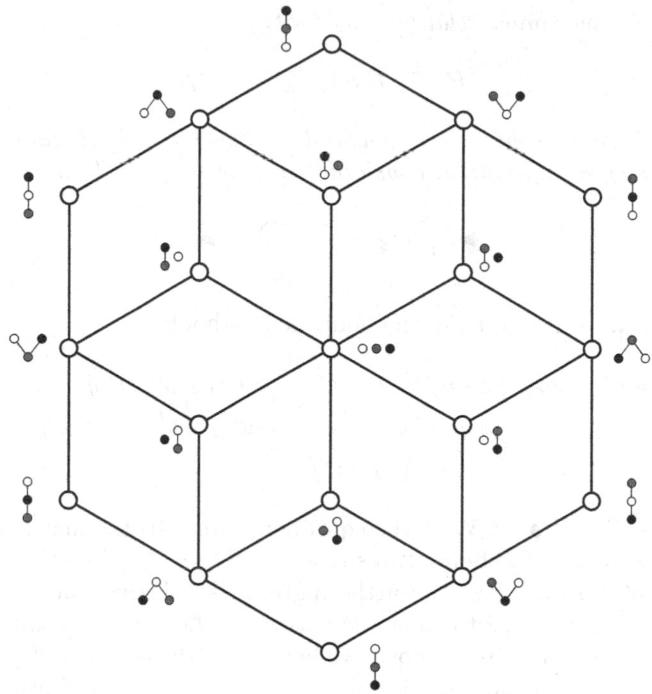

Fig. 3. The lattice of orders on a three element set. (Color figure online)

2 Directed Saturation

In preparation for the introduction of the lattices of orders we first need to introduce the following notion of directed saturation and propose it's basic properties.

Definition 1. *Let P and Q be quasiorder relations on a set X. We then call the relation*

$$P \blacktriangleright Q := \left(P^d \circ (P \cup Q)^{\complement} \circ P^d\right)^{\complement}$$

the ***q-saturation of*** *P towards Q. Furthermore, we put $P \triangleright Q := Q \cap (P \blacktriangleright Q)$.*

The letter q in q-saturation stands for quasiorder. As the following Lemma 1 shows, $P \blacktriangleright Q$ is a quasiorder relation again. Furthermore, it is the largest quasiorder relation one can get from P by just adding pairs from Q.

Lemma 1. *Let P and Q be quasiorder relations on a set X. Then the following statements hold:*

(i) For $a, b \in X$ the following equivalence holds:

$$(a, b) \in P \blacktriangleright Q \iff {}^{P}a \times b^{P} \subseteq P \cup Q.$$

(ii) $P \blacktriangleright Q$ *is a quasiorder relation on* X *satisfying*

$$P \subseteq P \blacktriangleright Q \subseteq P \cup Q.$$

Furthermore, it is the largest quasiorder extension of P *contained in* $P \cup Q$.

(iii) Let $(Q_t)_{t \in T}$ *be a family of quasiorder relations on* X. *Then*

$$P \blacktriangleright \Big(\bigcap_{t \in T} Q_t\Big) = \bigcap_{t \in T} (P \blacktriangleright Q_t).$$

Proof. (i) For $a, b \in X$ the following equivalences hold:

$$\begin{aligned}(a,b) \in P \blacktriangleright Q &\iff \nexists (x,y) \in (P \cup Q)^{\complement} : x\ P\ a \text{ and } b\ P\ y \\ &\iff \forall x, y \in X : \big(x \in {}^{P}a \text{ and } y \in b^{P} \Longrightarrow (x,y) \in P \cup Q\big) \\ &\iff {}^{P}a \times b^{P} \subseteq P \cup Q.\end{aligned}$$

(ii) We put $S := P \blacktriangleright Q$. With the equivalence from (i) the inclusion $P \subseteq S$ is a simple consequence of P being transitive and $S \subseteq P \cup Q$ directly follows from $(a,b) \in {}^{P}a \times b^{P}$ for all $a, b \in X$. Furthermore, S is reflexive since ${}^{P}a \times a^{P} \subseteq P$ for every $a \in X$. It remains to show the trasitivity. Let $(a,b) \in S$ and $(b,c) \in S$. Applying the equivalence from above, we have to show that ${}^{P}a \times c^{P} \subseteq P \cup Q$. So let $(x,y) \in {}^{P}a \times c^{P}$. In the following we show $(x,y) \in P \cup Q$. Since $b \in b^{P}$ and $b \in {}^{P}b$, it follows that $(x,b) \in {}^{P}a \times b^{P}$ and $(b,y) \in {}^{P}b \times c^{P}$ and hence (applying the equivalence from above) $(x,b) \in P \cup Q$ and $(b,y) \in P \cup Q$. In the trivial case of $x\ Q\ b\ Q\ y$ it directly follows that $(x,y) \in P \cup Q$. In case of $x\ P\ b$ it follows that ${}^{P}x \subseteq {}^{P}b$ and hence $(x,y) \in \big({}^{P}x \times c^{P}\big) \subseteq \big({}^{P}b \times c^{P}\big) \subseteq P \cup Q$. Analogously, in case of $b\ P\ y$ it follows that $y^{P} \subseteq b^{P}$ and hence $(x,y) \in \big({}^{P}a \times y^{P}\big) \subseteq \big({}^{P}a \times b^{P}\big) \subseteq P \cup Q$. Summing up, in all cases we have $(x,y) \in P \cup Q$. Hence, S is a quasiorder relation. Let E be another quasiorder relation with $P \subseteq E \subseteq P \cup Q$ and let $(a,b) \in E$. Let $(x,y) \in {}^{P}a \times b^{P}$. Since $P \subseteq E$, it follows that $x\ E\ a\ E\ b\ E\ y$ and hence $(x,y) \in E \subseteq P \cup Q$. With the equivalence from above it follows that $(a,b) \in S$ and hence $E \subseteq S$.

(iii) We put $Q := \bigcap_{t \in T} Q_t$, $F := P^{d} \circ (P \cup Q)^{\complement} \circ P^{d}$ and, accordingly, $F_t := P^{d} \circ (P \cup Q_t)^{\complement} \circ P^{d}$ for every $t \in T$. Since the relation product distributes[1] over the union of relations, we receive that

$$\begin{aligned}F &= P^{d} \circ \Big(P \cup \textstyle\bigcap_{t \in T} Q_t\Big)^{\complement} \circ P^{d} = P^{d} \circ \Big(\bigcup_{t \in T} (P \cup Q_t)^{\complement}\Big) \circ P^{d} \\ &= \textstyle\bigcup_{t \in T} \Big(P^{d} \circ (P \cup Q_t)^{\complement} \circ P^{d}\Big) = \displaystyle\bigcup_{t \in T} F_t.\end{aligned}$$

It easily follows that $P \blacktriangleright Q = F^{\complement} = \bigcap_t F_t^{\complement} = \bigcap_t (P \blacktriangleright Q_t)$. □

[1] This means that for relations A, B_t $(t \in T)$ and C the following two distributivity laws hold: $A \circ (\bigcup_t B_t) = \bigcup_t (A \circ B_t)$ and $(\bigcup_t B_t) \circ C = \bigcup_t (B_t \circ C)$.

Example 1. Let $X := \{1,2,3,4\}$ and let L be the canonic linear order on X given by $1\ L\ 2\ L\ 3\ L\ 4$. In this setting, Fig. 4 shows two suborders P_1 and P_2 of L together with $Q_1 = P_1 \rhd L^d$ and $Q_2 = P_2 \rhd L^d$. Since $P_1 \subseteq P_2$ but $Q_1 \not\supseteq Q_2$, the pair $((\cdot) \rhd L^d, (\cdot) \rhd L)$ is clearly not a Galois connection from $(\mathcal{S}_L, \subseteq)$ to $(\mathcal{S}_{L^d}, \subseteq)$.

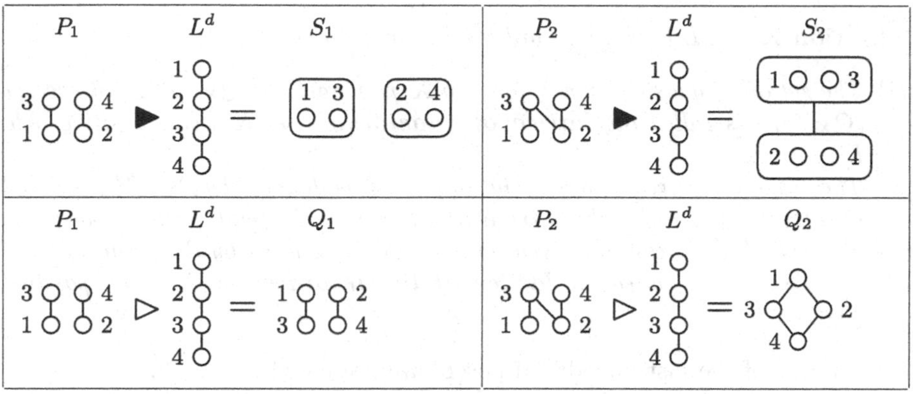

Fig. 4. Example of two suborders P_1 and P_2 of the canonic linear order L on the set $\{1,2,3,4\}$. Furthermore, the figure shows the q-saturations $S_i := P_i \blacktriangleright L^d$ as well as the largest subrelations $Q_i = P_i \rhd L^d$ of L^d such that $P_i \cup Q_i$ remains a quasiorder $(i = 1,2)$.

3 Lattices of Orders

We can now introduce the order relation that gives rise to the lattices of orders. The set of all binary relations on a set X is denoted by $\mathcal{R}_X$. For a given linear order relation L on X we call the binary relation $\sqsubseteq_L$ on $\mathcal{R}_X$ given by

$$R \sqsubseteq_L S \quad :\Longleftrightarrow \quad R \cap L \subseteq S \cap L \text{ and } R \cap L^d \supseteq S \cap L^d \qquad (R, S \in \mathcal{R}_X)$$

the **bipolar order** with **positive pole** L and **negative pole** L^d. Furthermore, we call $R^+ := R \cap L$ the **positive direction** and $R^- := R \cap L^d$ the **negative direction** of the binary relation R.

Proposition 1. *$(\mathcal{R}_X, \sqsubseteq_L)$ is a poset and for binary relations R and S on X with $R \sqsubseteq_L S$ the following statements hold:*

(i) $R \cap \Delta = S \cap \Delta$,
(ii) $R \sqsubseteq_L (R \cap S) \sqsubseteq_L S$ *and* $R \sqsubseteq_L (R \cup S) \sqsubseteq_L S$.

Proof. Straightforward. □

Accordingly, the ordered set $(\mathcal{R}_X, \sqsubseteq_L)$ consists of different connected components, each representing one of the $2^{|X|}$ different ways to choose subrelations of the diagonal Δ. Hence, if one restricts oneself to reflexive relations, one stays inside of the same connected component of $(\mathcal{R}_X, \sqsubseteq_L)$. In the following definition we entitle certain sub-posets of the ordered set $(\mathcal{R}_X, \sqsubseteq_L)$. In doing so, we make use of the following convention: For a poset $(Y, \leq)$ and a subset $S \subseteq Y$ we simply write $(S, \leq)$ when we refer to the sub-poset $(S, \leq \cap S^2)$.

Definition 2. *Let L be a linear order relation on a set X.*

*(i) The set of all quasiorder relations on X is denoted by $\mathcal{Q}_X$. The ordered set $(\mathcal{Q}_X, \sqsubseteq_L)$ is called the **lattice of quasiorders** on X (with positive pole L).*
*(ii) The set of all (partial) order relations on X is denoted by $\mathcal{P}_X$. The ordered set $(\mathcal{P}_X, \sqsubseteq_L)$ is called the **lattice of orders** on X (with positive pole L).*
*(iii) The set of all linear order relations on X is denoted by $\mathcal{L}_X$. The ordered set $(\mathcal{L}_X, \sqsubseteq_L)$ is called the **lattice of linear orders** on X (with positive pole L).*

Examples of the just introduced posets are presented in Figs. 3, 5 and 6.

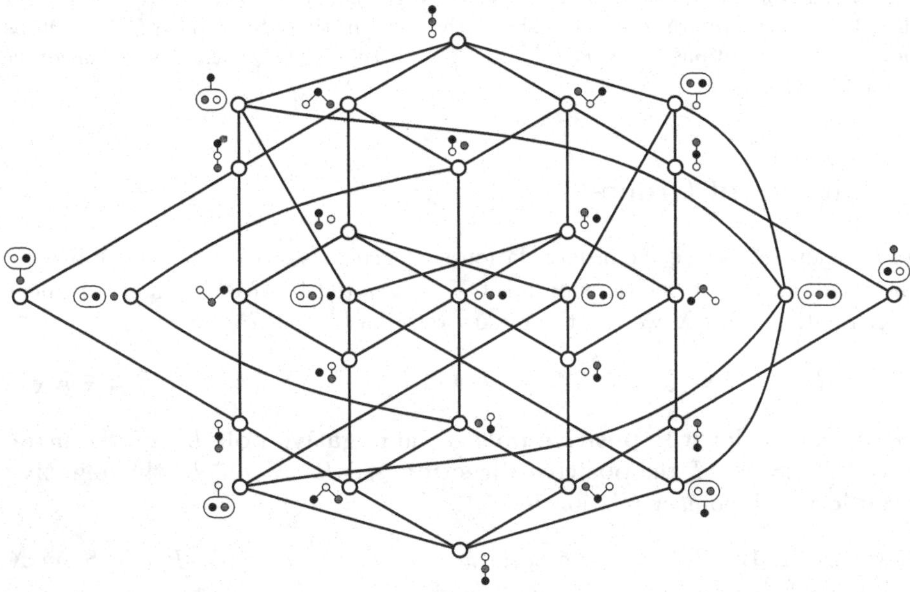

Fig. 5. The lattice of quasiorders on a three element set. (Color figure online)

Theorem 1. *Let L be a linear order relation on a set X. Then the following statements hold:*

(i) The lattice of quasiorder relations $(\mathcal{Q}_X, \sqsubseteq_L)$ forms a complete lattice with top element L and bottom element L^d. Supremum and infimum of a nonempty family $(R_t)_{t\in T}$ of quasiorder relations are given by[2]

$$\bigvee_{t\in T} R_t = \Big(\bigcup_{t\in T} R_t^+\Big)^\circ \blacktriangleright \bigcap_{t\in T} R_t^- \quad \text{and} \quad \bigwedge_{t\in T} R_t = \Big(\bigcup_{t\in T} R_t^-\Big)^\circ \blacktriangleright \bigcap_{t\in T} R_t^+,$$

where $R^+ = R\cap L$ is the positive direction and $R^- = R\cap L^d$ is the negative direction of $R \in \mathcal{Q}_X$. Furthermore, one receives the dual order $\sqsupseteq_L$ of $\sqsubseteq_L$ by replacing L by its dual L^d, i.e. $\sqsupseteq_L = \sqsubseteq_{L^d}$.

(ii) The lattice of orders is a complete sublattice of the lattice of quasiorders.

(iii) The lattice of linear orders is a complete sublattice of the lattice of orders.

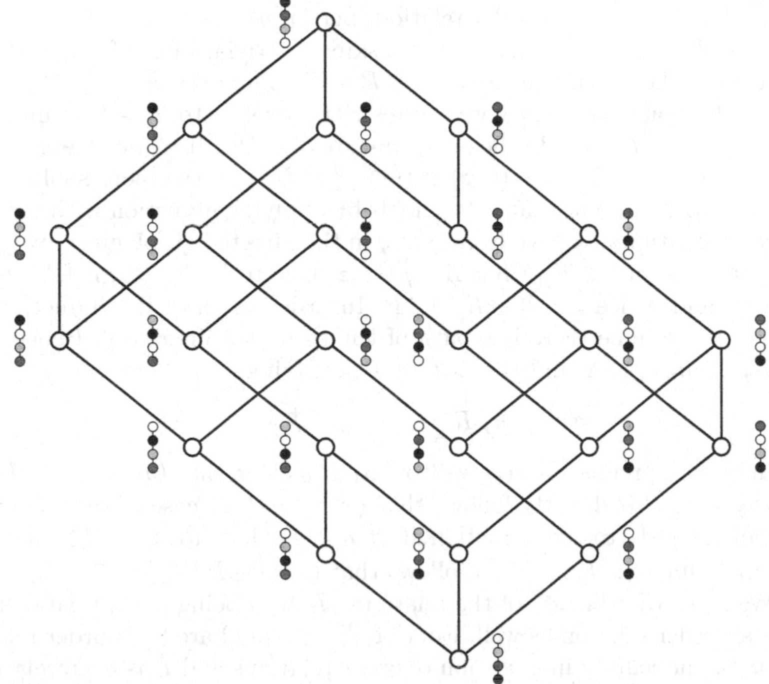

Fig. 6. The lattice of linear orders on a four element set. In the finite case, the lattice of linear orders is isomorphic to the lattice of permutations from [2]. The highly symmetric graph of the line diagram forms a permutahedron. (Color figure online)

[2] Please remember that $(\cdot)^\circ$ denotes the transitive closure and that $\blacktriangleright$ denotes the q-saturation introduced in Definition 1.

Proof. (i) It is obvious that L is the top element and that L^d is the bottom element of the ordered set $(\mathcal{Q}_X, \sqsubseteq_L)$. Hence, we can restrict ourselves to showing that suprema and infima exist for every non-empty selection of quasiorder relations on X. Furthermore, by the mentioned duality $\sqsupseteq_L = \sqsubseteq_{L^d}$, which is also obvious, we can restrict ourselves to the case of suprema. For the given non-empty family $(R_t)_{t\in T}$ we put $P := (\bigcup_t R_t^+)^\circ$ and $Q := \bigcap_t R_t^-$. Clearly P and Q are both quasiorder relations. By Lemma 1 (i) $S := P \blacktriangleright Q$ is a quasiorder relation with $S \cap L = P$ and $S \cap L^d = P \rhd Q$. For every $s \in T$ it now follows that

$$\begin{aligned} R_s \cap L &= R_s^+ \subseteq P = S \cap L \quad \text{and} \\ R_s \cap L^d &= R_s^- \supseteq Q \supseteq P \rhd Q = S \cap L^d. \end{aligned}$$

Hence, we have $R_s \sqsubseteq_L S$ for every $s \in T$, which makes S an upper bound of the family $(R_t)_{t\in T}$. Let U be another upper bound of the given family. We now have to show that $S \sqsubseteq_L U$. Hence, we have to show that (a) $P \subseteq U^+$ and that (b) $U^- \subseteq P \rhd Q$. Property (a) easily follows from $\bigcup_t R_t^+ \subseteq U^+$, which directly yields $P \subseteq (U^+)^\circ = U^+$. In order to prove the second property (b) we simply show that $P \cup U^-$ is a quasiorder relation. Let $(x,y),(y,z) \in P \cup U^-$. We show that $(x,z) \in P \cup U^-$. In doing so, we can skip the trivial cases of $x\ P\ y\ P\ z$ and $x\ U^-\ y\ U^-\ z$. The remaining cases (I) $x\ P\ y\ U^-\ z$ and (II) $x\ U^-\ y\ P\ z$ can be shown in a dual manner. Hence, we can restrict ourselves to show that in case (I) we receive $(x,z) \in P \cup U^-$. In the easy sub-case of $x\ L^d\ z$ it follows from $P \subseteq U^+$ that $x\ U^+\ y\ U^-\ z$ and hence $(x,z) \in U \cap L^d = U^-$. In the more sophisticated sub-case of $x\ L\ z$ we show that $x\ P\ z$ holds by applying induction with regard to the shortest length n of a path from x to y in the directed graph given by $\bigcup_t R_t^+$. If $n = 1$, there is an $s \in T$ with $x\ R_s^+\ y\ U^-\ z$. Due to $U^- \subseteq R_s^-$, it follows that $(x,z) \in R_s$ and hence $(x,z) \in R_s^+ \subseteq P$. In order to show the induction step $n \to n+1$, we assume there is a path of length n+1 from x to y. Hence, there are $a_0, a_1, \ldots, a_{n+1} \in X$ and $t_1, \ldots, t_n \in T$ satisfying

$$x = a_0\ R_{t_1}^+\ a_1\ R_{t_2}^+\ a_2 \ldots a_n\ R_{t_n}^+\ a_{n+1} = y.$$

By the induction premise we receive from $a_1\ P\ y\ U^-\ z$ that $(a_1, z) \in P \cup U^-$. In case of $(a_1,z) \in P$ it directly follows that $(x,z) \in P$. In case of $(a_1, z) \in U^-$ it follows (analogously to the induction start $n = 1$) that $(a_1, z) \in R_{t_1}^-$ and hence $(x,z) \in R_{t_1}$. Due to $x\ L\ z$ it then follows that $(x,z) \in R_{t_1}^+ \subseteq P$.

(ii) We show that in case of the relations $(R_t)_{t\in T}$ being order relations, $S = P \blacktriangleright Q$ is an order relation as well. First of all, P and Q are both order relations since Q is the nonempty intersection of order relations and P is a subrelation of the order relation L. Since $P \blacktriangleright Q$ is an order relation iff $(P \rhd Q) \cap P^d = \Delta$, we show the latter. Let $(x,y) \in P \rhd Q$ with $(y,x) \in P$. By the definition of P there are $a_0, a_1, \ldots, a_n \in X$ and $t_1, \ldots, t_n \in T$ with

$$y = a_0\ R_{t_1}^+\ a_1\ R_{t_2}^+\ a_2 \ldots a_{n-1}\ R_{t_n}^+\ a_n = x.$$

For all $i, j \in \{0, \ldots, n\}$ it follows that $a_i\ P\ x$ and $y\ P\ a_j$ and hence, by Lemma 1 (i), $(a_i, a_j) \in P \cup Q$. Accordingly, for $i = 1, \ldots, n$ it follows that

$(a_{i-1}, a_i) \in R_{t_i}^+$ and $(a_i, a_{i-1}) \in Q \subseteq R_{t_i}^-$. Thus, it follows that $a_i \; R_{t_i} \; a_{i-1} \; R_{t_i} \; a_i$ and hence $a_{i-1} = a_i$. Summing up, it follows that $x = y$, which proves $(P \rhd Q) \cap P^d = \Delta$ and shows (by the equivalence from above) that $S = P \blacktriangleright Q$ is an order relation.

(iii) We show that in case of the relations $(R_t)_{t \in T}$ being linear order relations, $S = P \blacktriangleright Q$ is a linear order as well. We put $P_0 := \bigcup_{t \in T} R_t^+$ and $Q_0 := Q \setminus P^d$. From the linearity of the R_t it easily follows that $L = P \,\dot{\cup}\, Q_0^d$. We show that $S = P \cup Q_0$. The linearity of S then is a direct consequence of $L = P \,\dot{\cup}\, Q_0^d$. The inclusion $S \subseteq P \cup Q_0$ is a direct consequence of $P \subseteq S \subseteq P \cup Q$ (Lemma 1 (ii)) and the fact that S is an order relation (see (ii)) and is hence antisymmetric. In order to prove the dual inclusion $P \cup Q_0 \subseteq S$, we first show the preparing statement $P \circ Q_0 \subseteq P \cup Q_0$. We do this by showing that for every natural number $n \in \mathbb{N}$ the following inclusion holds:

$$P_0^n \circ Q_0 \subseteq P \cup Q_0. \qquad (*)$$

The latter is shown by induction over n. Let $a \; P^n \; b \; Q_0 \; c$. In case of $n = 1$ there is an $s \in T$ with $a \; R_s^+ \; b \; Q_0 \; c$. Since $Q_0 \subseteq Q = \bigcap_{t \in T} R_t^-$ it follows that $a \; R_s \; b \; R_s \; c$ and hence $a \; R_s \; c$. In case of $a \; L \; c$ it directly follows that $(a, c) \in R_s^+ \subseteq P$. In the opposite case of $a \; L^d \; c$ we distinguish between the trivial case of $(a, c) \in Q_0$ and the non-trivial case of $(a, c) \in L^d \setminus Q_0$. In the latter case it follows from $L^d = P^d \,\dot{\cup}\, Q_0$ that $c \; P \; a \; R_s^+ \; b$ and hence $(b, c) \in P^d$ which contradicts $(b, c) \in Q_0$. Hence, we have now shown that $(*)$ holds for $n = 1$. In order to prove the induction step we assume that $(*)$ holds for an $n \in \mathbb{N}$. Let $a_1, \ldots, a_{n+1} \in X$ and $t_1, \ldots, t_{n+1} \in T$ with

$$a_{n+1} \; R_{t_{n+1}}^+ \; a_n \; R_{t_n}^+ \; a_{n-1} \ldots a_1 \; R_{t_1}^+ \; b \; Q_0 \; c.$$

By the induction hypothesis it follows that $(a_n, c) \in P \cup Q_0$. In case of $(a_n, c) \in P$ it directly follows that $a_{n+1} \; R_{t_{n+1}}^+ \; a_n \; P \; c$ and hence $(a_{n+1}, c) \in P$. Furthermore, in the remaining case of $(a_n, c) \in Q_0$ we have $a_{n+1} \; R_{t_{n+1}}^+ \; a_n \; Q_0 \; c$, which matches the situation of $n = 1$ already handled above. Hence, we have shown that $P \circ Q_0 \subseteq P \cup Q_0$. Dually, one can show that $Q_0 \circ P \subseteq P \cup Q_0$. Together this yields

$$P \circ Q_0 \circ P \;\subseteq\; (P \cup Q_0) \circ P \;=\; (P \circ P) \cup (Q_0 \circ P) \;\subseteq\; P \cup Q_0.$$

Now we can finally show the remaining inclusion $P \cup Q_0 \subseteq S$. Due to $S \cap L = P$ it suffices to show that $Q_0 \subseteq P \blacktriangleright Q$. Let us assume the opposite. Then (by Lemma 1 (i)) there are $(a, b) \in Q_0$ and $x, y \in X$ with $x \; P \; a \; Q_0 \; b \; P \; y$ and $(x, y) \notin P \cup Q$. But this contradicts $P \circ Q_0 \circ P \subseteq P \cup Q_0$. □

In order to better understand the lattice of orders, one has to understand the conditions under which for two order relations $P \subseteq L$ and $Q \subseteq L^d$ the union $P \cup Q$ is an order relation again. Clearly the mapping $\varphi : R \mapsto (R^+, R^-)$ delivers an order-embedding from the lattice of (quasi-)orders into the direct product

$$\mathbb{D} \;:=\; (\mathcal{S}_L, \subseteq) \times (\mathcal{S}_{L^d}, \supseteq),$$

where $\mathcal{S}_L$ and $\mathcal{S}_{L^d}$ denote the sets of suborder relations of L and of L^d, respectively. Accordingly, in this way the lattice of orders can be viewed as a sub-poset of the complete lattice $\mathbb{D}$. But φ does not preserve infima and suprema. Otherwise the lattice of orders would have been (isomorphic to) a subdirect product of $(\mathcal{S}_L, \subseteq)$ and $(\mathcal{S}_{L^d}, \supseteq)$. The fact that φ does not preserve infima or suprema can be seen in Fig. 3 showing the lattice of orders on a three-element set: In the example, the linear orders

1 3 2 and 2 1 3

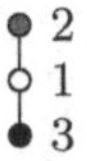

both contain the pair $(3,1) \in L^d$. Accordingly, the pair $(3,1)$ belongs to the intersection of the negative directions of the two relations. Hence, if φ was supremum-preserving, $(3,1)$ should belong to the supremum of the two relations. But it does not, since their supremum is the positive pole.

If one takes a look at the definition of the supremum in the lattice of orders, the previous observation is not very surprising. Whereas on the one hand, the positive direction of the supremum just depends on the positive directions R_t^+ of the input relations, the negative direction of the supremum, on the other hand, depends on both, the positive and the negative directions of the input relations. In other words, the negative direction of the supremum does not solely depend on the negative directions of the input relations and hence can not be deduced as a simple intersection, which is, of course, the supremum in $(\mathcal{S}_{L^d}, \supseteq)$.

Statement (iii) of Theorem 1 can be translated into the following: the orders of order dimension ≤ 1 form a complete sublattice of the lattice of orders. This gives rise to the question as to whether the statement holds true for other boundaries of the order dimension. The answer is no, as the following simple consideration shows: Let P be an order relation of order dimension $k > 2$ and let L be an arbitrary linear extension of L. If one chooses L to be the positive pole, P is the supremum of the trivial order relations $\Delta \cup \{(x,y)\}$ with $(x,y) \in P$. Accordingly, P is the supremum of order relations of order dimension 2.

In order to describe the covering relation of the lattice of orders, we need to recall some notions about closure systems. Let $\mathcal{C}$ be a closure system on a set A and let $\lceil\cdot\rceil : \mathfrak{P}(A) \to \mathfrak{P}(A)$ be the corresponding closure operator. The closure system $\mathcal{C}$ is called a **convex geometry** if for every closed set $C \in \mathcal{C}$ and for all points $a, b \in A \setminus C$ the following condition, called the **anti-exchange-property**, holds:

$$a \in \lceil C \cup \{b\}\rceil \quad \text{and} \quad b \in \lceil C \cup \{a\}\rceil \implies a = b. \tag{AXP}$$

A subset $B \subseteq Y$ is called a **base** of $Y \subseteq A$ if $\lceil B\rceil = \lceil Y\rceil$ and $\lceil D\rceil \subsetneq \lceil Y\rceil$ for every $D \subsetneq B$. Furthermore, a point $a \in Y$ is called an **extremal** point of Y if $\lceil Y \setminus \{a\}\rceil \subsetneq \lceil Y\rceil$. The set of all extremal points of Y is denoted by $\mathrm{Extr}(Y)$. Moreover, A closed set $C \in \mathcal{C}$ is called a **copoint** to a point $a \in A$ if $a \notin C$ and $a \in D$ for every closed set $D \in \mathcal{C}$ with $C \subsetneq D$.

Proposition 2. *Let (X, P) be a poset. Then the following statements hold:*

(i) The closure system $\mathcal{S}_P$ of all suborders of P is a convex geometry and has the closure operator $R \mapsto \Delta \cup R^\circ$.
(ii) For every $R \subseteq P$ it holds that $\mathrm{Extr}(R) = \mathrm{Cover}(\Delta \cup R^\circ)$.
(iii) Let B be a base of $R \subseteq P$. Then B is unique with $B = \mathrm{Cover}(\Delta \cup R^\circ)$.
(iv) Let Q_1 and Q_2 be suborders of P with $Q_1 \subseteq Q_2$. Then Q_1 is covered by Q_2 in $(\mathcal{S}_P, \subseteq)$ iff $|Q_2 \setminus Q_1| = 1$.
(v) Every copoint in $\mathcal{S}_P$ is copoint to exactly one point, i.e, if $Q \in \mathcal{S}_P$ is copoint to both $(a, b) \in P$ and to $(c, d) \in P$ it follows that $(a, b) = (c, d)$.

Proof. The statements are well-known and straightforward to show. Moreover, (iii) to (v) are consequences of $\mathcal{S}_P$ being a convex geometry; see for instance [9,10] □

Corollary 1. *In the lattice of orders $(\mathcal{P}_X, \sqsubseteq_L)$ two order relations P and Q are neighboured iff they differ in exactly one pair, i.e., iff $|P \triangle Q| = 1$.*

Proof. We start the proof with the backwards direction. Let $|P \triangle Q| = 1$. Then there is a pair $(a, b) \in X^2$ with either (I) $P = Q \cup \{(a, b)\}$ or (II) $Q = P \cup \{(a, b)\}$. In case (I) Q is the lower cover of P if $(a, b) \in L$ and Q is the upper cover of P if $(a, b) \in L^d$. And clearly in the dual case (II) P and Q are neighbours as well. In order to show the forward direction of the equivalence we assume that P is the lower cover of Q in $(\mathcal{L}_X, \sqsubseteq_L)$ and that $|P \triangle Q| > 1$. Since $P \sqsubseteq_L Q$ it follows that $P \sqsubseteq_L P \cap Q \sqsubseteq_L Q$. In case of P and Q being incomparable as subsets (i.e. $P \nsubseteq Q$ and $Q \nsubseteq P$) it even follows that[3] $P \sqsubset_L P \cap Q \sqsubset_L Q$, which contradicts P and Q being neighbours. Hence, it remains to show that the two remaining cases of $P \subseteq Q$ and $Q \subseteq P$ also lead to contradictions as well. In case of $P \subseteq Q$ it follows that $|Q \setminus P| > 1$. Hence, by Proposition 2 (iv) P and Q are no neighbours in $(\mathcal{S}_Q, \subseteq)$. Thus, there is a suborder relation R of Q with $P \subsetneq R \subsetneq Q$. But due to $P \sqsubseteq_L Q$ it follows that $P^- = R^- = Q^-$ and hence $P^+ \subsetneq R^+ \subsetneq Q^+$. The latter implies that $P \sqsubset_L R \sqsubset_L Q$, which contradicts P and Q being neighbours in the lattice of orders. Dually, the final remaining case $Q \subseteq P$ also leads to a contradiction. □

We want to point out that the previous Corollary 1 cannot be transferred to the lattice of quasiorders: As one can see in Fig. 5 there are neighboured quasiorder relations that differ in more than one pair.

Corollary 2. *The lattice of orders is a cover-preserving sublattice of the lattice of quasiorders, i.e., two order relations are neighboured in the lattice of orders iff they are neighboured in the lattice of quasiorders.*

Proof. Direct consequence of Corollary 1. □

---

[3] We write $P \sqsubset_L Q$ if $P \sqsubseteq_L Q$ and $P \neq Q$.

Corollary 1 helps us to understand the undirected graph that is described by the line diagram of the lattice of orders. As one can see in Fig. 3, one receives the line diagrams of the lattices of orders that result from different choices of the positive pole L by just rotating the depicted line diagram. In other words, the line diagram is the diagram for a multi-lattice structure. In the inside there is the discrete order Δ surrounded by the linear order relations on the outside.

M

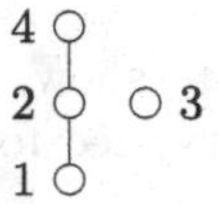

	1	2	3	4
1	×	×		×
2		×		×
3			×	
4				×

N

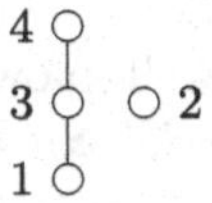

	1	2	3	4
1	×		×	×
2		×		
3			×	×
4				×

Fig. 7. Example of two order relations that are incomparable in the lattice of orders, regardless of the choice of the positive pole L.

One question that occurs in this context is the following: Given two order relations P and Q, is there always a linear order relation L satisfying $P \sqsubseteq_L Q$? In other words, is it always possible to choose a positive pole L such that two given partial order relations are comparable in the resulting lattice of orders? As the following example shows, the answer to this question is no:

Example 2. Let M and N be the order relations depicted in Fig. 7. Suppose there is a linear order relation L on the base set $X = \{1, 2, 3, 4\}$ with $M \sqsubseteq_L N$. The condition $M \cap L \subseteq N$ can be translated into $M \setminus N \subseteq L^{\complement}$. Since L is linear, it follows that $M \setminus N \subseteq L^d$. Dually, the condition $N \cap L^d \subseteq M$ yields $N \setminus M \subseteq L$. Since $(3, 4) \in N \setminus M$ and $(2, 4) \in M \setminus N$, it directly follows that $3 \; L \; 4 \; L \; 2$ and hence $3 \; L \; 2$. But since we also have $(1, 2) \in M \setminus N$ and $(1, 3) \in N \setminus M$, it furthermore follows that $2 \; L \; 1 \; L \; 3$ and hence $2 \; L \; 3$, which contradicts the already derived $3 \; L \; 2$. Thus, in our example there is no linear order L with $M \sqsubseteq_L N$.

4 Contextual Representation

Facing a class of complete lattices raises the question of how the class can be described by a of formal context construction. In this section we answer this question for the class of the lattices of orders. The most fundamental question that has to be answered in this regard is of course: What are the join- and the meet-irreducible elements of the lattice of orders? As it turns out, the answer requires an understanding of the class of orders introduced in the following definition. The basic idea is to take a linear order, choose an interval and reorder the elements inside the interval:

Definition 3. *We put* $V := \{\downarrow, \bullet, \uparrow\}$ *and define the relation* $\leq_V$ *to be the linear order relation on* V *given by* $\downarrow <_V \bullet <_V \uparrow$. *Let* L *be a linear order relation on a set* X, *let* $(x, y) \in L$ *and let* $\varphi : [x, y]_L \to V$. *We then define* $L_{|(x,y),\varphi}$ *to be the binary relation on* X *given by*

$$(g, h) \in L_{|(x,y),\varphi} \;:\Longleftrightarrow\; \begin{cases} (\{g, h\} \nsubseteq [x, y]_L \text{ and } g\, L\, h) & \text{or} \\ (\{g, h\} \subseteq [x, y]_L \text{ and } g\, L\, h \text{ and } \varphi g \leq_V \varphi h) & \text{or} \\ (\{g, h\} \subseteq [x, y]_L \text{ and } (\varphi g, \varphi h) = (\downarrow, \uparrow)). \end{cases}$$

The mapping φ *is called a* ***reordering*** *of* $[x, y]_L$. *Furthermore,* φ *is called a* ***proper*** *reordering if* $\varphi x = \uparrow$ *and* $\varphi y = \downarrow$. *It is called a* ***linear*** *reordering if it is proper and fulfils* $\varphi g \neq \bullet$ *for every* $g \in [x, y]_L$ *(Fig. 8).*

Proposition 3. *Let* L *be a linear order relation on a set* X, *let* $(x, y), (v, w) \in L_{\neq}$ *and let* $\varphi : [x, y]_L \to V$ *and* $\psi : [v, w] \to V$. *Then the following statements hold:*

(i) $L_{|(x,y),\varphi}$ *is a partial order relation.*
(ii) φ *is a proper reordering iff* $L^d \cap \mathrm{Cover}(L_{|(x,y),\varphi}) = \{(y, x)\}$.
(iii) If φ *is proper, then* $L_{|(x,y),\varphi}$ *is a linear order iff* φ *is a linear reordering.*
(iv) If φ *and* ψ *are proper reorderings, the following equivalence holds:*

$$[x, y]_L \subseteq [v, w]_L \text{ and } \varphi = \psi|_{[x,y]_L} \iff L_{|(v,w),\psi} \sqsubseteq_L L_{|(x,y),\varphi}.$$

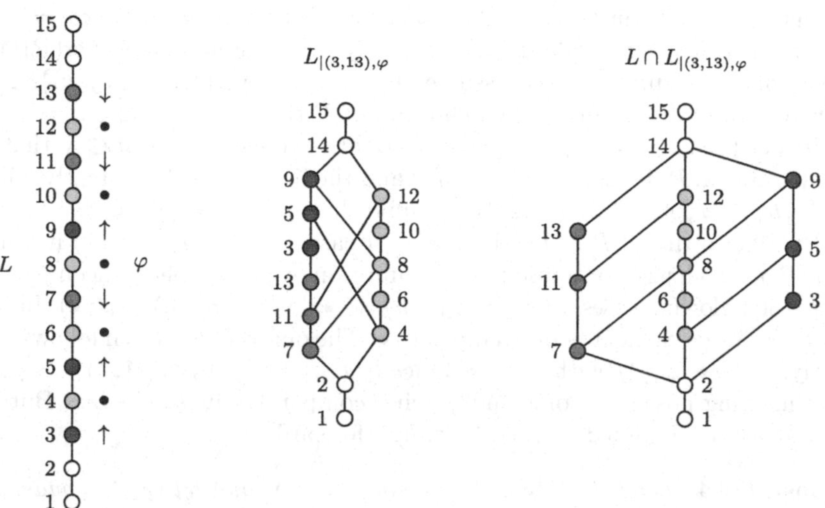

Fig. 8. An illustrating example to Definition 3 and to Proposition 3. (Color figure online)

Proof. We put $R := L_{|(x,y),\varphi}$. (i) Clearly, R is reflexive. Let $(g,h) \in R$ and $(h,g) \in R$. We show that $g = h$. Since L is linear we can w.l.o.g. assume that $g\ L\ h$. In case of $\{g,h\} \nsubseteq [x,y]_L$ it directly follows that $g\ L\ h\ L\ g$ and hence $g = h$. In case of $\{g,h\} \subseteq [x,y]_L$ it follows from $\downarrow\ \leq_V\ \uparrow$ that $\varphi g \leq_V \varphi h \leq_V \varphi g$ and hence $\varphi g = \varphi h$. But this again leads to $g\ L\ h\ L\ g$ and hence to $g = h$. It remains to show the transitivity of R. Let $(g,h) \in R$ and $(h,i) \in R$. We show that $g\ R\ i$ in a case-by-case analysis. Case (I) $\{g,h\} \nsubseteq [x,y]_L$ and $\{h,i\} \nsubseteq [x,y]_L$: It directly follows that $g\ L\ h\ L\ i$ and hence $g\ L\ i$. It furthermore follows that $\{g,i\} \nsubseteq [x,y]_L$ (since otherwise $\{g,h,i\} \subseteq [g,i]_L \subseteq [x,y]_L$) and hence $g\ R\ i$. Case (II) $\{g,h\} \nsubseteq [x,y]_L$ and $\{h,i\} \subseteq [x,y]_L$: It follows that $g \notin [x,y]_L$ and hence $\{g,i\} \nsubseteq [x,y]_L$. Furthermore, it follows that $g\ L\ h\ L\ y$ and hence $g\ L\ y$. Due to $g \notin [x,y]_L$ it follows that $(x,g) \notin L$ and hence $g\ L\ x\ L\ i$, which leads to $g\ L\ i$ and hence to $g\ R\ i$. Case (III) $\{g,h\} \subseteq [x,y]_L$ and $\{h,i\} \nsubseteq [x,y]_L$: analogous to case (II). Case (IV) $\{g,h,i\} \subseteq [x,y]_L$: In case of $\varphi g \leq_V \varphi h \leq_V \varphi i$ and $g\ L\ h\ L\ i$ it directly follows that $\varphi g \leq_V \varphi i$ as well as $g\ L\ i$ and hence $g\ R\ i$. In case of $\varphi g \leq_V \varphi h$ and $(\varphi h, \varphi i) = (\downarrow,\uparrow)$ it directly follows that $(\varphi g, \varphi i) = (\downarrow,\uparrow)$ and hence $g\ R\ i$. Analogously, the case $\varphi h \leq_V \varphi i$ and $(\varphi g, \varphi h) = (\downarrow,\uparrow)$ leads to $(\varphi g, \varphi i) = (\downarrow,\uparrow)$ and hence to $g\ R\ i$. The remaining case of $(\varphi h, \varphi i) = (\downarrow,\uparrow) = (\varphi g, \varphi h)$ obviously leads to a contradiction.

(ii) Since x and y are the bottom and the top element in the interval $[x,y]_L$, x is the bottom element of $\varphi^{-1}(\varphi x)$ and, dually, y is the top element of $\varphi^{-1}(\varphi y)$. In case of $(\varphi x, \varphi y) = (\uparrow,\downarrow)$ it directly follows that $y\ R\ x$ and, furthermore, that every $h \in [x,y]_L$ with $y\ R_{\neq}\ h\ R_{\neq}\ x$ has to be mapped to $\bullet$ (by φ, of course). But it then follows that $y\ L\ h\ L\ x$, which leads to the contradiction $x = y$. Hence, there is no element h between y and x, making y the lower cover of x w.r.t. R. And since for every pair $(v,w) \in L^d \cap R_{\neq}$ it follows that $\varphi v = \downarrow$ and $\varphi w = \uparrow$, one receives $v\ R\ y\ R\ x\ R\ w$, making (y,x) the only pair contained in $\mathrm{Cover}(R) \cap L^d$. For the opposite direction, we assume that $L^d \cap \mathrm{Cover}(R) = \{(y,x)\}$. From $y\ R\ x$ and $(y,x) \notin L$ it follows by definition of R that $\varphi y = \downarrow$ and $\varphi x = \uparrow$.

(iii) Let $(\varphi x, \varphi y) = (\uparrow,\downarrow)$. Suppose there are two elements $g,h \in X$ that are incomparable in R. By the definition of R and the linearity of L it directly follows that $\{g,h\} \subseteq [x,y]_L$. W.l.o.g we can assume that $g\ L\ h$. In case of $\varphi g \leq_V \varphi h$ it directly follows that $g\ R\ h$. Furthermore, in case of $(\varphi g, \varphi h) = (\uparrow,\downarrow)$ it follows that $h\ R\ g$, which also contradicts the assumption of g and h being incomparable. Hence, the following cases remain: $(\varphi g, \varphi h) = (\bullet,\downarrow)$ or $(\varphi g, \varphi h) = (\uparrow,\bullet)$. In both cases there is one element being mapped to $\bullet$. The opposite direction follows more directly: Let $g \in [x,y]_L$ with $\varphi g = \bullet$. Since $(\varphi x, \varphi g) = (\uparrow,\bullet) \notin \{(\uparrow,\downarrow)\} \cup \leq_V$, the only remaining possibility of x and g being comparable in R is $g\ L\ x$. But this yields the contradiction $x = g$. (iv): Straightforward. □

Proposition 4. *Let L be a linear order on a set X and let Q be a quasiorder on X. Then for $(x,y) \in L_{\neq}$ the following equivalences hold:*

(i) $(y,x) \in Q \setminus Q^d \iff \exists$ *proper reordering* φ *of* $[x,y]_L$ *with* $Q \sqsubseteq_L L_{|(x,y),\varphi}$.
(ii) $(x,y) \in Q \setminus Q^d \iff \exists$ *proper reordering* φ *of* $[x,y]_L$ *with* $(L_{|(x,y),\varphi})^d \sqsubseteq_L Q$.

(iii) $(x,y) \notin Q \iff \exists$ *linear reordering* ψ *of* $[x,y]_L$ *with* $Q \sqsubseteq_L L \cap L_{|(x,y),\psi}$.
(iv) $(y,x) \notin Q \iff \exists$ *linear reordering* ψ *of* $[x,y]_L$ *with* $(L \cap L_{|(x,y),\psi})^d \sqsubseteq_L Q$.

Proof. We just show statements (i) and (iii), since (ii) und (iv) follow by duality. The backwards directions of (i) and (iii) are direct consequences of the previous Proposition 3 (ii). For the forward directions we define the two reorderings φ and ψ of $[x,y]_L$ by

$$\varphi g := \begin{cases} \uparrow & \text{, if } x\ Q\ g, \\ \downarrow & \text{, if } g\ Q\ y, \\ \bullet & \text{, otherwise.} \end{cases} \qquad \text{and} \qquad \psi g := \begin{cases} \uparrow & \text{, if } x\ Q\ g, \\ \downarrow & \text{, otherwise.} \end{cases}$$

Since in both cases (i) and (iii), it holds that $(x,y) \notin Q$, both reorderings are well-defined and proper. Let $(g,h) \in L^d \cap L_{|(x,y),\varphi}$. By the definition of $L_{|(x,y),\varphi}$ it follows that $\varphi g = \downarrow$ and $\varphi h = \uparrow$, which yields $g\ Q\ y\ Q\ x\ Q\ h$, which again yields $(g,h) \in Q$. Let, on the other hand, $(g,h) \in L \cap Q$ and suppose that $(g,h) \notin L_{|(x,y),\varphi}$. It then follows that $\{g,h\} \subseteq [x,y]_L$ and $\varphi g \not\leq_V \varphi h$, which leads to the two cases $\varphi g = \uparrow$ or $\varphi h = \downarrow$. In the first case it follows that $x\ Q\ g$, which yields $x\ Q\ h$, which again translates into $\varphi h = \uparrow$, a contradiction to $\varphi g \not\leq_V \varphi h$. Analogously, the remaining case $\varphi h = \downarrow$ leads to a contradiction. Hence, we have shown (i) and it remains to show the forward direction of (iii): It suffices to show that $Q \cap L \subseteq L_{|(x,y),\psi}$. Let $(g,h) \in Q \cap L$ and assume that $(g,h) \notin L_{|(x,y),\psi}$. It follows that $\{g,h\} \subseteq [x,y]_L$ and $\psi g \not\leq_V \psi h$. But this yields $\psi g = \uparrow$ and $\psi h = \downarrow$, which translates into $x\ Q\ g$ and $(x,h) \notin Q$, a contradiction to $g\ Q\ h$. □

Lemma 2. *Let L be a linear order relation on a set X. Then for a given order relation P on X the following statements are equivalent:*

(a) P is meet-irreducible in the lattice of orders $(\mathcal{P}_X, \sqsubseteq_L)$.
(b) P^d is join-irreducible in the lattice of orders $(\mathcal{P}_X, \sqsubseteq_L)$.
(c) There is a pair $(x,y) \in L_{\neq}$ and a proper reordering φ of $[x,y]_L$ with either

$$(\alpha)\ P = L \cap L_{|(x,y),\varphi} \text{ and } \varphi \text{ linear} \qquad \text{or} \qquad (\beta)\ P = L_{|(x,y),\varphi}.$$

Furthermore, for a quasiorder relation Q the following statements are equivalent:

(a^*) Q is meet-irreducible in the lattice of quasiorders $(\mathcal{Q}_X, \sqsubseteq_L)$.
(b^*) Q^d is join-irreducible in the lattice of quasiorders $(\mathcal{Q}_X, \sqsubseteq_L)$.
(c^*) There is a pair $(x,y) \in L_{\neq}$ and a proper reordering φ of $[x,y]_L$ with either

$$(\alpha^*)\ Q = L \cap L_{|(x,y),\varphi} \text{ and } \varphi \text{ linear} \qquad \text{or} \qquad (\beta^*)\ Q = L_{|(x,y),\varphi} \cup \{(x,y)\}.$$

Proof. The equivalences (a)↔(b) and (a*)↔(b*) each follow by duality. (a)→(c): By premise P has a unique upper cover R. By Corollary 1 there is a pair $(x,y) \in L_{\neq}$ with either (I) $R = P \mathbin{\dot\cup} \{(x,y)\}$ or (II) $P = R \mathbin{\dot\cup} \{(y,x)\}$. In Case (I) by Proposition 4 (iii) there is a linear reordering φ of $[x,y]_L$ with $P \sqsubseteq_L L \cap L_{|(x,y),\varphi}$.

Suppose that $P \sqsubset_L L \cap L_{|(x,y),\varphi}$. By premise, it then follows that $R \sqsubseteq_L L \cap L_{|(x,y),\varphi}$ and hence $(x,y) \in L_{|(x,y),\varphi}$, a contradiction to Proposition 3 (i) and (ii). In Case (II) by Proposition 4 (i) there is a proper reordering φ of $[x,y]_L$ with $P \sqsubseteq_L L_{|(x,y),\varphi}$. Suppose that $P \sqsubset_L L_{|(x,y),\varphi}$. It then follows that $R \sqsubseteq_L L_{|(x,y),\varphi}$ and hence $(y,x) \notin L_{|(x,y),\varphi}$, a contradiction to Proposition 3 (ii).

(c)→(a): We prove that both cases (α) and (β) imply (a). Case (α): Let R be an upper cover of $P = L \cap L_{|(x,y),\varphi}$. Since $P \subseteq L$, by Corollary 1 there has to be a pair $(g,h) \in L \setminus P$ with $R = P \mathbin{\dot\cup} \{(g,h)\}$. It follows that $(g,h) \notin L_{|(x,y),\varphi}$ and hence $\{g,h\} \subseteq [x,y]_L$ and $\varphi g \not\leq_V \varphi h$. Since by ($\alpha$) the map φ does not map onto $\bullet$, it follows that $\varphi g = \uparrow$ and $\varphi h = \downarrow$. Since $\varphi g = \uparrow$ and $\varphi h = \downarrow$ as well as $x\ L\ g$ and $h\ L\ y$, it follows that $x\ P\ g\ R\ h\ P\ y$ and thus $x\ R\ y$. Since by Proposition 3 (ii), the pair (x,y) cannot belong to P, it follows that $(g,h) = (x,y)$. Accordingly, P can have at most one upper cover. And since for all $(v,x) \in P$ and $(y,w) \in P$ it follows that $(v,w) = (x,y)$ or $\{v,w\} \nsubseteq [x,y]_L$ and $v\ L\ w$. Thus, $R = P \cup {}^P x \times y^P$ is indeed an order relation and hence the only upper cover of P. Case (β): By Proposition 3 (ii) it follows that $L^d \cap \mathrm{Cover}(L_{|(x,y),\varphi}) = \{(y,x)\}$. Accordingly, $P \setminus \{(y,x)\}$ is an order relation and hence an upper cover of P. Furthermore, it follows that every other upper cover has to be of the form $R = P \cup \{(g,h)\}$ with $(g,h) \in L \setminus P$. Since $P = L_{|(x,y),\varphi}$ it follows that $\{g,h\} \subseteq [x,y]_L$ and $\varphi g \not\leq_V \varphi h$, leading to the following three cases of $(\varphi g, \varphi h) \in \{(\bullet,\downarrow),(\uparrow,\bullet),(\uparrow,\downarrow)\}$. In case of $(\varphi g, \varphi h) = (\bullet,\downarrow)$ it follows that $h\ P\ x$ and hence $g\ R\ h\ R\ x$. Since R is an order relation it follows that $g\ R\ x$, which yields either $x = h$ or $g\ P\ x$. Since $x \neq h$ (different images under φ) it follows that $g\ P\ x$. From $\varphi g = \bullet$ it then follows that $g\ L\ x$ and hence $g = x$, a contradiction to $\varphi x = \uparrow$. Analogously, the case $(\varphi g, \varphi h) = (\uparrow,\bullet)$ leads to a contradiction. In the remaining case $(\varphi g, \varphi h) = (\uparrow,\downarrow)$ it directly follows that $(h,g) \in P$, which contradicts $P \cup \{(g,h)\}$ being an order. Accordingly, all three cases lead to contradictions, which proves $P \setminus \{(y,x)\}$ being the only upper cover of P.

(a*)→(c*): The characterisation of the meet-irreducible elements of the lattice of quasiorders can now be deduced from the above. For every meet-irreducible quasiorder in $(\mathcal{Q}_X, \sqsubseteq_L)$ the following implication holds for $(g,h),(v,w) \in L_{\neq}$:

$$\{g,h\}^2 \subseteq Q \text{ and } \{v,w\}^2 \subseteq Q \implies (g,h) = (v,w). \qquad (**)$$

If the implication is not valid, one could easily construct two different upper covers of Q (by leaving away pairs from the negative direction of Q). As a consequence, if Q is meet-irreducible, then Q is an order relation or there is a unique pair $(g,h) \in L_{\neq}$ with $(g,h),(h,g) \in Q$ such that $Q \setminus \{(h,g)\}$ is an order relation. By Corollary 2, in case of Q being an order relation, Q is meet-irreducible not just in the lattice of quasiorders, but also in the lattice of orders, which leaves the two cases of (I) $Q \subseteq L$ and (II) $Q \nsubseteq L$. In Case (I) Q is indeed meet-irreducible in the lattice of quasiorders (since then every quasiorder S with $Q \sqsubseteq_L S$ is a suborder relation of L), whereas in Case (II) Q has two upper covers (namely $L_{|(x,y),\varphi} \cup \{(x,y)\}$ and $L_{|(x,y),\varphi} \setminus \{(y,x)\}$) and is hence not meet-irreducible in $(\mathcal{Q}_X, \sqsubseteq_L)$. Accordingly, in case of Q being an order relation, statement (α^*) follows. Let now Q be a proper quasiorder relation, i.e., not an order relation. By $(**)$ there is a unique pair $(g,h) \in L_{\neq}$ with $\{g,h\}^2 \subseteq Q$. Accordingly, the relations $P_l := Q \setminus \{(g,h)\}$, $P_m := Q \setminus \{(g,h),(h,g)\}$ and $P_u := Q \setminus \{(h,g)\}$ are order relations, which, together with Q, form the following cover-preserving sub-poset of the lattice of quasiorders:

$P_u = Q \setminus \{(h,g)\}$

Q $\quad$ $P_m = Q \setminus \{(g,h),(h,g)\}$

$P_l = Q \setminus \{(g,h)\}$

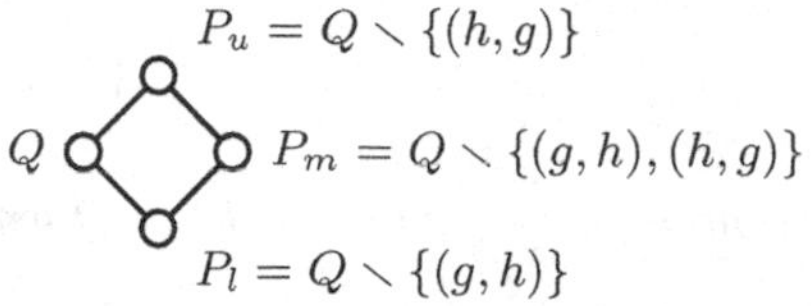

Since every upper cover U of P_l in $(\mathcal{P}_X, \sqsubseteq_L)$ yields an upper cover $U \cup \{(g,h)\}$ of Q in $(\mathcal{Q}_X, \sqsubseteq_L)$, it follows that P_l is meet-irreducible in $(\mathcal{P}_X, \sqsubseteq_L)$. Since $P_l \nsubseteq L$, it follows that (β) has to hold for P_l, which translates to (β^*) being valid for Q. (c*)→(a*): The case of (α^*)→(a*) directly follows from (α)→(a). Moreover, (β^*)→(a*) follows from (β)→(a), since in case of $Q \nsubseteq L$, upper covers of Q in $(\mathcal{Q}_X, \sqsubseteq_L)$ canonically correspond to upper covers of $Q \setminus \{(x,y)\}$ in $(\mathcal{P}_X, \sqsubseteq_L)$. □

In the finite case one can w.l.o.g. assume that $X = \{1, \ldots, n\}$ and that the positive pole L is the order induced by the canonic order on natural numbers. By the previous Lemma 2, for a given pair $x < y$ there are 2^{y-x} meet-irreducible order relations P with $P \cup \{(x,y)\}$ being the unique upper cover of P and there are 3^{y-x} meet-irreducible order relations Q with $Q \setminus \{(y,x)\}$ being the unique upper cover of Q. Accordingly, if one now wants to count the meet-irreducibles in the resulting lattice of orders one has to count zthe numbers in the following $(n \times n)$-square:

	1	2	3	4	…	$n-1$	n
1		1	2	4		2^{n-2}	2^{n-1}
2	1		1	2		2^{n-3}	2^{n-2}
3	3	1		1		2^{n-4}	2^{n-3}
4	9	3	1			2^{n-5}	2^{n-4}
⋮							
$n-1$	3^{n-2}	3^{n-3}	3^{n-4}	3^{n-5}			1
n	3^{n-1}	3^{n-2}	3^{n-3}	3^{n-4}		1	

Corollary 3. *The number of meet-irreducibles in the lattice of orders on an n-element set is*

$$\frac{2^{n+2} + 3^n - 6n - 5}{4}.$$

The same holds true for the lattice of quasiorders. And since both lattices are self-dual, the same number also holds for the join-irreducibles.

Proof. Follows from Lemma 2 as an application of finite geometric series. □

The following representation theorem of the lattices of orders is our second main result. An example is presented in Fig. 9.

Theorem 2. *Let L be a linear order relation on a set X. Let $V_\alpha := V \setminus \{\bullet\}$, let $V_\beta := V$ and let M_α and M_β be the sets of triples given by*

$$M_\tau := \Big\{ \big(\tau, (x,y), \varphi\big) \;\Big|\; (x,y) \in L_{\neq} \text{ and } \varphi : [x,y]_L \to V_\tau \text{ with } (\varphi x, \varphi y) = (\uparrow, \downarrow) \Big\},$$

where $\tau \in \{\alpha, \beta\}$. Furthermore, let $\mathbb{P}_L$ and $\mathbb{Q}_L$ be the formal contexts given by

$$\mathbb{P}_L := \begin{array}{c|c|c|} & M_\alpha & M_\beta \\ \hline M_\alpha & \times & I \\ \hline M_\beta & I^d & J \\ \hline \end{array} \quad \& \quad \mathbb{Q}_L := \begin{array}{c|c|c|} & M_\alpha & M_\beta \\ \hline M_\alpha & \times & I \\ \hline M_\beta & I^d & J \cup \Delta \\ \hline \end{array}$$

with $\times := M_\alpha \times M_\alpha$ being the full relation and with the other binary relations $I \subseteq M_\alpha \times M_\beta$ and $J \subseteq M_\beta \times M_\beta$ being defined by

$$\big(\alpha, (x,y), \varphi\big) \; I \; \big(\beta, (v,w), \psi\big) \;\; :\Longleftrightarrow \;\; \begin{cases} \text{there are no } i, j \in [x,y]_L \cap [v,w]_L \\ \text{with } i \; L \; j \text{ and} \\ (\varphi i, \varphi j) = (\uparrow, \downarrow) = (\psi i, \psi j), \end{cases}$$

$$\big(\beta, (x,y), \varphi\big) \; J \; \big(\beta, (v,w), \psi\big) \;\; :\Longleftrightarrow \;\; \begin{cases} \text{for all } i, j \in [x,y]_L \cap [v,w]_L \text{ with } i \; L \; j \\ \text{the following two implications hold:} \\ (\varphi i, \varphi j) = (\uparrow, \downarrow) \implies \psi i \leq_V \psi j \\ (\psi i, \psi j) = (\uparrow, \downarrow) \implies \varphi i \leq_V \varphi j. \end{cases}$$

Then the following statements holds:

(i) The concept lattice of $\mathbb{P}_L$ is isomorphic to the lattice of orders with positive pole L. An isomorphism $\Phi : \mathfrak{B}(\mathbb{P}_L) \to (\mathcal{P}_X, \sqsubseteq_L)$ is given by

$$\begin{aligned} \Phi(A,B) &:= \Delta \cup \pi_2\big[A \cap M_\beta\big] \cup \big(\pi_2\big[B \cap M_\beta\big]\big)^d \\ &= \Big(L \setminus \pi_2\big[B \cap M_\alpha\big]\Big) \cup \Big(L \setminus \pi_2\big[A \cap M_\alpha\big]\Big)^d, \end{aligned}$$

where π_2 denotes the projection into the second component, i.e., for a subset S of $M_\alpha \cup M_\beta$ it is $\pi_2[S] := \{(x,y) \mid (\tau, (x,y), \varphi) \in S\}$.

(ii) Both contexts, $\mathbb{P}_L$ and $\mathbb{Q}_L$, are reduced.

If, additionally, L fulfils the finite interval property, the following statements hold:

(iii) The concept lattice of $\mathbb{Q}_L$ is isomorphic to the lattice of quasiorders. An isomorphism $\Psi : \mathfrak{B}(\mathbb{Q}_L) \to (\mathcal{Q}_X, \sqsubseteq_L)$ is given by

$$\Psi(A,B) := \Big(L \setminus \pi_2[B \cap M_\alpha]\Big) \cup \Big(L \setminus \pi_2[A \cap M_\alpha]\Big)^d.$$

(iv) The incidence relation of $\mathbb{P}_L$ is a closed subrelation of the incidence of $\mathbb{Q}_L$.

Proof. (i) We put $M := M_\alpha \cup M_\beta$, denote the incidence relation of $\mathbb{P}_L$ by K and define $\tilde{\gamma} : M \to \mathcal{P}_X$ and $\tilde{\mu} : M \to \mathcal{P}_X$ by

$$\tilde{\gamma}(\cdot) := \tilde{\mu}^d(\cdot) \qquad \text{and} \qquad \tilde{\mu}(\tau,(x,y),\varphi) := \begin{cases} L_{|(x,y),\varphi} \cap L & \text{, if } \tau = \alpha, \\ L_{|(x,y),\varphi} & \text{, if } \tau = \beta. \end{cases}$$

We show that $\tilde{\mu}[M]$ is meet-dense in the lattice of orders. Let $P \in \mathcal{P}_X$. We put $\mathcal{M}_P := \{R \in \tilde{\mu}[M] \mid P \sqsubseteq_L R\}$. Then for every $R \in \mathcal{M}_P$ it follows that $P \cap L^d \supseteq R \cap L^d$ and $P \cap L \subseteq R \cap L$. By Proposition 4 (i) and (iii) it even follows that

$$P \cap L^d = \bigcup\{R \cap L^d \mid R \in \mathcal{M}_P\} \quad \text{and} \quad P \cap L = \bigcap\{R \cap L \mid R \in \mathcal{M}_P\},$$

which yields $P = \bigwedge \mathcal{M}_P$. Dually one shows that $\tilde{\gamma}[M]$ is join-dense. By the Basic Theorem on Concept Lattices [7] it remains to show that

$$\tilde{\gamma}(\tau,(x,y),\varphi) \sqsubseteq_L \tilde{\mu}(\nu,(v,w),\psi) \iff (\tau,(x,y),\varphi)\ K\ (\nu,(v,w),\psi),$$

which is an easy excercise. The fact that Φ is an isomorphism follows directly from Proposition 4.

(ii) By (i) and Lemma 2, $\mathbb{P}_L$ results from reducing the ordinal scale of the lattice of orders. Analogously, one can show that $\mathbb{Q}_L$ results from reducing the ordinal scale of the lattice of quasiorders.

(iii) We use the notions as in (i) with the only difference that for a proper reordering φ of $[x,y]_L$ we put $\tilde{\mu}(\beta,(x,y),\varphi) := L_{|(x,y),\varphi} \cup \{(x,y)\}$ (instead of just $L_{|(x,y),\varphi}$). We show that for a given $Q \in \mathcal{Q}_X$ and for $\mathcal{M}_Q := \{R \in \tilde{\mu}[M] \mid Q \sqsubseteq_L R\}$ it follows that $Q = \bigwedge \mathcal{M}_Q$. Analogously to (i), it directly follows that $Q \cap L = \bigcap\{R \cap L \mid R \in \mathcal{M}_Q\}$ and $Q \cap L^d \supseteq \bigcup\{R \cap L^d \mid R \in \mathcal{M}_Q\} =: S$ and hence $Q \cap L^d \supseteq S^\circ$. It remains to show that in the latter even $Q \cap L^d = S^\circ$ holds. Let $(y,x) \in Q \cap L^d$. In case of $(x,y) \notin Q$, the existence of a proper reordering of $[x,y]_L$ with $Q \sqsubseteq_L L_{|(x,y),\varphi}$ follows from Proposition 4 (i). So it remains the case of $\{x,y\}^2 \subseteq Q$. Since L has no infinite intervals, the same holds true for $Q \cap L^d$. Accordingly, there is a sequence $y = a_0, a_1, \ldots, a_n = x$ with $(a_{i-1}, a_i) \in \mathrm{Cover}(Q \cap L^d)$ for $i = 1, \ldots, n$. Moreover, for $i \in \{1, \ldots, n\}$ we define the proper reordering $\varphi_i : [a_i, a_{i-1}]_L \to V$ by

$$\varphi_i g := \begin{cases} \uparrow & \text{, if } y = a_i, \\ \downarrow & \text{, if } g = a_{i-1}, \\ \bullet & \text{, otherwise,} \end{cases} \quad \text{and put} \quad Q_i := \{(a_i, a_{i-1})\} \cup L_{|(a_i,a_{i-1}),\varphi_i}.$$

We show that $Q \sqsubseteq_L Q_i$. Since $L^d \cap Q_i = \{(a_{i-1}, a_i)\} \subseteq Q$ it remains to show that $L \cap Q \subseteq Q_i$. Let $(g,h) \in L \cap Q$ and assume that $(g,h) \notin Q_i$. In particular, it follows that $(g,h) \notin L_{|(a_i,a_{i-1}),\varphi_i}$, which yields $\{g,h\} \subseteq [a_i, a_{i-1}]_L$ and $(\varphi_i g, \varphi_i h) \in \{(\uparrow, \bullet), (\uparrow, \downarrow), (\bullet, \downarrow)\}$. In case of $\varphi_i g = \uparrow$ it follows that $a_{i-1}\ Q\ a_i = g\ Q\ h$ and hence $a_{i-1}\ (Q \cap L^d)\ h\ (Q \cap L^d)\ a_i$. Since a_i and a_{i-1} are neighboured in $Q \cap L^d$ it follows that $(g,h) = (a_i, a_i)$ or $(g,h) = (a_i, a_{i-1})$. In both cases it follows that $(g,h) \in Q_i$, which can also be deduced from the remaining case of $\varphi_i h = \downarrow$ in an analogous fashion. The fact that Ψ is an isomorphism follows from Proposition 4 (iii) and (iv). Statement (iv) directly follows from [7] Theorem 13. □

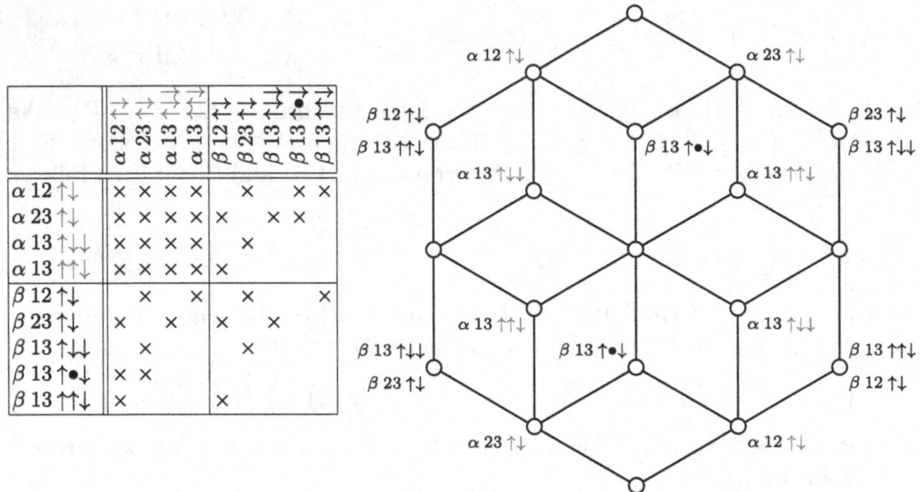

	α 12 ↑↓	α 23 ↑↓	α 13 ↑↓↓	α 13 ↑↑↓	β 12 ↑↓	β 23 ↑↓	β 13 ↑↓↓	β 13 ↑•↓	β 13 ↑↑↓
α 12 ↑↓	×	×	×	×		×		×	×
α 23 ↑↓	×	×	×	×	×		×	×	
α 13 ↑↓↓	×	×	×	×		×			
α 13 ↑↑↓	×	×	×	×	×				
β 12 ↑↓		×		×		×			×
β 23 ↑↓	×		×		×		×		
β 13 ↑↓↓		×				×			
β 13 ↑•↓	×	×							
β 13 ↑↑↓	×				×				

Fig. 9. The representing context $\mathbb{P}_L$ of the lattice of orders on the three element set $\{1, 2, 3\}$ with the positive pole L given by $1\ L\ 2\ L\ 3$ and it's concept lattice. For a better readability we leave away all brackets and commata. Furthermore, the mappings in the third component are represented by it's respective image-tuples (in L-order). For example, the label β 13 ↑•↓ means the tuple $(\beta, (1,3), \varphi)$ with $\varphi(1) = \uparrow$, $\varphi(2) = \bullet$ and $\varphi(3) = \downarrow$. (Color figure online)

5 Conclusion and Outlook

We presented an alternative way of ordering order relations. Even though we specified the fundamental properties and briefly described connections to existing work, there remain a lot of open questions: Are there relevant sublattices or intervals? How exactly do the lattices of mergings [6], which gave rise to our approach, fit into the picture? Does the new view offer aspects regarding existing problems in order theory? Furthermore, we introduced the q-saturation as a tool to describe infima und suprema in the lattices of orders. As it turns out, there is also a partial order version, the *p-saturation*, which seems to be a topic worth investigating on it's own.

References

1. Dushnik, B., Miller, E.W.: Partially ordered sets. Am. J. Math. **63**(3), 600–610 (1941)
2. Yanagimoto, T., Okamoto, M.: Partial orderings of permutations andmonotonicity of a rank correlation statistic. Ann. Inst. Stat. Math. **21**, 489–506 (1969)
3. Trotter, W.T.: Combinatorics and Partially Ordered Sets. The Johns Hopkins University Press, Baltimore (1992)
4. Avann, S.P.: The lattice of natural partial orders. Aequationes Math. **8**, 95–102 (1972)
5. Pouzet, M., Reuter, K., Rival, I., Zaguia, N.: A generalized permutahedron. Algebra Univers. **34**, 496–509 (1995)
6. Ganter, B., Meschke, C., Mühle, H.: Merging ordered sets. In: Valtchev, P., Jäschke, R. (eds.) ICFCA 2011. LNCS (LNAI), vol. 6628, pp. 183–203. Springer, Heidelberg (2011). https://doi.org/10.1007/978-3-642-20514-9_15
7. Ganter, B., Wille, R.: Formal Concept Analysis. Springer, Heidelberg (1999). https://doi.org/10.1007/978-3-642-59830-2
8. Davey, B.A., Priestley, H.A.: Introduction to Lattices and Order, 2nd edn. Cambridge University Press, Cambridge (2002)
9. Edelman, P.H., Jamison, R.E.: The theory of convex geometries. Geom. Dedicata **19**, 247–270 (1985)
10. Edelman, P.H., Saks, M.E.: Combinatorial representation and convex dimension of convex geometries. Order **5**, 23–32 (1988)

Methods and Applications

On-demand Relational Concept Analysis

Alexandre Bazin[1], Jessie Carbonnel[2(✉)], Marianne Huchard[2], Giacomo Kahn[3], Priscilla Keip[4,5], and Amirouche Ouzerdine[2]

[1] Université de Lorraine, CNRS, Inria, LORIA, F54000 Nancy, France
contact@alexandrebazin.com

[2] LIRMM, CNRS and Université de Montpellier, Montpellier, France
{jcarbonnel,huchard,ouzerdine}@lirmm.fr

[3] Université d'Orléans, INSA Centre Val de Loire, LIFO, Orléans, France
giacomo.kahn@isima.fr

[4] CIRAD, UPR AIDA, 34398 Montpellier, France
keip@lirmm.fr

[5] AIDA, Univ Montpellier, CIRAD, Montpellier, France

Abstract. Formal Concept Analysis (FCA) and its associated conceptual structures are used to support exploratory search through conceptual navigation. Relational Concept Analysis (RCA) is an extension of Formal Concept Analysis to process relational datasets. RCA and its multiple interconnected structures represent good candidates to support exploratory search in relational datasets, as they are enabling navigation within a structure as well as between the connected structures. However, building the entire structures does not present an efficient solution to explore a small localised area of the dataset, to retrieve the closest alternatives to a given query. In these cases, generating only a concept and its neighbour concepts at each navigation step appears as a less costly alternative. In this paper, we propose an algorithm to compute a concept, and its neighbourhood, in connected concept lattices. The concepts are generated directly from the relational context family, and possess both formal and relational attributes. The algorithm takes into account two RCA scaling operators and it is implemented in the RCAExplore tool.

Keywords: Relational Concept Analysis · Formal Concept Analysis · Exploratory search · On-demand generation · Local generation

1 Introduction

Many datasets in thematic areas like environment or product lines comprise databases complying with a relational data model. Typical applications in which we are currently involved concern issues relative to watercourse quality[1] (Fresqueau project), the inventory and use of pesticidal, antibacterial and antifungal

[1] http://dataqual.engees.unistra.fr/.

D. Cristea et al. (Eds.): ICFCA 2019, LNAI 11511, pp. 155–172, 2019.
https://doi.org/10.1007/978-3-030-21462-3_11

plants[2] (Knomana project), and the analysis and representation of complex product lines [6]. In these applications, there is a wide range of question forms, such as classical querying, establishing correlations between descriptions of objects from several categories or case based reasoning. These questions can be addressed by complementary approaches including conceptual classification, knowledge pattern and rule extraction, or exploratory search [23,26]. In the Knomana project, for example, one main purpose will be, after the ongoing plant inventory, to support farmers, their advisors, local entrepreneurs or researchers in selecting plants of immediate interest for agricultural crop protection and animal health. As users will face large amounts of data with no prior knowledge of the data, and will mainly formulate potentially imprecise queries, without an immediate answer, exploratory search seems to be a suitable approach.

Previous work [7,10,11,14,17] has shown that Formal Concept Analysis (FCA) is a relevant support for data exploration. FCA is a data analysis framework organising datasets composed of objects described by attributes in canonical structures called concept lattices, which have good properties for exploratory search. We expect Relational Concept Analysis (RCA), an FCA extension to handle several datasets connected by relationships, to be beneficial as well. Considering RCA for relational dataset exploration brings issues relative to (a) the use of scaling (logical) operators describing the relationships, (b) the iterative process to build the concept lattices, and (c) the presence of several concept lattices connected via relational attributes. Despite this additional complexity, RCA helps the user to concentrate on the classification of objects of several categories, where the object groups (concepts) are described by intrinsic attributes and by their relations to object groups of other categories. Besides, the relational attributes offer a support to navigate between the object groups of the different categories, while the concept lattices offer a (by-specialisation) navigation between object groups of the same category.

There are several complementary strategies to implement dataset exploration using RCA. One may consist in exhaustively computing concept lattices (and related artefacts like implication rules) at several steps, using several logical operators and considering only some of the object categories and some of the inter-categories relationships [5]. Another strategy, which is followed here, consists in an on-demand computation of a concept and its neighbourhood comprising its upper, lower and relational covers.

The next section presents the main principles of Relational Concept Analysis (Sect. 2). The on-demand computation of a concept and its neighbourhood is presented in Sect. 3, and its integration in the RCAExplore tool is presented in Sect. 4. Section 5 illustrates the algorithm with the example introduced in Sect. 2. Related work is exposed in Sect. 6. We conclude the paper with a few perspectives in Sect. 7.

[2] http://www.cirad.fr/en/news/all-news-items/articles/2017/science/identifying-plants-used-as-natural-pesticides-in-africa-knomana.

2 Relational Concept Analysis

Formal Concept Analysis (FCA) [15] is a data analysis framework to structure objects described by attributes in a canonical structure called a concept lattice. It revolves around the notions of formal *contexts* and *concepts*.

Definition 1. *A* formal context *is a triple* $\mathbb{K} = (\mathcal{O}, \mathcal{A}, \mathcal{I})$ *in which* $\mathcal{O}$ *is a set of* objects, $\mathcal{A}$ *a set of* attributes, *and* $\mathcal{I} \subseteq \mathcal{O} \times \mathcal{A}$ *an* incidence relation *stating "which objects possess which attributes".*

Two *derivation operators* (both denoted by $(\cdot)'$) are defined on a formal context $(\mathcal{O}, \mathcal{A}, \mathcal{I})$. Let O be a set of objects and A a set of attributes. Then:

$$O' = \{a \in \mathcal{A} \mid \forall o \in O, (o, a) \in \mathcal{I}\}$$

$$A' = \{o \in \mathcal{O} \mid \forall a \in A, (o, a) \in \mathcal{I}\}$$

The composition of these two operators forms a *Galois connection* and thus a closure operator. We say that a set S is *closed* when $S = S''$.

Definition 2. *The attribute (resp. object) set* X *is called a* generator *of the set* X''. *It is a* minimal generator *if none of its subsets have the same closure.*

Definition 3. *A* formal concept *is a pair* $(A, B) \in 2^{\mathcal{O}} \times 2^{\mathcal{A}}$ *such that* $A = B'$ *and* $B = A'$. *The set* A *is called the* extent *and* B *the* intent *of the concept.*

A concept is thus a maximal set of objects sharing a maximal set of attributes and corresponds to a maximal rectangle of crosses (up to permutation of rows and columns) when the context is represented as a crosstable. The set $\mathcal{C}_{\mathbb{K}}$ of all concepts of the context $\mathbb{K}$, together with the order induced by the set-inclusion relation (denoted by $\leq_s$) on either of the components (but usually the extents), forms a complete lattice called the *concept lattice* of $\mathbb{K}$. Figure 1 depicts two concept lattices. It uses a condensed representation in which an attribute (resp. an object) is introduced in the greatest (resp. the smallest) concept having this attribute in the lattice.

Thus, a concept inherits the attributes from its super-concepts, and the objects from its sub-concepts. For instance, concept *C_DM_tools_0* of the concept lattice on the left-hand side of Fig. 1 has for intent {*OS:Windows, DM:Conceptual*}, and for extent {*Astah, ErwinDM, Magic Draw, ER/Studio*}.

Definition 4. *Let* (A, B) *be a formal concept of a formal context* $\mathbb{K}$. *A formal concept* (A_2, B_2) *is called a* cover *of* (A, B) *if* A *and* A_2 *are comparable and there is no concept* (A_3, B_3) *with* A_3 *between* A *and* A_2. *It is an* upper cover *when* $A \subset A_2$ *and a* lower cover *otherwise.*

Definition 5. *A concept is an* object-concept *(resp.* attribute-concept*) if it is the smallest (resp. greatest) concept which extent (resp. intent) contains a particular object (resp. attribute).*

In Fig. 1 (left-hand side), *C_DM_tools_0* is an attribute-concept, while *C_DM_tools_2* is an object-concept, and *C_DM_tools_3* is both.

Relational Concept Analysis (RCA) [18,19] is an adaptation of FCA to process relational datasets. A relational dataset is composed of several categories of objects described by both their own (intrinsic) attributes and their relationships with objects of other categories. As input, RCA takes a Relational Context Family (RCF), gathering a set of formal contexts (representing the different categories of objects and their attributes) and a set of relational contexts (representing relationships between objects of different categories), where each relational context defines links between the objects of two formal contexts.

Definition 6 (Relational Context Family). *A Relational Context Family is a pair* $(\boldsymbol{K}, \boldsymbol{R})$ *such that:*

- $\boldsymbol{K} = \{\mathbb{K}_i = (\mathcal{O}_i, \mathcal{A}_i, \mathcal{I}_i)\}$ *is a set of formal contexts (object-attribute relations)*
- $\boldsymbol{R} = \{r_k\}, r_k \subseteq \mathcal{O}_i \times \mathcal{O}_j$ *is a set of relational contexts (object-object relations), with* $\mathcal{O}_i$ *being the set of objects of context* $\mathbb{K}_i$ *(the source context) and* $\mathcal{O}_j$ *being the set of objects of context* $\mathbb{K}_j$ *(the target context).*

Table 1. (top) Two formal contexts: (left-hand side) Data Modelling tools (DM_tools) and (right-hand side) DataBase Management Systems (DBMS). (bottom) Relational context stating which DM_tools *support* which DBMS

$\mathbf{K}_s =$

DM_tools	OS:Windows	OS:Mac OS	OS:Linux	DM:Conceptual	DM:Physical	DM:Logical	DM:ETL
Astah	×	×	×	×			
Erwin DM	×			×	×	×	
ER/Studio	×			×	×	×	×
Magic Draw	×	×	×	×	×	×	
MySQL Workbench	×	×	×		×		

DBMS	DT:Enum	DT:Set	DT:Geometry	DT:Spatial	DT:Audio	DT:Image	DT:Video	DT:XML	DT:JSON	DT:Period
MySQL	×	×	×							
Oracle				×	×	×	×	×		
PostgreSQL	×		×					×	×	
Teradata	×		×					×	×	×

$\mathbf{R}_s =$

support	MySQL	Oracle	PostgreSQL	Teradata
Astah	x	x		
Erwin DM	×	×		×
ER/Studio	×	×	×	×
Magic Draw	×	×	×	
MySQL Workbench	×			

The three contexts of Table 1 present an example of an RCF $(\mathbf{K}_s, \mathbf{R}_s)$ taken from the software product line domain. Table 1 (top) displays two formal contexts. The one on the left-hand side presents five Data Modelling tools (*DM_tools*) against seven attributes representing their compatible operating systems (*OS:*), and the data models (*DM:*) the tools may manage. The table on the right-hand side describes four DataBase Management Systems (*DBMS*) according to the data types (*DT:*) they may handle. Table 1 (bottom) presents a

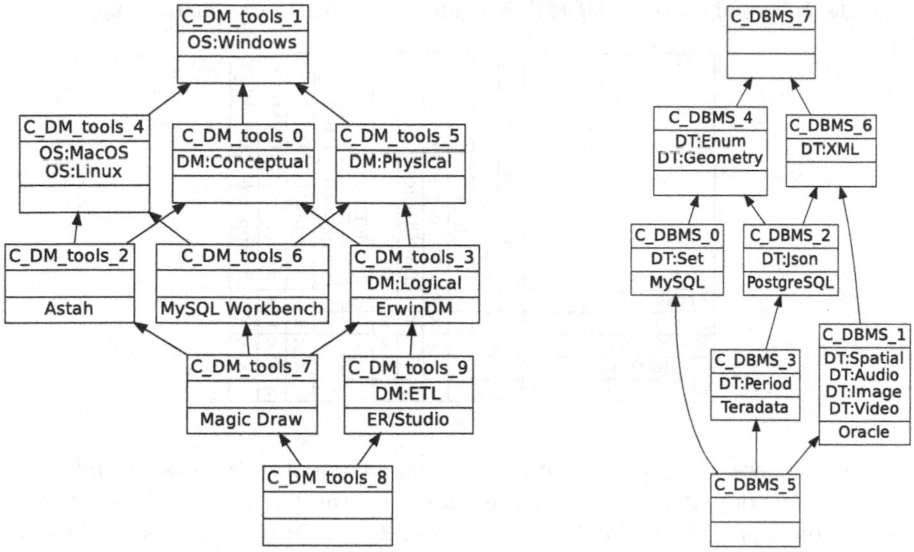

Fig. 1. (left) concept lattice of DM_tools, (right) concept lattice of DBMS

relational context stating which Data Modelling tools *support* which DataBase Management Systems.

The RCA process uses two steps iteratively. The first is building the concept lattices of all the formal contexts in $\mathbf{K}$, ignoring the relations contained in the relational contexts. The two concept lattices associated with Table 1 (top) are presented in Fig. 1. The second step is taking the relations into account by augmenting the formal contexts with new attributes that represent information about those relations. These *relational attributes* take the form of a $\rho r.C$ construct in which ρ is a *scaling operator* (sometimes called *quantifier*), r is a relation in the RCF and C is a formal concept of the lattice of the target of r built in the first step. In our example, we may introduce the relational attribute $\exists$ $support.(C\_DBMS\_2)$ to characterise the *DM_tools* that support at least one *DBMS* offering Json and XML. Given two formal contexts $\mathbb{K}_i, \mathbb{K}_j \in \mathbf{K}$ and a relational context $r \subseteq \mathcal{O}_i \times \mathcal{O}_j$, the application of RCA extends the set of attributes $\mathcal{A}_i$ with a set of relational attributes representing links to the concepts of $\mathbb{K}_j$. The extended attribute set is denoted by $\mathcal{A}_i^+$ and the incidence relation $\mathcal{I}_i$ is extended and denoted $\mathcal{I}_i^+$ to take these new attributes into account by associating them to objects of $\mathcal{O}_i$ depending on the relation r, the concept C involved in the relational attribute and the scaling operator ρ. In this paper, we focus on two scaling operators: the *existential* operator $\exists$, associating an object o to the relational attribute $\exists r.(C)$ if o is linked to at least one object of the extent of C by r, and the *universal strict* operator $\exists\forall$, associating an object o to $\exists\forall r.(C)$ if all the objects linked to o by r are included in the extent of C, and $r(o) \neq \emptyset$.

Table 2. Formal context *DM_tools* extended according to the relation *support*

DM_tools$^+$	OS:Windows	OS:Mac OS	OS:Linux	DM:Conceptual	DM:Physical	DM:Logical	DM:ETL	∃ support(C_DBMS_3)	∃ support(C_DBMS_1)	∃ support(C_DBMS_0)	∃ support(C_DBMS_2)
Astah	×	×	×	×					×	×	
Erwin DM	×			×	×	×		×	×	×	×
ER/Studio	×			×	×	×	×	×	×	×	×
Magic Draw	×	×	×	×	×	×			×	×	×
MySQL Workbench	×	×	×		×					×	

Table 2 shows *DM_tools*$^+$, the formal context about *DM_tools* extended by the relational context *support* and the scaling operator $\rho = exist$. According to the relation *support*, the Data Modelling tool `Magic Draw` supports the DBMS `PostgreSQL`. Now let us consider the concept *C_DBMS_2* in the concept lattice (right-hand side) of Fig. 1. Its extent contains `PostgreSQL`. As one object in the concept's extent is linked to `Magic Draw` through the relation *support*, then `Magic Draw` is associated with the relational attribute $\exists\ support(C\_DBMS\_2)$.

The concept lattice associated with a formal context $\mathbb{K}^+ = (\mathcal{O}, \mathcal{A}^+, \mathcal{I}^+)$ then structures the objects from $\mathcal{O}$ both by their attributes and their relations to other sets of objects through the relational attributes. Figure 2 presents the extended

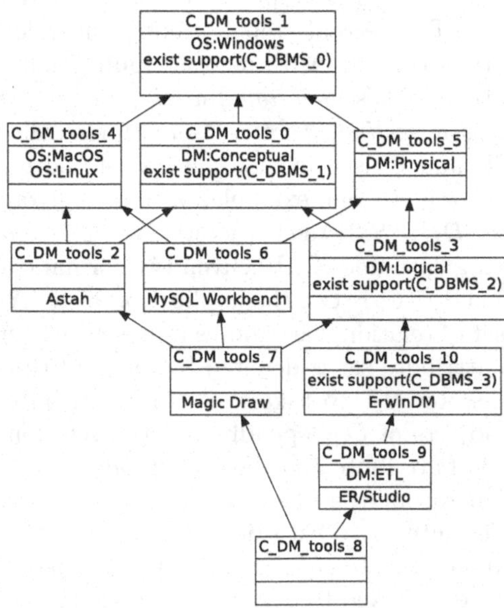

Fig. 2. Concept lattice of the extended context *DM_tools*$^+$

concept lattice corresponding to the extended formal context *DM_tools*$^+$, according to the relation *support* and the *existential* scaling operator.

In this way, for complex data models including more than one relation, RCA produces a succession of concept lattices, extended at each step by the new abstractions obtained at the previous step. Let $\mathbf{K}^0 = \mathbf{K} = \{\mathbb{K}_i^0\}$ denote the contexts of the initial RCF and $\mathbf{L}^0$ the family of their lattices. We will use $\mathbf{K}^n = \{\mathbb{K}_i^{(n-1)+}\}$ and $\mathbf{L}^n$ to denote the state of the contexts and lattices after n applications of the previously described two steps.

In the following, we introduce some exploration algorithms in a RCF. In order to do so, we define the notion of relational cover of a concept, and we also use the notion of *transversal* from hypergraph theory.

Definition 7. *Let $\mathbb{K}_i^n$ be a formal context in the RCF $(\boldsymbol{K}^n, \boldsymbol{R})$ and (A, B) a formal concept of $\mathbb{K}_i^n$. A* relational cover *of (A, B) is a concept C such that $\rho r.C \in B$.*

Definition 8. *Let $\mathcal{F}$ be a family of subsets of a ground set G. A* transversal *is a set of elements of G that intersects every set of $\mathcal{F}$. A transversal is* minimal *when none of its proper subsets is a hitting set (i.e. a transversal itself).*

3 The Exploration Algorithm

In this section, we present algorithms for computing a concept from connected lattices, along with its neighbouring concepts. It enables exploration by minimal steps, which show the closest alternatives to the concept representing a given query. Then, the user may choose one of these alternatives and pursue his/her exploration (i.e., taking a step in the exploration) by investigating the neighbouring concepts, which can be in the same lattice, or in a connected one.

With the example given on Data Modelling tools, let us consider a user who wants to find a tool for Linux (`OS:Linux`) dedicated to physical data models `DM:physical`. It is easy to compute independently the concept *C_DM_tools_6* from Fig. 2 gathering the corresponding tools. Let us suppose now that the user is not satisfied by the proposed tools, they can consider removing a constraint (like using Linux or MacOS) which leads them to consider *C_DM_tools_5*. Reversely, the user may want to add constraints like (1) a tool also considering logical data models (`DM:Logical`); (2) or a tool admitting Json data types (`DT:Json`) from concept *C_DBMS_2*. In these two cases, considering *C_DM_tools_3* is relevant. Thus, the purpose of the algorithm is to show *C_DM_tools_6* and all possible similar solutions present and gathered in the neighbouring concepts. It aims to avoid the inherent complexity and running time of computing the whole lattices, and it is also designed to assist users during their data exploration task.

The algorithm takes as input a RCF $(\mathbf{K}^n, \mathbf{R})$ and a formal concept $C = (A, B) \in \mathbf{L}^n$. Its output is the concept $(A, B^+) \in \mathbf{L}^{(n+1)}$ and its upper, lower and relational covers, where B^+ is B enhanced with relational attributes. Meanwhile, the RCF is updated with the relational attributes for a next step.

3.1 Redefining Derivation Operators

In order to compute the new concept and its covers in $\mathbf{L}^{(n+1)}$, we need the contexts of $\mathbf{K}^{n+1}$. The explicit knowledge of all the relational attributes of those contexts would require the computation of all the concepts in the contexts of the input $\mathbf{K}^n$. This is too time-consuming. We would prefer to manipulate only a minimal number of relational attributes allowing us to derive, on-the-fly, the other relational attributes.

Any object described by an attribute $\rho r.(X,Y)$ is also necessarily described by all the attributes of the form $\rho r.(X_2,Y_2)$ such that $(X_2,Y_2) \in \mathbf{L}^n$ with $X \subseteq X_2$ and/or $Y_2 \subseteq Y$. As such, intents can be represented without loss of information by their relational attributes constructed from attributes-wise maximal concepts. However, two problems arise with such a representation: (1) the set intersection cannot be used to compute the intent of a set of objects anymore, and (2) if only maximal relational attributes are explicitly present in the context, the extent of a set of attributes cannot be computed through a simple test of set inclusion. To remedy this, we provide three algorithms to use on sets of attributes (both intrinsic and relational) with only the maximal relational attributes given explicitly.

The Ex algorithm (Algorithm 1) computes the extent, in the formal context $\mathbb{K}_i$, of an attribute set A represented by its maximal relational attributes. For each object o and attribute $\rho r.(X,Y) \in A$, it checks whether $r(o)$ and X intersect in the correct way (depending on the scaling operator). The algorithm runs in $O(|\mathcal{O}_i| \times |A| \times |\mathcal{O}_j|)$ where $\mathcal{O}_j$ is the biggest set of objects in the RCF.

Algorithm 1. Ex($\mathbb{K}_i, A$)

Input: $\mathbb{K}_i = (\mathcal{O}_i, \mathcal{A}_i, \mathcal{I}_i)$ a formal context, $A \subseteq \mathcal{A}_i$ a set of attributes
Output: Computes the extent of a set of attributes A

$O \leftarrow \mathcal{O}_i$
foreach $a \in A$ **do**
  if $a \sim \exists\forall r.(X,Y)$ **then**
    $O \leftarrow \{o \in O \mid r(o) \subseteq X\}$
  else if $a \sim \exists r.(X,Y)$ **then**
    $O \leftarrow \{o \in O \mid r(o) \cap X \neq \emptyset\}$
  else
    $O \leftarrow \{o \in O \mid (o,a) \in \mathcal{I}_i\}$
return O

Let A and B be two attribute sets represented by their maximal relational attributes. A relational attribute $\exists r.(X,Y)$ is in the intersection of A and B if and only if there exist two attributes $\exists r.(X_2,Y_2) \in A$ and $\exists r.(X_3,Y_3) \in B$ such that $X \subseteq X_2$ and $X \subseteq X_3$. The same holds for the $\exists\forall$ scaling operator. The Intersect algorithm (Algorithm 2) uses this property to compute the maximal relational attributes of the intersection of A and B by recursively intersecting the intents of the concepts used to build the relational attributes in A and B. The fact that, at any given time, the depth of all the relational attributes is bounded ensures that the recursion ends. The algorithm runs in $O((|\mathcal{A}_i|^2 \times Q_{Ex}(\mathbb{K}_j, \mathcal{A}_j))^{n+1})$ with $\mathcal{A}_i$ the biggest attribute set in the RCF, $Q_{Ex}(\mathbb{K}_j, \mathcal{A}_j)$ the complexity of Algorithm 1 with, as inputs, $\mathbb{K}_j$ and $\mathcal{A}_j$ such that $\mathbb{K}_j$ is the context in the RCF for which $|\mathcal{O}_j| \times |\mathcal{A}_j|$ is the biggest and n the maximal depth of a relational attribute, i.e. the current number of times RCA's steps have been applied.

Algorithm 2. INTERSECT($\mathbb{K}_i, A, B$)

Input: $\mathbb{K}_i = (\mathcal{O}_i, \mathcal{A}_i, \mathcal{I}_i)$ a formal context, $A, B \subseteq \mathcal{A}_i$ two attribute sets
Output: The relational intersection of A and B

$R \leftarrow A \cap B$
$\mathcal{F} \leftarrow \emptyset$
foreach $a_1 \sim \exists r.(X_1, Y_1) \in B$ *with* $r \subseteq \mathcal{O}_i \times \mathcal{O}_j$ *and* $\mathbb{K}_j = (\mathcal{O}_j, \mathcal{A}_j, \mathcal{I}_j)$ **do**
  foreach $a_2 \sim \exists r.(X_2, Y_2) \in A$ **do**
    $\mathcal{F} \leftarrow \mathcal{F} \cup \{\exists r.($EX($\mathbb{K}_j$,INTERSECT$(\mathbb{K}_j, Y_1, Y_2))$,INTERSECT$(\mathbb{K}_j, Y_1, Y_2))\}$
$R \leftarrow R \cup$ MAX$(\mathcal{F}, \subseteq_{\mathcal{A}_i})$
$\mathcal{F} \leftarrow \emptyset$
foreach $a_1 \sim \exists\forall r.(X_1, Y_1) \in B$ *with* $r \subseteq \mathcal{O}_i \times \mathcal{O}_j$ *and* $\mathbb{K}_j = (\mathcal{O}_j, \mathcal{A}_j, \mathcal{I}_j)$ **do**
  foreach $a_2 \sim \exists\forall r.(X_2, Y_2) \in A$ **do**
    $\mathcal{F} \leftarrow \mathcal{F} \cup \{\exists\forall r.($EX($\mathbb{K}_j$,INTERSECT$(\mathbb{K}_j, Y_1, Y_2))$,INTERSECT$(\mathbb{K}_j, Y_1, Y_2))\}$
$R \leftarrow R \cup \mathcal{F}$
return R

IN (Algorithm 3) uses INTERSECT to compute the intent, in the formal context $\mathbb{K}_i$, of a set O of objects described by their maximal relational attributes. It starts with the set of all explicitly known attributes and intersects it with the description of each object in the context $\mathbb{K}_i$. It runs in $O(|O| \times Q_{Int})$ where Q_{Int} is the complexity of INTERSECT.

Algorithm 3. IN($\mathbb{K}_i, O$)

Input: $\mathbb{K}_i = (\mathcal{O}_i, \mathcal{A}_i, \mathcal{I}_i)$ a formal context, $O \subseteq \mathcal{O}_i$ a set of objects
Output: Computes the intent of a set of objects O

$A \leftarrow \mathcal{A}_i$
foreach $o \in O$ **do**
  $A \leftarrow$ INTERSECT$(A, \mathit{Intent}(\{o\}))$
return A

3.2 Computing the Closed Neighbourhood

Now that we have redefined the derivation operators on implicitly known relational contexts, we are able to compute the upper, lower and relational covers of a concept.

The easiest are the relational covers. A concept (X, Y) is a relational cover of a concept (U, V) if and only if $\rho r.(X, Y)$ is a maximal relational attribute in V. Upper covers are easy too. Candidates can be generated by adding an object – the set of which we have perfect knowledge of – to the current extent and computing the corresponding concept. The covers are the candidates that have the smallest extent. Computing the lower covers is more challenging. They could be computed by adding attributes to the intent but the full set of relational attributes is only known implicitly. We chose to, instead, remove objects. The lower covers of (X, Y) being the concepts with the maximal extents that are contained in X and do not contain any of the minimal generators of X, a simple way to compute them would be to remove minimal transversals of the minimal generators.

Algorithm 4. GROWCONTEXT($\mathbb{K}_i, r, \rho, o, OC_j$)

Input: $\mathbb{K}_i = (\mathcal{O}_i, \mathcal{A}_i, \mathcal{I}_i)$ a formal context, $r \subseteq \mathcal{O}_i \times \mathcal{O}_j$ a relational context, ρ a scaling operator, $o \in \mathcal{O}_i$ an object, OC_j the set of object-concepts of $\mathbb{K}_j = (\mathcal{O}_j, \mathcal{A}_j, \mathcal{I}_j)$

Output: Extends the context $\mathbb{K}_i$ and adds the crosses, giving $\mathbb{K}_i^+$

if $\rho == \exists$ **then**
    foreach $(X, Y) \in OC_j$ *such that* $r(o) \cap X \neq \emptyset$ **do**
        $\mathcal{A}_i \leftarrow \mathcal{A}_i \cup \exists r.(X, Y)$
        $\mathcal{I}_i \leftarrow \mathcal{I}_i \cup (o, \exists r.(X, Y))$
if $\rho == \exists\forall$ **then**
    $Y \leftarrow$IN$(\mathbb{K}_j, r(o))$
    $X \leftarrow$EX$(\mathbb{K}_i, Y)$
    $\mathcal{A}_i \leftarrow \mathcal{A}_i \cup \exists\forall r.(X, Y)$
    $\mathcal{I}_i \leftarrow \mathcal{I}_i \cup (o, \exists\forall r.(X, Y))$

Algorithm 4 (GROWCONTEXT) takes as input a context, a relation, a scaling operator, an object o and the set of object-concepts [2] of o. It constructs new relational attributes, adds them to the context and completes the incidence relation accordingly. For the scaling operator $\exists$, each relation r and each object-concept (X, Y) in the target of r give rise to a new relational attribute $\exists\, r.(X, Y)$ that is added to the context $\mathbb{K}_i$. As object-concepts contain the irreducible elements of the lattice, relational attributes constructed with them are enough to reconstruct all the other possible relational attributes. For the scaling operator $\exists\forall$, the new added relational attribute $\exists\forall\, r.C$ is built thanks to Algorithms 3 and 1, roughly speaking by computing the intent of $r(o)$ and the associated extent.

Algorithm 5 computes the closed neighbourhood of a concept C. It takes as input a set of formal contexts $\mathbf{K} = (\mathbb{K}_1, \ldots, \mathbb{K}_w)$ of a RCF, a strategy $\mathcal{S} = \{(r, \rho)_{lj}, \ldots\}$, $l, j \in \{1, \ldots, w\}$ and a starting concept C from a context $\mathbb{K}_i$. A strategy is an assignment of scaling operators to relations, which corresponds to a user choice. The goal is to compute (or complete) the intent corresponding to the extent of C, as well as its upper, lower and relational covers, in the extended context $\mathbb{K}_i^+$. For each $(r, \rho)_{ij} \in \mathcal{S}$ such that $r : \mathbb{K}_i \mapsto \mathbb{K}_j$, the first loop (Lines 1 to 4) computes OC_j the object-concepts of $\mathbb{K}_j$ and calls Algorithm 4 to obtain the extended context. In Line 5, the intent of concept C is extended with the relational attributes added during the previous loop. The next loop (Lines 6 to 8) computes the relational covers $\mathcal{R}$ of concept C. For each relational attribute in the intent of C, the corresponding concept (in the target context) is added to the cover. In Lines 9 to 11, the lower covers $\mathcal{L}$ of C are computed by removing from the extent of C a minimal transversal of the set of minimal generators of C's extent. Finally, the upper covers $\mathcal{U}$ of C are computed in Lines 12 to 14. Candidates are created by adding an object o to the extent of C. Only the extent-wise minimal resulting concepts are kept.

Algorithm 5. INCREMENTAL($\mathbf{K}, \mathcal{S}, C, \mathbb{K}_i$)

Input: $\mathbf{K} = \{\mathbb{K}_1, \ldots, \mathbb{K}_w\}$, $\mathcal{S} = \{(r, \rho)_{lj}, \ldots\}, l, j \in \{1, \ldots, w\}$ a strategy, $C = (O, A)$ a concept of $\mathbb{K}_i = (\mathcal{O}_i, \mathcal{A}_i, \mathcal{I}_i)$

Output: $C, \mathcal{U}, \mathcal{R}, \mathcal{L}$ the completed concept C and its closed relational neighbourhood

1 **foreach** $(r, \rho)_{ij} \in \mathcal{S}$ **do**
2     $OC_j \leftarrow$ OBJECTPOSET($\mathbb{K}_j$)
3   **foreach** $o \in \mathcal{O}_i$ **do**
4   GROWCONTEXT($\mathbb{K}_i, r, \rho, o, OC_j$)
5 $A \leftarrow$ IN($\mathbb{K}_i, O$)
6 $\mathcal{R} \leftarrow \emptyset$
7 **foreach** $a \sim \rho r.(X_1, Y_1) \in A$ **do**
8     $\mathcal{R} \leftarrow \mathcal{R} \cup \{(X_1, Y_1)\}$
9 $\mathcal{L} \leftarrow \emptyset$
10 **foreach** $T \in minTrans(minGen(O))$ **do**
11     $\mathcal{L} \leftarrow \mathcal{L} \cup \{(O \setminus T, \text{IN}(\mathbb{K}_i, O \setminus T))\}$
12 $\mathcal{U} \leftarrow \emptyset$
13 **foreach** $o \in \mathcal{O}_i \setminus O$ **do**
14     $\mathcal{U} \leftarrow \mathcal{U} \cup \{(\text{EX}(K_i, \text{IN}(\mathbb{K}_i, O \cup \{o\})), \text{IN}(\mathbb{K}_i, O \cup \{o\}))\}$
15 $\mathcal{U} \leftarrow$ MIN($\mathcal{U}, \subseteq_{\mathcal{O}_i}$)
16 **return** $C, \mathcal{U}, \mathcal{R}, \mathcal{L}$

4 Implementation

In this section we present RCAExplore[3], the tool in which the proposed algorithms are integrated. We quote the different algorithms that helped for the implementation, and we explain how they concretely work in the RCAExplore tool. A version of RCAExplore with the implemented algorithms is currently available on demand to the authors and will soon be available online.

RCAExplore is a framework for FCA and RCA. It provides to the user the ability to finely tune the RCA process. At each step, the user can select the formal contexts and the relational contexts that interest them and modify their content if they want. The user can also select one or more quantifiers to process each relation, and choose for each selected formal context a particular algorithm to build the associated conceptual structure. RCAExplore implements 5 different algorithms (*fca, iceberg, ares, acposet* and *ocposet*) to build the corresponding conceptual structures. *fca* builds a concept lattice, *iceberg* builds an iceberg lattice [21], *ares, acposet* and *ocposet* respectively build the partially ordered sets induced by the concepts that introduce objects or attributes (AOC-poset), attributes (AC-poset) or objects (OC-poset). The *ocposet* algorithm is called for each relation to extend the source formal context (Algorithm 5, Line 2).

The on-demand algorithms proposed in this paper build only a part of the concept lattice (i.e., a concept and its neighbours) on which the user wants to focus on. It is integrated into RCAExplore as a new way of navigating, where the user gradually progresses into the lattice without getting lost. First, the user selects the formal context from which they want to start the navigation. This formal context must have at least one attribute or one relation toward another formal context of the RCF. Then the user selects the attribute set around which they want to build the first concept. For this concept, the relational, upper and lower covers are computed as mentioned in Sect. 3. To compute the lower cover, we needed to compute minimal transversals of minimal generators (Algorithm 5, Lines 10,11). For this purpose, we implemented the MTMiner algorithm [20]. Once covers are calculated, the user can choose another concept among those calculated to continue the navigation. If this new concept contains a relational attribute, the user can either switch to the corresponding lattice and build what is around the concept introducing this relational attribute, or stay in the same lattice and resume the process.

5 Illustrative Example

In this section, we illustrate the defined algorithms and we briefly introduce the real case studies on which we tested them. We consider the RCF $(\mathbf{K}_s, \mathbf{R}_s)$ with $\mathbf{K}_s = \{DM\_tools, DBMS\}$ and $\mathbf{R}_s = \{support\}$ presented in Sect. 2. We apply the strategy $\{(support, \exists)\}$.

[3] http://dataqual.engees.unistra.fr/logiciels/rcaExplore.

User Request (step 1). Let us consider a user who initially wants to select a data modelling tool that runs on Windows (*OS:Windows*) and that handles conceptual data models (*DM:Conceptual*). Traditional FCA may compute the formal concept associated with these 2 attributes (i.e., *C_DM_tools_0*, left-hand side of Fig. 1) and inform the user that the corresponding tools are **Astah**, **Erwin DM**, **Magic Draw** and **ER/Studio**. Now the user wants to get insight into the neighbourhood of the proposed solution, e.g. to know what are the specificities of the tools in case the functional needs or the working environment evolves.

Algorithm (run 1). Let us apply our algorithms on this concept to (1) retrieve the supported DBMS (relational cover) and (2) find the closest alternatives to the query (lower and upper covers). We call Algorithm 5 with the following parameters: $\text{INCREMENTAL}(\mathbf{K}_s, \{(support, \exists)\}, C\_DM\_tools\_0, DM\_tools)$.

In Algorithm 5, Lines 1 to 4 extend the context of *DM_tools* with the relational attributes representing the object-concepts of *DBMS* (*support*'s target context). In our case, we have only one relation (*support*, ∃) visited at Line 1. In Line 2, OC_j takes the object-concepts of *DBMS*, i.e., concepts 0, 1, 2 and 3 from the concept lattice on the right-hand side of Fig. 1: the whole OC-poset is shown in the right-hand side of Fig. 3. Then, the loop on Lines 3 and 4 considers the 5 objects of *DM_tools*, on which GROWCONTEXT is called. Each object o_i of *DM_tools* is associated to the relational attributes representing the concepts of OC_j having in their extents at least one object linked to o_i.

As *support*(**Astah**) $= \{MySQL, Oracle\}$, the relational attributes $\exists support\text{-}(C\_DBMS\_0)$ (corresponding to *MySQL* object-concept) and $\exists support(C\_DBMS\_1)$ (corresponding to *Oracle* object-concept) are added to *DM_tools* and associated to **Astah**. We apply the same process with the other objects. After Line 4 execution, we obtain the extended context presented in Table 2.

Line 5 updates the intent of the input concept to add the relational attributes: the intent of *C_DM_tools_0* is now $\{OS : Windows, DM : Conceptual, \exists support(C\_DBMS\_0), \exists support(C\_DBMS\_1)\}$. The concepts of *DBMS* corresponding to the added relational attributes, i.e. the concepts C_DBMS_0 and C_DBMS_1, form the relational cover of the input concept (Lines 6 to 8).

Then, (Lines 9 to 11) we compute the minimal generators of the extent of *C_DM_tools_0*, which are {**Erwin DM**, **Astah**} and {**ER/Studio**, **Astah**}. Their minimal transversals are {**Astah**} and {**Erwin DM**,**ER/Studio**}. The two concepts having {**Astah**,**Magic Draw**} and {**Erwin DM**,**ER/Studio**,**Magic Draw**} for extent represent the lower cover of *C_DM_tools_0* (respectively *C_DM_tools_2* and *C_DM_tools_3* in Fig. 2).

Finally, in Lines 12 to 14, we consider the objects of *DM_tools* that are not in *C_DM_tools_0*'s extent, i.e., **MySQL Workbench**. For each one of them, we compute the concept corresponding to their union with *C_DM_tools_0*'s extent, and we obtain in this case the concept *C_DM_tools_1* of Fig. 2, which represents the upper cover of *C_DM_tools_0*.

With the implementation of the algorithms in RCAExplore, we obtain, as output, the concept $C\_DM\_tools\_0$ and its relational, upper and lower covers as shown in left-hand side of Fig. 3, where the input concept is highlighted.

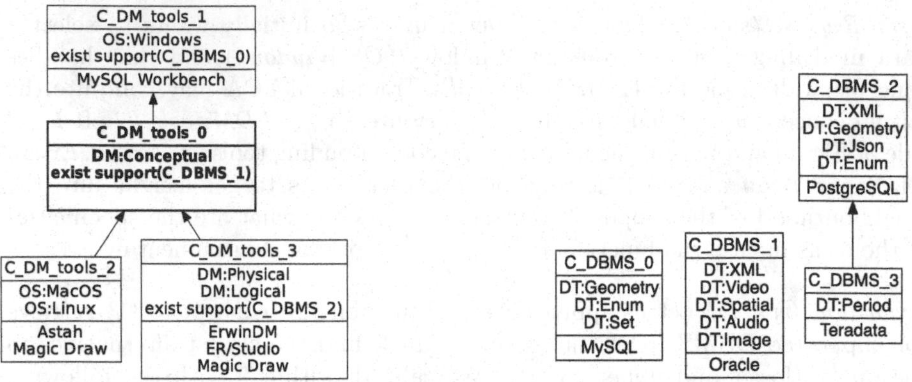

Fig. 3. (left) First iteration on DM_tools, (right) OC-poset on DBMS (results obtained with RCAExplore extension to on-demand algorithm)

User Request (step 2). The user can now select another concept from which they want to continue the navigation, e.g. *C_DM_tools_*1, because they think that the constraints of handling conceptual data models (*DM:Conceptual*) may not be so important.

Algorithm (run 2). In this case, the incremental algorithm is called again on the chosen concept, and computes its covers as shown in left-hand side of Fig. 4. In this example, the concept *C_DM_tools_*1 has no upper cover, being the top concept of the lattice, as we can confirm thanks to Fig. 2. The concept *C_DM_tools_*2 is a sub-concept of *C_DM_tools_*4 but their connection is not visible in Fig. 4 because their covers have not been computed yet. By selecting one of those concepts, their covers will be computed and the connection will be visible.

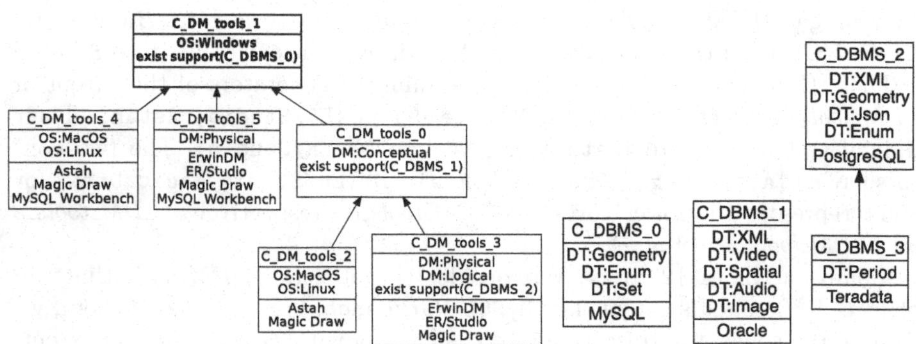

Fig. 4. (left) Second iteration on DM_tools, (right) OC-poset on DBMS (results obtained with RCAExplore extension to on-demand algorithm)

User Request (step 3) and Algorithm (run 3). In the follow-up, the user could continue exploring the concepts of *DM_tools*, or they could jump to a relational cover, which is one concept of *DBMS*, e.g. from *C_DM_tools_1* to *C_DBMS_0*. In this case, the user focuses on the kind of DBMS that are supported by the group of considered *DM_tools*. The algorithm will build the covers of *C_DBMS_0*, that are *C_DBMS_4* and *C_DBMS_5*, from Fig. 1 (right-hand side). There will be no relational cover, as *DBMS* is not the source of an object-object relation.

6 Related Work

Lattice structures are among the first structures used as a support for exploratory search [17], and this task has later attracted a lot of attention in Formal Concept Analysis theory [9]. Many works focus on *conceptual neighbourhood* to present both information related to a query and its closest variants [1,10,16]. Another approach proposes to retrieve cousin concepts [8] which are similar yet not comparable concepts. In this paper, we consider RCA to retrieve the conceptual neighbourhood in interconnected lattices, structuring both intrinsic and relational attributes.

The exponential growth of concept lattices is well-known [15]. As a consequence, the main limitation of FCA-based exploratory search lies in the complexity and computation of the structures [7]. Many solutions have been proposed to reduce the complexity of conceptual navigation. Some authors propose to prune the concept lattice to restrict the explorable dataspace, by computing iceberg concept lattices [27], or by applying constraints to bound the final structure [7]. To ease the navigation, the authors of [24] seek to extract more simplified browsable structures; they first extract a tree from the concept lattice, and then reduce the obtained tree using clustering and fault-tolerance methods. The tool `SearchSleuth` [10] enables FCA-based exploratory search for web queries, a field where the domain cannot be entirely processed using FCA and concept lattices. To tackle this issue, they generate a new formal context specific to a query at each navigation step. In a previous work [3], we proposed to compute the conceptual neighbourhood of a query in a sub-order of the concept lattice restricted to the attribute- and object-concepts (attribute-object-concept poset), a condensed alternative to concept lattices. At each step, only the conceptual neighbourhood is computed. In the present work, we also generate the conceptual neighbourhood on-the-fly, but this time in interconnected concept lattices.

Mimouni et al. [25] use RCA to structure, query and browse a collection of legal documents. First, they build interconnected lattices representing different types of legal documents referring to each other. Then, their approach allows for the retrieval of the concept corresponding to a user query, and to explore variations of this query by navigation in the neighbour concepts. In their approach, they compute all the lattices during the first step.

Ferré and Hermann [12] propose *Query-based Faceted Search* and an implementation in the tool `SEWELIS`, that allows to browse relational datasets in the form of RDF files. Also, Ferré et al. [13] propose RLCA, a relational extension of *Logical Formal Analysis*, an adaptation of FCA to describe objects by formulas of ad-hoc logics instead of binary attributes. While RCA computes connected yet separate concept lattices, one per sort of objects, RLCA gathers the objects, their descriptions and their relations to other objects in one structure, giving two complementary points of view on data. Besides, Ferré [14] introduces *abstract conceptual navigation* as an abstraction describing user guidance in a dynamic space of concepts connected by navigation links. Our approach fits into this vision.

7 Conclusion

In this paper, we proposed algorithms to compute the conceptual neighbourhood of a query in connected concept lattices generated with RCA. First, we redefined the traditional FCA derivation operators to take into account relational attributes. Then, we presented a way to compute the relational, upper and lower covers of a given concept in extended lattices, without computing all structures. Two RCA scaling operators, i.e., existential and universal strict, may be used. We illustrated how the algorithms work on a running example from the domain of software product line engineering. Also, we implemented our algorithms in RCAExplore, an existing tool to handle relational context families with relational concept analysis.

In the future, we plan to perform a systematic scalability study on real datasets from the projects Fresqueau and Knomana and from available product descriptions [4]. As the computation of minimal transversals of minimal generators may be a limitation while applying our algorithms on large datasets, we envision to develop metaheuristics to reduce the time complexity of their computation. We will also work on the presentation of the selectable attributes to the user. Our tests of the on-demand algorithm on the Fresqueau dataset showed that the extended formal contexts can have an important number of relational attributes, and it is difficult for the user to choose one of them. We also collected concrete questions from the Knomana project partners as they have some real exploration tasks in their domain. We evaluated the potentiality of constructing the whole conceptual structure on their data [22]. As a next step, we will qualitatively evaluate the benefits of the on-demand approach. In this evaluation, we will consider partial contexts to reduce the user focus with regard to their request, and we will implement strategies to keep records of the data exploration.

Acknowledgement. The authors warmly thank Xavier Dolques who helped us during the implementation in RCAexplore. This work was supported by the INRA-CIRAD Glofoods metaprogramme (Knomana project) and by the French National Research Agency under the Investments for the Future Program, referred as ANR-16-CONV-0004.

References

1. Alam, M., Le, T.N.N., Napoli, A.: LatViz: a new practical tool for performing interactive exploration over concept lattices. In: Proceedings of CLA 2016, pp. 9–20 (2016)
2. Arévalo, G., Berry, A., Huchard, M., Perrot, G., Sigayret, A.: Performances of galois sub-hierarchy-building algorithms. In: Kuznetsov, S.O., Schmidt, S. (eds.) ICFCA 2007. LNCS (LNAI), vol. 4390, pp. 166–180. Springer, Heidelberg (2007). https://doi.org/10.1007/978-3-540-70901-5_11
3. Bazin, A., Carbonnel, J., Kahn, G.: On-demand generation of AOC-posets: reducing the complexity of conceptual navigation. In: Kryszkiewicz, M., Appice, A., Ślęzak, D., Rybinski, H., Skowron, A., Raś, Z.W. (eds.) ISMIS 2017. LNCS (LNAI), vol. 10352, pp. 611–621. Springer, Cham (2017). https://doi.org/10.1007/978-3-319-60438-1_60
4. Ben Nasr, S., et al.: Automated extraction of product comparison matrices from informal product descriptions. J. Syst. Softw. **124**, 82–103 (2017)
5. Braud, A., Dolques, X., Huchard, M., Ber, F.L.: Generalization effect of quantifiers in a classification based on relational concept analysis. Knowl.-Based Syst. **160**, 119–135 (2018)
6. Carbonnel, J., Huchard, M., Nebut, C.: Towards the extraction of variability information to assist variability modelling of complex product lines. In: Proceedings of VAMOS 2018, pp. 113–120 (2018)
7. Carpineto, C., Romano, G.: Exploiting the potential of concept lattices for information retrieval with CREDO. J. Univers. Comp. Sci. **10**(8), 985–1013 (2004)
8. Codocedo, V., Lykourentzou, I., Napoli, A.: A semantic approach to concept lattice-based information retrieval. Ann. Math. Artif. Intell. **72**(1–2), 169–195 (2014)
9. Codocedo, V., Napoli, A.: Formal concept analysis and information retrieval – a survey. In: Baixeries, J., Sacarea, C., Ojeda-Aciego, M. (eds.) ICFCA 2015. LNCS (LNAI), vol. 9113, pp. 61–77. Springer, Cham (2015). https://doi.org/10.1007/978-3-319-19545-2_4
10. Ducrou, J., Eklund, P.W.: SearchSleuth: the conceptual neighbourhood of an web query. In: Proceedigs of CLA 2007, pp. 249–259 (2007)
11. Dunaiski, M., Greene, G.J., Fischer, B.: Exploratory search of academic publication and citation data using interactive tag cloud visualizations. Scientometrics **110**(3), 1539–1571 (2017)
12. Ferré, S., Hermann, A.: Reconciling faceted search and query languages for the semantic web. Int. J. Metadata Semant. Ontol. **7**(1), 37–54 (2012)
13. Ferré, S., Ridoux, O., Sigonneau, B.: Arbitrary relations in formal concept analysis and logical information systems. In: Dau, F., Mugnier, M.-L., Stumme, G. (eds.) ICCS-ConceptStruct 2005. LNCS (LNAI), vol. 3596, pp. 166–180. Springer, Heidelberg (2005). https://doi.org/10.1007/11524564_11
14. Ferré, S.: Reconciling Expressivity and Usability in Information Access - From Filesystems to the Semantic Web. Habilitation thesis, Matisse, Univ. Rennes 1 (2014). habilitation à Diriger des Recherches (HDR), defended on November 6th
15. Ganter, B., Wille, R.: Formal Concept Analysis. Springer, Heidelberg (1999). https://doi.org/10.1007/978-3-642-59830-2
16. Godin, R., Gecsei, J., Pichet, C.: Design of a browsing interface for information retrieval. In: Proceedings of SIGIR 1989, pp. 32–39 (1989)

17. Godin, R., Saunders, E., Gecsei, J.: Lattice model of browsable data spaces. Inf. Sci. **40**(2), 89–116 (1986)
18. Rouane, M.H., Huchard, M., Napoli, A., Valtchev, P.: A proposal for combining formal concept analysis and description logics for mining relational data. In: Kuznetsov, S.O., Schmidt, S. (eds.) ICFCA 2007. LNCS (LNAI), vol. 4390, pp. 51–65. Springer, Heidelberg (2007). https://doi.org/10.1007/978-3-540-70901-5_4
19. Huchard, M., Hacene, M.R., Roume, C., Valtchev, P.: Relational concept discovery in structured datasets. Ann. Math. Artif. Intell. **49**(1–4), 39–76 (2007)
20. Hébert, C., Bretto, A., Crémilleux, B.: A data mining formalization to improve hypergraph minimal transversal computation. Fundam. Informaticae **80**, 415–433 (2007)
21. Jäschke, R., Hotho, A., Schmitz, C., Ganter, B., Stumme, G.: TRIAS - an algorithm for mining iceberg tri-lattices. In: Proceedings of ICDM 2006, pp. 907–911 (2006)
22. Keip, P., et al.: Effects of input data formalisation in Relational Concept Analysis for a data model with a ternary relation. In: Proceedings of ICFCA 2019 (2019, to appear)
23. Marchionini, G.: Exploratory search: from finding to understanding. Commun. ACM **49**(4), 41–46 (2006)
24. Melo, C.A., Grand, B.L., Aufaure, M.: Browsing large concept lattices through tree extraction and reduction methods. Int. J. Intell. Inf. Technol. **9**(4), 16–34 (2013)
25. Mimouni, N., Nazarenko, A., Salotti, S.: A conceptual approach for relational IR: application to legal collections. In: Baixeries, J., Sacarea, C., Ojeda-Aciego, M. (eds.) ICFCA 2015. LNCS (LNAI), vol. 9113, pp. 303–318. Springer, Cham (2015). https://doi.org/10.1007/978-3-319-19545-2_19
26. Palagi, É., Gandon, F.L., Giboin, A., Troncy, R.: A survey of definitions and models of exploratory search. In: ACM Workshop ESIDA@IUI, pp. 3–8 (2017)
27. Stumme, G., Taouil, R., Bastide, Y., Pasquier, N., Lakhal, L.: Computing iceberg concept lattices with Titanic. Data Knowl. Eng. **42**(2), 189–222 (2002)

Mining Formal Concepts Using Implications Between Items

Aimene Belfodil[1,2(✉)], Adnene Belfodil[1(✉)], and Mehdi Kaytoue[1,3]

[1] Univ Lyon, INSA Lyon, CNRS, LIRIS UMR 5205, 69621 Lyon, France
{aimene.belfodil,adnene.belfodil,mehdi.kaytoue}@insa-lyon.fr
[2] Mobile Devices Ingénierie, 100 Avenue Stalingrad, 94800 Villejuif, France
[3] Infologic, 99 Avenue de Lyon, 26500 Bourg-Lès-Valence, France

Abstract. Formal Concept Analysis (FCA) provides a mathematical tool to analyze and discover concepts in Boolean datasets (i.e. Formal contexts). It does also provide a tool to analyze complex attributes by transforming them into Boolean ones (i.e. items) thanks to *conceptual scaling.* For instance, a numerical attribute whose values are $\{1, 2, 3\}$ can be transformed to the set of items $\{\leq 1, \leq 2, \leq 3, \geq 3, \geq 2, \geq 1\}$ thanks to interordinal scaling. Such transformations allow us to use standard algorithms like *Close-by-One* (CbO) to look for concepts in complex datasets by leveraging a closure operator. However, these standard algorithms do not use the relationships between attributes to enumerate the concepts as for example the fact that ≤ 1 implies ≤ 2 and so on. For such, they can perform additional closure computations which substantially degrade their performance. We propose in this paper a generic algorithm, named CbOI for *Close-by-One using Implications*, to enumerate concepts in a formal context using the inherent implications between items provided as an input. We show that using the implications between items can reduce significantly the number of closure computations and hence the time effort spent to enumerate the whole set of concepts.

1 Introduction

Formal Concept Analysis (FCA) [8,25] provides a mathematical tool to analyze and discover concepts in Boolean datasets (i.e. Formal contexts). It does also provide a tool to analyze complex attributes by transforming them to Boolean ones (i.e. items) thanks to *conceptual scaling* [11]. For instance, a numerical attribute whose values are $\{1, 2, 3\}$ can be transformed to a set of items $\{\leq 1, \leq 2, \leq 3, \geq 3, \geq 2, \geq 1\}$ thanks to *interordinal scaling* (see Fig. 1). Such transformations allow to use standard algorithms for enumerating all formal concepts in a formal context [5,9,17,18] by leveraging a closure operator. However, these standard algorithms do not take advantage of the relationships between attributes to enumerate the concepts as for example the fact that "≤ 1 implies ≤ 2" and so on. For such, they perform additional closure computations which substantially degrade their performance. Some algorithms [2,6,14] have been proposed in the literature to handle some particular instances of contexts

D. Cristea et al. (Eds.): ICFCA 2019, LNAI 11511, pp. 173–190, 2019.
https://doi.org/10.1007/978-3-030-21462-3_12

with implications. For instance, interordinal scaled contexts are directly linked to interval patterns as investigated by [14]. On the other hand, ordinal scaled contexts are linked to datasets augmented with a taxonomy (i.e hierarchy) between items (e.g. *cats* are *animals*) [2,6]. Yet, to the best of our knowledge, when a formal context is provided with an arbitrary set of implications (i.e. forming some directed graph between items), no generic algorithm is provided.

In this paper, we propose a generic algorithm, named `CbOI` for *Close-by-One using Implications*, to enumerate formal concepts using the inherent implications between items provided as an input. In other words, provided a pair (formal context, directed graph of implications between attributes), `CbOI` uses at its best the provided implications between items to enumerate exhaustively and non-redundantly formal concepts. The proposed algorithm relies on a *Divide & Conquer* scheme to enumerate closed sets in a strongly accessible set system [4,12,17]. In fact, we show that closed sets are upsets (i.e. upward closed) in an equivalent poset of the input directed graph. We use then the fact that the set of these upsets forms a strongly accessible set system since it is an anti-matroid [7]. Building on these notions, we elaborate algorithm `CbOI`.

This paper is organized as follow: Sect. 2 recalls basic definitions about binary relations and partially ordered sets as well as definitions from Formal Concept Analysis. Section 3 introduces the *Divide & Conquer* Algorithm following [4] and explain the drawbacks of not using the implications to enumerate concepts. Next, Sect. 4 presents the formalization of the problem as well as the newly proposed algorithm dubbed `CbOI`. Details about implementation[1] of `CbOI` as well as its experimental evaluation are presented and discussed in Sect. 5.

2 Preliminary Definitions

In this paper, given an arbitrary set E, $\wp(E)$ denotes the powerset of E (i.e. $\wp(E) = \{A \mid A \subseteq E\}$). Moreover, for any application $f : E \to F$, and for any subset $A \subseteq E$, $f[A]$ denotes the image of A by f (i.e. $f[A] = \{f(a) \mid a \in A\}$).

2.1 Binary Relations, Pre-orders and Partial Orders

This section recalls basic definitions on binary relations, pre-ordered sets and partially ordered sets. More details can be found in [22]. Let E be an arbitrary set. A *binary relation* R over a set E is an arbitrary subset of $E \times E$. For any element (a, b) in R we denote aRb. A binary relation R is said to be: *(1) reflexive:* $(\forall a \in E)\ a\,R\,a$. *(2) transitive:* $(\forall a, b, c \in E)$ if $a\,R\,b$ and $b\,R\,c$ then $a\,R\,c$. *(3) symmetric:* $(\forall a, b \in E)$ if $a\,R\,b$ then $b\,R\,a$. and *(4) anti-symmetric:* $(\forall a, b \in E)$ if $a\,R\,b$ and $b\,R\,a$ then $a = b$.

A binary relation $\to$ on E is said to be a *pre-order* on E if it is *reflexive* and *transitive*. The pair $(E, \to)$ is said to be a *preordered-set* or a *proset* for short.

A binary relation $\leftrightarrow$ on E is said to be an *equivalence relation* on E if it is a *symmetric preorder* on E. The equivalence class of $a \in E$ the set of its equivalent

[1] **Source Code.** https://github.com/BelfodilAimene/CbOImplications.

element $\{b \in E \mid b \leftrightarrow a\}$. The quotient set is the partition of E on equivalent classes and we denote it $E/\leftrightarrow$.

Partially Ordered Sets. A binary relation $\leq$ on E is said to be a *partial order* on E if it is an *anti-symmetric preorder* on E. The pair $(E, \leq)$ is said to be a *partially ordered set* or a *poset* for short.

A mapping $\sigma : E \to E$ is said to be a closure operator on a poset $(E, \leq)$ iff it is: *(1) extensive:* $(\forall a \in E)\ a \leq \sigma(a)$, *(2) monotonous:* $(\forall a, b \in E)$ if $a \leq b$ then $\sigma(a) \leq \sigma(b)$ and *(3) idempotent:* $(\forall a \in E)\ \sigma(\sigma(a)) = \sigma(a)$. The set $\sigma[E] = \{\sigma(e) \mid e \in E\}$ is then called the set of fixpoints of σ.

Up-sets and Up-closure. Given a poset $(E, \leq)$, a subset $S \subseteq E$ is said to be an up-set (or upper-ideal) *iff*: $(\forall x \in S, \forall y \in E)\ x \leq y \Rightarrow y \in S$. There is an operator $\uparrow$ on $\wp(E)$ that associates to any subset $S \subseteq E$, the smallest up-set enclosing it. It is given by: $\uparrow S = \{e \in E \mid (\exists s \in S)\ s \leq e\}$. Operator $\uparrow$ is called *up closure* and is a closure operator on $(\wp(E), \subseteq)$. The set of all up-sets, denoted $\mathcal{U}(E)$, is given by the set of all fix-points of $\uparrow$. It is closed under arbitrary intersection and arbitrary union (i.e. $(\mathcal{U}(E), \subseteq)$ does form a complete sublattice of $(\wp(E), \subseteq)$). Down-sets and down closure $\downarrow$ can be defined analogically. Please note that for $s \in S$, we denote $\uparrow s$ and $\downarrow s$ rather than $\uparrow \{s\}$ and $\downarrow \{s\}$. Sets $\uparrow s$ and $\downarrow s$ are called principal filter and principal ideal respectively. In fact, we have $\uparrow S = \bigcup_{s \in S} \uparrow s$ and $\downarrow S = \bigcup_{s \in S} \downarrow s$.

For a subset $S \subseteq E$, the set of minimal elements $min(S)$ and $max(S)$ are given respectively by the lower and the upper borders of S. Formally:

$$min(S) = \{s \in S \mid \downarrow s \cap S = \{s\}\} \quad \text{and} \quad max(S) = \{s \in S \mid \uparrow s \cap S = \{s\}\}$$

One important remark for a finite posets $(E, \leq)$ is that for any up-set $S \in \mathcal{U}(E)$ we have $S = \uparrow min(S)$. Moreover, $min(S)$ represents the smallest subset C in S (w.r.t. $\subseteq$) such that $S = \uparrow C$.

Last but not least, any finite poset $(E, \leq)$ can be represented by its Hasse Diagram [22]: the transitive reduction [1] of directed acyclic graph (dag) $(E, \leq)$. It does represent for each element $e \in E$ the set of its direct lower (resp. upper) neighbors $lowers(E)$ (resp. $uppers(E)$). For $e \in E$, we have:

$$lowers(e) = \{l \in E \mid e \text{ covers } l\} = max(\downarrow e \backslash \{e\})$$
$$uppers(e) = \{u \in E \mid u \text{ covers } e\} = min(\uparrow e \backslash \{e\})$$

For $e_1, e_2 \in E$, we say that e_2 *covers* e_1 iff $e_1 \neq e_2$, $e_1 \leq e_2$ and there is no element e lying strictly between e_1 and e_2: i.e. $(\forall e \in E)\ e_1 \leq e \leq e_2 \Rightarrow e = e_1$ or $e = e_2$.

2.2 Formal Concept Analysis

Formal Concept Analysis (FCA) was introduced in [25] as a mathematical framework to analyze and manipulate concepts in (binary) databases. *FCA* starts by a *formal context* [8]. A *(formal) context* is a triple $\mathbb{K} = (\mathcal{G}, \mathcal{M}, \mathcal{I})$ consisting of two sets $\mathcal{G}$ and $\mathcal{M}$ and an incidence relation $\mathcal{I}$ between $\mathcal{G}$ and $\mathcal{M}$. Elements of

Table 1. A formal context.

$\mathcal{G}$	Any	Hotel	Restaurant	Chinese restaurant	Italian restaurant
$place_1$	×	×			
$place_2$	×	×	×	×	
$place_3$	×		×		×

$\mathcal{G}$ are called objects and elements of $\mathcal{M}$ are called attributes or items. In order to express that an object $g \in \mathcal{G}$ *has* an attribute $m \in \mathcal{M}$, we write $g\mathcal{I}m$ or $(g, m) \in \mathcal{I}$. A formal context $\mathbb{K}$ is said to be *finite* if $\mathcal{G}$ and $\mathcal{M}$ are both finites. Note that subsets of items $B \subseteq \mathcal{M}$ are called *patterns* or *itemsets*. Table 1 depicts a *formal context* $(\mathcal{G}, \mathcal{M}, \mathcal{I})$ where objects in $\mathcal{G}$ are places while attributes in $\mathcal{M}$ are place tags. The incidence relation $\mathcal{I}$ is represented by the crosses in the table and it represents the fact that a *place* is tagged by a *tag*. For instance, we have $place_3\ \mathcal{I}\ Restaurant$ that can be read as: "*$place_3$ is tagged as a Restaurant*".

Two fundamental operators, namely *extent* and *intent*, are defined on a formal context $\mathbb{K} = (\mathcal{G}, \mathcal{M}, \mathcal{I})$. The *extent operator*, denoted *ext*, associates to each itemset $B \subseteq \mathcal{M}$ the set of objects $g \in \mathcal{G}$ having all items in B. Formally, it is given by: $ext : \wp(\mathcal{M}) \to \wp(\mathcal{G}), B \mapsto \{g \in \mathcal{G} \mid (\forall m \in B)\, g\mathcal{I}\, m\}$. The set of all possible extents of a context $\mathbb{K}$ is denoted $\mathbb{K}_{ext}$ and is given by $\mathbb{K}_{ext} = ext[\wp(\mathcal{M})]$. Dually, the *intent operator*, denoted *int*, associates to each subset of objects $A \subseteq \mathcal{G}$ the set of items $m \in \mathcal{M}$ common to the objects in A. Formally, it is given by: $int : \wp(\mathcal{G}) \to \wp(\mathcal{M}), A \mapsto \{m \in \mathcal{M} \mid (\forall g \in A)\, g\mathcal{I}\, m\}$. The set of all possible intents of a context $\mathbb{K}$ is denoted $\mathbb{K}_{int}$ and is given by $\mathbb{K}_{int} = int[\wp(\mathcal{G})]$.

For ease of notations, for $g \in \mathcal{G}$ and $m \in \mathcal{M}$, we denote respectively $int(g)$ and $ext(m)$ rather than $int(\{g\})$ and $ext(\{m\})$. Note that, for $B \subseteq \mathcal{M}$ and $A \subseteq \mathcal{G}$, we have: $ext(B) = \bigcap_{m \in B} ext(m)$ and $int(A) = \bigcap_{g \in A} int(g)$. A key theorem in FCA (Proposition 10 in [8]) is given below:

Theorem 1. *The pair of functions (ext, int) form a Galois connection between the powerset lattices $(\wp(\mathcal{G}), \subseteq)$ and $(\wp(\mathcal{M}), \subseteq)$. That is $ext \circ int$ and $int \circ ext$ are closure operators on $(\wp(\mathcal{G}), \subseteq)$ and $(\wp(\mathcal{M}), \subseteq)$ respectively. Hence, $\mathbb{K}_{ext}$ and $\mathbb{K}_{int}$ are* Moore Families *(i.e. closed under arbitrary intersection).*

This theorem allows one to build what is called a *concept lattice* $(\mathfrak{B}(\mathbb{K}), \leq)$ which is a complete lattice. Elements of $\mathfrak{B}(\mathbb{K})$ are called (formal) concepts and are of the form $(A, B) \in \wp(\mathcal{G}) \times \wp(\mathcal{M})$ such that $A = ext(B)$ and $B = int(A)$. In other words: $\mathfrak{B}(\mathbb{K}) = \{(ext(B), B) \mid B \in \mathbb{K}_{int}\} = \{(A, int(A)) \mid A \in \mathbb{K}_{ext}\}$. For two concepts (A_1, B_1) and (A_2, B_2) in $\mathfrak{B}(\mathbb{K})$, we say that (A_1, B_1) is a *subconcept* of (A_2, B_2) and we denote $(A_1, B_1) \leq (A_2, B_2)$ if $A_1 \subseteq A_2$ (or equivalently $B_2 \subseteq B_1$). From now on, elements of $\mathbb{K}_{int}$ are called *closed patterns* since they are the fixpoints of the closure operator $int \circ ext$.

As a matter of example, consider again the formal context depicted in Fig. 2, we have $ext(\{hotel, restaurant\}) = \{place_2\}$, meaning that the only place that is both tagged *hotel* and *restaurant* is $place_2$. Dually $int(\{place_2, place_3\}) = \{Any, Restaurant\}$, that is what is common to $place_2$ and $place_3$ is that they are both tagged as *Any* and as *Restaurant*. Since $ext(\{Any, Restaurant\}) = \{place_2, place_3\}$, the pair $(\{place_2, place_3\}, \{Any, Restaurant\})$ is a concept.

Another important notion in FCA tightly linked to formal concepts and closed sets are implications (see Definition 1). For example, we have in the context depicted in Table 1 that *Italian Restaurant* $\rightarrow$ *Restaurant*. Indeed, all *Italian Restaurant* are *Restaurant*.

$\mathcal{G}$	x	y
g_1	1	4
g_2	2	2
g_3	2	2
g_4	3	2

$\mathcal{G}$	$x \geq 1$	$x \geq 2$	$x \geq 3$	$x \leq 3$	$x \leq 2$	$x \leq 1$	$y \geq 2$	$y \geq 4$	$y \leq 4$	$y \leq 2$
g_1	×			×	×	×	×	×	×	
g_2	×	×		×	×		×		×	×
g_3	×	×		×	×		×	×	×	
g_4	×	×	×	×			×	×	×	

Fig. 1. **(left)** A numerical dataset with 2 numerical attributes. **(right)** A formal context that is the result of an *interordinal scaling* of the numerical dataset.

Definition 1 ***Item-Implications.*** *Let $A, B \subseteq \mathcal{M}$ be two itemsets. We say that A implies B and we denote $A \rightarrow B$ iff: $ext(A) \subseteq ext(B)$. In other words, if an object has all items in the set of attributes A then it has all items in itemset B. For two items $a, b \in \mathcal{M}$, we call an implication $\{a\} \rightarrow \{b\}$* item-implication *and we denote it $a \rightarrow b$ for ease of notation.*

Handling Complex Data in Formal Concept Analysis. While basic FCA provides a tool to analyze (formal) contexts (i.e. Boolean datasets), several techniques are proposed in FCA literature to handle more complex datasets like numerical ones. The most straightforward way, is called *(conceptual) scaling* [8,11]. It is the action of transforming a dataset to a formal context w.r.t. the nature of patterns we are looking for.

For instance, if we want to look for (closed) interval patterns in numerical datasets [14] (i.e. (closed) n-dimensional hyper-rectangle in the euclidean space), we use *interordinal scaling*. Figure 1 depicts such a transformation. It can be shown that the set of closed patterns (i.e. intents) with a non-empty extent in the interordinal scaled context encoding a numerical dataset represents the set of non-empty extent closed interval patterns [14].

Other example of datasets are those augmented by a taxonomy (i.e. hierarchy of items) [6]. For instance Fig. 2 depicts such a dataset. In general, patterns we are looking for in such datasets are (closed) itemsets using antichains of the taxonomy. *Ordinal scaling* can be used to handle such a type of datasets. The ordinal scaling of the dataset presented in Fig. 2 is presented in Table 1. Note that FCA offers other more sophisticated tools to handle complex data, we can cite among other techniques pattern structures [10,19] and pattern setups [3,21].

$\mathcal{G}$	*tags*
$place_1$	{Hotel}
$place_2$	{Chinese Restaurant, Hotel}
$place_3$	{Italian Restaurant}

Fig. 2. **(left)** Point-of-interests annotated with tags and **(right)** a taxonomy of tags.

3 Formal Concepts Enumeration

The enumeration of formal concepts is an equivalent task to the enumeration of closed patterns in a formal context. Indeed, for a formal context $\mathbb{K}$, we have $\mathfrak{B}(\mathbb{K}) = \{(ext(C), C) \mid C \in \mathbb{K}_{int}\}$ where $\mathbb{K}_{int}$ is no more than the set of fixpoints of the closure operator $int \circ ext$. Many algorithms are proposed in the literature [5,9,17,18] (among others) to enumerate exhaustively and non-redundantly closed sets of some closure operator σ on $(\wp(E), \subseteq)$ with E an arbitrary finite set. A. Gély showed that many of those algorithms can be seen as an instance of a more general *divide-&-conquer* algorithm proposed in [12]. Later, Boley et al. [4] proved that such a generalization can be used to enumerate exhaustively and non-redundantly closed sets (i.e. fix-points) of a closure operator $\sigma : \mathcal{F} \to \mathcal{F}$ in a *finite strongly accessible set system* $(E, \mathcal{F})$ rather than only for the case $\mathcal{F} = \wp(E)$. We recall below some definitions about set systems.

A *set system* on an arbitrary finite set E is a pair $(E, \mathcal{F})$ where $\mathcal{F}$ is a set of subsets of E (i.e. $\mathcal{F} \subseteq \wp(E)$) called the set of *feasible sets*. For a set system $(E, \mathcal{F})$, we present below some properties:

- *[P0]* $\emptyset \in \mathcal{F}$.
- *[P1]* $\forall S \in \mathcal{F} \backslash \{\emptyset\}\ \exists e \in E$ s.t. $S \backslash \{e\} \in \mathcal{F}$.
- *[P2]* $\forall S, T \in \mathcal{F}$ s.t. $S \subsetneq T$ we have $(\exists e \in T \backslash S)\ S \cup \{e\} \in \mathcal{F}$.
- *[P3]* $\forall S, T \in \mathcal{F}$ s.t. $|S| < |T|$ we have $(\exists e \in T \backslash S)\ S \cup \{e\} \in \mathcal{F}$.
- *[P4]* $\forall S, T \in \mathcal{F}$ s.t. $T \nsubseteq S$ we have $(\exists e \in T \backslash S)\ S \cup \{e\} \in \mathcal{F}$.
- *[P5]* $\forall S, T \in \mathcal{F}$, $S \cup T \in \mathcal{F}$ (i.e. $\mathcal{F}$ is closed under set union).

Implications between properties are depicted in Fig. 3. A set system $(E, \mathcal{F})$ is said to be *accessible* if it has property *[P1]* and *strongly accessible* [4] if it has both *[P1]* and *[P2]* properties. It is said to be a *greedoid* [15] if it has both *[P1]* and *[P3]* and an *anti-matroid* [7] if it has properties *[P0]* and *[P4]* (or equivalently *[P1]* and *[P5]*). Note that anti-matroids have the 6 aforementioned properties. *Anti-matroids* are *greedoids* and *greedoids* are *strongly accessibles*.

Given a finite strongly accessible set system $(E, \mathcal{F})$ and a closure operator $\sigma : \mathcal{F} \to \mathcal{F}$. Boley et al. [4] showed that Algorithm 1, dubbed *Divide & Conquer Closed Set Listing* (D&C for short), enumerates exhaustively and non-redundantly fixpoints of σ (i.e. $\sigma[\mathcal{F}]$). Algorithm 1 starts from the smallest element $\sigma(\emptyset)$ in $\sigma[\mathcal{F}]$ (Line 8) then enumerates in depth-first fashion concepts in $\sigma[\mathcal{F}]$ by performing closure computations (Line 3) then checking, thanks to *canonicity test* (Line 4), if the closed set is already generated making D&C non-redundant.

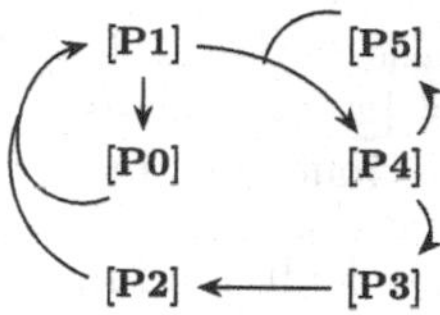

Fig. 3. Implications between set system properties (in *finite* set systems). Note that *[P0] and [P2]* together imply *[P1]*; and *[P1] and [P5]* together imply *[P4]*.

Algorithm 1. D&C (Divide & Conquer Closed Set Listing) Algorithm

---

Input: Finite strongly accessible set system $(E, \mathcal{F})$ and
a closure operator σ on $(\mathcal{F}, \subseteq)$.
Output: Elements of $\sigma[\mathcal{F}]$

1 **procedure** D&C(C, B)
2   **for** $e \in E \backslash (B \cup C)$ *s.t.* $C \cup \{e\} \in \mathcal{F}$ **do**
3     $C_{new} \leftarrow \sigma(C \cup \{e\})$ // Compute the new closed element C_{new}
4   **if** $C_{new} \cap B = \emptyset$ **then**
5   D&C(C_{new}, B)
6     $B \leftarrow B \cup \{e\}$ // Update the set of banned elements B
7   **Print**(C) // Output the closed element $C \in \sigma[\mathcal{F}]$
8 D&C$(\sigma(\emptyset), \emptyset)$

Concepts Enumeration and Eventual Problems. Algorithm D&C can be instantiated easily for instance to enumerate closed patterns (or concepts) of a formal context $\mathbb{K} = (\mathcal{G}, \mathcal{M}, \mathcal{I})$ in a top-down fashion. Indeed, the considered set system can be the general one $(\mathcal{M}, \wp(\mathcal{M}))$ and the closure operator is $int \circ ext$. Rather than outputting only closed patterns $C \in \mathbb{K}_{int}$ (Line 7), one can print the concept $(ext(C), C)$. Such an algorithm is in fact Close-by-One (CbO) [17,18]. Algorithm CbO has a delay time complexity[2] of $\mathcal{O}(|\mathcal{G}||\mathcal{M}|^2)$, a total time complexity of $\mathcal{O}(|\mathcal{G}||\mathcal{M}|^2|\mathbb{K}_{int}|)$. Please note that the time complexity of the closure computation is given by $\mathcal{O}(|\mathcal{G}||\mathcal{M}|)$.

However, such an enumeration does not use the inherent implications between items. For instance, let us reconsider the *interordinal scaled context* presented in Fig. 1, we present below some steps of execution of CbO:

1. **Begins:** $C_0 = \{x \geq 1, x \leq 3, y \geq 2, y \leq 4\}$ and $B = \emptyset$.
2. **Add item $\mathbf{x \geq 2}$ to** C_0: $C_{new} = \{x \geq 1, \mathbf{x \geq 2}, x \leq 3, y \geq 2, y \leq 4\}$ at Line 3. The canonicity test does not fail since $B = \emptyset$ and algorithm continues by enumerating all subconcepts of $(ext(C_{new}), C_{new})$. Further, at line 6, we have $B = \{\mathbf{x \geq 2}\}$.

---

[2] **Delay time:** maximum time between two outputs, between the beginning and the first output and between the last output and the ending of an enum. algo. (cf. [13]).

3. **Add item $\mathbf{x \geq 3}$ to** C_0: $C_{new} = \{x \geq 1, \underline{x \geq 2}, \mathbf{x \geq 3}, x \leq 3, y \geq 2, y \geq 4, y \leq 4\}$ at Line 3. Since $B = \{\underline{x \geq 2}\}$, canonicity test fails since $C_{new} \cap B = \{\underline{x \geq 2}\}$. The enumeration continues until the end.

The problem shown beforehand after adding $x \geq 3$ is the fact that there was a useless closure computation that led to a certain failure. One could avoid this closure computation if the inherent implication $x \geq 3 \rightarrow x \geq 2$ is used. Indeed, since $x \geq 3 \rightarrow x \geq 2$, any closed itemset containing $x \geq 3$ contains $x \geq 2$. This shows that one can avoid some closure computations if implications are used properly, or in other words, avoid visiting some non valid closed itemsets. Moreover, such implications are sometime known from the user since they are *inherent* to the attributes and not derived from the incidence relation of the context. This is the case of interordinal and ordinal scaled datasets for example. While some state-of-the-art algorithms try to use this knowledge (i.e. implications between some items of the context) in some particular datasets as it is the case of numerical datasets (interordinal scaled contexts) [14] or datasets augmented with a taxonomy of items (ordinal scaled contexts) [2,6]; no general algorithm has been proposed to enumerate concepts in a context while taking benefit from an arbitrary provided set of item-implications (cf. Definition 1).

4 Using Item-Implications to Enumerate Concepts

In this section, we start by formalizing the user inputs, that is: the formal context and the implications between items provided by the user. Next, we show how to transform this pair of inputs to an equivalent pair with a poset between (sets of) items. We show then that the closed patterns of the input context are, up to the aforementioned transformation, in the anti-matroid of up-sets of the poset (cf. Propositions 2 and 4). The proposed algorithm is then a straightforward implementation of `D&C` since anti-matroids are strongly accessibles.

Let $\mathbb{K} = (\mathcal{G}, \mathcal{M}, \mathcal{I})$ be a finite formal context.

Definition 2. *The* item-implication basis *associated to a context* $\mathbb{K}$*, denoted* $\rightarrow$*, is given below:*

$$\rightarrow := \{(a, b) \in \mathcal{M} \times \mathcal{M} \mid a \rightarrow b\} = \{(a, b) \in \mathcal{M} \times \mathcal{M} \mid ext(a) \subseteq ext(b)\}$$

The definition given beforehand regroups all item-implications existing in the context. Moreover, pair $(\mathcal{M}, \rightarrow)$ forms a pre-ordered set. We model now the item-implication basis known/provided by the user. We can say informally that such a set of implications are those that are inherent to the attributes (not necessarily derived) from the incidence relation.

Definition 3. *A* valid Item-Implication basis *for* $\mathbb{K}$ *is any* sub relation $\mathfrak{I}$ *of* $\rightarrow$.

User Inputs. The input is then a pair $(\mathbb{K}, \mathfrak{I})$ where $\mathbb{K}$ is a finite formal context and $\mathfrak{I}$ is a valid item-implication basis for $\mathbb{K}$. Figure 4 (left) depicts an example of the input pair $(\mathbb{K}, \mathfrak{I})$. It is clear that $\mathfrak{I}$ provides a partial information about a pre-order. Relation $\mathfrak{I}$ can be augmented to a sub pre-order $\rightarrow_{\mathfrak{I}}$ of $\rightarrow$ thanks to *reflexive closure* (i.e. adding (m, m) to $\mathfrak{I}$ for all $m \in \mathcal{M}$) and *transitive closure* (i.e. the smallest transitive relation containing $\mathfrak{I}$). Thus we will be dealing from now on with an equivalent pair $(\mathbb{K}, \rightarrow_{\mathfrak{I}})$ where $(\mathcal{M}, \rightarrow_{\mathfrak{I}})$ is a pre-ordered set.

4.1 Building a Partial Order

Pre-order $\rightarrow_{\mathfrak{I}}$ could contain some cycles (i.e. not anti-symmetric). One can define an equivalence relation $\leftrightarrow_{\mathfrak{I}}$ on $\mathcal{M}$ such that $(\forall a, b \in \mathcal{M})\ a \leftrightarrow_{\mathfrak{I}} b$ iff $a \rightarrow_{\mathfrak{I}} b$ and $b \rightarrow_{\mathfrak{I}} a$. Please note that if $a \leftrightarrow_{\mathfrak{I}} b$ then $ext(a) = ext(b)$. With $\mathcal{M}' = \mathcal{M}/\leftrightarrow_{\mathfrak{I}}$ the quotient set of $\mathcal{M}$ on $\leftrightarrow_{\mathfrak{I}}$ and the following relation $\leq_{\mathfrak{I}}$:

$$(\forall S_1, S_2 \in \mathcal{M}')\ S_1 \leq_{\mathfrak{I}} S_2 \text{ iff } (\exists a \in S_1, \exists b \in S_2)\ a \rightarrow_{\mathfrak{I}} b$$

One can show that $(\mathcal{M}', \leq_{\mathfrak{I}})$ does form a partially ordered set (poset). Accordingly, context $\mathbb{K} = (\mathcal{G}, \mathcal{M}, \mathcal{I})$ is also transformed to $\mathbb{K}' = (\mathcal{G}, \mathcal{M}', \mathcal{I}')$ where:

$$(\forall g \in \mathcal{G}, \forall S \in \mathcal{M}')\ \ g\,\mathcal{I}'\,S\,\textbf{iff}\,(\forall m \in S)\ g\,\mathcal{I}\,m$$

Figure 4 gives an example of such a transformation from $\mathbb{K}$ to $\mathbb{K}'$ and from $(\mathcal{M}, \rightarrow_{\mathfrak{I}})$ to $(\mathcal{M}, \leq_{\mathfrak{I}})$. Note that this transformation is a *partial column-clarification* (see [8]) in the sense that it does concern only the item-implication basis $\rightarrow_{\mathfrak{I}}$ provided by the user since we want to use only the *user inputs*. If the total item-implication basis $\rightarrow$ associated with the context is used, the beforehand transformation will be equivalent to a column clarification. Proposition 1 provides that looking for concepts (or extents) in $\mathbb{K}$ is equivalent to look for concepts in the partially clarified context $\mathbb{K}'$. As such, from now on, we will consider the pair $(\mathbb{K}, \leq_{\mathfrak{I}})$ such that $(\mathcal{M}, \leq_{\mathfrak{I}})$ is a partial order. If not so, the context and the item-implication basis are transformed as shown beforehand (cf. Fig. 4).

Proposition 1. *We have* $\mathbb{K}_{ext} = \mathbb{K}'_{ext}$.

Proof. Recalling that $\mathbb{K}_{ext}$ and $\mathbb{K}'_{ext}$ are closed under arbitrary intersection. We prove the double inclusion:

- $\mathbb{K}'_{ext} \subseteq \mathbb{K}_{ext}$: Let $A \in \mathbb{K}'_{ext}$, that is $\exists \mathbb{S} \subseteq \mathcal{M}'$ such that $ext_{\mathbb{K}'}(\mathbb{S}) = A$, that is: $A = \bigcap_{S \in \mathbb{S}} ext_{\mathbb{K}'}(\{S\})$. However, it is clear that: $ext_{\mathbb{K}'}(\{S\}) = \{g \in \mathcal{G} \mid g\,\mathcal{I}'\,S\} = \{g \in \mathcal{G} \mid (\forall m \in S)\ g\,\mathcal{I}\,m\} = ext_{\mathbb{K}}(S)$. Hence, $A = \bigcap_{S \in \mathbb{S}} ext_{\mathbb{K}}(S)$. In other words, $A \in \mathbb{K}_{ext}$ since it is the intersection of some elements in $\mathbb{K}_{ext}$.
- $\mathbb{K}_{ext} \subseteq \mathbb{K}'_{ext}$: Let $A \in \mathbb{K}_{ext}$, that is: $\exists B \subseteq \mathcal{M}$ such that $ext_{\mathbb{K}}(B) = A$, that is: $A = \bigcap_{m \in B} ext_{\mathbb{K}}(\{m\})$. For $m \in B$, let $S_m \in \mathcal{M}'$ be the unique set containing m. We have $ext_{\mathbb{K}'}(\{S_m\}) = ext_{\mathbb{K}}(\{m\})$. Hence, $A = \bigcap_{m \in B} ext_{\mathbb{K}'}(\{S_m\})$; that is $A \in \mathbb{K'}_{ext}$ since it is an intersection of some elements in $\mathbb{K'}_{ext}$. $\square$

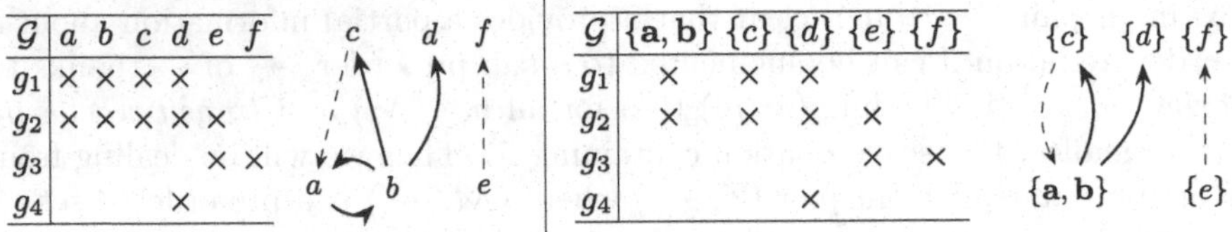

$\mathcal{G}$	a	b	c	d	e	f
g_1	×	×	×	×		
g_2	×	×	×	×	×	
g_3					×	×
g_4				×		

$\mathcal{G}$	$\{\mathbf{a,b}\}$	$\{c\}$	$\{d\}$	$\{e\}$	$\{f\}$
g_1	×	×	×		
g_2	×	×	×	×	
g_3				×	×
g_4			×		

Fig. 4. **(left)** Input context $\mathbb{K} = (\mathcal{G}, \mathcal{M}, \mathcal{I})$, the input-item implications $\mathfrak{I}$ (continuous arrows) and the item-implications of the context that are valid but not provided (dashed arrows). **(right)** The result of transformation to the context $\mathbb{K}' = (\mathcal{G}, \mathcal{M}', \mathcal{I}')$ and the Hasse diagram of the poset $(\mathcal{M}', \leq_{\mathfrak{I}})$ (continuous arrows).

4.2 Closed Patterns are Up-Sets

Consider now the obtained pair $(\mathbb{K}, \leq_{\mathfrak{I}})$ where $\mathbb{K} = (\mathcal{G}, \mathcal{M}, \mathcal{I})$ is a context and $(\mathcal{M}, \leq_{\mathfrak{I}})$ is a poset s.t. $\forall a, b \in \mathcal{M}$, we have if $a \leq_{\mathfrak{I}} b$ then $ext(a) \subseteq ext(b)$.

Lemma 1. $\forall C \subseteq \mathcal{M} : \uparrow C \subseteq int \circ ext(C)$

Proof. Let $m \in \uparrow C$, thus $\exists c \in C$ such that $c \rightarrow_{\mathfrak{I}} m$ or in other words, $ext(c) \subseteq ext(m)$. Since int is order-reversing ((ext, int) is a Galois connection (see Theorem 1)), $int \circ ext(m) \subseteq int \circ ext(c)$. Since $int \circ ext$ is monotonous then $int \circ ext(c) \subseteq int \circ ext(C)$. Hence, $int \circ ext(m) \subseteq int \circ ext(C)$. By extensivity of $int \circ ext$ we conclude that $m \in int \circ ext(C)$; that is $\uparrow C \subseteq int \circ ext(C)$. □

A straightforward corollary of Lemma 1 is given below:

Proposition 2. *Closed patterns are up-sets on* $(\mathcal{M}, \leq_{\mathfrak{I}})$ *that is* $\mathbb{K}_{int} \subseteq \mathcal{U}(\mathcal{M})$ *where* $\mathcal{U}(\mathcal{M}) = \{S \subseteq \mathcal{M} \mid \uparrow S = S\}$ *is the set of upsets on* $(\mathcal{M}, \leq_{\mathfrak{I}})$.

Proof. Let $C \in \mathbb{K}_{int}$, we have $C \subseteq \uparrow C$ by extensivity of $\uparrow$. In the other hand, we have $\uparrow C \subseteq int \circ ext(C) = C$ according to Proposition 1 and by using the impotence of $int \circ ext$. Hence, $C = \uparrow C$ or in other words $C \in \mathcal{U}(\mathcal{M})$. □

Proposition 2 shows that all closed patterns are up-sets. It should be noticed that if the provided item-implication basis $\mathfrak{I}$ is equal to the complete item-implication basis $\rightarrow$ associated to context (see Definition 2), then all principal filters in poset $(\mathcal{M}, \leq_{\mathfrak{I}})$ are in $\mathbb{K}_{int}$ (i.e. $(\forall m \in \mathcal{M})\ int \circ ext(m) = \uparrow m$).

We conclude that: rather than enumerating elements of $\wp(\mathcal{M})$ to look for closed itemsets, as for instance Close-by-One does, one can consider only elements of $\mathcal{U}(\mathcal{M})$. For that, one can use Algorithm D&C (see Algorithm 1) to enumerate concepts of $\mathbb{K}$ if the set system $(\mathcal{M}, \mathcal{U}(\mathcal{M}))$ is strongly accessible which is fortunately the case according to Proposition 4. We state in Proposition 3 a characterization of the upper neighbors of $S \in \mathcal{U}(\mathcal{M})$ in the poset $(\mathcal{U}(\mathcal{M}), \subseteq)$. This proposition is crucial to build the algorithm and to prove Proposition 4.

Proposition 3. *Let* $(\mathcal{M}, \leq)$ *be a* <u>*finite*</u> *poset, we have:*

$$(\forall S \in \mathcal{U}(\mathcal{M}), \forall a \in \mathcal{M} \backslash S)\ \ S \cup \{a\} \in \mathcal{U}(\mathcal{M}) \textit{ iff } a \in max(\mathcal{M} \backslash S)$$

Proof. Let $S \in \mathcal{U}(\mathcal{M})$ and let $a \in \mathcal{M} \backslash S$. We show below both implications:

- ($\Leftarrow$) Let $a \in max(\mathcal{M} \backslash S)$, that is: $(\uparrow a) \cap (\mathcal{M} \backslash S) = \{a\}$. Hence, $\uparrow a \subseteq S \cup \{a\}$. We have $\uparrow (S \cup \{a\}) = \uparrow S \cup \uparrow \{a\} = S \cup \uparrow \{a\}$. Since, $\uparrow \{a\} \subseteq S \cup \{a\}$, then $S \cup \uparrow \{a\} \subseteq S \cup \{a\}$. By extensivity of $\uparrow$ we have $S \cup \{a\} \subseteq S \cup \uparrow \{a\}$. Thus, $\uparrow (S \cup \{a\}) = S \cup \{a\}$ or in other words $S \cup \{a\} \in \mathcal{U}(\mathcal{M})$.
- ($\Rightarrow$) Let $a \notin max(\mathcal{M} \backslash S)$, that is $\exists b \in \mathcal{M} \backslash S$ such that $a \leq b$ and $a \neq b$. Hence, $b \in \uparrow (S \cup \{a\})$ (since $b \in \uparrow a$) but in the same time $b \notin S \cup \{a\}$. In other words $\uparrow (S \cup \{a\}) \neq S \cup \{a\}$. Hence, $S \cup \{a\} \notin \mathcal{U}(\mathcal{M})$. □

Figure 5 depicts for $C \in \mathcal{U}(\mathcal{M})$ the set $max(\mathcal{M} \backslash C)$. Hence, the only direct upper neighbors of $C = \{\mathbf{a}, \mathbf{c}, \mathbf{d}\}$ in $(\mathcal{U}(\mathcal{M}), \subseteq)$ according to Proposition 3 are $\{\mathbf{a}, \mathbf{c}, \mathbf{d}, \underline{b}\}$, $\{\mathbf{a}, \mathbf{c}, \mathbf{d}, \underline{e}\}$ and $\{\mathbf{a}, \mathbf{c}, \mathbf{d}, \underline{f}\}$ since $max(\mathcal{M} \backslash C) = \{\underline{b}, \underline{e}, \underline{f}\}$.

Proposition 4. *For $(\mathcal{M}, \leq)$ a finite poset, $(\mathcal{M}, \mathcal{U}(\mathcal{M}))$ is an anti-matroid [7].*

Proof. We have $\emptyset \in \mathcal{U}(\mathcal{M})$. It remains to show now that $(\mathcal{M}, \mathcal{U}(\mathcal{M}))$ has property [P4] (see Sect. 3). Let be two upsets $A, C \in \mathcal{U}(\mathcal{M})$ s.t. $C \nsubseteq A$, we need to show that: $\exists a \in C \backslash A$ s.t. $A \cup \{a\} \in \mathcal{U}(\mathcal{M})$.

Let us show before that $max(C \backslash A) \subseteq max(\mathcal{M} \backslash A)$. Let $e \in max(C \backslash A)$, hence $e \in \mathcal{M} \backslash A$. Let $f \in \mathcal{M}$ s.t. $e \leq f$ and $e \neq f$. In one hand, since $e \in C$ we have $f \in C$ since $C \in \mathcal{U}(\mathcal{M})$. In the other hand, since $e \in max(C \backslash A)$ then $f \notin C \backslash A$. Therefore, $f \in A$ or in other words $f \notin \mathcal{M} \backslash A$. We conclude that $e \in max(\mathcal{M} \backslash A)$. Thus $max(C \backslash A) \subseteq max(\mathcal{M} \backslash A)$.

According to Proposition 3, $\forall a \in max(\mathcal{M} \backslash A)$, we have $A \cup \{a\} \in \mathcal{U}(\mathcal{M})$. Moreover, we have $max(C \backslash A) \neq \emptyset$ since $\mathcal{M}$ is finite and $C \backslash A \neq \emptyset$ (since $C \nsubseteq A$). Since $\emptyset \neq max(C \backslash A) \subseteq max(\mathcal{M} \backslash A)$ then $\exists a \in C \backslash A$ s.t. $A \cup \{a\} \in \mathcal{U}(\mathcal{M})$. □

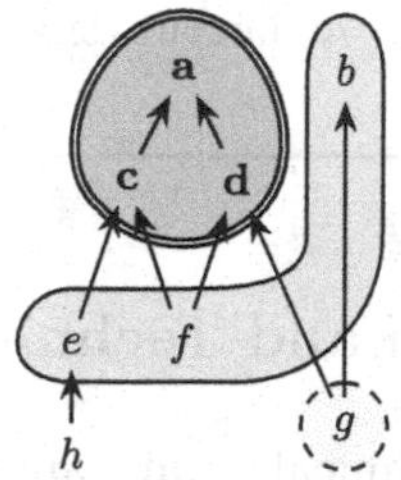

Fig. 5. Poset $(\mathcal{M}, \leq_{\mathfrak{I}})$ with $\mathcal{M} = \{a, b, c, d, e, f, g, h\}$. The set *contoured with two lines* $C = \{\mathbf{a}, \mathbf{c}, \mathbf{d}\} \in \mathcal{U}(\mathcal{M})$ is an upset and $min(C) = \{c, d\}$. The set *contoured with one line* `addables`$(C) = max(\mathcal{M} \backslash C) = \{b, e, f\}$ regroups addable items. Indeed, all upper neighbors of items b, e and f are in C (note that item b has no upper-neighbor). The set *contoured with one dashed line* `potential_addables`$(C) = \{g\}$ contains potential addable items. Indeed, $uppers(g) = \{b, \mathbf{d}\}$ contains at least one element in C.

4.3 Close-by-One Using Implications (CbOI) Algorithm

We have shown in Proposition 2 that, for a $(\mathbb{K}, \leq_{\mathfrak{J}})$ with $\mathbb{K} = (\mathcal{G}, \mathcal{M}, \mathcal{I})$ s.t. $(\mathcal{M}, \leq_{\mathfrak{J}})$ is a poset, all closed patterns in $\mathbb{K}_{int}$ are upsets in $\mathcal{U}(\mathcal{M})$. Since $(\mathcal{M}, \mathcal{U}(\mathcal{M}))$ is strongly accessible (Proposition 4), we can use Algorithm 1 (D&C) to enumerate concepts in $\mathbb{K}$ using closure operator $int \circ ext$ on $(\mathcal{M}, \mathcal{U}(\mathcal{M}))$. Algorithm 2 dubbed *Close-by-One using Implications (CbOI for short)* is then a straightforward implementation of Algorithm 1 where somehow only line 2 is modified according to Proposition 3: For $C \in \mathbb{K}_{int}$, the set $\texttt{addables}(C)$ denotes the set of items to add to build the next closed itemsets. It is given by:

$$\texttt{addables}(C) = max(\mathcal{M} \backslash C) = \{a \in \mathcal{M} \backslash C \mid uppers(a) \subseteq C\}$$

In the next section, we show that some optimizations can be made to compute and maintain efficiently the set of addable items for each generated closed sets (Line 2). Moreover, one can partially compute the closure (Line 3–4) in order to perform canonicity test (Line 5).

Algorithm 2. Algorithm *Close-by-One using Implications* (CbOI)

Input: Pair $(\mathbb{K}, \leq_{\mathfrak{J}})$ with $\mathbb{K} = (\mathcal{G}, \mathcal{M}, \mathcal{I})$ a formal context and $(\mathcal{M}, \leq_{\mathfrak{J}})$ is a poset s.t. if $m_1 \leq_{\mathfrak{J}} m_2$ then $ext(m_1) \subseteq ext(m_2)$.

Output: Elements of $\mathfrak{B}(\mathbb{K})$

1 **procedure** CbOI(A, C, B)
2   **for** $m \in \texttt{addables}(C) \backslash B$ **do**
3     $A_{new} \leftarrow A \cap ext(m)$ // Compute the new extent A_{new}
4     $C_{new} \leftarrow int(A_{new})$ // Compute the new intent C_{new}
5   **if** $C_{new} \cap B = \emptyset$ **then**
6   CbOI(A_{new}, C_{new}, B)
7     $B \leftarrow B \cup \{m\}$ // Update the set of banned attributes B
8   Print(A, C)   // Output the extent A and the intent C
9 CbOI($\mathcal{G}$, $int(\mathcal{G})$, $\emptyset$)

---

5 Empirical Evaluation and Technical Details

We start by explaining some technical details around CbOI implementation provided in https://github.com/BelfodilAimene/CbOImplications.

5.1 Implementation Details

Computing the Partial Order from the Input Item-Implications. As explained in Sect. 4.1, the user input is a pair $(\mathbb{K}^{(0)}, \mathfrak{J}^{(0)})$ where context $\mathbb{K}^{(0)} = (\mathcal{G}, \mathcal{M}^{(0)}, \mathcal{I}^{(0)})$ a finite context and $\mathfrak{J} \subseteq \mathcal{M}^{(0)} \times \mathcal{M}^{(0)}$ such that for all $(m_1, m_2) \in \mathfrak{J}$ we have $ext(m_1) \subseteq ext(m_2)$ (i.e. valid item-implication basis). One can model this provided item-implication basis as a directed graph $(\mathcal{M}^{(0)}, \mathfrak{J}^{(0)})$. The aim at the beginning, is to compute the associated partial order and the partial scaling of the context w.r.t. to the item implication basis. To do so we start by computing the set of strongly connected components $\mathcal{M}$ on the directed graph $(\mathcal{M}^{(0)}, \mathfrak{I}^{(0)})$. This can be done for instance using Tarjan's Algorithm [24] whose complexity is $\mathcal{O}(|\mathcal{M}^{(0)}| + |\mathfrak{I}^{(0)}|)$. Once done, we can build the associated directed acyclic graph (DAG) $(\mathcal{M}, \mathfrak{I}^{(1)})$ where there is an arc (S_1, S_2) in $\mathfrak{I}^{(1)}$ iff $\exists m_1 \in S_1$ and $m_2 \in S_2$ such that $(m_1, m_2) \in \mathfrak{I}^{(0)}$. Such an operation is called *graph condensation*. Once $\mathcal{M}$ computed, the context $\mathbb{K}^{(0)}$ is transformed to an equivalent context $\mathbb{K} = (\mathcal{G}, \mathcal{M}, \mathcal{I})$ (i.e. $\mathbb{K}_{ext} = \mathbb{K}^{(0)}_{ext}$ according to Proposition 1).

The DAG $(\mathcal{M}, \mathfrak{I}^{(1)})$ represents the partial order $\leq_{\mathfrak{I}}$ (i.e. a reflexive and transitive closure of $\mathfrak{I}$ creates $\leq_{\mathfrak{I}}$). A more usual and efficient way to store $\leq_{\mathfrak{I}}$ is to compute the *transitive reduction* of $(\mathcal{M}, \mathfrak{I}^{(1)})$ [1] to obtain the Hasse Diagram of $\leq_{\mathfrak{I}}$. We obtain then $(\mathcal{M}, \mathfrak{I})$ where we store for each $m \in \mathcal{M}$ both sets of its direct lower and upper neighbors (i.e. $lowers(m)$ and $uppers(m)$) w.r.t. $(\mathcal{M}, \leq_{\mathfrak{I}})$.

Computing and Maintaining Addable Items. Now that the partial order $\leq_{\mathfrak{I}}$ is encoded by the list of lower and upper neighbors for each item $m \in \mathcal{M}$ (i.e. the Hasse Diagram). One solution is that at each step of the algorithm, for a closed itemset C, we computed the set $\texttt{addables}(C) = \{a \in \mathcal{M} \backslash C \mid uppers(a) \subseteq C\}$ whose complexity is $\mathcal{O}(\mathcal{G}^2)$. This computation of addable items could lessen the performances of the implementation. To address this drawback, we propose to keep for each generated itemset $C \in \mathcal{U}(\mathcal{M})$, the set of its addable items $\texttt{addables}(C)$, the set of potential addable items $\texttt{potential\_addables}(C)$ and its minimal elements $min(C)$ (only if we want to output them). An item $p \in \mathcal{M}$ is said to be *potentially addable* if $p \in \mathcal{M} \backslash (C \cup \texttt{addables}(C))$ and it has at least one element of its direct upper neighbors $uppers(p)$ in C. Formally: $\texttt{potential\_addables}(C) = (\bigcup_{c \in min(C)} lowers(c)) \backslash \texttt{addables}(C)$. Figure 5 gives an example about addable and potential addable items for a closed itemset C.

These three aforementioned sets of addable items, potential addable items and minimal items can be maintained incrementally as follow. Given an up-set $C \in \mathcal{U}(\mathcal{M})$ and an item $a \in \texttt{addables}(C)$, the following steps are performed to compute the three sets associated to the up-set $C \cup \{a\}$:

1. $min(C \cup \{a\}) := (min(C) \backslash uppers(a)) \cup \{a\}$.
2. $\texttt{potential\_addables}(C \cup \{a\}) := \texttt{potential\_addables}(C) \cup lowers(a)$
3. Initialize addable items $\texttt{addables}(C \cup \{a\})$ by $\texttt{addables}(C) \backslash \{a\}$.
4. For each item p in the computed $\texttt{potential\_addables}(C \cup \{a\})$, if we have $uppers(p) \subseteq min(C \cup \{a\})$ then remove it from $\texttt{potential\_addables}(C \cup \{a\})$ and add it to $\texttt{addables}(C \cup \{a\})$. This can be further optimized by maintaining for each potentially addable item the number of direct upper neighbors that are not already in C. Whenever, element a is add, we subtract 1 from the values associated to elements in $lowers(a)$. Once a potentially addable element sees its value become 0, he is considered as addable item and no longer potentially addable.

Going back to Fig. 5, adding the addable item $\underline{b}$ updates the different sets as follow: $C_{new} = \{\mathbf{a}, \mathbf{c}, \mathbf{d}, \underline{b}\}$, $min(C_{new}) = \{\mathbf{c}, \mathbf{d}, \underline{b}\}$, $\texttt{addables}(C_{new}) = \{e, f, \underline{\underline{g}}\}$ and $\texttt{potential\_addables}(C_{new}) = \emptyset$.

Computing Next Closure and Performing Canonicity Test. Line 3–5 in Algorithm 2 are dedicated to closure computation of the newly generated set and checking if such a closed pattern is already generated. Some optimizations can be made here. For instance, vertical representation of the context (i.e. keeping for each item, its extent) can be held in memory in order to compute efficiently the new pattern extent (Line 3). For closure computation (Line 4) and canonicity test (Line 5), one can use the optimizations explained below:

1. We have a canonicity test fail (i.e. $int \circ ext(C \cup \{m\}) \cap B \neq \emptyset$) iff $\exists b \in B \cap \texttt{addables}(C)$ such that $ext(C \cup \{m\}) \subseteq ext(b)$. Hence, we do not need to compute the closure $int \circ ext(C \cup \{m\})$ to perform the canonicity test. Note that to ensure a fair comparison between `CbOI` and `CbO`, this same optimization has been used for `CbO` implementation.
2. To maintain both sets of addable and potential addable items as explained beforehand, closure computation is computed incrementally by adding item per item until there is no addable item a s.t. $ext(C \cup \{m\}) \subseteq ext(a)$.

Outputting Minimal Elements. If the item-implications in $\leq_{\mathfrak{I}}$ are well-known by the user, one should output only minimal element of a closed pattern C w.r.t. $\leq_{\mathfrak{I}}$ (i.e. $min(C)$) since C contains some redundant information [6,14].

5.2 Experimental Evaluation

Experiment Settings. Experiments were conducted in a machine with an Intel Core i7-7700HQ 2.80 GHz CPU and 7.7 GiB memory space and the implementation was done using Python 2.7.12. Table 2 reports the benchmark input contexts and their associated item-implications basis. *Europarl*[3] and *Yelp*[4] are datasets augmented with a taxonomy. Hence, their corresponding contexts $\mathbb{K}_1$ and $\mathbb{K}_2$ are obtained via an ordinal scaling and their associated implication basis are derived from the hierarchy of items induced by the provided taxonomy. *Basketball*[5], *Airport* (see Footnote 5) and *Iris*[6] are numerical datasets. Analogously, their corresponding contexts $\mathbb{K}_3$, $\mathbb{K}_4$ and $\mathbb{K}_5$ are the result of an interordinal scaling and the associated implication basis are constituted with two chains of implications per attribute (i.e. if the domain of the numerical attribute is $\{1, 2, 3\}$ then the item implications basis associated to the inter-ordinal scaling is given by $\leq 1 \rightarrow \leq 2 \rightarrow \leq 3$ and $\geq 3 \rightarrow \geq 2 \rightarrow \geq 1$). *Mushroom* (see Footnote 6) features only nominal attributes. Hence, its associated context $\mathbb{K}_6$ represents the

[3] EPD8 (last accessed on 04 Octobre 2018): http://parltrack.euwiki.org/.
[4] Yelp (last accessed on 25 April 2017): www.yelp.com/dataset/challenge.
[5] Bilkent repository: http://funapp.cs.bilkent.edu.tr/.
[6] UCI repository: https://archive.ics.uci.edu/ml/index.php.

Table 2. Benchmark Inputs and their characteristics: the number of objects $|\mathcal{G}|$, the number of attributes $|\mathcal{M}|$, the size of the concept lattice $|\mathbb{K}_{int}|$ of the context $\mathbb{K}$, the number of strict (irreflexive) item-implications $| \rightarrow_{\mathfrak{I}}^{*} |$ in the pre-order associated to the corresponding input implication basis, the number of strict item-implications in the context $| \rightarrow^{*} |$ (see Definition 2) and the density given by $| \rightarrow_{\mathfrak{I}}^{*} |/| \rightarrow^{*} |$.

	context $\mathbb{K} = (\mathcal{G}, \mathcal{M}, \mathcal{I})$				implications $\mathfrak{I}$		
$(\mathbb{K}, \mathfrak{I})$	Name	$\|\mathcal{G}\|$	$\|\mathcal{M}\|$	$\|\mathbb{K}_{int}\|$	$\| \rightarrow_{\mathfrak{I}}^{*} \|$	$\| \rightarrow^{*} \|$	Density
$(\mathbb{K}_1, \mathfrak{I}_1)$	Europarl	4 742	357	1 307	709	1 034	68.57%
$(\mathbb{K}_2, \mathfrak{I}_2)$	Yelp	127 162	1 174	63 300	1 514	2 111	71.72%
$(\mathbb{K}_3, \mathfrak{I}_3)$	Basketball	40	272	272 223	4 716	13 724	34.36%
$(\mathbb{K}_4, \mathfrak{I}_4)$	Iris	150	246	6 516 292	3 964	11 704	33.87%
$(\mathbb{K}_5, \mathfrak{I}_5)$	Airport	135	1 348	82 467 125	90 182	313 432	28.77%
$(\mathbb{K}_6, \mathfrak{I}_6)$	Mushrooms	8 124	119	238 710	0	949	0.00%

result of nominal scaling of all attributes. Note that the set of implications $\mathfrak{I}_6$ is empty, yet there are some implications between items that are context-dependent (i.e. $\rightarrow$ is not empty).

Evaluation Results. Table 3 reports the number of closures and the performance of `CbO` and `CbOI` on the different benchmark inputs. For each benchmark context $\mathbb{K}_i$, we run both algorithms on the provided implication basis (i.e. input $(\mathbb{K}_i, \mathfrak{I}_i)$) as well as on the total one that is associated to the context (i.e. $(\mathbb{K}_i, \rightarrow)$). For a fair comparison, we report the context load/preparation time into the memory for `CbO` as well as the load/preparation time of the pair (context, implication basis) for `CbOI`. When $\rightarrow$ implication basis is used, the load time includes the time spent to compute it.

It is clear that the number of closures performed by `CbOI` is much less than the ones performed by `CbO` in all tests excepts when no implications are provided. This corresponds to the case $(\mathbb{K}_6, \mathfrak{I}_6)$ where the number of closures performed by `CbOI` is supposed to be equivalent to the number of closures performed by `CbO` if the same order of choice of items to add is followed.

Concerning the execution time, it is clear that the load time for `CbO` is lesser than the load/compute time for `CbOI` since `CbOI` does load and prepare additionally the item-implication basis. However, even if `CbOI` has this drawback, one can see that the enumeration time is much faster than `CbO` (i.e. up to 15× faster for input $(\mathbb{K}_3, \mathfrak{I}_3)$ or even more for input $(\mathbb{K}_4, \mathfrak{I}_4)$). This compensates the overhead induced by the implication-basis load time in `CbOI`. One could notice that `CbO` perform better than `CbOI` when no implication is provided as it is the case in test $(\mathbb{K}_6, \mathfrak{I}_6)$. This is due to the fact that `CbOI` manages more structures than `CbO` during enumeration. It is worth noting that even if the implication basis is computed then used to enumerate concepts (see tests $(\mathbb{K}_i, \rightarrow)$), `CbOI` performs faster than `CbO` (up to 6× faster for input $(\mathbb{K}_3, \mathfrak{I}_3)$). Still, we can observe that `CbOI` is less efficient when the underlying implication basis is huge (case $(\mathbb{K}_3, \rightarrow)$,

Table 3. CbO and CbOI performance comparison on the benchmark inputs

$(\mathbb{K}, \mathfrak{I})$	CbO nb closure	load (ms)	enum (ms)	total (ms)	CbOI nb closure	load (ms)	enum (ms)	total (ms)
$(\mathbb{K}_1, \mathfrak{I}_1)$	185 418	**86**	184	270	**17 020**	220	**48**	**268**
$(\mathbb{K}_1, \rightarrow)$					**13 409**	191	**46**	**237**
$(\mathbb{K}_2, \mathfrak{I}_2)$	24 437 659	**8 715**	30 360	39 075	**3 317 590**	18 084	**16 107**	**34 191**
$(\mathbb{K}_2, \rightarrow)$					**2 974 130**	19 976	**15 030**	**35 006**
$(\mathbb{K}_3, \mathfrak{I}_3)$	13 340 233	**3**	57 286	57 289	**703 999**	20	**3 628**	**3 648**
$(\mathbb{K}_3, \rightarrow)$					**445 735**	39	**9 114**	**9 153**
$(\mathbb{K}_4, \mathfrak{I}_4)$	170 615 166	**10**	709 517	709 527	**9 618 493**	26	**77 586**	**77 612**
$(\mathbb{K}_4, \rightarrow)$					**8 383 741**	73	**141 016**	**141 089**
$(\mathbb{K}_5, \mathfrak{I}_5)$	NA	**53**	>12 h	>12 h	**122 717 962**	268	**1 496 175**	**1 496 443**
$(\mathbb{K}_5, \rightarrow)$					**106 409 230**	1 400	**8 221 648**	**8 223 048**
$(\mathbb{K}_6, \mathfrak{I}_6)$	**4 363 487**	**155**	**13 800**	**13 955**	4 363 511	184	15 985	16 169
$(\mathbb{K}_6, \rightarrow)$	4 363 487	**155**	13 800	13 955	**1 338 245**	244	**10 003**	**10 247**

$(\mathbb{K}_4, \rightarrow)$ and $(\mathbb{K}_5, \rightarrow)$). This can be explained by the fact that CbOI spends more time to handle a huge and complex system of item-implications but the gain obtained from these base of implications does not compensate this effort.

6 Conclusion

In this paper, we have investigated how to incorporate and leverage the inherent implications between items in some given context so as to enumerate more efficiently its formal concepts. Experimental studies demonstrated that the proposed algorithm dubbed CbOI for *Close-by-One using Implications* is far more efficient than its concurrent CbO in most configurations. Indeed, many aspects of the devised algorithm can be considerably improved. For instance, including FCbO optimizations [16] during the enumeration process can significantly reduce the number of falsely generated closed patterns. Moreover, the load/preparation time of CbOI can be more enhanced by, for example, computing more efficiently the transitive reduction of the implication basis [20]. Another important remark is that the same notions here can be used to look for minimal generators. In fact, Kaytoue et al. [14] showed that there is no one-to-one correspondence between the minimal generators in the interval pattern structure and the minimal generators in the interordinal scaled contexts conversely to closed interval patterns. However, if we consider only up-sets w.r.t. the poset induced by the implications, there is a one-to-one correspondence between interval patterns and up-sets. Hence, minimal generators for interval patterns can be mined efficiently by algorithm DeFMe [23] since it considers strongly accessible set systems.

Aknowledgement. This work has been partially supported by the project *ContentCheck* **ANR-15-CE23-0025** funded by the French National Research Agency, the **ANRt** French program and the **APRC Conf Pap-CNRS** project. The authors would like to thank the reviewers for their valuable remarks. They also warmly thank Anes Bendimerad for interesting discussions.

References

1. Aho, A.V., Garey, M.R., Ullman, J.D.: The transitive reduction of a directed graph. SIAM J. Comput. **1**(2), 131–137 (1972)
2. Belfodil, A., Cazalens, S., Lamarre, P., Plantevit, M.: Flash points: discovering exceptional pairwise behaviors in vote or rating data. In: Ceci, M., Hollmén, J., Todorovski, L., Vens, C., Džeroski, S. (eds.) ECML PKDD 2017. LNCS (LNAI), vol. 10535, pp. 442–458. Springer, Cham (2017). https://doi.org/10.1007/978-3-319-71246-8_27
3. Belfodil, A., Kuznetsov, S.O., Kaytoue, M.: Pattern setups and their completions. In: CLA, pp. 243–253 (2018)
4. Boley, M., Horváth, T., Poigné, A., Wrobel, S.: Listing closed sets of strongly accessible set systems with applications to data mining. Theor. Comput. Sci. **411**(3), 691–700 (2010)
5. Bordat, J.P.: Calcul pratique du treillis de galois d'une correspondance. Mathématiques et Sciences humaines **96**, 31–47 (1986)
6. Cellier, P., Ferré, S., Ridoux, O., Ducassé, M.: An algorithm to find frequent concepts of a formal context with taxonomy. In: CLA, pp. 226–231 (2006)
7. Dietrich, B.L.: Matroids and antimatroids-a survey. Discrete Math. **78**(3), 223–237 (1989)
8. Ganter, B., Wille, R.: Formal Concept Analysis. Springer, Heidelberg (1999). https://doi.org/10.1007/978-3-642-59830-2
9. Ganter, B.: Two basic algorithms in concept analysis. Technical report, Technische Hoschule Darmstadt (1984)
10. Ganter, B., Kuznetsov, S.O.: Pattern structures and their projections. In: Delugach, H.S., Stumme, G. (eds.) ICCS-ConceptStruct 2001. LNCS (LNAI), vol. 2120, pp. 129–142. Springer, Heidelberg (2001). https://doi.org/10.1007/3-540-44583-8_10
11. Ganter, B., Wille, R.: Conceptual scaling. In: Roberts, F. (ed.) Applications of Combinatorics and Graph Theory to the Biological and Social Sciences, pp. 139–167. Springer, New York (1989). https://doi.org/10.1007/978-1-4684-6381-1_6
12. Gély, A.: A generic algorithm for generating closed sets of a binary relation. In: Ganter, B., Godin, R. (eds.) ICFCA 2005. LNCS (LNAI), vol. 3403, pp. 223–234. Springer, Heidelberg (2005). https://doi.org/10.1007/978-3-540-32262-7_15
13. Johnson, D.S., Papadimitriou, C.H., Yannakakis, M.: On generating all maximal independent sets. Inf. Process. Lett. **27**(3), 119–123 (1988)
14. Kaytoue, M., Kuznetsov, S.O., Napoli, A.: Revisiting numerical pattern mining with formal concept analysis. In: IJCAI, pp. 1342–1347 (2011)
15. Korte, B., Lovász, L.: Mathematical structures underlying greedy algorithms. In: Gécseg, F. (ed.) FCT 1981. LNCS, vol. 117, pp. 205–209. Springer, Heidelberg (1981). https://doi.org/10.1007/3-540-10854-8_22
16. Krajca, P., Outrata, J., Vychodil, V.: Advances in algorithms based on CbO. In: CLA, pp. 325–337 (2010)

17. Kuznetsov, S.O.: A fast algorithm for computing all intersections of objects in a finite semi-lattice. Nauchno-Tekhnicheskaya Informatsiya ser. **2**(1), 17–20 (1993)
18. Kuznetsov, S.O.: Learning of simple conceptual graphs from positive and negative examples. In: Żytkow, J.M., Rauch, J. (eds.) PKDD 1999. LNCS (LNAI), vol. 1704, pp. 384–391. Springer, Heidelberg (1999). https://doi.org/10.1007/978-3-540-48247-5_47
19. Kuznetsov, S.O.: Pattern structures for analyzing complex data. In: Sakai, H., Chakraborty, M.K., Hassanien, A.E., Ślęzak, D., Zhu, W. (eds.) RSFDGrC 2009. LNCS (LNAI), vol. 5908, pp. 33–44. Springer, Heidelberg (2009). https://doi.org/10.1007/978-3-642-10646-0_4
20. Le Gall, F.: Powers of tensors and fast matrix multiplication. In: Proceedings of the 39th International Symposium on Symbolic and Algebraic Computation, pp. 296–303. ACM (2014)
21. Lumpe, L., Schmidt, S.E.: Pattern structures and their morphisms. In: CLA, vol. 1466, pp. 171–179 (2015)
22. Roman, S.: Lattices and Ordered Sets. Springer, New York (2008). https://doi.org/10.1007/978-0-387-78901-9
23. Soulet, A., Rioult, F.: Efficiently depth-first minimal pattern mining. In: Tseng, V.S., Ho, T.B., Zhou, Z.-H., Chen, A.L.P., Kao, H.-Y. (eds.) PAKDD 2014. LNCS (LNAI), vol. 8443, pp. 28–39. Springer, Cham (2014). https://doi.org/10.1007/978-3-319-06608-0_3
24. Tarjan, R.E.: Depth-first search and linear graph algorithms. SIAM J. Comput. **1**(2), 146–160 (1972)
25. Wille, R.: Restructuring lattice theory: an approach based on hierarchies of concepts. In: Rival, I. (ed.) Ordered Sets, vol. 83, pp. 445–470. Springer, Dordrecht (1982). https://doi.org/10.1007/978-94-009-7798-3_15

Effects of Input Data Formalisation in Relational Concept Analysis for a Data Model with a Ternary Relation

Priscilla Keip[1,2(✉)], Alain Gutierrez[3], Marianne Huchard[3], Florence Le Ber[4], Samira Sarter[6], Pierre Silvie[1,2,5], and Pierre Martin[1,2]

[1] CIRAD, UPR AIDA, 34398 Montpellier, France
{priscilla.keip,pierre.silvie,pierre.martin}@cirad.fr
[2] AIDA, Univ Montpellier, CIRAD, Montpellier, France
[3] LIRMM, Université de Montpellier, CNRS, Montpellier, France
marianne.huchard@lirmm.fr
[4] ICube, Université de Strasbourg, CNRS, ENGEES, Illkirch-Graffenstaden, France
florence.leber@engees.unistra.fr
[5] IRD, UMR EGCE, 91198 Gif-sur-Yvette, France
[6] ISEM, Univ Montpellier, CIRAD, CNRS, EPHE, IRD, Montpellier, France
samira.sarter@cirad.fr

Abstract. Today pesticides, antimicrobials and other pest control products used in conventional agriculture are questioned and alternative solutions are searched out. Scientific literature and local knowledge describe a significant number of active plant-based products used as bio-pesticides. The Knomana (KNOwledge MANAgement on pesticide plants in Africa) project aims to gather data about these bio-pesticides and implement methods to support the exploration of knowledge by the potential users (farmers, advisers, researchers, retailers, etc.). Considering the needs expressed by the domain experts, Formal Concept Analysis (FCA) appears as a suitable approach, due do its inherent qualities for structuring and classifying data through conceptual structures that provide a relevant support for data exploration. The Knomana data model used during the data collection is an entity-relationship model including both binary and ternary relationships between entities of different categories. This leads us to investigate the use of Relational Concept Analysis (RCA), a variant of FCA on these data. We consider two different encodings of the initial data model into sets of object-attribute contexts (one for each entity category) and object-object contexts (relationships between entity categories) that can be used as an input for RCA. These two encodings are studied both quantitatively (by examining the produced conceptual structures size) and qualitatively, through a simple, yet real, scenario given by a domain expert facing a pest infestation.

Keywords: Biopesticides · Data exploration · Formal Concept Analysis · Relational Concept Analysis

D. Cristea et al. (Eds.): ICFCA 2019, LNAI 11511, pp. 191–207, 2019.
https://doi.org/10.1007/978-3-030-21462-3_13

1 Introduction

Today pesticides, antimicrobials and other pest control products used in conventional agriculture are questioned and alternative solutions are searched out, including active plant-based products. The Knomana (KNOwledge MANAgement on pesticides plants in Africa) project aims to identify plants used as biopesticides, currently from the scientific literature, and to implement methods to support the exploration of knowledge by the potential users (farmers, advisers, researchers, retailers, etc.). About 30000 descriptions of plant uses have been collected and recorded according to a data model designed through meetings with domain experts. Each plant use is described using 36 attributes such as the plant taxonomy, the protected system (i.e. crop, animal and human being), or the preparation method.

Considering the data exploitation needs expressed by the domain experts, Formal Concept Analysis (FCA) appears as a suitable approach, due do its inherent qualities for structuring and classifying data through conceptual structures that provide a relevant support for data exploration. To exploit the recorded data, and extract knowledge about alternative protection systems, we rely on Relational Concept Analysis (RCA) [11], one of the possible extensions of Formal Concept Analysis [9] for relational data. RCA input is a so-called *relational context family*, composed with object-attribute (formal) contexts, which describe objects from various categories, and object-object (relational) contexts, which describe relationships between objects of several categories. It outputs a set of conceptual structures (such as concept lattices, AOC-posets or Iceberg lattices) connected through *relational attributes* which point to concepts. Each conceptual structure classifies the objects of one category according to the (initial) attributes and the *relational attributes*, thus according to the relations that the objects of this category have with objects and object groups (concepts) of another (or the same) category.

Considering a real-word context such as the Knomana project dataset, although a data model has been set for data collection purpose, there are still many ways to encode the data model into a *relational context family*. In this paper, we study the impact of the definition of a *relational context family* on the practicability of RCA on the Knomana project dataset. The first question is raised by the fact that relational contexts represent binary relations between objects, but the data model we deal with contains a ternary relation. The model has thus to be converted into a model with binary relations, while respecting the original semantics of the ternary relation. Two encodings are envisaged: the first one considers a reification of the ternary relation (i.e. a specific object-attribute context represents the 3-tuples), while the second one projects the ternary relation into three binary relations, one for each pair of the linked object sets. The second question is related with the possibility, for each relationship, to consider one direction only or both directions. It is connected to the potential explorations that the domain experts may have in mind. The proposed encodings are studied quantitatively (by examining the produced conceptual structures size and running time) and qualitatively, through a simple, yet real, scenario given by a domain expert facing a pest infestation.

Section 2 presents Relational Concept Analysis and the RCAExplore tool which is used in our evaluations. In Sect. 3, we propose two encodings of the initial data model into *relational context families*. Then, in Sect. 4, we first show the size and computation time of several conceptual structures for the two encodings on an excerpt of the Knomana dataset restricted to a few key plants designated by the domain experts. Then we study a simple, yet real, exploration scenario using both encodings, to show their relevancy with respect to the studied question. Section 5 exposes related work. We conclude and give perspectives of this work in Sect. 6.

2 Background

RCA extends the purpose of Formal Concept Analysis (FCA, [9]) to relational data. RCA applies iteratively FCA on a Relational Context Family (RCF), that is a pair $(\mathcal{K}, \mathcal{R})$, where $\mathcal{K}$ is a set of object-attribute contexts and $\mathcal{R}$ is a set of object-object contexts. $\mathcal{K}$ contains n object-attribute contexts $K_i = (G_i, M_i, I_i), i \in \{1, ..., n\}$. $\mathcal{R}$ contains m object-object contexts $R_j = (G_k, G_l, r_j)$, $j \in \{1, ..., m\}$, where $r_j \subseteq G_k \times G_l$ is a binary relation with $k, l \in \{1, ..., n\}$, $G_k = dom(r_j)$ the domain of the relation, and $G_l = ran(r_j)$ the range of the relation. RCA relies on a relational scaling mechanism that is used to transform a relation r_j into a set of *relational attributes* that extends the object-attribute context describing the objects of $dom(r_j)$. A relational attribute $\exists r_j(C)$, where $\exists$ is the existential quantifier, $C = (X, Y)$ is a concept, and $X \subseteq ran(r_j)$, is owned by an object $g \in dom(r_j)$ if $r_j(g) \cap X \neq \emptyset$. Other quantifiers can be found in [11]. RCA process consists in applying FCA first on each object-attribute context of an RCF, and then iteratively on each object-attribute context extended by the relational attributes created using the concepts from the previous step. The RCA process stops when the families of lattices of two consecutive steps are isomorphic and the extended object-attribute contexts are unchanged.

Table 1. A relational context family about (a) `Pests` and (b) `Plants` able to treat them, as indicated in `treatedBy` relation (c)

Pests	Chrysomelidae	Noctuidae	Liposcelididae	Trichocomaceae
CallosobruchusC	×			
SpodopteraL		×		
LiposcelisB			×	
AspergillusF				×
PenicilliumE				×
CallosobruchusM	×			

Plants	Rhizome	Root	Leaf	Seed	Fruit
AcorusC	×	×			
AlpiniaO	×				
AgeratumC			×		
LaphangiumL			×		
PelargoniumR			×		
HelianthusA				×	
CitrusL					×

treatedBy	AcorusC	AlpiniaO	AgeratumC	LaphangiumL	PelargoniumR	HelianthusA	CitrusL
CallosobruchusC	×						
SpodopteraL	×				×		×
LiposcelisB		×					
AspergillusF			×				
PenicilliumE				×			
CallosobruchusM						×	

In the following, we consider a small example from the Knomana dataset. Object-attribute contexts `Plants` and `Pests` (Table 1(a) and (b)) respectively represent a set of plants and a set of pests. Pests are described by their family (e.g. *Callosobruchus maculatus* –`CallosobruchusM` in Table 1– is a member of the leaf beetle family, *Chrysomelidae*), while plants are described by the parts (e.g. fruit or leaf) that are used for treating the pests. The object-object context `treatedBy` (Table 1(c)) represents the link between pests and plants they can be treated by. At the first step of the RCA process, FCA is applied on `Plants` and `Pests` and results in two lattices $\mathcal{L}_{\texttt{Plants}}$ and $\mathcal{L}_{\texttt{Pests}}$ (Fig. 1).

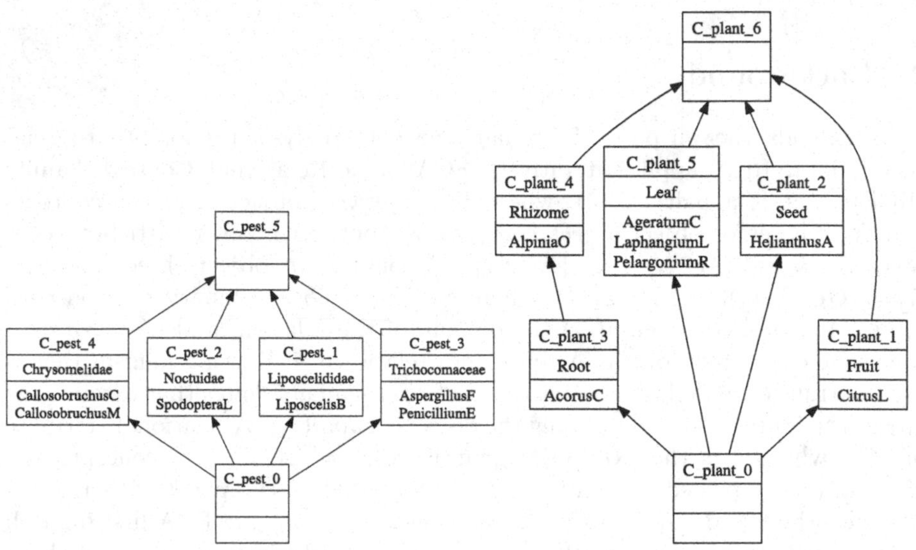

Fig. 1. Lattices $\mathcal{L}_{\texttt{Pests}}$ (left) and $\mathcal{L}_{\texttt{Plants}}$ (right)

At the second step of the process, `Pests` context is extended with relational attributes built from context `isTreatedBy` and concepts of $\mathcal{L}_{\texttt{Plants}}$ (Table 2). For instance, the relational attribute $\exists$`treatedBy(C_plant_2)` is added to `CallosobruchusM` since this pest is related to `HelianthusA` which is an object of `C_plant_2`. A new lattice is built, that is represented in Fig. 2. In this last lattice, we can observe that pests grouped in `C_pest_8` are both treated by *Acorus Calamus* `AcorusC` (sweet flag), using its root or rhizome (see `C_plant_3`).

The tool RCAexplore[1] was developed during project ANR 11 MONU 14 Fresqueau, in order to explore relational hydroecological data. RCAExplore is an implementation of the RCA process where several choices can be made before each iteration: the algorithm to be used, the scaling operator, and the considered contexts.

[1] http://dataqual.engees.unistra.fr/logiciels/rcaExplore.

Table 2. Extended context from `Pests`.

Pests*	Chrysomelidae	Noctuidae	Liposcelididae	Trichocomaceae	∃ treatedBy(C_plant_0)	∃ treatedBy(C_plant_1)	∃ treatedBy(C_plant_2)	∃ treatedBy(C_plant_3)	∃ treatedBy(C_plant_4)	∃ treatedBy(C_plant_5)	∃ treatedBy(C_plant_6)
CallosobruchusC	×							×	×		×
SpodopteraL		×				×		×	×	×	×
LiposcelisB			×						×		×
AspergillusF				×						×	×
PenicilliumE				×						×	×
CallosobruchusM	×						×				×

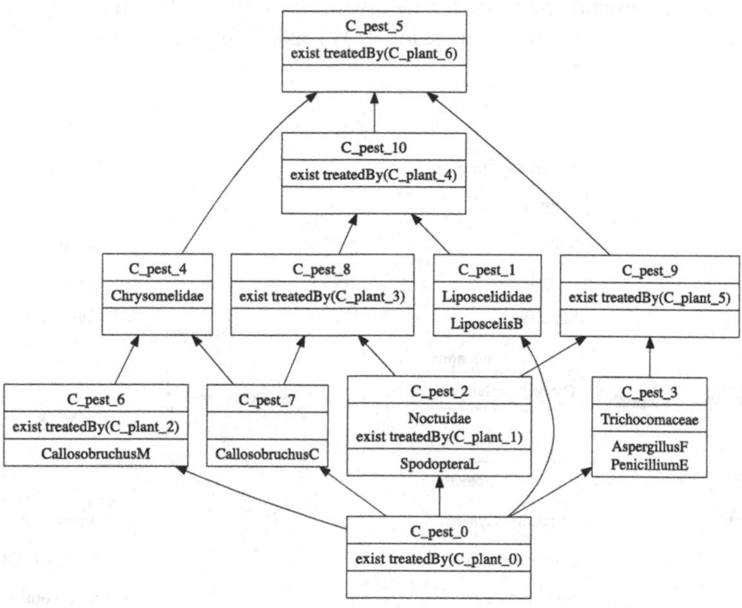

Fig. 2. Lattice built on the extended context `Pest*` of Table 2

3 From the Knomana Model to a Relational Context Family and Conceptual Structures

The Knomana database gathers descriptions of plant uses, each one characterized using 36 data types, including the protecting plant, the protected organism, the controlled aggressor (also called pest and disease), the method adopted to prepare the product to be applied, or the reference to the document describing the use. Currently, the Knomana database comprises 28700 plant use descriptions manually entered from 250 documents (mainly scientific publications).

The descriptions include 966 plant species, originated from 60 territories, used to protect 39 species of organism (animal, vegetal, and human) against 253 species of aggressors (Bacteria, Chromista, Eukaryota, Fungi, Insecta, and Virus).

In the data model, data types are grouped as data classes to represent the three main entity categories of the system: biopesticide, protected system and targeted organism. To represent the biological system, these three main entity categories (or data classes) are linked through a ternary relationship. As the relational contexts of RCA are binary relationships, two different data models have been designed. The first one, called M1 (see left-hand side of Fig. 3), consists in reifying the ternary relationship as a specific data class, the latter supporting binary relationships with each of the three main data classes of the biological system. The second one, called M2 (see right-hand side of Fig. 3), consists in establishing binary relationships between the data classes of the biological system, corresponding to projections of the ternary relationship. The relation directions have been determined in order to obtain classifications and propagation of relations according to the first question (Q1) asked by the experts (introduced below).

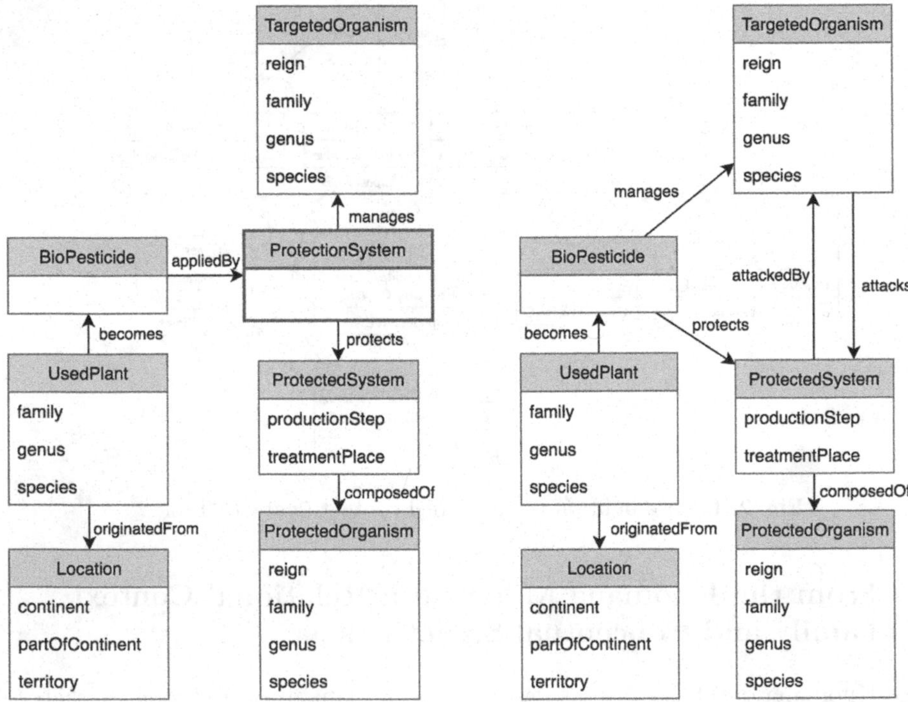

Fig. 3. Models M1 (left-hand side) and M2 (right-hand side). Implementations of the data model: (M1) with the ternary relationship as a data class, (M2) without the ternary relationship, which is transformed by establishing binary relationships between the main data classes

The encoding of M1 and M2 as relational context families (RCF) consists in converting each data class as a formal context (object-attribute context) and each arrow as a relational context (object-object context). To measure the effect of the encodings on the resulting conceptual structures, two encodings of the M1 and M2 arrows are considered: the "original" encoding implements the arrows presented in Fig. 3, while the "enhanced" encoding includes the arrows and their opposite (making a sort of symmetric closure at the model level).

M1 and M2 encodings are evaluated on a dataset reduced to the descriptions associated to six protecting plant species, i.e. *Cymbopogon citratus*, *Hyptis suaveolens*, *Lantana camara*, *Moringa oleifera*, *Ocimum gratissimum*, and *Thymus vulgaris*. These plants have been selected by domain experts of the Knomana project for their first investigations, according to these plant efficacy in contrasted situations, diversity of applications, and high presence in most of the West African territories. The dataset comprises 225 descriptions composed of 16 pieces of information: the protecting plant (name of the species, genus and family), the plant origin (territory, part of continent, and continent), the targeted organism (name of the species, genus, family, and reign), and the protected system. The latter is described using the protected organism (name of the species, genus, family, and reign), the production step (e.g. crop, or cattle) and the treatment place (e.g. field, or stock). Table 3 presents the number of objects and attributes of the formal contexts of M1 and M2 on the reduced dataset, and details the number of binary attributes generated for each data type of the model.

Table 3. Description of formal contexts of M1 and M2

Formal context (Class name)	Number of objects	Number of attributes	Included data types	Number of values
BioPesticide	38	0	(empty class)	-
UsedPlant	6	16	family	4
			genus	6
			species	6
Location	20	32	continent	5
			partOfContinent	7
			territory	20
ProtectedSystem	14	19	productionStep	7
			treatmentPlace	12
ProtectedOrganism	21	48	reign	3
			family	11
			genus	15
			species	19
TargetedOrganism	111	234	reign	5
			family	43
			genus	78
			species	108
ProtectionSystem	225	0	ternary relation	-

RCAExplore software enables to evaluate four kinds of conceptual structures, and algorithms that allow to build them: concept lattices built with addIntent/addExtent [16] (FCA), AOC-posets built with Ares [4] (ARES), and ICEBERG lattices [21] for support 30 and 40 (ICEBERG30, ICEBERG40). An AOC-poset is a restriction of the concept lattice to the concepts introducing objects or attributes. ICEBERG is a restriction of the concept lattice to the concepts having a minimal support (i.e. extent size), here to concepts with a minimal support of 30% and 40%. In the next section, we conduct an evaluation of the dataset along two dimensions: quantitatively, by assessing the possibility of building the conceptual structures and their size, if appropriate, and qualitatively, by analyzing the ability of M1 and M2 to answer a specific case of the following generic biological question Q1 raised by our domain experts: "Given a plant able to protect an organism against an aggressor, which other plants can alternatively be used with the same benefits?". This question corresponds to a so-called "replacement" scenario: replace a plant by another one with supposed similar ability to deal with the observed aggressor on the attacked organism.

4 Results

In this section, we first present the effects of the two proposed encodings on the conceptual structure construction for our six key plant dataset (Sect. 4.1), and then on a real replacement scenario for an *Aspergillus*[2] attack (Sect. 4.2).

4.1 Six Key Plant Dataset: Conceptual Structure Variants

Tables 4 and 5 respectively show for model M1 and model M2 the numbers of concepts that were built for each algorithm, and for the enhanced case (where we take the symmetric closure of the model) and the original cases (as shown in Fig. 3). Tables 6 and 7 respectively show for model M1 and model M2 the numbers of relational attributes that were built. Step numbers until RCA stops and execution times are compared in Table 8.

As a first remark, some computations failed (see italics figures between parentheses) on a laptop in the enhanced case. For enhanced M1 (resp. enhanced M2), concept lattices and Iceberg30 (resp. concept lattices) could not be computed because of lack of memory. Computing AOC-posets and Iceberg40 was always possible, but Iceberg40 shows very few concepts in the original models, thus we suspect it will not be very useful for experts in this case.

In Tables 4 and 5, the AOC-posets for both enhanced models M1 and M2 show similar concept numbers, except for `ProtectionSystem`, which is specific to M1. The concept number of `ProtectionSystem` AOC-poset can be roughly obtained from the sum of the concept numbers of the AOC-posets of the neighbour contexts (`Biopesticide`, `ProtectedSystem` and `TargetedOrganism`). The concept numbers of `Location` and `ProtectedOrganism` do not change between

[2] *Aspergillus* genus groups several species of microscopic fungi.

Table 4. M1 model: number of concepts for each algorithm (italics figures between parentheses are for failed computations because of lack of memory)

	M1 enhanced				M1 original			
	FCA	ARES	ICEBERG30	ICEBERG40	FCA	ARES	ICEBERG30	ICEBERG40
ProtectionSystem	*(1660066)*	1151	*(2750032)*	619	415	238	9	3
BioPesticide	*(16212)*	359	*(3625)*	57	576	113	15	3
UsedPlant	*(51)*	36	*(16)*	3	51	37	12	3
Location	*(191)*	63	*(31)*	5	29	27	4	3
ProtectedSystem	*(1585)*	154	*(771)*	20	42	32	5	4
ProtectedOrganism	*(300)*	95	*(38)*	17	33	29	4	3
TargetedOrganism	*(380084)*	560	*(9386)*	61	153	151	5	2
TOTAL	*(2058489)*	2418	*(2763899)*	782	1299	627	54	21

Table 5. M2 model: number of concepts for each algorithm (italics figures between parentheses are for failed computations because of lack of memory)

	M2 enhanced				M2 original			
	FCA	ARES	ICEBERG30	ICEBERG40	FCA	ARES	ICEBERG30	ICEBERG40
BioPesticide	*(307618)*	354	9563	84	23650	178	60	6
UsedPlant	*(55)*	36	16	3	57	38	12	3
Location	*(191)*	63	31	5	29	27	4	3
ProtectedSystem	*(2270)*	153	316	20	747	81	21	5
ProtectedOrganism	*(495)*	93	42	108	33	29	4	3
TargetedOrganism	*(1363817)*	555	48556	108	7186	216	35	4
TOTAL	*(1674446)*	1254	58524	328	31702	569	136	24

original models M1 and M2 because they are sinks in the model graph. In the enhanced case, `Location` and `UsedPlant` concept numbers are almost the same so that we can assume that they do not influence each other too much. For both enhanced and original models and considering the cases where the computation finished, M2 Iceberg40 lattices contain more concepts, what suggests that M2 concepts are more populated than M1 concepts; nevertheless there is no significant change in scaling factor. Whole concept lattices are built only for the original M1 and M2 models. For enhanced M2 model, the number of concepts for `Biopesticide` and `TargetedOrganism` explodes, likely due to the circuit between `TargetedOrganism` and `ProtectedSystem`.

Tables 6 and 7 show the numbers of relational attributes and inform us about the grouping factor provided by the conceptual structures. The formal contexts that are sinks in the model (no outgoing relation) have no relational attributes. While observing the enhanced models, we can notice that M2 gives rise to less relational attributes (but the difference is more significant for Iceberg40 than for AOC-posets). For the original models, this is the reverse, there are more relational attributes for M2 than for M1, which could be explained by the fact that M2 original contains a circuit, which is not the case of M1 original. We also observe a significant difference between Iceberg30 and Iceberg40 in M2.

Table 6. M1 model: number of relational attributes added at each formal context for each algorithm (italics figures between parentheses are for failed computations because of lack of memory)

	M1 enhanced				M1 original			
	FCA	ARES	ICEBERG30	ICEBERG40	FCA	ARES	ICEBERG30	ICEBERG40
ProtectionSystem	*(3330)*	1087	*(2025)*	138	195	183	10	6
BioPesticide	*(47959)*	1173	*(423074)*	622	415	238	9	3
UsedPlant	*(1881)*	428	*(457)*	78	605	140	19	6
Location	*(51)*	36	*(16)*	35	0	0	0	0
ProtectedSystem	*(48126)*	1230	*(23091)*	655	33	29	4	3
ProtectedOrganism	*(289)*	156	*(219)*	68	0	0	0	0
TargetedOrganism	*(47908)*	1137	*(423058)*	853	0	0	0	0
TOTAL	*(149544)*	5247	*(871940)*	2449	1248	590	42	18

Table 7. M2 model: number of relational attributes added at each formal context for each algorithm (italics figures between parentheses are for failed computations because of lack of memory)

	M2 enhanced				M2 original			
	FCA	ARES	ICEBERG30	ICEBERG40	FCA	ARES	ICEBERG30	ICEBERG40
BioPesticide	*(24777)*	744	48888	131	7933	297	56	9
UsedPlant	*(5323)*	417	9594	89	23679	205	64	9
Location	*(51)*	36	16	3	0	0	0	0
ProtectedSystem	*(29560)*	1002	58161	209	7219	245	39	7
ProtectedOrganism	*(585)*	153	316	20	0	0	0	0
TargetedOrganism	*(5789)*	507	9879	104	747	81	21	5
TOTAL	*(66085)*	2859	126854	556	39578	828	180	30

Table 8 shows the running times and the step numbers. The step numbers are similar in original M1 and M2. Computing concept lattices for original M2 needs more steps due to the existing circuit and the creation of many non-introducer concepts. The 16 steps for obtaining the AOC-poset of enhanced M1 are noticeable and correspond to a running time relatively high, compared to the others. The different running time for original M1 and M2 (from 127 ms to about 9 s) allows to envisage online work for experts. For enhanced models (AOC-posets), it can be preferable to compute them offline.

Table 8. (left) Final step number and (right) computation time (milliseconds) for each algorithm and each model

	enhanced models		original models	
	M1	M2	M1	M2
FCA	*(5)*	*(4)*	5	9
ARES	16	10	5	6
ICEBERG30	*(7)*	10	6	6
ICEBERG40	11	8	5	6

	enhanced models		original models	
	M1	M2	M1	M2
FCA	-	-	351	9722
ARES	311149	28288	1195	1677
ICEBERG30	-	29864	137	144
ICEBERG40	796	223	127	166

In the light of the above evaluation, AOC-poset and Iceberg40 are appropriate for the dataset on both M1 and M2 original models. They will be used in the next section on a real question raised by the experts. Iceberg40 gives incomplete information, but allows us to focus on frequent situations. AOC-poset, as it holds all the introducer concepts, contains the whole initial information. It can be used to build the entire concept lattice. A concept which appears in the concept lattice and not in the AOC-poset represents a group Ext of objects and a group Int of attributes such that (1) each object of Ext is introduced in a sub-concept because it has an attribute which is not in Int, and (2) each attribute of Int is introduced in a super-concept because it is owned by an object which is not in Ext. These concepts are useful to reveal data regularities. In our context, they could be connection points between different exploration paths. In the future, we will evaluate in which extent they are useful during exploration, as they could be built on the fly based on the introducer concepts. Let us notice that the algorithm running time does not cover all the needed time for a concrete analysis. In a real scenario, the analyst also needs to select and extract or focus on presumed relevant data.

4.2 Aspergillus Attack: Answering a Concrete Replacement Scenario

To assess the pertinence of our approach, we have selected a smaller dataset from Knomana base and have explored it with both models M1 and M2 with AOC-posets. This smaller dataset contains the same 6 plants, but only targeted organisms of *Aspergillus* family. The aim is then to answer an instantiation of general question Q1: "knowing recognized benefits of *Hyptis suaveolens* in the management of *Arachis hypogaea* against *Aspergillus parasiticus*, which other plants could alternatively be used?".

Figure 4 gives a simplified version of an excerpt from the AOC-posets built from `BioPesticide` and `ProtectionSystem` contexts, according to M1 model. In this figure, arrows represent the navigation links (the relational attributes) which allowed to find and highlight the concepts and the hierarchy we want to give to our domain expert to explore data around their question. The bold numbers are the concept numbers, the italic text is for objects of the AOC-poset and normal text is for relational attributes.

Starting from the introducer concept of *Hyptis suaveolens*, in `UsedPlant` AOC-poset (not shown), we can navigate to concepts 0, 10, 9, 7 and 5 of `BioPesticide` AOC-poset (see left of Fig. 4). The most specific concept among them is concept 5 which introduces H_B, i.e. the biopesticide produced from *Hyptis suaveolens* coming from Benin. Following the relational attributes of the concept 5, we can navigate to concepts 9 and 10 of ProtectionSystem AOC-poset (see right of Fig. 4). Concepts 9 and 10 are together non comparable; their relational attributes show that concept 9 groups plant uses that protect *Arachis hypogaca* (`PeS 0`) from *Aspergillus ochraceus* (`TO 1`), while concept 10 groups plant uses that protect *Arachis hypogaea* from *Aspergillus parasiticus* (`TO 2`), the last biological system being the one we want to manage.

In `BioPesticide` AOC-poset, we also notice that concept 5 owns a subconcept, concept 1, that introduces OG_B, i.e. a biopesticide produced from *Ocimum gratissimum* coming from Benin. According to the lattice order, which is preserved in AOC-posets, it can be deduced, thanks to the inheritance of the attributes, that the biopesticide produced from *Ocimum gratissimum* coming from Benin allows to manage at least one of the same biological systems as presented in concepts 9 and 10 of `ProtectionSystem` AOC-poset. *Ocimum gratissimum* can thus be used instead of *Hyptis suaveolens* in order to protect *Arachis hypogaea* against *Aspergillus parasiticus*, but also against *Aspergillus ochraceus*, and *Aspergillus flavus*. These facts can be checked in Knomana knowledge base.

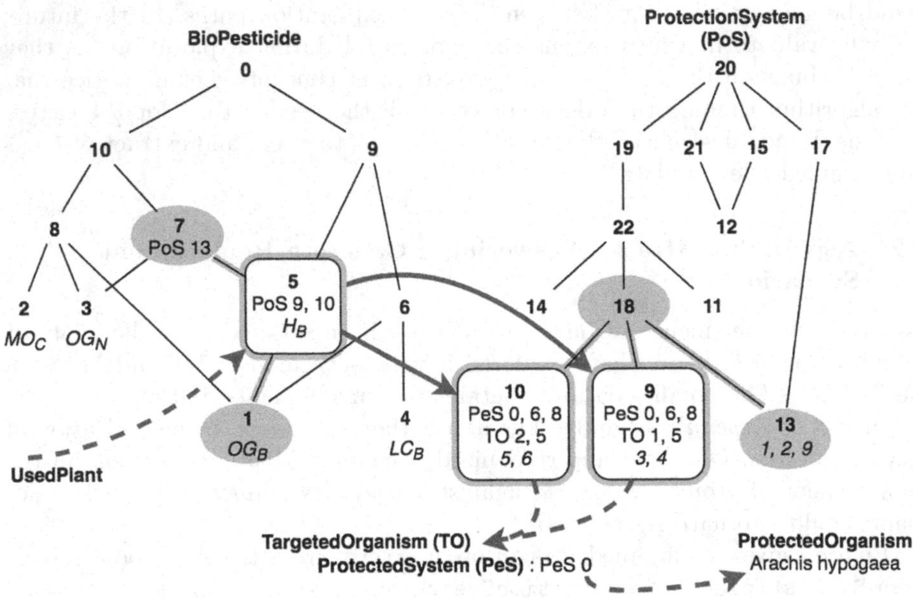

Fig. 4. Simplified version of an excerpt from the AOC-posets built from `BioPesticide` and `ProtectionSystem` contexts with M1 to answer Q1 in the case study: navigated concepts and their links are highlighted

Besides, concept 5 inherits a relational attribute PoS 13 from concept 7, that leads to concept 13 in `ProtectionSystem` AOC-Poset. This concept 13 is not comparable with concepts 9 and 10, but all these three concepts are subconcepts of concept 18. Concept 13 groups same plant uses as concepts 9 and 10 (protecting *Arachis hypogaea* against *Aspergillus flavus*), but also a different use (protecting *Oryza sativa* against *Aspergillus flavus*) due to the chosen encoding. Actually, following model M1, protected systems are first classified with respect to the production step and the treatment location, and then with respect to the protected organism.

Furthermore, concept 1 is a subconcept of concept 5 but also of other concepts in `BioPesticide` AOC-poset. Based on these hierarchical links we can infer that the biopesticide produced from *Ocimum gratissimum* coming from Benin can protect other biological systems than the ones previously described. The hierarchical organization highlights these facts for the domain experts.

Let us now consider the analysis based on M2 model; a simplified excerpt of the resulting AOC-posets is shown in Fig. 5. Starting from the introducer concept of *Hyptis suaveolens* in UsedPlant AOC-poset (not shown), we can navigate to concept 5 of BioPesticide AOC-poset (see middle of Fig. 5) that introduces H_B, i.e. the biopesticide produced from *Hyptis suaveolens* coming from Benin. As for model M1, concept 5 has a subconcept introducing OG_B. Both models give currently the same result. Going further, we see that relational attributes of concept 5 are of two types: eight of them lead to concepts of `ProtectedSystem` AOC-poset (see right of Fig. 5) while the seven others lead to concepts of `TargetedOrganism` AOC-poset (see left of Fig. 5). The attributes leading to `TargetedOrganism` concepts reveal, by looking at the most specific concepts (number 1, 2 and 0), that the biopesticides produced from *Hyptis suaveolens* and *Ocimum gratissimum* are used to fight against *Aspergillus ochraceus*, *Aspergillus parasiticus* and *Aspergillus flavus*. The attributes leading to `ProtectedSystem` concepts allow to find again *Arachis hypogaea* that is introduced by the most specific among the targeted concepts, concept 0, thanks to relational attribute `PeO 0`.

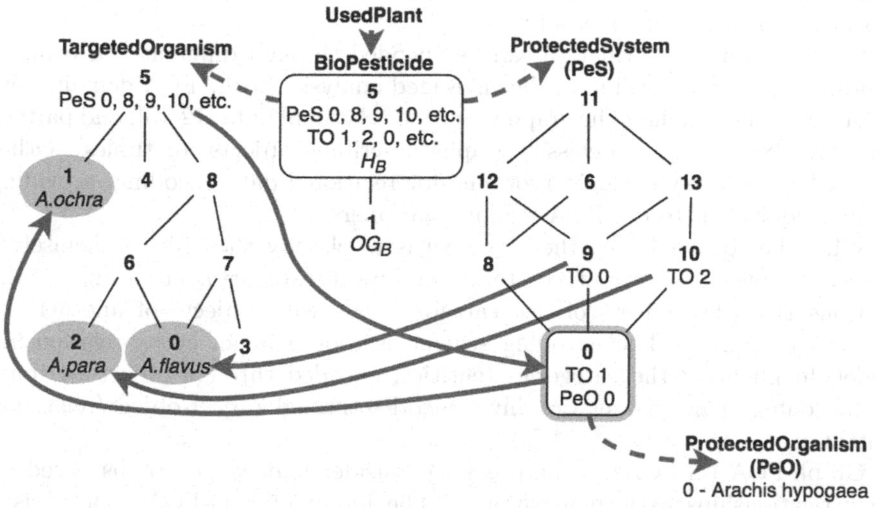

Fig. 5. Simplified version of an excerpt from the AOC-posets built from `TargetedOrganism` and `ProtectedSystem contexts` with M2 to answer Q1 in the case study: navigated concepts and their links are highlighted

To summarize, for the case study of this small dataset, both models M1 and M2 allow to find *Ocimum gratissimum* as an alternative plant to *Hyptis suaveolens* for protecting *Arachis hypogaea* against *Aspergillus parasiticus*. In addition,

a query will not give more information, contrary to RCA. The formed concepts not only give one or more answers to the initial question, but they also show how these answers are classified, and which additional (not included in the initial question) description they share. Indeed, both models also show that *Hyptis suaveolens* can be replaced by *Ocimum gratissimum* to protect *Arachis hypogaea* against *Aspergillus ochraceus*, and *Aspergillus flavus*, which is an additional information. Besides, by examining the neighborhood of the concepts which give the searched answer, the experts can formulate hypotheses for new research. E.g., if they notice that a plant protects a targeted organism against a specific aggressor, they may design experiments for evaluating if other plants with similar characteristics (grouped in the same concept) may also have the same effect. However, a set of three binary relations is not equivalent to one ternary relation. Model M2 will thus be sometimes less precise because it lacks the ternary relation. Furthermore, the navigation is more difficult in M2 than in M1 lattice family because of the greater number of concepts and relational attributes in M2 lattices.

5 Related Work

As for any data analysis method, studying the data encoding for Formal Concept Analysis and its impact on the analysis results is an essential phase. In our case, we need to take into account two specific features of our dataset: multi-relational information and ternary relations.

Multi-relational information can be encoded through different and complementary schemes, according to the envisaged analysis. In the FCA domain, several approaches highlight the graph nature of relational data [12,15], and pattern structures [8] are used to classify graphs describing objects (or tuples). Other approaches [1,7] rely on logical formula for relational data encoding, providing features equivalent to the RCA scaling quantifiers.

With the RCA scheme, the objective is to classify the objects themselves in several conceptual structures (one per object category), according to the relations that the objects of one category have with objects of another (or the same) category. The encoding scheme is rooted in an entity-relationship model, highlighting the categories (entities, encoded through object-attribute formal contexts) and the relationships (encoded through object-object/relational contexts).

Graph-FCA (G-FCA) [5] proposes to consider knowledge graphs based on n-ary relationships as formal contexts. The intent of a G-FCA concept is a projected graph pattern and the extent is an object relation. In the same vein, triadic concept analysis [14] (resp. more generally polyadic concept analysis [22]) has been introduced to deal with 3-dimensional (resp. n-dimensional) formal contexts. Both proposals could be a solution for giving additional views and highlighting more specific information on our data and we will consider them as future work.

Reading and interpreting RCA structures is known to be difficult. To facilitate this interpretation, [19] proposed to synthesize the concepts of a main lattice and their related concepts from the other lattices within a hierarchy of closed partially-ordered (sequential) patterns, i.e. directed acyclic graphs. This idea has been generalized by [6], where a family of concept lattices built by RCA is summarized through a hierarchy of concept graphs. Each concept graph is a set of concepts (potentially coming from several lattices) whose intents are mutually dependent, allowing to highlight relational patterns. Concept graphs are then organized according to the specialization order between concepts they include.

Overlays on RCAExplore have been proposed in [20] to help the analyst choices, e.g. by forecasting the number of concepts and rules resulting from a relational concept family and a quantifier or an algorithm, and several configurations are studied on an environmental dataset. Encoding legal document description in an RCF is presented in [17], where a relation links legal documents representing orders to documents representing legislative texts. The resulting conceptual structures are analyzed through relational queries and exploration strategies. The effect of several encodings of the UML meta-model in a relational context family (RCF), that includes or not the navigability and unnamed roles, has been studied in [10], allowing to conclude which RCFs are practicable, and in general about the applicability of RCA in class model normalization. Later on, using concept lattices versus AOC-posets has been studied on 15 UML class models and 15 Java code models in [18], to conclude to the superiority of AOC-posets in performance and relevancy of the produced structures.

6 Conclusion and Perspectives

The Knomana project provides a valuable collection of information about biopesticide plants in Africa. The project comes with many challenges, including information gathering, moving from raw information to knowledge associated with a stable vocabulary and ontology, and exploitation of the gathered information. In this paper, we investigate the information exploitation dimension through the application of relational concept analysis. We analyze variants for encoding the initial data model into a relational context family, and the effect of several encoding options, both quantitatively and qualitatively on a few key plants designated by the domain experts of the Knomana project as their first investigation focus.

The Knomana project is intended to extend its geographical scope to the whole world. The information collection is a continuous task, involving master students and researchers from several countries. Answering the expert questions will benefit from other approaches, such as using an on-demand algorithm [2,3], exploring other scaling quantifiers, as well as applying metrics evaluating the interest of formal concepts [13]. We envisage to define strategies for formalizing the expert questions and automatize, at least partially, the construction of appropriate relational context families. To save execution time, we plan to implement in RCAExplore other AOC-poset building algorithms, that we previously implemented in a more specific tool[3]. Besides, RCAExplore is currently moving to the COGUI platform[4] in order to pool knowledge processing activities.

Acknowledgement. This work was supported by the French National Research Agency under the Investments for the Future Program, referred as ANR-16-CONV-0004 and by INRA-CIRAD GloFoodS metaprogram (KNOMANA project).

References

1. Baader, F., Distel, F.: A finite basis for the set of $\mathcal{EL}$-implications holding in a finite model. In: Medina, R., Obiedkov, S. (eds.) ICFCA 2008. LNCS (LNAI), vol. 4933, pp. 46–61. Springer, Heidelberg (2008). https://doi.org/10.1007/978-3-540-78137-0_4
2. Bazin, A., Carbonnel, J., Huchard, M., Kahn, G.: On-demand relational concept analysis. CoRR abs/1803.07847 (2018). http://arxiv.org/abs/1803.07847
3. Bazin, A., Carbonnel, J., Huchard, M., Kahn, G., Keip, P., Ouzerdine, A.: On-demand relational concept analysis. In: Cristea, D., et al. (eds.) ICFCA 2019, LNAI 11511, pp. 155–172. Springer, Cham (2019)
4. Dicky, H., Dony, C., Huchard, M., Libourel, T.: Ares, adding a class and restructuring inheritance hierarchy. In: Onzièmes Journées Bases de Données Avancées, Nancy, France (Informal Proceedings), pp. 25–42 (1995)
5. Ferré, S.: A proposal for extending formal concept analysis to knowledge graphs. In: Baixeries, J., Sacarea, C., Ojeda-Aciego, M. (eds.) ICFCA 2015. LNCS (LNAI), vol. 9113, pp. 271–286. Springer, Cham (2015). https://doi.org/10.1007/978-3-319-19545-2_17
6. Ferré, S., Cellier, P.: How hierarchies of concept graphs can facilitate the interpretation of RCA lattices? In: 14th International Conference CLA 2018, Olomouc, Czech Republic, pp. 69–80 (2018)
7. Ferré, S., Ridoux, O., Sigonneau, B.: Arbitrary relations in formal concept analysis and logical information systems. In: Dau, F., Mugnier, M.-L., Stumme, G. (eds.) ICCS-ConceptStruct 2005. LNCS (LNAI), vol. 3596, pp. 166–180. Springer, Heidelberg (2005). https://doi.org/10.1007/11524564_11
8. Ganter, B., Kuznetsov, S.O.: Pattern structures and their projections. In: Delugach, H.S., Stumme, G. (eds.) ICCS-ConceptStruct 2001. LNCS (LNAI), vol. 2120, pp. 129–142. Springer, Heidelberg (2001). https://doi.org/10.1007/3-540-44583-8_10
9. Ganter, B., Wille, R.: Formal Concept Analysis - Mathematical Foundations. Springer, Heidelberg (1999). https://doi.org/10.1007/978-3-642-59830-2
10. Guédi, A.O., Huchard, M., Miralles, A., Nebut, C.: Sizing the underlying factorization structure of a class model. In: 17th IEEE International Conference EDOC 2013, Vancouver, BC, Canada, pp. 167–172 (2013)
11. Hacene, M.R., Huchard, M., Napoli, A., Valtchev, P.: Relational concept analysis: mining concept lattices from multi-relational data. Ann. Math. Artif. Intell. **67**(1), 81–108 (2013)

[3] http://www.lirmm.fr/AOC-poset-Builder/.
[4] https://www.lirmm.fr/cogui/.

12. Kötters, J.: Concept lattices of a relational structure. In: Pfeiffer, H.D., Ignatov, D.I., Poelmans, J., Gadiraju, N. (eds.) ICCS-ConceptStruct 2013. LNCS (LNAI), vol. 7735, pp. 301–310. Springer, Heidelberg (2013). https://doi.org/10.1007/978-3-642-35786-2_23
13. Kuznetsov, S.O., Makhalova, T.P.: On interestingness measures of formal concepts. CoRR abs/1611.02646 (2016). http://arxiv.org/abs/1611.02646
14. Lehmann, F., Wille, R.: A triadic approach to formal concept analysis. In: 3rd International Conference ICCS 1995, Santa Cruz, California, USA, pp. 32–43 (1995)
15. Liquière, M., Sallantin, J.: Structural machine learning with galois lattice and graphs. In: ICML, Madison, Wisconsin, pp. 305–313 (1998)
16. van der Merwe, D., Obiedkov, S., Kourie, D.: AddIntent: a new incremental algorithm for constructing concept lattices. In: Eklund, P. (ed.) ICFCA 2004. LNCS (LNAI), vol. 2961, pp. 372–385. Springer, Heidelberg (2004). https://doi.org/10.1007/978-3-540-24651-0_31
17. Mimouni, N., Nazarenko, A., Salotti, S.: A conceptual approach for relational IR: application to legal collections. In: Baixeries, J., Sacarea, C., Ojeda-Aciego, M. (eds.) ICFCA 2015. LNCS (LNAI), vol. 9113, pp. 303–318. Springer, Cham (2015). https://doi.org/10.1007/978-3-319-19545-2_19
18. Miralles, A., Molla, G., Huchard, M., Nebut, C., Deruelle, L., Derras, M.: Class model normalization - outperforming formal concept analysis approaches with aoc-posets. In: 12th International Conference on CLA 2015, Clermont-Ferrand, France, pp. 111–122 (2015). http://ceur-ws.org/Vol-1466/paper09.pdf
19. Nica, C., Braud, A., Dolques, X., Huchard, M., Le Ber, F.: Extracting hierarchies of closed partially-ordered patterns using relational concept analysis. In: Haemmerlé, O., Stapleton, G., Faron Zucker, C. (eds.) ICCS 2016. LNCS (LNAI), vol. 9717, pp. 17–30. Springer, Cham (2016). https://doi.org/10.1007/978-3-319-40985-6_2
20. Ouzerdine, A., Braud, A., Dolques, X., Huchard, M., Le Ber, F.: Régler le processus d'exploration dans l'analyse relationnelle de concepts. le cas de données hydroécologiques. In: Actes de la 19e conférence sur l'extraction et la gestion de connaissances (EGC 2019). Nouvelles Technologies de l'Information (2019)
21. Stumme, G., Taouil, R., Bastide, Y., Pasquier, N., Lakhal, L.: Computing iceberg concept lattices with Titanic. Data Knowl. Eng. **42**(2), 189–222 (2002)
22. Voutsadakis, G.: Polyadic concept analysis. Order **19**(3), 295–304 (2002). https://doi.org/10.1023/A:1021252203599

Parallelization of the GreConD Algorithm for Boolean Matrix Factorization

Petr Krajča(✉) and Martin Trnecka

Department of Computer Science, Palacký University Olomouc, 17. listopadu 12, 77146 Olomouc, Czech Republic
petr.krajca@upol.cz, martin.trnecka@gmail.com

Abstract. Boolean matrix factorization (BMF) is a well established and widely used tool for data analysis. Vast majority of existing algorithms for BMF is based on some greedy strategy which makes them highly sequential, thus unsuited for parallel execution. We propose a parallel variant of well-known BMF algorithm—GreConD, which is able to distribute workload among multiple parallel threads, hence can benefit from modern multicore CPUs. The proposed algorithm is based on formal concept analysis, intended for shared memory computers, and significantly reducing computation time of BMF via parallel execution.

Keywords: Boolean matrix factorization · Parallel algorithm · GreConD · Formal concept analysis

1 Introduction

Boolean Matrix Factorization (BMF) also known as Boolean matrix decomposition [3] is a problem of decomposing a Boolean matrix into two Boolean matrices such that the Boolean matrix product of the two matrices exactly or approximately equals the given input matrix. This decomposition can be seen as a concise representation of the original data, and thus, it is desirable to find output matrices with the least dimension possible.

The problem of finding the least dimension for which an exact decomposition of a Boolean matrix exists—known as the Boolean rank or Schein rank—is NP-hard [13]. Therefore, existing BMF algorithms seek for a sub-optimal decomposition with the dimension as close to the Boolean rank as possible, typically utilizing some heuristic approach.

Many efficient BMF algorithms exist, e.g. GreConD [4], GreEss [3], Asso [11], PaNDa$^+$ [10] and Hyper [15]. All of them are applicable on a reasonably large data [3]. On the other hand, their performance suffers on (very) large data. Moreover, almost all of them lack an efficient implementation. This is a consequence of two facts: (i) they are good enough for a basic experimental evaluation

Petr Krajča was supported by the grant JG 2019 of Palacký University Olomouc, No. JG_2019_008.

D. Cristea et al. (Eds.): ICFCA 2019, LNAI 11511, pp. 208–222, 2019.
https://doi.org/10.1007/978-3-030-21462-3_14

which is usually performed on small data; and (ii) for such evaluation the factorization needs to be computed only once. Unfortunately, practical applications of these methods usually require several iteration on large data, i.e. an efficient and fast implementation is necessary. Hence, it may be cumbersome to use existing algorithms for a practical analysis of large data.

With the development and growing affordability of multi-core processors, the interest in parallel computing increases and parallel algorithms are preferred to better utilize modern hardware and to handle large datasets. Interestingly, to the best of our knowledge, in existing literature, there is no attempt to parallelize any BMF algorithm.

The adjective 'Boolean' needs to be emphasized here. There are many parallel algorithms for some of the classical existing factorization methods, e.g. non-negative matrix factorization (NNMF) or singular value decomposition (SVD). These algorithms are originally intended for real-valued matrices, see for instance [7], and can be applied to Boolean matrices as well. However, as [14] and others point out, results of these methods lack meaningful interpretation when applied to Boolean matrices. Since the interpretation of data is crucial from the knowledge discovery point of view, it is more appropriate to use BMF for Boolean data than the methods originally designed for real-valued data.

A pioneering work [12] is an exception. In [12] authors consider a parallelization of BMF where a general parallelization scheme is proposed. The proposed scheme is applicable to any sequential heuristic BMF algorithm and use a parallelization to improve quality of BMF. Basically, it improves quality of obtained factors via construction of several close to optimal final decompositions in more processes running simultaneously.

The main reason for the absence of parallel BMF algorithms is their nature. A typical algorithm for BMF is based on some greedy strategy which makes the algorithm inherently sequential, hence difficult to parallelize. However, we show that the GreConD algorithm, according to many studies (e.g. [2,3]) the fastest BMF algorithm, provides enough opportunities for efficient parallelization. This means, unlike the algorithm proposed in [12], the parallel algorithm we propose is able to utilize multiple processor cores to speed up the computation.

The rest of the paper is organized as follows. Section 2 provides basic notions of Boolean matrix factorization and its connection to formal concept analysis [6]—which is widely utilized in BMF. Then in Sect. 3 a description including full pseudocodes of our new parallel BMF algorithm is provided. Section 4 presents results from various experiments and some implementation notes as well. Finally, Sect. 5 concludes the paper.

2 Preliminaries

Through the paper matrices are denoted by upper-case bold letters ($\mathbf{I}$). $\mathbf{I}_{ij}$ denotes the entry corresponding to the row i and the column j of $\mathbf{I}$. The set of all $m \times n$ Boolean (binary) matrices is denoted by $\{0,1\}^{m\times n}$. The number of 1s in Boolean matrix $\mathbf{I}$ is denoted by $\|\mathbf{I}\|$, i.e $\|\mathbf{I}\| = \sum_{i,j} \mathbf{I}_{ij}$.

2.1 A Brief Introduction to BMF

A general aim in BMF is for a given Boolean matrix $\mathbf{I} \in \{0,1\}^{m \times n}$ to find matrices $\mathbf{A} \in \{0,1\}^{m \times k}$ and $\mathbf{B} \in \{0,1\}^{k \times n}$ for which

$$\mathbf{I} \approx \mathbf{A} \circ \mathbf{B} \tag{1}$$

where $\circ$ is Boolean matrix multiplication, i.e. $(\mathbf{A} \circ \mathbf{B})_{ij} = \max_{l=1}^{k} \min(\mathbf{A}_{il}, \mathbf{B}_{lj})$, and $\approx$ represents approximate equality. This approximate equality is assessed by $||\cdot||$ (i.e. by number of 1s) and with the corresponding metric E which is defined for matrices $\mathbf{I} \in \{0,1\}^{m \times n}$, $\mathbf{A} \in \{0,1\}^{m \times k}$, and $\mathbf{B} \in \{0,1\}^{k \times n}$ by

$$E(\mathbf{I}, \mathbf{A} \circ \mathbf{B}) = ||\mathbf{I} \ominus (\mathbf{A} \circ \mathbf{B})||, \tag{2}$$

where $\ominus$ is *Boolean subtraction* which is the normal matrix subtraction with an alternative definition $0 - 1 = 0$. In words, function E is a number of 1s in $\mathbf{I}$ that are not in $(\mathbf{A} \circ \mathbf{B})$. Note, the metric (2), or its variant, is generally used to assess the quality of factorization.

A decomposition of $\mathbf{I}$ into $\mathbf{A} \circ \mathbf{B}$ may be interpreted as a discovery of k factors that exactly or approximately describe the data: interpreting $\mathbf{I}$, $\mathbf{A}$, and $\mathbf{B}$ as the object-attribute, object-factor, and factor-attribute matrices.

The model (1) can be interpreted as follows: the object i has the attribute j, i.e. $\mathbf{I}_{ij} = 1$, if and only if there exists factor l such that l applies to i and j is one of the particular manifestations of l.

Note also an important geometric view of BMF: a decomposition $\mathbf{I} \approx \mathbf{A} \circ \mathbf{B}$ with k factors represents a coverage of the 1s in $\mathbf{I}$ by k rectangular areas—rectangles for short—in $\mathbf{I}$ full of 1s. The lth rectangle is the Boolean product of the lth column in matrix $\mathbf{A}$ and the lth row in matrix $\mathbf{B}$. Let us consider the following Boolean matrix factorization of $\mathbf{I}$:

$$\mathbf{I} = \begin{pmatrix} 10111 \\ 01101 \\ 01001 \\ 10110 \end{pmatrix} = \begin{pmatrix} 110 \\ 011 \\ 001 \\ 100 \end{pmatrix} \circ \begin{pmatrix} 10110 \\ 00101 \\ 01001 \end{pmatrix} = \mathbf{A} \circ \mathbf{B}$$

The geometric view of BMF tell us that the matrix $\mathbf{I}$ can be described as the *Boolean sum* $\oplus$ which is the normal matrix sum where $1 + 1 = 1$ of three matrices depicted below.

$$\begin{pmatrix} 10111 \\ 01101 \\ 01001 \\ 10110 \end{pmatrix} = \begin{pmatrix} 1&0&1&1&0 \\ 0&0&0&0&0 \\ 0&0&0&0&0 \\ 1&0&1&1&0 \end{pmatrix} \oplus \begin{pmatrix} 0&0&1&0&1 \\ 0&0&1&0&1 \\ 0&0&0&0&0 \\ 0&0&0&0&0 \end{pmatrix} \oplus \begin{pmatrix} 0&0&0&0&0 \\ 0&1&0&0&1 \\ 0&1&0&0&1 \\ 0&0&0&0&0 \end{pmatrix}$$

Each of these matrices contains one rectangle which is a *Boolean product* of a particular column of the matrix A and a row of the matrix B. If the rectangular areas cover only non zero elements in the matrix $\mathbf{I}$, then the $\mathbf{A} \circ \mathbf{B}$ is called the *from-below matrix decomposition* [3].

2.2 BMF with Help of Formal Concept Analysis

The problem of BMF is closely connected to *formal concept analysis* (FCA) [6]. The main notion of FCA is *formal context* which is defined as a triple $\langle X, Y, I\rangle$ where X is a nonempty set of objects, Y is a nonempty set of attributes, and I is a binary relation between X and Y. Hence the formal context $\langle X, Y, I\rangle$ with m objects and n attributes is in a fact a Booolean matrix $\mathbf{I} \in \{0,1\}^{m\times n}$ where $\mathbf{I}_{ij} = 1$ if $\langle x, y\rangle \in I$, and vice versa.

This one-to-one correspondence allows to apply means of formal concept analysis to Boolean matrices. Namely, to every Boolean matrix $\mathbf{I} \in \{0,1\}^{n\times m}$ one may associate a pair $\langle {}^{\uparrow}, {}^{\downarrow}\rangle$ of arrow operators assigning to sets $C \subseteq X = \{1,\ldots,m\}$ and $D \subseteq Y = \{1,\ldots,n\}$ the sets $C^{\uparrow} \subseteq Y$ and $D^{\downarrow} \subseteq X$ defined by

$$C^{\uparrow} = \{j \in Y \mid \forall i \in C : \mathbf{I}_{ij} = 1\},$$
$$D^{\downarrow} = \{i \in X \mid \forall j \in D : \mathbf{I}_{ij} = 1\},$$

where $C^{\uparrow}$ is a set of all attributes (columns) shared by all objects (rows) in C and $D^{\downarrow}$ is a set of all objects sharing all attributes in D.

The pair $\langle C, D\rangle$ for which $C^{\uparrow} = D$ and $D^{\downarrow} = C$ is called a *formal concept* and C and D are called *extent* and *intent* of a formal concept $\langle C, D\rangle$, respectively. Note that formal concepts correspond to maximal rectangles full of 1s. The set of all formal concepts of a formal context $\langle X, Y, I\rangle$ we denote by

$$\mathcal{B}(X, Y, I) = \{\langle C, D\rangle \mid C \subseteq X, D \subseteq Y, C^{\uparrow} = D, D^{\downarrow} = C\}.$$

The set of all concepts can be equipped with a partial order $\leq$ such that $\langle A, B\rangle \leq \langle C, D\rangle$ iff $A \subseteq C$ (or $D \subseteq B$), a pair $\langle A, B\rangle$ is a subconcept of $\langle C, D\rangle$, while $\langle C, D\rangle$ is a superconcept of $\langle A, B\rangle$. The whole set of partially ordered formal concepts is called the *concept lattice* of $\mathbf{I}$.

Now, we explain the connection between a set of formal concepts and the Boolean matrix factorization.

Every set $\mathcal{F} = \{\langle C_1, D_1\rangle, \ldots, \langle C_k, D_k\rangle\} \subseteq \mathcal{B}(X, Y, I)$ (with a fixed indexing of the formal concepts $\langle C_l, D_l\rangle$) induces the $m \times k$ and $k \times n$ Boolean matrices $\mathbf{A}_{\mathcal{F}}$ and $\mathbf{B}_{\mathcal{F}}$ such that

$$(\mathbf{A}_{\mathcal{F}})_{il} = \begin{cases} 1, \text{if } i \in C_l, \\ 0, \text{if } i \notin C_l, \end{cases} \tag{3}$$

and

$$(\mathbf{B}_{\mathcal{F}})_{lj} = \begin{cases} 1, \text{if } j \in D_l, \\ 0, \text{if } j \notin D_l, \end{cases} \tag{4}$$

for $l = 1,\ldots,k$. That is, the lth column of $\mathbf{A}_{\mathcal{F}}$ and lth row $\mathbf{B}_{\mathcal{F}}$ are the characteristic vectors of C_l and D_l, respectively. The set $\mathcal{F}$ is also called a set of *factor concepts*. Clearly, $\mathbf{A}_{\mathcal{F}} \circ \mathbf{B}_{\mathcal{F}}$ is the from-below matrix decomposition.

2.3 GreConD Algorithm

One of the most successful algorithms for BMF is the GRECOND[1] algorithm [4]. To produce matrices $\mathbf{A}_{\mathcal{F}}$ and $\mathbf{B}_{\mathcal{F}}$ it uses a particular greedy search for factor concepts which allows to compute factor formal concepts "on demand", i.e. without the need to compute the whole concept lattice for the input binary matrix first. This strategy leads to a magnitude time saving while the quality of decomposition compared to the direct adoption of the set cover algorithm is almost retained.

GRECOND is designed to compute an exact from-below factorization, i.e. it provides an approximate solution. It stops when first k factors are discovered or when the error E does not exceed ϵ. Note that if k is unspecified (i.e. the number of factors does not matter) and $\epsilon = 0$, then the algorithm provides an exact solution.

Note, GRECOND according to various experimental evaluations (e.g. [2,3]) is one of the fastest BMF algorithm which produces some of the best results.

In the following section we provide a description of our parallelization of the GRECOND algorithm which we call PARAGRECOND.

3 Algorithm

The algorithm, we propose, consists of four basic building blocks, we shall call them procedures. The main one is the PARAGRECOND procedure which bears the name of the algorithm and represents the core of the algorithm. Further, there is a NEXTFACTOR procedure which takes care of the factor discovery. Note that factors are discovered in independent threads which are described with the NEXTFACTORTHREAD procedure. Furthermore, to compute factors efficiently, every thread uses an auxiliary procedure EXTENDCONCEPT. Figure 1 provides an overview of these procedures and their relationships to threads. The rest of this section is devoted to a more detailed description of our algorithm and these procedures.

3.1 Core of the Algorithm

The PARAGRECOND procedure (see Algorithm 1 for the pseudo-code) is by its very nature similar to the original GRECOND algorithm as proposed in [4]. It takes a formal context $\langle X, Y, I \rangle$ and a number t as arguments and returns a set of factor concepts. The argument t indicates the number of threads used to compute matrix decomposition.

Remark 1. We describe our algorithm using the FCA terminology. Therefore, if there is no risk of confusion, we use notions of a formal context $\langle X, Y, I \rangle$, the corresponding relation I, and the corresponding binary matrix $\mathbf{I}$ interchangeably.

[1] GRECOND is abbreviation for Greedy Concepts on Demand.

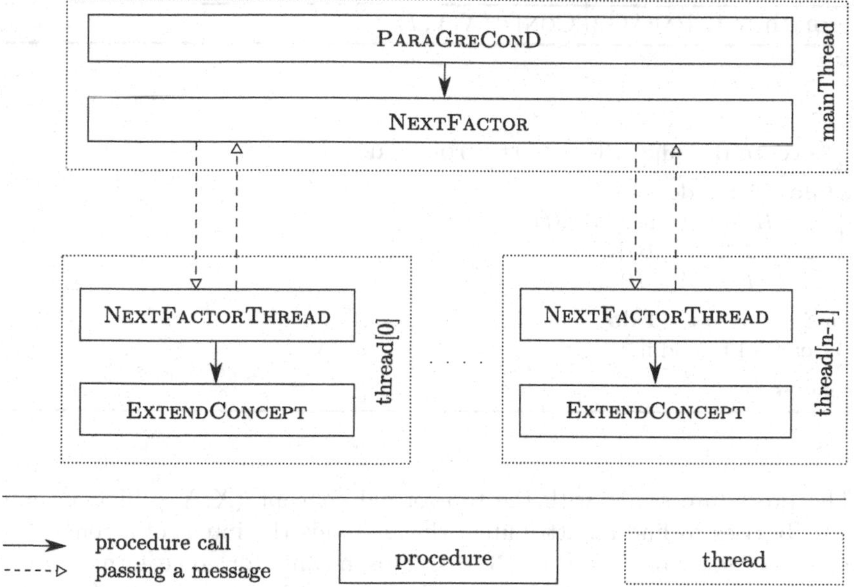

Fig. 1. Conceptual view of the algorithm

First, PARAGRECOND creates an empty set of factor concepts $\mathcal{F}$ (line 1) and creates a copy I' of the relation I (line 2). The relation I' shall represent ones not covered yet by any factor concept. For convenience, we assume that I' is a global variable accessible from other procedures without any restrictions. Further, during initialization t threads are spawned (lines 3–4). These threads serve as workers which are used to find factor concepts. Their role is explained shortly.

The main part of the algorithm (lines 5–8) consists of three steps: (i) A formal concept covering presumably the largest amount of ones in the I' is identified. Next, (ii) the set $\mathcal{F}$ is extended with this formal concept. Subsequently, (iii) ones covered by this concept are removed from I'. These steps are repeated until I' is empty, i.e., it contains no 1s. When the algorithm stops, it remains to cancel all threads (lines 9–10) and return the result.

3.2 Parallel Factor Discovery

The crucial part of the algorithm is the NEXTFACTOR procedure (see Algorithm 2) which uses a greedy strategy to identify a formal concept covering presumably maximal number of ones in I'. The amount of covered ones is given by:

$$\text{COVERAGE}(I, A, B) = |I \cap (A \times B)|.$$

where A and B are extent and intent of a formal concept, respectively, and I is a binary relation $I \subseteq X \times Y$.

Algorithm 1. ParaGreConD($\langle X, Y, I\rangle, t$)

```
F ← ∅
I' ← I
for i ← 0 to t − 1 do
    create thread[i] with NextFactorThread
while |I'| > 0 do
    ⟨A, B⟩ ← NextFactor(t)
    F ← F ∪ {⟨A, B⟩}
    I' ← I' − (A × B)
for i ← 0 to t − 1 do
    cancel thread[i]
return F
```

The procedure starts with the top formal concept $\langle X, X^{\uparrow}\rangle$ (line 1) and it repeatedly tries to find an attribute j that extends the intent of a concept and increases coverage of ones (lines 3–15). This means, for some formal concept $\langle A, B\rangle$ it tries to find an attribute $j \notin B$ and formal concept $\langle (B \cup \{j\})^{\downarrow}, (B \cup \{j\})^{\downarrow\uparrow}\rangle$ which covers more ones than $\langle A, B\rangle$. If there is no such attribute, the procedure stops and returns the given formal concept $\langle A, B\rangle$. Otherwise, formal concept with the highest coverage is passed as an input for another iteration where the next attempt to extend the intent of the concept is made.

Remark 2. This approach is in its nature highly sequential. Apparently, the initial formal concept is extended with the individual attributes in consecutive steps. Hence, it might seem there is no opportunity for parallelization. However, in each individual step one has to compute $\langle (B \cup \{j\})^{\downarrow}, (B \cup \{j\})^{\downarrow\uparrow}\rangle$ and its corresponding coverage for all attributes $j \notin B$. In fact, this is a very time demanding task, fortunately, these formal concepts and their corresponding coverages can be computed independently, hence in separate threads of execution.

The NextFactor procedure, we propose, utilizes threads spawned earlier in the ParaGreConD procedure and observation from Remark 2. Each thread (described in Algorithm 3) receives a formal concept $\langle A, B\rangle$ and a minimal coverage the output formal concept has to satisfy. These values are send from the NextFactor procedure to each thread at the beginning of each iteration (Algorithm 2, line 6). Then, each thread tries to find an attribute j such that Coverage($I', (B \cup \{j\})^{\downarrow}, (B \cup \{j\})^{\downarrow\uparrow}$) is maximal and greater or equal than the given minimal coverage (Algorithm 3, line 3). Note that this task is delegated to the auxiliary procedure ExtendConcept.

In order to distribute workload equally, all threads shares a global variable *attrIndex* holding an attribute (or its index) to be processed next. At the beginning of each iteration, this global variable is initialized to 1 (Algorithm 2, line 4) and all threads are using atomic *fetch-and-add*[2] operation to obtain an attribute j to work with.

Algorithm 2. NextFactor(t)

```
1  F ← ⟨X, X↑⟩
2  minCoverage = 0
3  repeat
4      AtomicStore(attrIndex, 1)
5      for i ← 0 to t − 1 do
6          send to thread[i] value ⟨F, minCoverage⟩
7      cov′ ← 0
8      for i ← 0 to t − 1 do
9          ⟨F′, cov⟩ ← receive
10         if (F′ ≠ None) and (cov ≥ minCoverage) and (cov > cov′) then
11             F ← F′
12             cov′ ← cov
13     changed ← (cov′ ≥ minCoverage)
14     minCoverage ← cov′
15 until not changed;
16 return F
```

Algorithm 3. NextFactorThread()

```
1  while not cancelled do
2      ⟨F, minCoverage⟩ ← receive;
3      ⟨F′, m⟩ ← ExtendConcept(F, minCoverage)
4      send to mainThread value ⟨F′, m⟩
```

When all attributes are processed (i.e., $attrIndex > |Y|$), each thread sends the obtained result back to the NextFactor procedure (Algorithm 3, line 4) which collects these partial results from all threads and picks the formal concept with the highest coverage (Algorithm 2, lines 8–12). Afterwards, in a loop, the procedure tries to find another attribute that extends the intent of a given formal concept while increasing its coverage as discussed earlier.

3.3 Computation of a Factor Concept

The major burden of computation lies in the ExtendConcept procedure. As mentioned earlier, its role is to find an attribute j and a corresponding formal

[2] This operation atomically increments value of the variable and returns the original value, i.e., its meaning is equivalent to `i++`.

concept $\langle (B \cup \{j\})^{\downarrow}, (B \cup \{j\})^{\downarrow\uparrow} \rangle$ having higher coverage than the original formal concept $\langle A, B \rangle$. From the computational point of view this task consists of three steps. It is necessary to compute (i) extent $(B \cup \{j\})^{\downarrow}$ of a formal concept, (ii) its intent $(B \cup \{j\})^{\downarrow\uparrow}$, and (iii) determine its coverage w.r.t. I'. In fact, all three steps are time demanding tasks. Thus, we use the following observations to make the EXTENDCONCEPT procedure more efficient.

Let $\langle X, Y, I \rangle$ be a formal context then for each formal concept $\langle A, B \rangle \in \mathcal{B}(X, Y, I)$, any attribute $j \in Y$, and each relation $I' \subseteq I$ holds

(Ob1) $(B \cup \{j\})^{\downarrow} = A \cap \{j\}^{\downarrow}$,
(Ob2) $X \cap \{j\}^{\downarrow} = \{j\}^{\downarrow}$,
(Ob3) $\text{COVERAGE}(I', A, B) \leq |A| \cdot |Y|$,
(Ob4) $\text{COVERAGE}(I', A, B) \leq |A| \cdot |B|$.

All these observations are easy to see, however, they have significant impact on efficiency of the algorithm. The observation (Ob1) allows us to compute an extent of the extended concept more efficiently, since the otherwise slow operation $^{\downarrow}$ can be replaced with an intersection $A \cap \{j\}^{\downarrow}$ which can be implemented more efficiently. It is just necessary to precompute and keep all sets $\{j\}^{\downarrow}$, $j \in Y$, for instance, in an array. Notice that this a quite common technique used, for example, in algorithms from the Close-by-One family, see [9] and [8].

Let us recall that each invocation of the NEXTFACTOR starts with the top formal concept, i.e., with $\langle X, X^{\uparrow} \rangle$. From the observations (Ob2) and (Ob1) follows that in its first iteration (see Algorithm 2, lines 3–15), NEXTFACTOR always deals with attribute concepts only. Since NEXTFACTOR is invoked multiple times, it makes sense to avoid redundant computations and keep these attribute concepts in the memory and reuse them.

Observations (Ob3) and (Ob4) provides upper bounds on number of ones each formal concept may cover. This means, that if some minimal coverage *cov* is required and for an extent A holds $|A| \cdot |Y| < cov$, then it is possible to immediately abandon computation. It is apparent that such formal concept does not satisfy minimal coverage condition. Analogously, if $|A| \cdot |B| < cov$ for some formal concept $\langle A, B \rangle$ and minimal coverage *cov*, then it is possible to skip the computation of a coverage, because it is with no doubt smaller than *cov*.

We incorporated all these observations and their implications into the EXTENDCONCEPT procedure, see Algorithm 4, to enhance its efficiency. The procedure accepts a formal concept $\langle A, B \rangle$ and a minimal coverage the extended formal concept has to satisfy. This value is kept in the auxiliary variable *cov* (line 1). Further, we keep a formal concept with the highest coverage in the variable *result*. In case there is no such formal concept, we indicate this with the special value *None* (see line 2).

As discussed above, EXTENDCONCEPT iterates over all attributes by increasing global shared variable *attrIndex* (line 3). If the given attribute is in B, procedure immediately proceeds with the next available attribute (line 4).

If $\langle A, B \rangle$ is the top formal concept, i.e., $A = X$, we employ the observation (Ob2). This means, the algorithm checks global array *attrConcepts* if the given attribute concept was already computed. If so, it is retrieved from the array (line 7), otherwise attribute concept is computed and stored to the array *attrConcepts* (lines 9 and 10). In both cases obtained formal concept is assigned to variables C and D, where C denotes the extent and D intent of a formal concept.

In case $\langle A, B\rangle$ is not the top concept (lines 12–16), the extent is determined first (line 12). Notice that observation (Ob1) is used here. Subsequently, if the upper bound of the coverage is greater or equal than the minimal coverage *cov*, see observation (Ob3), procedure proceeds to computation of an intent D (line 14). Otherwise, intent D is considered to be an empty set (line 16).

Afterwards, no matter what was the input, the coverage is checked (lines 17 and 18). If the upper bound of the coverage given by $|C| \cdot |D|$ is smaller than *cov*, the computation of a coverage is skipped, see observation (Ob4). If the coverage is greater than the minimal coverage *cov*, formal concept $\langle C, D\rangle$ is considered as a result and its coverage as minimal (lines 19–21). Then, procedure continues with the next available attribute. If there are no more attributes to process, procedure returns the formal concept stored in the variable *result* along with its coverage *cov*. Subsequently, these values are processed in the NEXTFACTOR procedure, see Algorithm 3 (line 4) and Algorithm 2 (lines 9–12).

Algorithm 4. EXTENDCONCEPT($\langle A, B\rangle$, *minCoverage*)

```
1  cov ← minCoverage
2  result ← None
3  while (j ← AtomicFetchAndAdd(attrIndex, 1) ≤ |Y| do
4      if j ∉ B then
5          if A = X then
6              if attrConcepts[j] is set then
7                  ⟨C, D⟩ ← attrConcepts[j]
8              else
9                  ⟨C, D⟩ ← ⟨{j}↓, {j}↓↑⟩
10                 attrConcepts[j] ← ⟨C, D⟩
11         else
12             C = A ∩ {j}↓
13             if |C| · |Y| ≥ cov then
14                 D ← C↑
15             else
16                 D ← ∅
17         if |C| · |D| > cov then
18             cov' ← Coverage(I', C, D)
19             if cov' > cov then
20                 cov ← cov'
21                 result ← ⟨C, D⟩
22 return ⟨result, cov⟩
```

4 Implementation and Evaluation

We implemented the ParaGreConD algorithm in order to evaluate its properties. Our referential implementation is in C++ and is intended for the GNU/Linux operating system. However, the algorithm makes no special requirements, hence can be implemented in any suitable programming language and for almost any operating system supporting multithreading.

Remark 3. The ParaGreConD algorithm, namely the NextFactor and NextFactorThread procedures, is described using the message passing programming style requiring a *send* and *receive* operations to be non-blocking and blocking, respectively. If these operations are unavailable, they can be easily emulated with traditional means like semaphores which are available in almost every appropriate operating system, see, for instance, [1]. We decided to use this approach also in our referential implementation to make the code more general.

Remark 4. Workload distribution in our algorithm requires the global variable *attrIndex* to be changed atomically. Thus, it is essential to choose proper means provided by the programming languages. For instance, in Java one should use the `AtomicInteger` class, in C++ there is a `std::atomic` template, or contemporary C/C++ compilers provide builtin operations[3] like `__atomic_fetch_add` which can be used to perform demanded operations atomically.

Evaluation of the algorithm was performed on otherwise idle computer which was equipped with two Intel Xeon E5-2680, 2.80 GHz CPUs (40 physical CPU cores in total), with 64 GB RAM, running Debian Linux 9.6. Source codes were compiled with GNU GCC 6.3.0. Each measurement was taken three times and an average value was considered.

The first set of experiments is focused on standard datasets from the FIMI Dataset Repository (T10I4D100K, T40I10D100K, chess, connect, accidents), UCI ML Dataset Repository (mushrooms, anonymous web), UCI KDD Repository (NSF), from [5] (americas large), and our own dataset (debian tags). Their characteristics are listed in Table 1. Besides, Table 1 contains number of discovered factors and time it took to obtain these factors with a single thread and forty threads. Apparently, the parallel algorithm is significantly faster than its sequential counterpart. Namely, for the datasets T10I4D100K, T40I10D100K, or NSF the parallel version is more than 18× faster.

The ability to utilize multiple processor cores to speed up computation is often called *scalability* and it is a crucial property of parallel algorithms. This property is typically expressed as a ratio $S(t) = \frac{T(1)}{T(t)}$ where $T(1)$ denotes time necessary to obtain results with a sequential algorithm and $T(t)$ denotes time of the computation if t threads is used.

Figure 2 shows performance improvement for selected datasets. One can see that for small numbers of threads performance improves almost linearly. However, at some point overall speedup slowdowns (see Fig. 2, left) or even stops

[3] Documented but not a part of the language specification.

Table 1. Dataset characteristics

dataset	rows	cols.	density of 1's (%)	factors	time (1 thread)	time (40 threads)	speedup
T40I10D100K	100,000	1,000	3.96	953	106.66 s	5.90 s	18.1×
T10I4D100K	100,000	1,000	1.01	906	31.47 s	1.73 s	18.1×
NSF	12,841	4,894	0.90	4911	253.33 s	13.44 s	18.8×
americas large	10,127	3,485	0.53	539	8,301 ms	620.46 ms	13.4×
accidents	340,183	468	7.22	690	236.82 s	20.21 s	11.7×
connect	67,557	129	33.3	214	5,933 ms	574.52 ms	10.3×
debian tags	14,315	476	0.99	496	597.46 ms	92.29 ms	6.5×
an. web.	32,710	296	1.02	281	412.48 ms	73.55 ms	5.6×
mushrooms	8124	120	19.1	120	220.05 ms	33.08 ms	6.6×
chess	3,196	74	50	124	220.05 ms	41.56 ms	5.2×

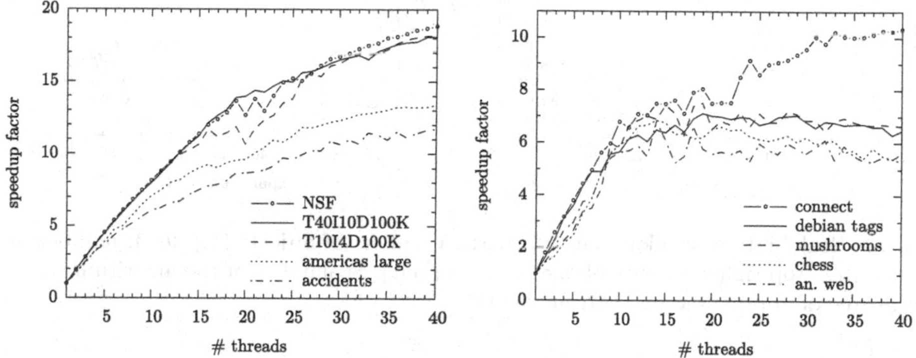

Fig. 2. Scalability of the algorithm for real datasets

(see Fig. 2, right). This is a consequence of so called Amdahl's law which sets the theoretical upper bound of a speedup, see, for instance, [1]. The theoretical speedup S_t can be expressed as:

$$S_t(t) = \frac{1}{(1-p) + \frac{p}{t}}, \tag{5}$$

where p denotes the proportion of the algorithm that can utilize multiple threads to reduce computation time and t is number of threads used to optimize this part of the algorithm. As a corollary of (5) maximal theoretical speedup S_{max} is given by

$$S_{max} = \frac{1}{1-p}. \tag{6}$$

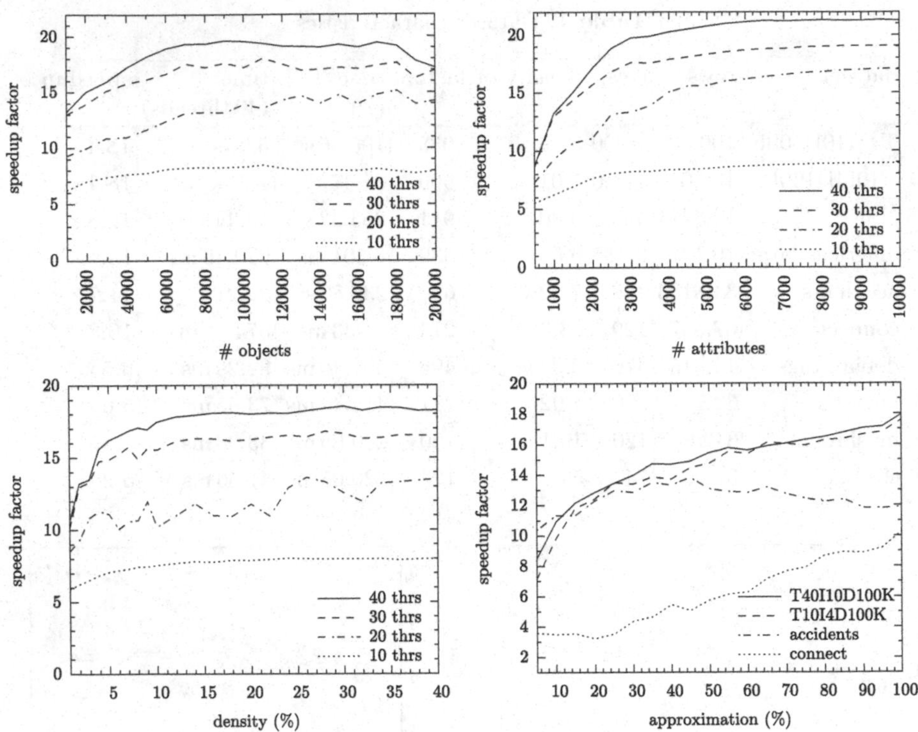

Fig. 3. Scalability of random datasets w.r.t. numbers of objects (top left), number of attributes (top right), density of ones (bottom left). Scalability of the algorithm w.r.t. to degree of approximation (bottom right).

This means, for instance, if 5 % of a task have to be performed sequentially, it is impossible to achieve better speedup factor than 20 no matter how many threads/CPUs is used.

The ParaGreCond algorithm is bounded by the Amdahl's law and by the fact that some parts of the algorithm are processed sequentially, namely, computation of an extent, intent, and coverage. Therefore, Eqs. (5) and (6) clearly explains why the speedup slowdowns or stops with higher numbers of threads. Note that the value p in (5) and (6) depends not only on the given algorithm but also on the input data. Further, Fig. 2 suggests that large datasets can benefit from parallel execution more than smaller ones.

We prepared a set of experiments involving artificial datasets to confirm this assumption. Outcomes of these experiments are presented in Fig. 3 which shows how varying numbers of objects, attributes, and ones affect scalability for 10, 20, 30, and 40 threads. The top left chart in Fig. 3 shows increasing scalability for random datasets with the increasing number of objects, one thousand of attributes, and 2 % density of ones. The top right chart in Fig. 3 depicts speedup for random datasets consisting of ten thousand objects, varying numbers of attributes, and 2 % density of ones. Finally, how scalability grows with the increasing density of ones in random datasets of size 10000×1000 can be seen in bottom left chart of Fig. 3. Clearly, these observations support assumption that the ParaGreConD algorithm scales better with large datasets.

The algorithm we present provides an exact solution of the BMF problem. However, it can be easily turned into an algorithm returning an approximate solution. Basically, it suffices to relax the $|I'| > 0$ condition in the ParaGreConD procedure to obtain results which correspond to the from-below approximation of the input matrix. Therefore, we decided to investigate if the degree of approximation (percentage of covered ones) and scalability are related. The result is presented in Fig. 3 (bottom right). Apparently, the speedup factor tends to grow with the increasing degree of approximation.

5 Conclusion and Future Research

The parallel algorithm ParaGreConD which is a parallelization of the well-know GreConD algorithm for Boolean matrix factorization was proposed. The algorithm itself is the first parallel algorithm for the from-below matrix decomposition problem. We provide a detailed description of the algorithm including detailed pseudocodes and implementation notes. Our experimental evaluation on real and synthetic data shows that the parallel algorithm significantly outperforms the original one in terms of runtime. Moreover, our experiments shows that the algorithm scales well especially for large datasets.

The future research should include the following topics. An application of our approach to more complex BMF algorithms, e.g. GreEss which provides a better results than GreConD from the quality of matrix factorization viewpoint. An examination if the parallel approach can bring a novel BMF algorithms, i.e. can be used to develop a new parallel strategy to Boolean matrix factorization.

References

1. Andrews, G.R.: Foundations of Multithreaded, Parallel, and Distributed Programming. Addison-Wesley, Reading (2002)
2. Belohlavek, R., Outrata, J., Trnecka, M.: Toward quality assessment of Boolean matrix factorizations. Inf. Sci. **459**, 71–85 (2018). https://doi.org/10.1016/j.ins.2018.05.016
3. Belohlavek, R., Trnecka, M.: From-below approximations in Boolean matrix factorization: geometry and new algorithm. J. Comput. Syst. Sci. **81**(8), 1678–1697 (2015). https://doi.org/10.1016/j.jcss.2015.06.002
4. Belohlavek, R., Vychodil, V.: Discovery of optimal factors in binary data via a novel method of matrix decomposition. J. Comput. Syst. Sci. **76**(1), 3–20 (2010). https://doi.org/10.1016/j.jcss.2009.05.002
5. Ene, A., Horne, W.G., Milosavljevic, N., Rao, P., Schreiber, R., Tarjan, R.E.: Fast exact and heuristic methods for role minimization problems. In: Ray, I., Li, N. (eds.) 13th ACM Symposium on Access Control Models and Technologies, SACMAT 2008, Estes Park, CO, USA, 11–13 June 2008, Proceedings, pp. 1–10. ACM (2008). https://doi.org/10.1145/1377836.1377838

6. Ganter, B., Wille, R.: Formal Concept Analysis Mathematical Foundations. Springer, Heidelberg (1999). https://doi.org/10.1007/978-3-642-59830-2
7. Kannan, R., Ballard, G., Park, H.: A high-performance parallel algorithm for nonnegative matrix factorization. SIGPLAN Not. **51**(8), 9:1–9:11 (2016). https://doi.org/10.1145/3016078.2851152
8. Krajca, P., Outrata, J., Vychodil, V.: Advances in algorithms based on CbO. In: Kryszkiewicz, M., Obiedkov, S.A. (eds.) Proceedings of the 7th International Conference on Concept Lattices and Their Applications, Sevilla, Spain, 19–21 October 2010. CEUR Workshop Proceedings, vol. 672, pp. 325–337. CEUR-WS.org (2010)
9. Kuznetsov, S.O.: Learning of simple conceptual graphs from positive and negative examples. In: Żytkow, J.M., Rauch, J. (eds.) PKDD 1999. LNCS (LNAI), vol. 1704, pp. 384–391. Springer, Heidelberg (1999). https://doi.org/10.1007/978-3-540-48247-5_47
10. Lucchese, C., Orlando, S., Perego, R.: A unifying framework for mining approximate top-k binary patterns. IEEE Trans. Knowl. Data Eng. **26**(12), 2900–2913 (2014). https://doi.org/10.1109/TKDE.2013.181
11. Miettinen, P., Mielikäinen, T., Gionis, A., Das, G., Mannila, H.: The discrete basis problem. IEEE Trans. Knowl. Data Eng. **20**(10), 1348–1362 (2008). https://doi.org/10.1109/TKDE.2008.53
12. Outrata, J., Trnecka, M.: Parallel exploration of partial solutions in Boolean matrix factorization. J. Parallel Distrib. Comput. **123**, 180–191 (2019). https://doi.org/10.1016/j.jpdc.2018.09.014
13. Stockmeyer, L.J.: The Set Basis Problem is NP-complete. Research reports, IBM Thomas J. Watson Research Division (1975)
14. Tatti, N., Mielikäinen, T., Gionis, A., Mannila, H.: What is the dimension of your binary data? In: Proceedings of the 6th IEEE International Conference on Data Mining (ICDM 2006), Hong Kong, China, 18–22 December 2006, pp. 603–612. IEEE Computer Society (2006). https://doi.org/10.1109/ICDM.2006.167
15. Xiang, Y., Jin, R., Fuhry, D., Dragan, F.F.: Summarizing transactional databases with overlapped hyperrectangles. Data Min. Knowl. Discov. **23**(2), 215–251 (2011). https://doi.org/10.1007/s10618-010-0203-9

Simultaneous, Polynomial-Time Layout of Context Bigraph and Lattice Digraph

Tim Pattison(✉) and Aaron Ceglar

Defence Science and Technology Group, Adelaide, Australia
{tim.pattison,aaron.ceglar}@dst.defence.gov.au

Abstract. Formal Concept Analysis (FCA) takes as input the bipartite context graph and produces a directed acyclic graph representing the lattice of formal concepts. Excepting possibly the supremum and infimum, the set of formal concepts corresponds to the set of proper maximal bicliques in the context bigraph. This paper proposes polynomial-time graph layouts which emphasise maximal bicliques in the context bigraph and facilitate "reading" directed paths in the lattice digraph. These layouts are applied to sub-contexts of the InfoVis 2004 data set which are indivisible by the CARVE divide-and-conquer FCA algorithm. The paper also investigates the relationship between vertex proximity in the bigraph layout and co-membership of maximal bicliques, and demonstrates the significant reduction of edge crossings in the digraph layout.

Keywords: Lattice drawing · Bigraph clustering · Resistance distance

1 Introduction

CARVE [28] is a divide-and-conquer algorithm which recursively partitions the bipartite context graph and lattice digraph of amenable formal contexts. It exploits structure found in some empirical datasets, such as the `InfoVis 2004` [30] and `FCA 2012` [8] bibligraphic data sets, and in software systems. The leaf nodes of the resultant partition tree [29] correspond to connected sub-graphs of the context bigraph which cannot be further divided by CARVE. An external algorithm performs Formal Concept Analysis (FCA) on each non-trivial indivisible sub-context and returns the corresponding directed acyclic graph (DAG). The vertices of these DAGs are formal concepts and the arcs represent the transitive reduction of the set inclusion relationship between their extents. The CARVE algorithm then assembles the lattice DAG for the original context from these constituent DAGs.

The CARVE software prototype [28,29] uses the Sugiyama method for layered graph drawing [34] to visualise these DAGs. It assigns vertices to layers, routes edges which span multiple layers via waypoints in each intervening layer, and attempts to minimise edge crossings between adjacent layers. Cycle removal is not required because the lattice digraph is acyclic. Figure 1 shows the resultant layout for an indivisible sub-context of `InfoVis 2004` representing 39 of the 614

D. Cristea et al. (Eds.): ICFCA 2019, LNAI 11511, pp. 223–240, 2019.
https://doi.org/10.1007/978-3-030-21462-3_15

attributes – papers published in, or cited by, the proceedings of the IEEE InfoVis Symposium – and 49 of the 1036 objects – authors of these papers.

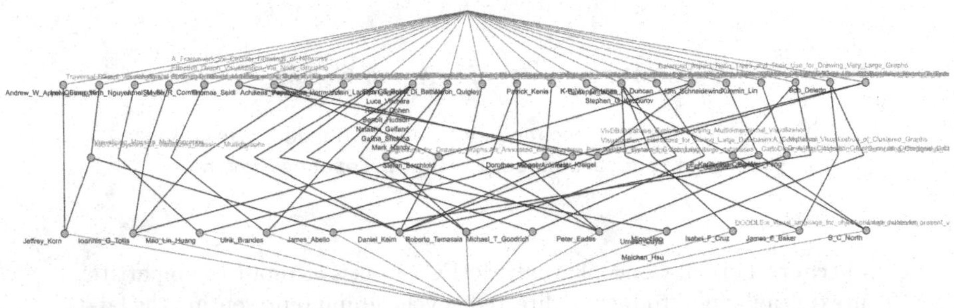

Fig. 1. Layout of the lattice digraph produced by CARVE for an indivisible sub-context of the `InfoVis` 2004 data set [30].

Figure 1 emphasises two important shortcomings of the current prototype: the object and attribute labels on vertices are long and in many cases hard to read because no attempt has been made to abbreviate them or otherwise prevent them from overlapping with each other or with edges; and it is difficult to "read" the ancestor-descendant relations from the diagram because long edges, edge crossings and waypoints interfere with the process of visually tracing the corresponding paths. Path tracing and label legibility are vital for the proper interpretation of line diagrams: concepts are comparable iff there exists an upward path between them, and the intent [extent] of a concept must be inferred from the set of attribute [object] labels appearing on its ancestors [descendants][1] [27].

The CARVE algorithm ensures that the digraphs to be visualised have two desirable properties which are not yet exploited by the CARVE prototype:

1. Neither the supremum nor infimum are object or attribute concepts.
2. The sets of *atoms* – upper neighbours of the infimum – and *co-atoms* – lower neighbours of the supremum – are disjoint.

There is no need to portray the infimum [supremum] or the edges it shares with atoms [co-atoms], since these vertices have no labels to display as a consequence of Property 1, and their existence and connections are implied, since concept lattices are complete [13]. Removing these uninformative vertices and adjacent edges creates room for vertically- or diagonally-oriented labels on the atoms and co-atoms, at least, which, as per Fig. 9, significantly reduces the number of overlapping labels. Property 2 ensures that a bigraph can be properly defined whose vertex classes are the atoms and co-atoms, and whose edges correspond to directed paths in the lattice digraph. Finding a good layered drawing for this bigraph as the basis of a good layout for the lattice DAG is a focus of this paper.

[1] A sentence containing square brackets is true both when read without the bracketed terms, and when read with each bracketed term substituted for the preceding term.

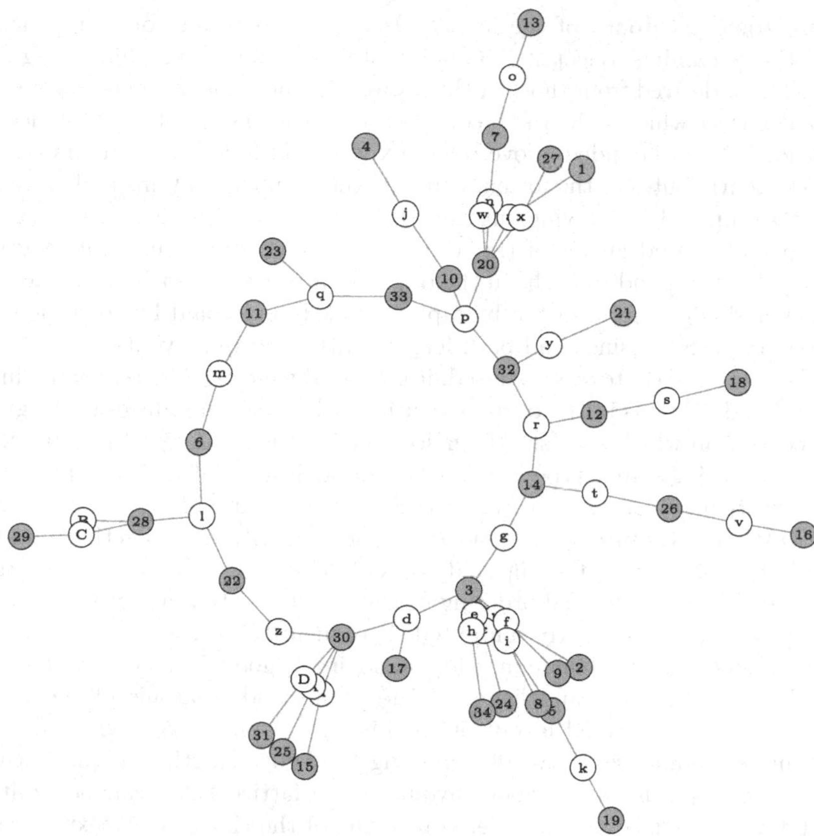

Fig. 2. Force-directed layout of the clarified context bigraph for an indivisible sub-context of the `InfoVis` 2004 data set [30].

The CARVE prototype also uses force-directed graph layout [6] to support exploratory visualisation of each indivisible sub-context bigraph. For the indivisible sub-context of `InfoVis 2004` whose lattice DAG is depicted in Fig. 1, Fig. 2 shows a force-directed layout of the corresponding clarified context bigraph. Objects and attributes are shown with grey and white fill respectively. Figure 2 was produced by the `Tikz` drawing package [31] using its `spring` layout with default parameters. A similar layout can be achieved within the CARVE prototype, but evolves slowly from a random initial layout and requires user experimentation with the layout parameters. A good initial layout which is easily computed, and more suitable defaults for the force-directed layout parameters, would improve the user experience. This paper proposes an initial layout; suitable default parameters and managing overlapping labels are beyond its scope.

Pattison and Ceglar [27] proposed a layered layout of the context bigraph in which vertex ordering was permuted to minimise edge crossings. They further proposed that the resultant ordering of objects [attributes] could be used to lay out the atoms [co-atoms] of the lattice DAG in preparation for incrementally adding the remaining concepts. The horizontal position of each remaining concept was then derived from those of the atoms of which it is a super-concept and the co-atoms of which it is a sub-concept. Whereas the number of vertices in the lattice DAG is bounded above by an exponential function of the number of objects and attributes in the formal context, the number of atoms and co-atoms is linear. Compared with laying out the full lattice DAG, it is therefore relatively cheap to find a good layout of the atom–co-atom bigraph. And since edges in the bigraph correspond to paths in the digraph, investing effort in reducing edge crossings and edge length in the bigraph will likely be repaid by corresponding decreases in path crossings and path length in the digraph layout.

Before a formal context is analysed, it is typically *clarified* by removing duplicate rows and columns in the context relation. Clarification induces a sub-graph of the context in which each set of vertices having the same set of neighbours is merged into a single archetype vertex whose label is a list of their labels. Importantly, clarifying a context does not change the structure of the corresponding lattice DAG [13]. Closure of the context relation constitutes a bijective mapping of attributes [objects] in the clarified context to attribute [object] concepts in the lattice. The elements and ordering of the resultant attribute object concept (AOC) poset can be efficiently computed [1] to identify the co-atoms and atoms.

This paper explores techniques for achieving a good layered layout of the clarified context bigraph and hence of the atoms and co-atoms of the lattice DAG. By a "good" layered layout of the bigraph, we mean one which at least approximately minimises both edge crossings and edge length. We qualitatively verify the hypothesis that a good layout of the lattice DAG can be built on the solid foundation of a good layered drawing of the clarified context bigraph. We largely avoid the question of digraph edge routing, which is often needed to prevent arcs passing through vertices to which they are not adjacent. We demonstrate that a good layered layout of the clarified context bigraph can be produced as a by-product of computing, in polynomial time, an unconstrained two-dimensional layout of the context bigraph. The latter is proposed as a means of bootstrapping the slower, force-directed layout used by CARVE. This bootstrap layout is based on the resistance distance metric on graphs [21] which has not previously been applied to formal contexts, but which we argue is well suited.

The remainder of this paper is organised as follows. In Sect. 2, we briefly review previous work relevant to: upward drawings of concept lattices; automated layout and edge crossing minimisation for layered graphs including bigraphs; the barycenter heuristic and its relationship to graph clustering and the resistance distance measure on graphs; two-dimensional drawings of bigraphs; and graph layout based on multi-dimensional scaling (MDS) of graph distance measures. A general introduction to the fields of FCA [13], graph theory [17] and graph drawing [6] are beyond the scope of this paper. In Sect. 3, we then describe the resistance distance measure, its suitability for FCA, its calculation, and its use in bigraph and digraph layout. Examples of laying out the context bigraph and lattice DAG using MDS of resistance distance are then provided in Sect. 4, followed by a brief discussion in Sect. 5 and proposals for future work.

2 Previous Work

Di Battista et al. [6] describe both force-directed methods for laying out general graphs and techniques for the layered drawing of directed graphs. Implementations of various algorithms in these two classes were included by Pohlmann [31] in the Tikz environment for vector graphics [35] in TEX, and have been used throughout this paper. Rival [32] surveyed theoretical results and research directions regarding the properties of upward line drawings of partially-ordered sets. Ganter and Wille [13] cite several earlier articles on automated lattice drawing, and describe the automated drawing of additive line diagrams. An additive line drawing assigns to each supremum-irreducible element a vector having a positive vertical component, and to each concept a position given by the sum of the vectors corresponding to the supremum-irreducible objects in its extent. Techniques have been proposed [14,37] to mitigate the problem of vertices overlapping with each other, or with edges to which they are not incident. Before applying one such technique, Zschalig [37] used a force-directed method to order the irreducible elements so as to reduce the number of edge crossings. Force-directed methods have been used to lay out the vertices of a lattice DAG in three dimensions [12], with Hannan and Pogel [19] choosing the resting spring length to be the symmetric difference between concept extents.

Gibson et al. [16] surveyed the state of the art in 2D graph drawing and graph visualisation, including force-directed methods, explicit optimisation of aesthetic criteria such as edge crossings and edge length, and MDS of graph measures such as minimum path length and resistance distance. Their taxonomy included a class called "constraint-based" layouts, of which layered layouts constituted a sub-class. For bipartite graphs such as the context bigraph, layout constraints are typically imposed on the basis of vertex type, with vertices of the same type – objects or attributes – spatially constrained to a line, curve or cluster, to emphasise the fact that edges are between vertices of different types. The class of bipartite graphs having planar drawings is restrictive; when their vertices are arranged in the two-dimensional plane, many bipartite graphs of practical interest will exhibit edge crossings. This problem is exacerbated by constraining the placement of vertices to emphasise the bipartite nature of the graph.

Despite the challenge to bigraph interpretability posed by edge crossings in a layered layout, we restrict consideration initially to the class of layered layouts in order to provide the basis for a layered layout of the lattice DAG. Layered bigraph layouts are a special case of layered digraph layouts in which there are only two layers and layer assignment is on the basis of vertex type. Unfortunately, the problem of minimising edge crossings in a layered drawing of a lattice DAG is NP-hard [12] in general, and NP-complete when there are only two layers and the ordering of vertices in one of the layers is fixed [10]. Minimising edge crossings in interactive timescales therefore becomes infeasible for large graphs.

A common technique for approximately minimising edge crossings in a layered drawing is the barycenter heuristic [10], whereby a bigraph vertex is placed at the mean horizontal position of its fixed neighbours. Kunegis [23] showed that each vertex of a bipartite graph could be placed in k dimensions at approximately the mean position of its adjacent vertices using the eigenvectors corresponding to the k smallest non-zero eigenvalues of the Laplacian matrix of the graph. The approximation resulted from the imposition of an orthogonality constraint to eliminate the trivially coincident placement of all vertices. In order to emphasise the bipartite nature of the graph in a 2D drawing, Kunegis synthesised a layered layout from his $k = 1$ dimensional, barycentered layout by using the second dimension to separate vertices into layers according to their type [23].

Compared with the explicit minimisation of edge crossings, complete eigen decomposition of the real, symmetric Laplacian matrix has only polynomial complexity, requiring $\mathcal{O}((n+m)^3)$ computation for n objects and m attributes [36]. The smallest non-zero eigenvalues of the Laplacian matrix are the dominant eigenvalues of its Moore-Penrose pseudo-inverse. Ho and van Dooren [20] described the efficient approximation of the corresponding eigenvectors for the current special case of a bipartite graph.

Gutman and Xiao [18] showed that for $k = n + m$ the vertex placement described by Kunegis [23] resulted in inter-vertex Euclidean distances whose squares are the corresponding resistance distances. Also known as commute or linear-network distance, resistance distance is a graph distance measure defined between all vertex pairs in a simple, connected, weighted graph [3,11,21]. Resistance distance is bounded above by the length of the shortest weighted path between the vertices of interest [21], but unlike path length, it decreases with each additional path. Cohen [3] observed that drawing a graph in two dimensions such that the Euclidean distances between vertices optimally approximated the corresponding resistance distances, "serves to call attention to clusters of vertices, such as cliques, that bear multiple or stronger connections". After neatly summarising the popularly-held view that, "Vertices in the same cluster of the graph have a small commute distance, whereas two vertices in different clusters of the graph have a 'large' commute distance," von Luxburg et al. [26] showed that this property weakens with increasing size for random geometric graphs.

When applied to bigraphs, it might be possible to infer co-membership of maximal bicliques from vertex proximity in a layout which preserves resistance distance. A *biclique* is a complete bipartite subgraph of the context bigraph. It is *proper* iff it contains at least one vertex of each type, and *maximal* iff no other biclique contains a proper superset of its vertices. Unless indicated otherwise, "biclique" will henceforth be used as shorthand for "proper biclique". Each maximal biclique is a formal concept, and – with the exception of the supremum and infimum as a consequence of Property 1 above – each formal concept is a maximal biclique. Since maximal bicliques are biclusters [25], graph layout techniques which emphasise clusters might also serve to highlight maximal bicliques.

Explicit knowledge of the maximal bicliques is not available until an efficient FCA algorithm has undertaken the computationally-intensive task of enumerating them. Thereafter, conventional information visualisation techniques (see e.g. [2]) – such as brushing and linking between the lattice DAG and the context bigraph – could be used to highlight the maximal bicliques. Here we investigate whether a layout based on resistance distance may form a suitable substrate for subsequent application of such techniques.

Given some measure of inter-vertex similarity or distance, multidimensional scaling (MDS) techniques [4] optimally preserve this measure while reducing the number of dimensions used to layout the graph vertices. MDS of graph metrics has been widely used for graph layout. Although graph theoretic distance (shortest path) is the most common metric used for this purpose [16,22], MDS has also been used to preserve resistance distance [3,5]. Force-directed approaches can also be used for this purpose, with the target distances typically corresponding to the relaxed length of the simulated springs between vertices [3].

3 Resistance Distance in Formal Concept Analysis

Resistance distance is a distance metric on simple, connected, weighted graphs. It is derived by considering a graph $\mathcal{G} = (\mathcal{V}, \mathcal{E})$, with vertex set $\mathcal{V}$ and edge set $\mathcal{E}$, to be an electrical network in which the edges are resistors whose conductances (inverse resistances) are specified by their edge weights [21]. The resistance distance or effective resistance $r_{st} \in \mathbb{R}$ is determined by injecting unit current at vertex $s \in \mathcal{V}$, withdrawing it at vertex $t \in \mathcal{V}$, and measuring the resultant voltage drop from s to t. Let external currents

$$i_j = \begin{cases} 1 & j = s \\ -1 & j = t \\ 0 & \text{otherwise} \end{cases} \tag{1}$$

be applied to the vertices $j \in \mathcal{V}$ of this network. Any assignment of currents to the edges of the network which satisfies Kirchoff's current law – viz. the total current at each node is zero – is called a unit (s, t) flow. The unit *electrical* (s, t) flow is one which additionally obeys Ohm's law – viz. the voltage drop across a resistor is proportional to the current flowing through it – and Kirchoff's voltage law – viz. the voltage drop around any closed loop is zero. Thomson's Principle (see e.g. [9]) states that the unit electrical (s, t) flow is the unit (s, t) flow which minimises the power dissipation in the network.

Observation 1. *The effective resistance $r_{st} \in \mathbb{R}$ is equal to the power dissipated by the unit electrical (s, t) flow.*

Proof. Let $\mathbf{v}, \mathbf{i} \in \mathbb{R}^n$ be the vectors of nodal voltages v_j and applied currents i_j respectively. For the applied nodal currents specified in Eq. 1, the power $P = \mathbf{v}'\mathbf{i} = v_s - v_t$.

Observation 2. *Let a network $\mathcal{G} = (\mathcal{V}, \mathcal{E})$ be extended by adding edges and optionally also vertices such that the extended network remains connected. Define r_{st}, r'_{st} to be the effective resistance between vertices $s, t \in \mathcal{V}$ in the original and extended networks respectively. Then $r'_{st} \leq r_{st}$.*

Proof. The unit electrical (s,t) flow in the original network is a unit (s,t) flow in the extended network. If no current flows in any of the added edges, then the latter flow is also a unit electrical (s,t) flow in the extended network, and $r'_{st} = r_{st}$. If on the other hand current flows in one or more of the added edges, then the unit electrical (s,t) flow in the original network is not a unit electrical (s,t) flow in the extended network. According to Thomson's Principle, the power dissipation in the latter is lower, and by Observation 1, so too is r'_{st}.

Resistance distance in bipartite graphs has been used as an inverse measure of similarity in recommender systems [24]. Let $K_{p,q}$ denote the complete bipartite graph having p object vertices, q attributes vertices and unit edge weights.

Observation 3 ([15]). *For distinct vertices s and t in $K_{p,q}$*

$$r_{st} = \begin{cases} \frac{2}{p} & s, t \text{ attributes} \\ \frac{2}{q} & s, t \text{ objects} \\ \frac{p+q-1}{pq} & s, t \text{ adjacent} \end{cases} \tag{2}$$

The resistance distances between the vertices of a complete bipartite graph are therefore strictly less than the corresponding shortest path lengths provided $p, q \geq 2$. Between vertices of the same type within $K_{p,q}$, resistance distance decreases as the number of vertices of the opposite type increases. Similarly between vertices of opposite type, resistance distance decreases, except where $p = 1$ or $q = 1$, as vertices of either type are added to $K_{p,q}$.

Corollary 1. *Let $K_{p,q}$ be extended by adding vertices and edges such that the extended network remains a connected bigraph. For distinct vertices s and t in $K_{p,q}$ within this extended bigraph, $r'_{st} \leq r_{st}$, where r_{st} is given by Eq. 2.*

Thus the exact resistance distances between the vertices of $K_{p,q}$ become upper bounds when $K_{p,q}$ is embedded as a maximal biclique within a larger, connected bigraph. The more vertices within this maximal biclique, the smaller the upper bound on the resistance distance between them. The restriction to a connected bigraph ensures that resistance distance remains defined, and is justified because the CARVE algorithm ensures that for each sub-context passed to the external FCA algorithm, the bigraph is connected.

Figure 3 compares the distributions of resistance distance within (red) and external to (blue) maximal bicliques for an indivisible sub-context of `InfoVis 2004`. Intra- and extra-biclique pairs are those separated by path lengths ≤ 2 and > 2 respectively. 100% of intra-biclique and only 3.3% of extra-biclique distances satisfy $0 < r_{ij} \leq 2$, while $r_{ij} > 2$ for 96.7% of extra-biclique and 0% of intra-biclique distances. This observation provides empirical support for the

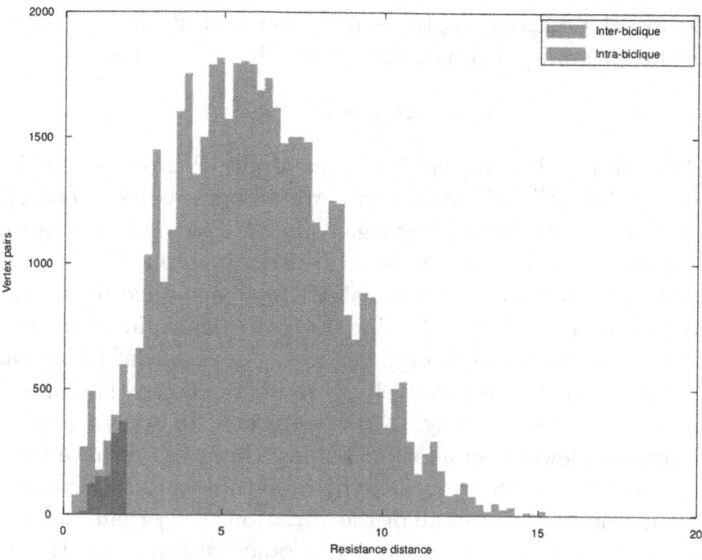

Fig. 3. Distributions of intra- and extra-biclique resistance distances for an indivisible sub-context of the `InfoVis 2004` data set having 151 maximal bicliques. (Color figure online)

use of resistance distance in clustering maximal bicliques, and demonstrates that the concerns of von Luxburg et al. [26] are not significant for the sub-context bigraphs of interest in this application.

The same vertex can belong to multiple maximal bicliques, with the intersecting membership of maximal bicliques described by the lattice DAG. For each maximal biclique to which a vertex belongs, Corollary 1 ensures that its resistance distance from its fellow vertices is small. Thus a layout which preserves resistance distance should place a vertex simultaneously close to its fellows in each such maximal biclique, and intersecting maximal bicliques should occupy partially-overlapping regions.

The context bigraph having bi-adjacency matrix $\mathbf{B} \in \{0,1\}^{n \times m}$ can be represented as a simple graph with adjacency matrix

$$\mathbf{A} = \begin{bmatrix} \mathbf{0} & \mathbf{B} \\ \mathbf{B}' & \mathbf{0} \end{bmatrix} \in \{0,1\}^{N \times N} \tag{3}$$

where $N \triangleq n + m$. Treating the unit edge weights as electrical conductances, the matrix $\mathbf{R}$ of effective resistances can be defined with the aid of the Laplacian matrix

$$\mathbf{L} \triangleq \begin{bmatrix} \mathbf{D}_o & -\mathbf{B} \\ -\mathbf{B}' & \mathbf{D}_a \end{bmatrix} \tag{4}$$

Here $\mathbf{D}_o, \mathbf{D}_a$ are the diagonal matrices of object and attribute degrees respectively. Then the resistance distance matrix $\mathbf{R} \in \mathbb{R}^{N \times N}$ is given by

$$\mathbf{R} \triangleq -2\mathbf{L}^+ + \mathbf{1}\delta' + \delta\mathbf{1}' \tag{5}$$

where δ is the vector of diagonal entries of the Moore-Penrose pseudo-inverse $\mathbf{L}^+$ of $\mathbf{L}$. The vector $\mathbf{1} \in \mathbb{R}^N$, whose elements are all ones, is an eigenvector of the Laplacian matrix with corresponding eigenvalue 0. The algebraic multiplicity of this zero eigenvalue is 1 for our connected context bigraph [17].

For a clarified formal context, classical MDS [4] is applied to the inter-vertex resistance distances r_{ij} calculated using Eq. 5 to determine the one- or two-dimensional coordinates of the bigraph vertices. The resultant layouts are henceforth denoted `R1D` and `R2D` respectively. It is easily shown that preserving the first few eigenvectors of $\mathbf{L}^+$ (see e.g. [23]) is equivalent to applying classical MDS to $\sqrt{r_{ij}}$ and involves fewer computational steps. However, we have found empirically that 1D and 2D MDS layouts of the indivisible sub-contexts of `InfoVis 2004` account for significantly more of the variation in r_{ij} than in $\sqrt{r_{ij}}$. To better preserve the clustering properties of resistance distance, we therefore limit attention to MDS of r_{ij}.

Having been assigned horizontal positions using `R1D`, the object and attribute vertices can be distinguished by assigning them distinct vertical coordinates to produce a layered layout of the clarified context bigraph. This is converted to a layered layout of the lattice digraph by re-assigning the subset of object and attribute vertices – and hence also concepts – which are neither atoms nor co-atoms to intervening layers which have been inserted to accommodate both these and the remaining (abstract) concepts. Layer assignment should respect the DAG structure. Whereas the horizontal position of each object and attribute concept is already assigned by MDS, the abstract concepts are positioned horizontally at the barycenters of their atomic descendants and co-atomic ancestors.

In both the 2D bigraph layout and 1D digraph layout, the benefits of clustering are in conflict with the requirement to avoid occlusion of vertices and their labels. In the case of the bigraph, a force-directed layout using the specified resistance distances as preferred spring lengths could be used to adjust the vertex positions. In the case of the digraph, vertex occlusion can be prevented by regularly spacing the vertices within each layer while retaining the ordinal position of each vertex within its assigned layer. Preserving vertex order leaves the number of edge crossings – at least those in the atom–co-atom bigraph – unchanged, but may increase edge length.

4 Examples

Figures 4 and 5 show the clarified context bigraphs for indivisible sub-contexts of `InfoVis 2004`, containing 42 and 46 maximal bicliques respectively, laid out using `R2D`. Figure 4 exhibits the same basic "hub and spoke" structure as the force-directed layout of the same sub-context in Fig. 2, but with less separation between some vertices lying on the spokes. This comparison confirms that vertex

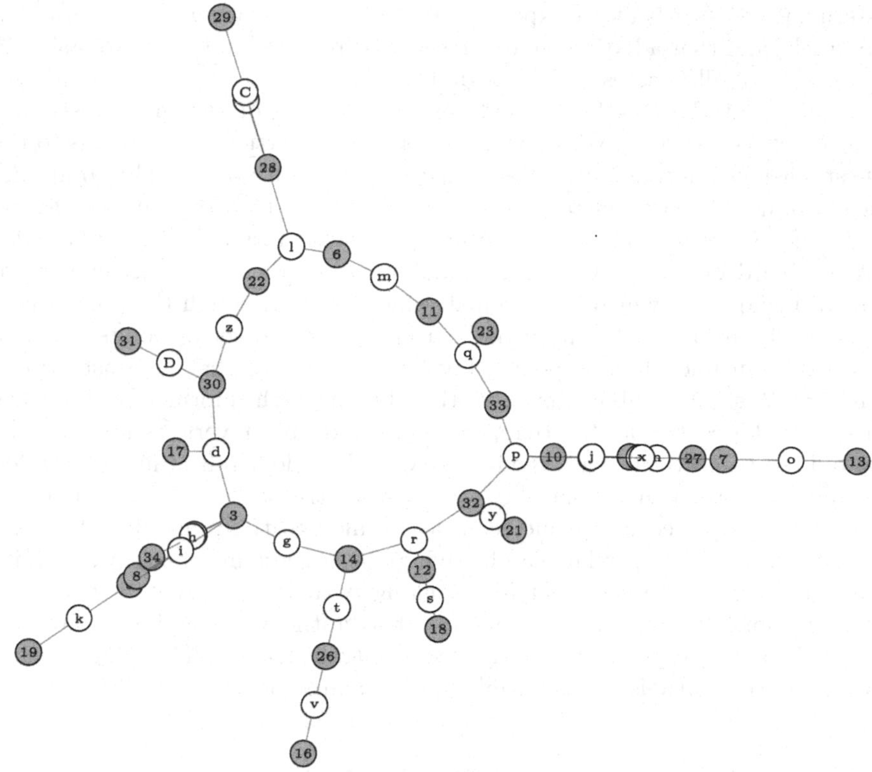

Fig. 4. Two dimensional resistance distance layout of the clarified context bigraph for the indivisible sub-context of the **InfoVis** 2004 data set shown in Fig. 2.

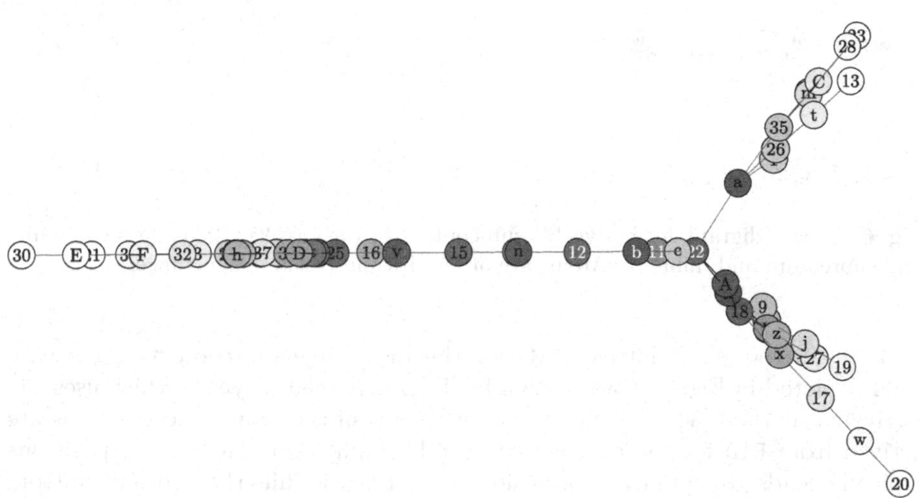

Fig. 5. 2D resistance distance layout of the clarified context bigraph for an indivisible sub-context of the **InfoVis** 2004 data set having 46 maximal bicliques.

clustering is necessarily at the expense of the spatial separation of vertices and of their labels, and that additional measures will be required in CARVE to maintain the visibility of all vertices and the legibility of their labels.

Estrada and Hatano [11] showed that the inverse of the mean resistance distance between a graph vertex and the remaining vertices corresponds to the well-established information centrality measure [33] on graphs. The total, and hence also mean, resistance distance from a nominated vertex decreases as the number of vertices to which it is structurally adjacent increases. Importantly for FCA, it also decreases as the number and – as per Eq. 2 – size of non-trivial maximal bicliques in which it participates increases. Although the point which minimises the mean Euclidean distance from a specified set of points is not in general the centroid of those points, Fig. 5 demonstrates that, at least for one of the `InfoVis 2004` sub-contexts, vertices having high information centrality lie towards the centre of `R2D`. Here, attribute and object vertices are coloured blue and red respectively, with the intensity of the colour reflecting information centrality. The calculated centrality values were translated and scaled to the range $[0 - 1)$, squared to enhance contrast, scaled to the range $[0 - 256)$ and rounded down. The paper labelled `b` having the highest information centrality is located towards the centre of Fig. 5, along with the – and its – two most central authors: `22` and `12`. Its remaining two authors – `2` and `11` – are also centrally located, despite significantly lower centrality scores. This suggests that information centrality is not uniformly approximated in `R2D`.

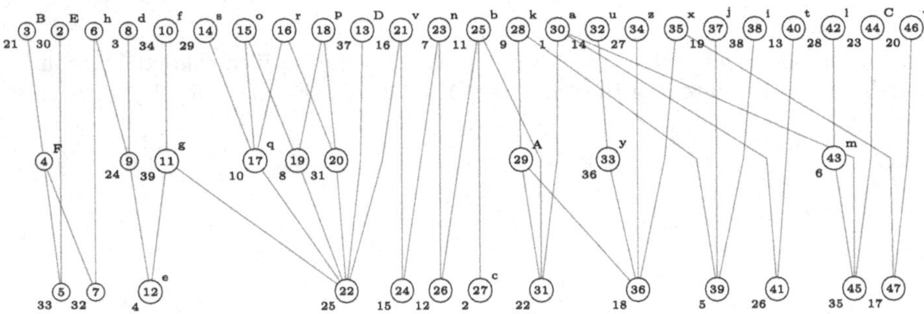

Fig. 6. Lattice digraph for indivisible sub-context of the `InfoVis 2004` data set, omitting supremum and infimum. Atoms are ordered using resistance distance.

Figure 6 shows the lattice DAG for the indivisible sub-context of `InfoVis 2004` depicted in Fig. 5. It was drawn by Tikz' `layered layout`, which uses the Sugiyama method [34]. The horizontal orderings of the atoms and co-atoms are derived from `R1D` for the clarified context bigraph; using their exact positions typically leads to significant occlusion of vertices within the top and bottom layers. The concepts are numbered, and were initially sorted, in descending lexicographic order of their extents based on the ascending order of objects in `R1D`. This total order is: consistent with the resistance-assigned ordering of the atoms;

approximately[2] consistent with the resistance-assigned ordering of the co-atoms; and a good starting point for within-layer ordering of the remaining concepts. Whilst crossing minimisation is desirable when routing those edges which bypass the middle layer, it has the potential – and for the present purposes undesirable – side-effect of permuting the pre-assigned ordering of the atoms and co-atoms. Here, however, the resistance-assigned ordering of the atoms is unchanged, and only minor permutation of the co-atoms has occurred. In particular, co-atoms 2 and 3 have been inexplicably swapped to *increase* the number of edge crossings, and 13 has been moved right, to reduce the length of, or bend in, its edge to atom 22. Minor improvements to the resultant layout are possible: for example, the edge from 25 to 31 should be routed to the left of 29 to eliminate two crossings.

Figure 7 shows a layout of the lattice DAG for an indivisible sub-context of `InfoVis 2004` containing 151 maximal bicliques. The atoms and co-atoms are ordered strictly according to `R1D`. Crossing minimisation – and hence also sensible edge routing – was disabled to prevent permutation of these layers. Vertices in the intervening layers were placed at the horizontal barycentre of their co-atomic ancestors and atomic descendants [27], and all edges drawn as straight lines. Some adjustment will be required to avoid occlusion by adjacent vertices and to route edges around vertices to which they are not structurally adjacent. The vertices are numbered in decreasing object-based lexicographic order as per Fig. 6, revealing that both the resistance-assigned ordering of the co-atoms, and the ordering of the intervening concepts induced by their barycentric positions, closely approximate this lexicographic order.

Dedicating a bottom layer to atoms and a top layer to co-atoms artificially increases edge length in the lattice DAG, because the vertical length of each directed atom–co-atom path is proportional to the number of arcs in the longest chain. Figure 8 shows the result of removing this constraint on the atoms (red). Compared with Fig. 7, there are significantly fewer edges spanning all 5 layers. Some vertices and adjacent edges are also vertically displaced out of the path of edges they would otherwise cross. To avoid crossing the green edges towards the left of Fig. 8, for example, the red edge could be routed via a waypoint four places to the left.

Table 1. Edge crossing count for different permutations of the atom–co-atom bi-adjacency matrix and for three different `InfoVis 2004` sub-contexts.

Permutation	Sub-context		
	42	46	151
Attributes-on-demand (baseline)	185	203	2756
Resistance distance (1D)	34	23	405
Object-based lexicographic attributes	37	21	418
Attribute-based lexicographic objects	35	24	456

[2] Table 1 demonstrates that this lexicographic permutation of the resistance-assigned ordering of the co-atoms has only a minor effect on edge crossings.

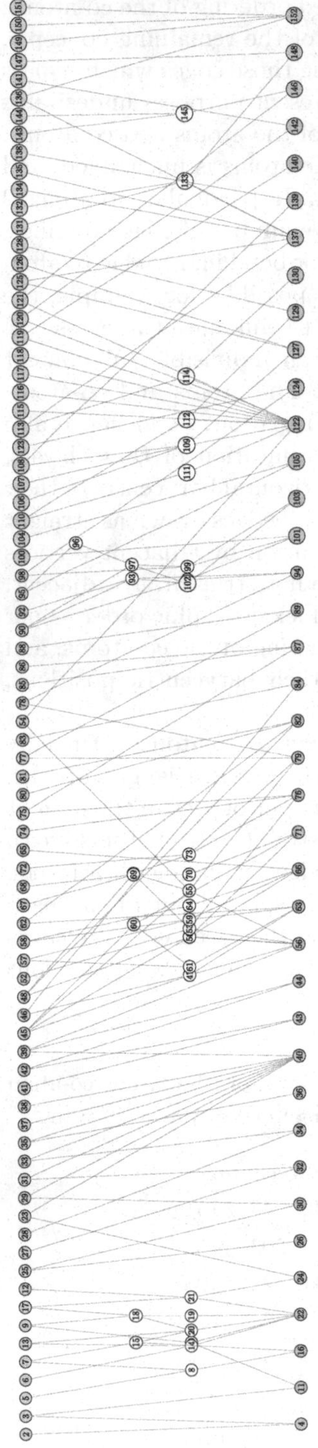

Fig. 7. Lattice digraph for an indivisible sub-context of `InfoVis 2004` containing 151 maximal bicliques. Atoms (red) and co-atoms (green) are ordered using resistance distance, and the remaining concepts placed at the horizontal barycenter of their co-atomic ancestors and atomic descendants. (Color figure online)

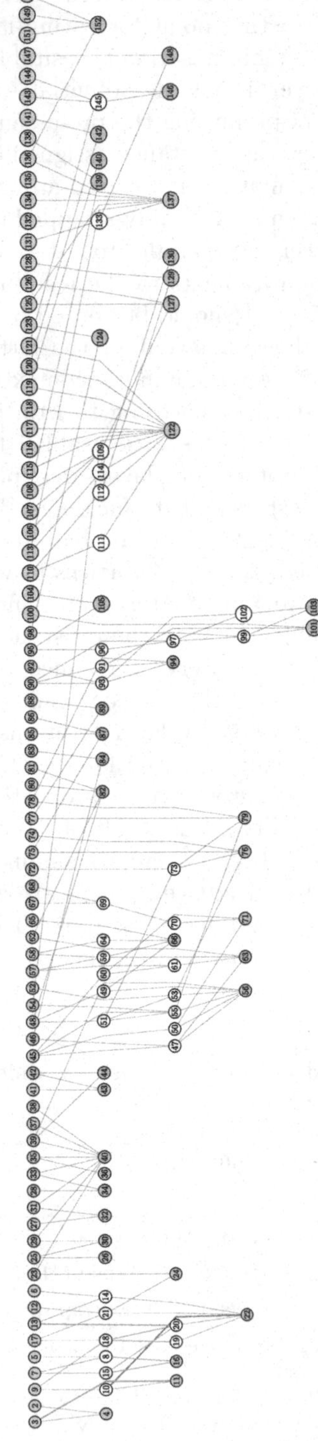

Fig. 8. As for Fig. 7 but using `Tikz layered layout`. Atoms are declared in the same resistance-based order, but are not constrained to the same layer. (Color figure online)

Table 1 shows the number of edge crossings in layered drawings of the atom–co-atom bigraph for three different indivisible sub-contexts of `InfoVis 2004` and different permutations of its bi-adjacency matrix. The baseline permutation results from ingesting the objects in order from the data set, and appending to the list of known attributes each new attribute encountered. For all three sub-contexts, the permutation based on `R1D` – requiring polynomial vice combinatorial computation – significantly reduced the number of edge crossings. The principal eigenvector for `R1D` contained several pairs of identical values, whose relative ordering is consequently arbitrary, but may nevertheless affect the number of edge crossings. These variants were not systematically evaluated. Preserving the resultant ordering of the atoms [co-atoms] and using object- [attribute-] based lexicographic ordering of the co-atoms [atoms] resulted in relatively small changes – typically an increase – in the number of edge crossings.

5 Discussion and Future Work

The CARVE algorithm constrains each lattice DAG corresponding to an indivisible sub-context to its own container, which is itself a leaf node in a hierarchy of nested containers. It further ensures that edge crossings in the DAG for the overall context can only occur between edges within the same leaf-node container [29]. In this paper, with the aid of example sub-contexts from `InfoVis 2004`, we have demonstrated the use of `R1D` to reduce edge crossings within these containers. In addition to improving the intelligibility of these constituent DAGs, this technique will thereby also improve the intelligibility of the overall DAG.

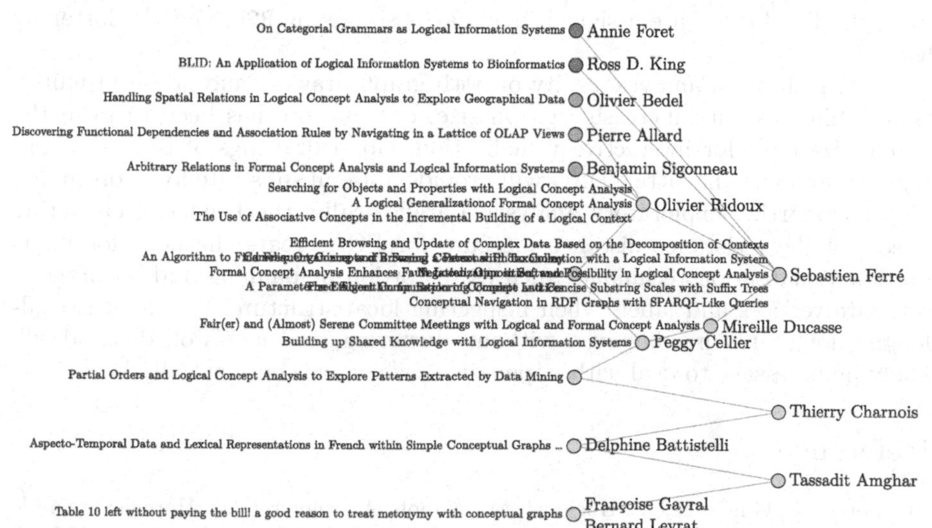

Fig. 9. `R1D` position equivalence classes for an FCA-related bibliographic data set [8]. The lattice is drawn rightward, and the infimum and supremum are omitted.

Similarly, each indivisible sub-context bigraph can also be laid out within its own leaf-node nested container [28] using R2D. In this case, however, edge crossings involving edges crossing the container boundary cannot be ruled out.

Table 1 demonstrates a significant reduction in edge crossings for the atom–co-atom bigraph resulting from permutation of the vertices from their original ordering using R1D. Comparison of Figs. 1 and 6 indicates that the layered layout of the full lattice DAG based on R1D also has significantly fewer edge crossings than the current implementation of the Sugiyama method within CARVE. These promising empirical observations warrant a more systematic evaluation of the resultant reductions in edge crossings, including comparison with the global minima. The generalisation of these observations to formal contexts featuring more densely connected context bigraphs also warrants investigation.

To deal with repeated entries in the principal MDS eigenvector, combinatoric optimisation of edge crossings could be performed over all permutations of these repeated entries within the broader R1D. The space of such permutations is considerably smaller than that of the original full layout. Figure 9 shows the DAG for an indivisible component of a bibliographic data set extracted from FCA-related conferences [8]. To improve the legibility of vertex labels, the lattice is drawn rightward rather than upward, and the infimum and supremum are again omitted. The ordering of the atoms and the co-atoms is derived from R1D. Vertices assigned identical positions are grouped into orange and purple equivalence classes, separated vertically and ordered arbitrarily. Here, the concepts within each resistance equivalence class occupy identical structural positions in the DAG. In this and limited other examples examined to date, attribute [object] vertices which are automorphically equivalent in the context bigraph correspond to attribute [object] concepts which are automorphically equivalent in the lattice digraph. The former are assigned identical positions by R2D, and the latter by R1D.

Noting that the analytic utility of both graph drawing and graph visualisation diminishes with increasing graph size, our purpose has been to push the viable size limit for interactive visualisation. Good drawings of both the context bigraph and the lattice DAG will provide a useful substrate for coordinated views supporting exploratory analysis. Paradoxically, the beneficial clustering effects of the resistance distance measure for FCA create the need for information visualisation techniques such as distortion or panning and zooming to separate vertices and labels when inspecting local structure. Labelling is challenging for graph drawing even without clustering, and labels on demand will likely be necessary to deal with paper titles, which can be especially long.

References

1. Berry, A., Gutierrez, A., Huchard, M., Napoli, A., Sigayret, A.: Hermes: a simple and efficient algorithm for building the AOC-poset of a binary relation. Ann. Math. Artif. Intell. **72**(1), 45–71 (2014). https://doi.org/10.1007/s10472-014-9418-6
2. Card, S., Mackinlay, J., Shneiderman, B.: Readings in Information Visualization: Using Vision to Think. Morgan Kaufmann, San Francisco (1999)

3. Cohen, J.D.: Drawing graphs to convey proximity: an incremental arrangement method. ACM Trans. Comput.-Hum. Interact. **4**(3), 197–229 (1997). https://doi.org/10.1145/264645.264657
4. Cox, T., Cox, M.: Multidimensional Scaling. Chapman Hall, London (1994)
5. Cuffe, P., Keane, A.: Visualizing the electrical structure of power systems. IEEE Syst. J. **11**(99), 1810–1821 (2017)
6. Di Battista, G., Eades, P., Tamassia, R., Tollis, I.: Graph Drawing: Algorithms for the Visualization of Graphs. Prentice Hall, New Jersey (1999)
7. Didimo, W., Patrignani, M. (eds.): GD 2012. LNCS, vol. 7704. Springer, Heidelberg (2013). https://doi.org/10.1007/978-3-642-36763-2
8. Doerfel, S., Jäschke, R., Stumme, G.: Publication analysis of the formal concept analysis community. In: Domenach, F., Ignatov, D.I., Poelmans, J. (eds.) ICFCA 2012. LNCS (LNAI), vol. 7278, pp. 77–95. Springer, Heidelberg (2012). https://doi.org/10.1007/978-3-642-29892-9_12
9. Doyle, P., Snell, J.: Random Walks and Electric Networks. The Mathematical Association of America (1984)
10. Eades, P., Wormald, N.C.: Edge crossings in drawings of bipartite graphs. Algorithmica **11**, 379–403 (1994)
11. Estrada, E., Hatano, N.: Resistance distance, information centrality, node vulnerability and vibrations in complex networks. In: Estrada, E., Fox, M., Higham, D.J., Oppo, G.L. (eds.) Network Science: Complexity in Nature and Technology, pp. 13–29. Springer, London (2010). https://doi.org/10.1007/978-1-84996-396-1_2
12. Freese, R.: Automated lattice drawing. In: Eklund, P. (ed.) ICFCA 2004. LNCS (LNAI), vol. 2961, pp. 112–127. Springer, Heidelberg (2004). https://doi.org/10.1007/978-3-540-24651-0_12
13. Ganter, B., Wille, R.: Formal Concept Analysis. Springer, Heidelberg (1999). https://doi.org/10.1007/978-3-642-59830-2
14. Ganter, B.: Conflict avoidance in additive order diagrams. J. Univers. Comput. Sci. **10**(8), 955–966 (2004)
15. Gervacio, S.V.: Resistance distance in complete n-partite graphs. Discrete Appl. Math. **203** (2016).https://doi.org/10.1016/j.dam.2015.09.017
16. Gibson, H., Faith, J., Vickers, P.: A survey of two-dimensional graph layout techniques for information visualisation. Inf. Vis. **12**(3–4), 324–357 (2012)
17. Gross, J.L., Yellen, J., Zhang, P. (eds.): Discrete Mathematics and its Applications. Handbook of Graph Theory, 2nd edn. Chapman and Hall/CRC Press, New York (2013)
18. Gutman, I., Xiao, W.: Generalized inverse of the Laplacian matrix and some applications. Bulletin: Classe des sciences mathématiques et naturelles **129**(29), 15–23 (2004). https://doi.org/10.2298/BMAT0429015G
19. Hannan, T., Pogel, A.: Spring-based lattice drawing highlighting conceptual similarity. In: Missaoui, R., Schmidt, J. (eds.) ICFCA 2006. LNCS (LNAI), vol. 3874, pp. 264–279. Springer, Heidelberg (2006). https://doi.org/10.1007/11671404_18
20. Ho, N.D., van Dooren, P.: On the pseudo-inverse of the Laplacian of a bipartite graph. Appl. Math. Lett. **18**, 917–922 (2005)
21. Klein, D., Randić, M.: Resistance distance. J. Math. Chem. **12**, 81–85 (1993)
22. Klimenta, M., Brandes, U.: Graph drawing by classical multidimensional scaling: new perspectives. In: Didimo, W., Patrignani, M. (eds.) GD 2012. LNCS, vol. 7704, pp. 55–66. Springer, Heidelberg (2013). https://doi.org/10.1007/978-3-642-36763-2_6

23. Kunegis, J.: Exploiting the structure of bipartite graphs for algebraic and spectral graph theory applications. Internet Math. **11**(3), 201–321 (2015). https://doi.org/10.1080/15427951.2014.958250
24. Kunegis, J., Schmidt, S., Albayrak, Ş., Bauckhage, C., Mehlitz, M.: Modeling collaborative similarity with the signed resistance distance kernel. In: Ghallab, M. (ed.) European Conference on Artificial Intelligence. IOS Press (2008)
25. Kuznetsov, S.O., Poelmans, J.: Knowledge representation and processing with Formal Concept Analysis. Wiley Interdisc. Rev. Data Min. Know. Disc. **3**(3), 200–215 (2013). https://doi.org/10.1002/widm.1088
26. von Luxburg, U., Radl, A., Hein, M.: Getting lost in space: large sample analysis of the resistance distance. In: Advances in Neural Information Processing Systems 23: 24th Annual Conference on Neural Information Processing Systems 2010. pp. 2622–2630. Curran, New York (2010)
27. Pattison, T., Ceglar, A.: Interaction challenges for the dynamic construction of partially-ordered sets. In: Bertet, K., Rudolph, S. (eds.) Proceedings of 11th International Conference on Concept Lattices and their Applications, pp. 23–34. CEUR Workshop Proceedings, Košice, Slovakia (2014). http://ceur-ws.org/Vol-1252/
28. Pattison, T., Ceglar, A., Weber, D.: Efficient Formal Concept Analysis through recursive context partitioning. In: Ignatov, D.I., Nourine, L. (eds.) Proceedings of 14th International Conference on Concept Lattices & Their Applications, vol. 2123. CEUR Workshop Proceedings. Czech Republic (2018). http://ceur-ws.org/Vol-2123/
29. Pattison, T., Weber, D., Ceglar, A.: Enhancing layout and interaction in Formal Concept Analysis. In: Proceedings of 2014 IEEE Pacific Visualization Symposium (PacificVis), pp. 248–252 (2014). https://doi.org/10.1109/PacificVis.2014.21
30. Plaisant, C., Fekete, J.D., Grinstein, G.: Promoting insight-based evaluation of visualizations: from contest to benchmark repository. IEEE Trans. Vis. Comput. Graph. **14**(1), 120–134 (2008). https://doi.org/10.1109/TVCG.2007.70412
31. Pohlmann, J.: Configurable Graph Drawing Algorithms for the TikZ Graphics Description Language. Master's thesis, Univerisität zu Lübeck (2011). http://www.tcs.uni-luebeck.de/downloads/papers/2011/
32. Rival, I.: Reading, drawing, and order. In: Rosenberg, I.G., Sabidussi, G. (eds.) Algebras and Orders, vol. 389, pp. 359–404. Springer, Dordrecht (1993). https://doi.org/10.1007/978-94-017-0697-1_9
33. Stephenson, K., Zelen, M.: Rethinking centrality: methods and applications. Soc. Netw. **11**, 1–37 (1989)
34. Sugiyama, K., Tagawa, S., Toda, M.: Methods for visual understanding of hierarchical system structures. IEEE Trans. Syst. Man Cybern. **11**(2), 109–125 (1981)
35. Tantau, T.: Graph drawing in TikZ. In: Didimo and Patrignani [7], pp. 517–528. https://doi.org/10.1007/978-3-642-36763_246
36. Wilkinson, J.: The Algebraic Eigenvalue Problem. Oxford University Press, Oxford (1965)
37. Zschalig, C.: An FDP-algorithm for drawing lattices. In: Diatta, J., Eklund, P., Liquire, M. (eds.) Proceedings of CLA 2007, vol. 331, pp. 58–71. CEUR-WS.org (2007). http://ceur-ws.org/Vol-331/Zschalig.pdf

Using Redescriptions and Formal Concept Analysis for Mining Definitions in Linked Data

Justine Reynaud(✉), Yannick Toussaint, and Amedeo Napoli

Université de Lorraine, CNRS, Inria, LORIA, 54000 Nancy, France
{justine.reynaud,yannick.toussaint,amedeo.napoli}@loria.fr

Abstract. In this article, we compare the use of Redescription Mining (RM) and Association Rule Mining (ARM) for discovering class definitions in Linked Open Data (LOD). RM is aimed at mining alternate descriptions from two datasets related to the same set of individuals. We reuse RM for providing category definitions in DBpedia in terms of necessary and sufficient conditions (NSC). Implications and AR can be jointly used for mining category definitions still in terms of NSC. In this paper, we firstly, recall the basics of redescription mining and make precise the principles of definition discovery. Then we detail a series of experiments carried out on datasets extracted from DBpedia. We analyze the different outputs related to RM and ARM applications, and we discuss the strengths and limitations of both approaches. Finally, we point out possible improvements of the approaches.

Keywords: Redescription Mining · Association Rule Mining · Concept Analysis · Linked Open Data · Definition of categories

1 Introduction

The Linked Open Data (LOD) cloud is a very large reservoir of data based on elementary triples *(subject, predicate, complement)*[1], where a triple is denoted as $\langle s, p, c \rangle$, with s, p and c denoting resources. These triples can be related to form a (huge) directed graph $G = (V, E)$ where vertices in V correspond to resources –also termed as individuals– and edges in E corresponds to relations or predicate linking resources. A specific ordering can be integrated within this graph structure. Individuals can be grouped in a class using the Resource Description Framework (RDF) thanks to the special predicate `rdf:type`, and then individuals are "instances" of this class. In turn, using RDF Schema (RDFS), the

[1] The elements of a triple are usually referred as *(subject, predicate, object)*. For avoiding any confusion with objects from FCA, we adopt here the term *complement* instead of *object*.

Supported by "Région Lorraine" and "Délégation Générale de l'Armement".

D. Cristea et al. (Eds.): ICFCA 2019, LNAI 11511, pp. 241–256, 2019.
https://doi.org/10.1007/978-3-030-21462-3_16

set of classes can be organized within a poset thanks to the partial ordering `rdfs:subClassOf`.

A class can be defined through an *extension* by enumeration of all individuals composing this extension. For example, the extension of the `Smartphone` class would include the set of all "known" smartphones in a given universe. Dually, a class may also be defined through an *intension* by enumeration of all characteristics common to individuals in the class. For example, the intension of the `Smartphone` class could be described as "a small computer equipped with a cellular antenna". It should be noticed that "extensions" and "intensions" are not necessarily closed sets as extents and intents are in Formal Concept Analysis (FCA [11]).

A basic classification problem, related to clustering and FCA, is to provide a suitable definition to a class of individuals, i.e. a description based on a set of characteristics which are common to all individuals. This problem arises whenever there is a need for building classes for an ontology or a knowledge base related to a particular domain. In the LOD cloud, this classification problem takes a specific form. There are classes defined by an extension but usually without any corresponding intension. More concretely, we may consider individuals as subjects s whose description is composed of the set of available pairs (p, c). An application of this classification problem is related to the mining of definitions of DBpedia categories, in the line of the work in [2].

Actually, DBpedia categories are automatically extracted from Wikipedia. In Wikipedia, a category is a specific page which lists all the pages related to itself, as this is the case for example for the page `Category:Smartphones`[2]. In DBpedia, a category is a resource appearing in the range of the predicate `dct:subject`. Moreover, categories are widespread as there is more than one million of categories but, most of the time, a category does not have any "processable" description and there does not exist any ordering or structure among categories.

Accordingly, given a class defined by a set of instances, the classification problem aims at finding a corresponding definition in terms of a description made of sets of characteristics or properties related to all these instances. Afterwards, the class can be defined in terms of necessary and sufficient conditions for an individual to be a member of the class. The necessary condition means that all instances share the characteristics of the description while the sufficient condition means that any individual having those characteristics should be an instance of the class.

Actually, the present work is a continuation of a work initiated in [2] and in [19]. In the first paper [2], authors rely on FCA [11] and implication between concepts for discovering definitions in LOD. These definitions are based on pairs of implications, i.e. $C \implies D$ and $D \implies C$, which stand for necessary and sufficient conditions. A double implication is considered as a definition $C \equiv D$. Most of the time $C \implies D$ is an implication while $D \longrightarrow C$ is an association rule whose confidence is less than 1. This means that a plausible definition can be set provided that the data at hand are completed. In the second paper [19],

[2] https://en.wikipedia.org/wiki/Category:SmartPhones.

authors propose a preliminary comparison between three approaches for mining definitions in LOD, (i) FCA, (ii) redescription mining and (iii) translation rule mining.

In the present paper, we make precise and discuss the mining of definitions in LOD using FCA and Redescription Mining (RM) [8,9]. RM aims at discovering alternate characterisations of a set of individuals from two sets of characteristics. The characterisations can be expressed thanks to Boolean connectors within propositional logic formulas. As experiments demonstrate it, FCA and RM are able to discover definitions which can be quite different, showing that both methods are indeed complementary and also the interest of such a comparison.

The paper is organised as follows. Problem statement and FCA are introduced in Sect. 2. Redescription Mining is detailed in Sect. 3, while Sect. 4 includes experiments which were carried out to evaluate the comparison and a discussion on the quality of the results. Section 5 discusses the Related work. Finally the conclusion ends the paper and sketches future work.

2 Problem Statement

2.1 Basics of FCA and Association Rules

We rely on Formal Concept Analysis (FCA) [11] in order to represent our data. Given G a set of objects, M a set of attributes and $I \subseteq G \times M$ a binary relation between G and M, (G, M, I) is a formal context. Derivation operators (denoted $.'$) for a set of objects $A \subseteq G$ and a set of attributes $B \subseteq M$ are $A' = \{m \in M \mid \forall a \in A, aIm\}$ and $B' = \{g \in G \mid \forall b \in B, gIb\}$. The two compositions of the both derivation operators, denoted by $(.)''$, are closure operators.

A pair (A, B) is a "concept" whenever $A' = B$ and $B' = A$, where A and B are closed. A and B are called the "extent" and the "intent" of the concept respectively. The set of concepts is organized within a "concept lattice" thanks to the partial ordering defined by $(A_1, B_1) \leq (A_2, B_2)$ when $A_1 \subseteq A_2$ or dually $B_2 \subseteq B_1$.

An association rule between two sets of attributes A and B, denoted $A \rightarrow B$ means that "if we observe A, then we observe B" with a *confidence* which can be considered as a conditional probability:

$$\mathrm{conf}(A \rightarrow B) = \frac{|A' \cap B'|}{|A'|}$$

where $(.)'$ corresponds to the derivation operator. An association rule is valid if its confidence is superior to a given threshold θ. When $conf(A \rightarrow B) = 1$, the rule is an implication, denoted by $A \Rightarrow B$. Moreover, if $B \Rightarrow A$, then A and B form a definition, denoted by $A \equiv B$.

2.2 Defining Categories in DBpedia

The content of *DBpedia* [17] is built with information extracted from *Wikipedia*, an online encyclopedia. In *Wikipedia*, a category, say X, is a specific kind of

Wikipedia page listing all pages related to X (see page `Category:Smartphones`[3] for example). These categories are annotated by the users of *Wikipedia*. In *DBpedia*, a category appears in RDF triples in the range of the relation `dct:subject`. For example, the triple $\langle$`x, dct : subject, Smartphones`$\rangle$ states that the `x` subject belongs to the `Smartphones` "category".

Moreover, speaking in terms of knowledge representation and reasoning, the name of a category is a purely syntactic expression, and thus a category does not have any formal definition as one could expect (see discussion in [2] on this aspect). Then it is impossible to perform any classification within the set of categories as the latter are not defined in terms of necessary and sufficient conditions. This is precisely what we want to deal with, i.e. providing a definition to a category. This amounts to find pairs of the form $(C, \{d_1, \ldots, d_n\})$ where C denotes a category, such as `Nokia_Mobile_Phone` for example, and d_i denotes a pair (p, c), such as `(manufacturer,Nokia)` for example. Then the whole set of d_i will stand for a possible description of C in terms of a list of $(attributes, values)$ pairs. A parallel can be drawn with concept definitions in Description Logics [3], where a form of definition is given by $C \equiv d_1 \sqcap \cdots \sqcap d_n$, such as:

$$\texttt{Nokia\_Mobile\_Phone} \equiv \texttt{Phone} \sqcap \exists \texttt{manufacturer.Nokia}$$

These definitions are useful for a practitioner aiming at contributing to *DBpedia*. Indeed, providing descriptions and then definitions to categories allows to be in agreement with knowledge representation principles, i.e. building sound and complete definitions of individual classes, as categories should be. In particular, this would help to find missing triples. For example, suppose that the definition `Nokia_Mobile_Phone` $\equiv$ `Phone` $\sqcap$ $\exists$`manufacturer.Nokia` is lying in *DBpedia*. Then, if an element `x` belongs to `Nokia_Mobile_Phone`, then this element should be a phone with manufacturer Nokia, i.e. `x` is an instance of `Phone` $\sqcap$ $\exists$`manufacturer.Nokia` ("necessary condition"). Conversely, if an element is an instance of `Phone` $\sqcap$ $\exists$`manufacturer.Nokia`, then `x` should be an instance of `Nokia_Mobile_Phone` ("sufficient condition"). This allows to complete incomplete triples if required.

2.3 A Practical Approach in FCA

Following the lines of [2] in the FCA framework, the discovery of category definitions relies on the construction of a context (G, M, I) from a set of triples denoted by ST. Given ST, G is the set of subjects, i.e. $G = \{s \mid \langle s, p, c\rangle \in ST\}$) and M is a set of pairs predicate-complement, i.e. $M = \{(p, c) \mid \langle s, p, c\rangle \in ST\}$). The incidence relation is defined as $sI(p, c) \iff \langle s, p, c\rangle \in ST$.

The set of pairs is partitioned into two sets: $M = M_{subj} \cup M_{descr}$ and $M_{subj} \cap M_{descr} = \emptyset$. The set M_{subj} is the set of all pairs (p, c) such that `p` = `dct:subject`. Since all the resources in the range of `dct:subject` are categories, hereafter, a pair (`dct:subject`, C) will simply be denoted C, where C corresponds to the

[3] https://en.wikipedia.org/wiki/Category:Smartphones.

label of a category we are trying to define and will be referred as a *category*. The set M_{descr} is the set of all pairs (p, c) such that $\texttt{p} \neq \texttt{dct:subject}$. Hereafter, a pair $(p, c) \in M_{descr}$ will be referred as a *description* and denoted $\exists p{:}c$ where c is an abbreviation of an abstract class containing only c.

Then the discovery process is based on a search for implications of the form $B_1 \Longrightarrow B_2$ where $B_1, B_2 \subseteq M$. Whenever an implication $B_1 \Longrightarrow B_2$ is discovered, the converse rule is checked. If $B_2 \Longrightarrow B_1$ is also an implication, then we have the definition $B_1 \equiv B_2$. If this is not the case, the set of triples involved in the context should be checked for potential incompleteness. One of the drawbacks of this approach is that it relies on implications. However, due to the incompleteness of LOD, a large number of definitions may be missed. To overcome this issue, an association rule $B_i \rightarrow B_j$ can be considered together with its converse $B_j \rightarrow B_i$, and we can wonder how far they are from being implications. Accordingly, in a previous work [19], we introduced the notion of a quasi-definition which is to a definition what an association rule is to an implication.

Definition 1 (Quasi-definition). *Given two sets of attributes B_i, B_j and a user-defined threshold θ, a quasi-definition $B_i \leftrightarrow B_j$ holds if $B_i \rightarrow B_j, B_j \rightarrow B_i$ and*

$$\min(\mathrm{conf}(B_i \rightarrow B_j), \mathrm{conf}(B_j \rightarrow B_i)) \geqslant \theta$$

The algorithm `Eclat` [21] is one of the existing algorithms used to compute association rules. It exhaustively enumerates all the frequent itemsets, i.e. itemsets whose support is above a given threshold. Here, we rely on `Eclat`, implemented in the `Coron` platform [12], to mine association rules.

Since we want to provide definitions of categories, we are interested in rules $c \rightarrow \{d_1, \ldots, d_n\}$ or, conversely, $\{d_1, \ldots, d_n\} \rightarrow c$ such that $c \in M_{subj}$ and $d_i \in M_{descr}$. Given an association rule R: $B_1 \rightarrow B_2$, the consequent can be decomposed into two rules R_C: $B_1 \rightarrow B_C$ and R_D: $B_1 \rightarrow B_D$ where $B_C = B_2 \cap M_{subj}$ and $B_D = B_2 \cap M_{descr}$ respectively. Since $B_C \subseteq B_2$, $B_2' \subseteq B_C'$, thus $|B_1' \cap B_2'| \leqslant |B_1' \cap B_C'|$, which means that if R holds, then R_C holds. Similarly, if R holds, then R_D holds. We take advantage of this property to build the quasi-definitions we are interested in, that is rules of the form $c \leftrightarrow \{d_1, \ldots, d_n\}$. For example, $\{\exists r_1{:}x_1, C_0\} \rightarrow \exists r_2{:}x_2$ is not kept because the antecedent include both categories and descriptions. On the other hand, $\exists r_1{:}x_1 \rightarrow \{\exists r_2{:}x_2, C_2\}$ can be decomposed into R_1: $\exists r_1{:}x_1 \rightarrow \exists r_2{:}x_2$ and R_2: $\exists r_1{:}x_1 \rightarrow C_2$. The rule R_2 is kept. If its converse is valid, we obtain the quasi-definition $C_2 \leftrightarrow \exists r_1{:}C_1$.

In the following, we present an alternative search for category definition based on "Redescription Mining" (RM), where the name of the category appears on the left hand side of the $\equiv$ symbol and a set of characteristics (composed of $\exists$*predicate.complement* expressions) appears on the right hand side.

3 Redescription Mining

3.1 Definitions

Redescription mining aims at searching for data subsets with multiple descriptions, as different views on the same set of objects [9]. Redescription mining takes as input a set of objects G and a set of attributes M partitioned into *views* V_i such as $M = V_1 \cup \cdots \cup V_n$ and $V_i \cap V_j = \emptyset$ if $i \neq j$. For example, the attributes can be partitioned w.r.t. the sources of the data (two different databases for example) or w.r.t. some criteria defined by a user. A value is associated to each pair (*object*, *attribute*), which can be Boolean, numerical or nominal, and which depends on the domain of the attribute. An example of such a dataset is provided in Fig. 1.

Views	V_1			V_2
Attributes	a_1	a_2	a_3	a_4
f_1		2	3	Triangle
f_2		3	3	Triangle
f_3	×	0	3	Triangle
f_4	×	2	3	Triangle
f_5	×	2	4	Rectangle

a_1: Has a right angle (Boolean)
a_2: Max number of equal sides (numerical)
a_3: Total number of sides (numerical)
a_4: Type (nominal)

Fig. 1. An example of dataset for redescription mining, with objects $\{f_1, \ldots, f_5\}$ and attributes $\{a_1, a_2, a_3, a_4\}$.

Given a set of objects G, a partition of a set of attributes M, redescription mining aims at finding a pair of "queries" (q_1, q_2), where q_1 and q_2 correspond to logical statements involving attributes and their values. These statements are expressed in propositional logic with the conjunction, disjunction and negation connectors. Below, a redescription say *RD* based on the pair (q_1, q_2) is denoted by $RD = q_1 \longleftrightarrow q_2$ or $RD = (q_1, q_2)$.

Given a redescription $RD = q_1 \longleftrightarrow q_2$, the set of objects G can be partitioned w.r.t. the queries which are satisfied by a subset of objects. There are four possible components in the partition, denoted by E_{ij} with $i, j \in \{0, 1\}$, depending on the fact that q_1 and/or q_2 are satisfied. For example, $E_{10}(RD)$ denotes the set of objects satisfying q_1 but not q_2 and $E_{11}(RD)$ denotes the set of objects satisfying both q_1 and q_2.

Redescriptions are mined w.r.t. a support, the Jaccard coefficient, and a p-value. The support of a redescription $RD = (q_1, q_2)$ is the proportion of objects in the dataset satisfying both queries q_1 and q_2, i.e. support$(RD) = \frac{|E_{11}(RD)|}{|G|}$.

The similarity between two datasets corresponding to two queries q_1 and q_2 is measured thanks to the Jaccard coefficient:

$$\text{jacc}(q_1 \leftrightarrow q_2) = \frac{|E_{11}(RD)|}{|E_{11}(RD)| + |E_{10}(RD)| + |E_{01}(RD)|}$$

Let us consider for example the redescription

$$RD : (a_2 = 2) \longleftrightarrow (a_4 = Triangle)$$

which is based on $q_1 = (a_2 = 2)$ and $q_2 = (a_4 = Triangle)$ w.r.t. the dataset in Fig. 1. We have that $|E_{11}(RD)| = |\{f_1, f_4\}| = 2$, $|E_{10}(RD)| = |\{f_5\}| = 1$, $|E_{01}(RD)| = |\{f_2, f_3\}| = 2$ and $|E_{00}(RD)| = |\emptyset| = 0$. Then it comes that support$(RD) = \frac{2}{5}$ and jacc$(RD) = \frac{2}{2+1+2} = \frac{2}{5}$. If the threshold for the Jaccard coefficient is $\frac{1}{2}$, then the redescription cannot be retained. By contrast, the redescription $(a_2 = 2) \wedge (a_3 = 3) \longleftrightarrow (a_4 = Rectangle)$ returns a Jaccard coefficient of $\frac{1}{2}$, meaning this time it can be accepted.

The significance of a redescription is computed w.r.t. the p-value. For a redescription $RD = (q_1, q_2)$, the p-value is the probability that $|E_{11}(RD)|$ is at least as much as computed, knowing the support of q_1 and q_2 and assuming they are random independant sets. In other words, we should answer the question "is the Jaccard computed due to random chance?". The p-value varies between 0 and 1, the lower it is, the more significant is the redescription. A p-value under the threshold 0.05 means that the computed $|E_{11}(RD)|$ is not due to random chance and that the redescription can be accepted as such [9,18].

3.2 A Redescription Mining Algorithm

In this paper, we reuse the `ReReMi` algorithm to mine redescriptions [9]. `ReReMi` takes two files F_1 and F_2 as input, which correspond to two subsets of attributes or "views" V_1 and V_2 in the dataset, and returns a set of redescriptions.

Firstly, a "candidate redescription" based on a given set of pairs (q_1, q_2), where q_1 contains only one attribute $\{a_1\} \subseteq V_1$ and q_2 only one attribute $\{a_2\} \subseteq V_2$, is checked. The checking is not necessarily systematic for all possible pairs or combinations of pairs of attributes, as a set of initial pairs can be specified by an analyst. Doing so, the set of candidate redescriptions is progressively extended, i.e. one attribute is added at a time to one of the queries of the candidate redescription.

A query q can be extended with a new attribute a in four possible ways: $q_1 \wedge a$, $q_1 \vee a$, $q_1 \wedge \neg a$ or $q_1 \vee \neg a$. The redescription with the best Jaccard coefficient is added to the candidate redescriptions. However, this extension can be customised using for example only one of the possibilities, e.g. $q_1 \wedge a$. The algorithm is based on a beam search: at each step, the top k pairs of queries with the higher Jaccard coefficient are extended. The algorithm continues until there is no more candidate available, i.e. until there is no way to increase the Jaccard coefficient of the current candidate redescription or there is no more attributes available to extend the query. A maximal depth can also be specified by the analyst. Since the algorithm is based on a greedy approach, there is no guarantee to obtain the best redescriptions: the algorithm may stop at a local maxima. Finally, the set of the candidate redescriptions is returned to the analyst.

Table 1. Datasets extracted.

	Persons	Objects	Films
Small	Turing_Award	Samsung_Galaxy	Hospital_films
Medium	Women_Mathematicians	Smartphones, Sports_cars	Road_movies
Large	Mathematicians	—	French_films

Table 2. Statistics on the datasets extracted.

Dataset	Triples	$\|G\|$	$\|M\|$	$\|M_{subj}\|$	$\|M_{descr}\|$	Predicates	Density
Samsung_Galaxy	940	59	277	30	247	33	5.2e−2
Turing_Award_laureates	2642	65	1360	503	857	35	2.2e−2
Hospital_films	1984	71	1265	490	775	46	1.6e−2
Women_mathematicians	9652	552	4243	1776	2467	98	2.9e−3
Smartphones	8418	598	2089	359	1730	98	5.8e−3
Sports_cars	9047	604	2730	435	2295	61	4.7e−3
Road_movies	20056	689	9314	2652	6662	103	2.4e−3
Mathematicians	32536	1660	12279	3848	8431	202	1.2e−3
French_films	121496	6039	25487	6028	19459	111	6.4e−4

3.3 Redescription Mining in Linked Open Data

For applying redescription mining to a set of linked data, i.e. a set of related RDF triples, we need first to transform this set of triples into a format that can be processed by the ReReMi algorithm. This operation is similar to the building of a context in the FCA framework. The attributes correspond to the predicates of the triples and they are separated into views.

We build an input "context" as described in Sect. 2.3. The two views correspond to the two sets of the partition of attributes, that is, $M = M_{subj}$ and M_{desc}. Based on that, searching for category definitions can be achieved by searching for redescriptions (q_1, q_2) where $q_1 = a$ with $a \in M_{subj}$ and q_2 is a query based on a set of one or more attributes from M_{desc}. Actually, this search should output a definition based on characteristics shared by all the resources of the category, that is, a set of necessary and sufficient conditions for being a member of the category.

4 Experiments

4.1 Datasets

We extracted 9 different subsets of triples with various sizes[4], which cover three domains : Persons, Manufactured objects and Films (see in Table 1).

[4] The datasets and the results of the experiments are available online, see https://gitlab.inria.fr/jreynaud/icfca19.

The small datasets have less than 100 objects and 1500 attributes, meaning that there are less than 1500 unique pairs (*predicate*, *complement*) in the extracted triples. The medium datasets have around 600 objects and between 2000 and 10000 attributes. Finally, the large datasets have more than 1500 objects and 10000 attributes. There is no manufactured object dataset of this size, therefore only two large datasets about persons and films are provided.

Statistics are given in Table 2. Overall, all the datasets are sparse, meaning that attributes have a low support. However, the density of manufactured objects seems to be higher than the density of persons and films. The number of predicates is low regarding the number of pairs (predicate, complement). This means that a lot of attributes share the same predicate and differ only on the complement.

4.2 Inputs

For mining association rules we used the Coron platform [12] with the `Eclat` algorithm. We set the minimum confidence to 0.5 and the minimum support to 3, 5 or 10, depending whether the dataset is small, medium or large respectively. The input file is a context built from all the triples, with all the attributes in M whenever they are categories or descriptions.

For mining the redescriptions, the attributes are partitioned. For each of the datasets, the partition of the attributes is built as follows: M_{subj} is constructed from the subset of triples whose predicate is `dct:subject` whereas M_{desc} is the complementary set. Here, there are only Boolean attributes and only conjunction is used in RM. From M_{subj} and M_{desc}, two tabular files compliant with `ReReMi` input are created, namely D_{subj} which contains attributes of the view M_{subj}, D_{desc} which contains attributes of the view M_{desc}. The thresholds used are 0.5 for Jaccard similarity (jacc $\geqslant 0.5$) and 3 for support (support $\geqslant 3$).

4.3 Extraction of Definitions

The `ReReMi` algorithm returns a set of redescriptions with their respective Jaccard coefficients. The `Eclat` algorithm returns a set of association rules that need to be processed. From the set of mined association rules, we build quasi-definitions $c \leftrightarrow \{d_1, \ldots, d_n\}$.

For measuring the precision of the algorithms, each rule (redescription or quasi-definitions) is manually evaluated by an analyst. Hereafter, a rule which is considered as "valid" by the expert is called a definition. This allows us to compute the precision as follows:

$$Prec_{RD} = \frac{|D_{RD}|}{|RD|} \quad \text{and} \quad Prec_{QD} = \frac{|D_{QD}|}{|QD|}$$

where RD (resp. QD) corresponds to the set of redescriptions (resp. association rules) and D_{RD} (resp. D_{QD}) corresponds to the number of redescriptions (resp. association rules) evaluated as valid by the domain expert. The set of definitions obtained from redescriptions is denoted D_{RD} and the set of definitions obtained from association rules is denoted D_{AR}.

Table 3 presents some redescriptions and quasi-definitions along with Jaccard coefficient and confidence for the datasets Turing_Award_laureates and Smartphones[5].

Table 3. Redescriptions and Association Rules extracted by ReReMi and Eclat for each the datasets Turing_Award_laureates and Smartphones, written in a Description Logics-like formalism. If the rule is valid (i.e. considered as true by the evaluator), the symbol ≡ is used. Otherwise, the symbol ≢ is used. The confidence corresponds to the minimal confidence between the two association rules $A \rightarrow B$ and $B \rightarrow A$.

N.	Redescriptions	jacc
	Turing_Award_laureates	
R1	Harvard_University_alumni ≡ ∃almaMater.Harvard_University	.89
R2	Stanford_University_alumni ≡ ∃almaMater.Stanford_University	.56
R3	National_Medal_of_Science_laureates ≡ ∃award.National_Medal_of_Science	1
R4	British_computer_scientists ≢ ∃award.Fellow_of_the_Royal_Society	.63
	Smartphones	
R5	Nokia_mobile_phones ≡ ∃manufacturer.Nokia	.82
R6	Samsung_Galaxy ≡ ∃manufacturer.Samsung_Electronics ⊓ ∃operatingSystem.Android_OS	.66
R7	Mobile_operating_systems ≡ Software ⊓ Work	.58
R8	MeeGo_Devices ≢ ∃operatingSystem.Sailfish_OS	.73
N.	**Association Rules**	**conf**
	Turing_Award_laureates	
R9	Harvard_University_alumni ≡ ∃almaMater.Harvard_University ⊓ Agent ⊓ Person ⊓ Scientist	.88
R10	Harvard_University_alumni ≡ ∃almaMater.Stanford_University ⊓∃award.Turing_Award ⊓ Agent ⊓ Person ⊓ Scientist	.75
R11	National_Medal_of_Science_laureates ≡ ∃award.National_Medal_of_Science ⊓ Agent ⊓ Person ⊓ Scientist	1
R12	Massachusetts_Institute_of_Technology_faculty ≢ ∃award.Turing_Award ⊓ Agent ⊓ Person ⊓ ∃birthPlace.New_York_City	.50
	Smartphones	
R13	Nokia_mobile_phones ≡ ∃manufacturer.Nokia ⊓ Device	.85
R14	Samsung_Galaxy ≡ ∃manufacturer.Samsung_Electronics ⊓ Smartphone ⊓ Device	.53
R15	Mobile_operating_systems ≡ Software ⊓ Work	.58
R16	Sony_mobile_phones ≢ Device ⊓∃ input.Capacitive_sensing ⊓∃ input.Proximity_sensor ⊓ ∃ input.Touchscreen	.64

There are many more quasi-definitions extracted than redescriptions, especially in the French_films dataset. In the domain of films, the rules extracted are about directors, actors and distributors. In the domain of persons, they are about the universities they come from or the award they won. Finally, in the domain of objects, rules are about manufacturers and brands.

[5] The entire set of redescriptions and quasi-definitions extracted are available online, see https://gitlab.inria.fr/jreynaud/icfca19.

Most of the "invalid" mined redescriptions are based on a description which is too "approximate", i.e. there are possibly too many exceptions to the rule. For example, a large proportion of British computer scientists are also fellows of the Royal Society, but not all are award winners (see rule R4). In some other cases, there are not enough counter-examples in the dataset. For example, in redescription R8, there are too few Meego smartphones which are not running Sailfish in the dataset.

4.4 Discussion

The number of extracted category definitions is reported in Table 4. The extracted rules depend on the data domain, and thus cannot be generalised to the whole *DBpedia*. For discovering more general definitions, we probably need to process larger datasets, e.g. considering a dataset about Person instead of Turing_Award_laureates. This would bring at the same time scalability issues that may be overcome with sampling. Further experiments in this direction could be considered in the future.

Table 4. Results of the experiments for each dataset. In the redescription mining settings, the number of extracted redescriptions ($|RD|$) and evaluated as true ($|D_{RD}|$) are reported. In the association rules mining settings, in addition to the number of association rules extracted ($|AR|$), the number of quasi-definitions ($|QD|$) is reported. For both redescriptions and quasi-defintions, the number of categories that have been defined ($|Cat|$) is reported along with the precision.

	Redescriptions				Association rules				
	$\|RD\|$	$\|D_{RD}\|$	$\|Cat\|$	P	$\|AR\|$	$\|QD\|$	$\|D_{QD}\|$	$\|Cat\|$	P
Turing_Award_laureates	33	16	16	0.48	563803	57	34	18	0.60
Women_Mathematicians	2	2	2	1.00	20483	6	5	5	0.83
Mathematicians	12	12	7	1.00	96807	29	24	14	0.83
Samsung_Galaxy	6	4	2	0.67	47004	20	20	5	1.00
Smartphones	24	22	17	0.92	3558380	34	26	12	0.76
Sports_cars	25	18	15	0.72	75030	49	35	17	0.71
Hospital_films	31	11	9	0.35	4345921	18	5	2	0.28
Road_movies	13	9	9	0.69	333491	34	18	13	0.53
French_films	6	6	4	1.00	371771	186	165	106	0.89

In Table 5, the number of predicates involved in definitions show that only a few predicates are involved in the definitions. Most of the time, there is only one attribute in the right side of a redescription, meaning that such an attribute is very discriminant and that redescriptions do not have any attribute in common. Then, it can be difficult to build a partial ordering between the defined categories.

By contrast, with association rules, all attributes that may be added without loss of confidence are included. In the Turing_Award_laureates dataset for

Table 5. Number of predicates involved in the definitions extracted by ReReMi and Eclat.

	Pred.	Pred (D_{RD})	Pred (D_{QD})
Turing_Award_laureates	35	4	5
Women_Mathematicians	98	2	3
Mathematicians	202	4	5
Samsung_Galaxy	33	2	7
Smartphones	98	5	8
Sports_cars	61	3	5
Hospital_films	46	5	2
Road_movies	103	3	7
French_films	111	4	10

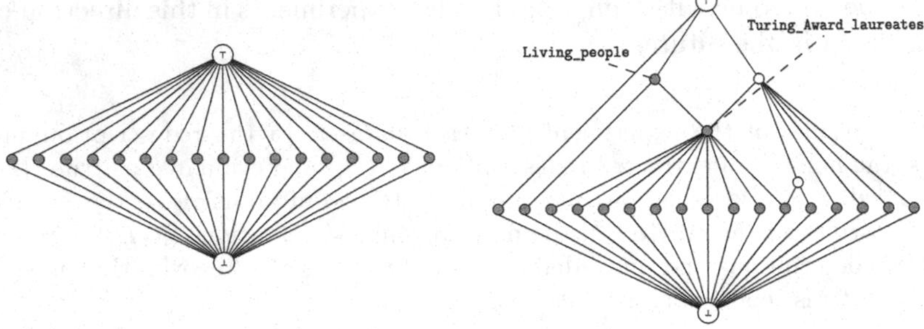

Fig. 2. Lattices build from the definitions obtained with redescriptions (left) and association rules (right) for the dataset Turing_Award_laureates. Gray nodes are object concepts, i.e. they define a category.

example, the redescription R1 and the quasi-definition R9 define the same category and both are valid. However, whereas R1 has only one attribute, R9 has four of them.

Simpler definitions like redescriptions may be useful and easier to understand. However, the attributes provided by quasi-definitions allow to build a classification. From the extracted rules, we built a context where G is the set of defined categories, and M is the set of attributes involved in a definition, and then we build the associated concept lattice. The results for the dataset Turing_Award_laureates are provided in Fig. 2.

The Fig. 3 presents the number and the precision of rules extracted. Compared to association rules, the number of redescriptions is 2 to 10 times less. In contrast with a previous assumption, there is no correlation between the number of rules extracted and the density of the context [19]. However, Smartphones and Sports_cars, which are two similar datasets, have approximately the same

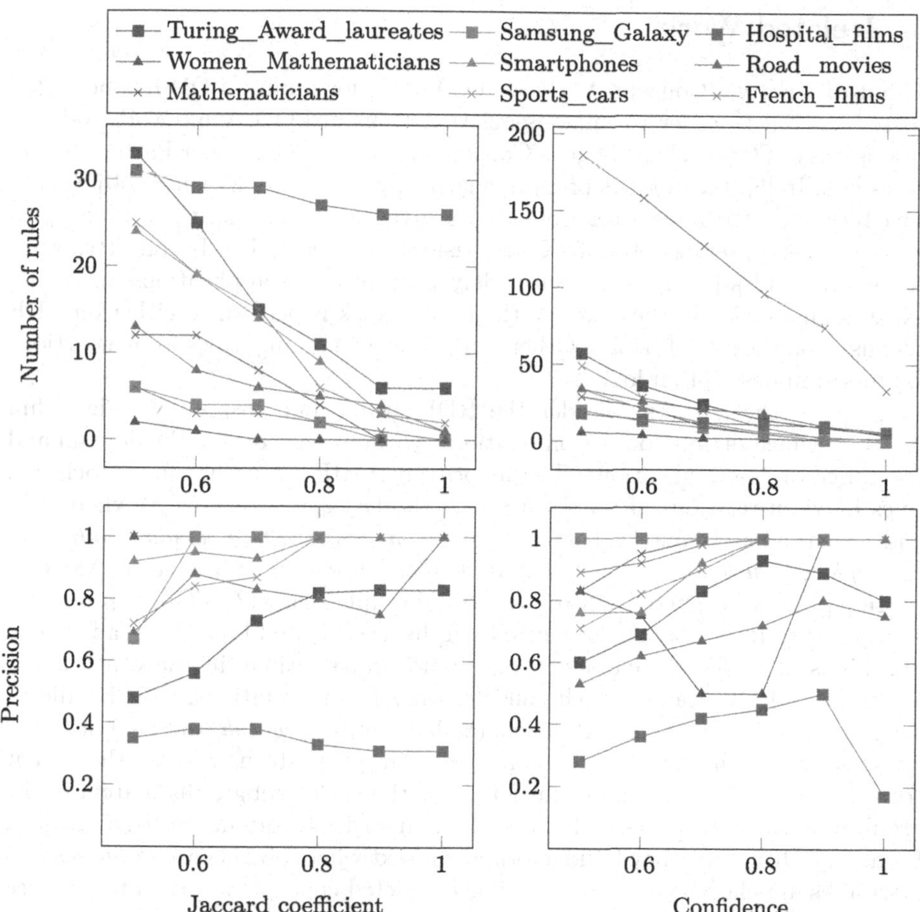

Fig. 3. Number and precision of the rules (redescriptions or quasi-defintions) extracted w.r.t. the Jaccard coefficient and the confidence.

number of extracted rules. This number could represent the "diversity" of the dataset: the more categories are defined, the more diverse is the dataset.

The precision increases w.r.t. the Jaccard coefficient threshold, meaning that the Jaccard coefficient is a suitable measure for redescription mining in LOD. Except for the dataset `Road_movies`, the Jaccard coefficient and confidence seem to return similar results for our task. The precision depends on the datasets and seems to be correlated to the number of extracted redescriptions. This would explain why redescriptions have a better precision than association rules.

The low precision of the `Hospital_films` dataset is hard to explain regarding its characteristics. However, from the extracted rules, it looks like this dataset suffers from an over-representation of some of its instances.

5 Related Work

The use of Formal Concept Analysis (FCA) [11] for mining LOD has been discussed in [13]. In order to mine association rules in LOD, some works rely on extensions of FCA, such as Logical Concept Analysis (LCA) [5] or Pattern Structures [10]. In [5], the authors propose a generalization of FCA where objects are variables and attributes are formulae in a given logic. In [10], the authors propose another generalisation of FCA, considering a partial order on attributes. Association rules [1,14] have been widely used in many applications, including LOD mining tasks. In these works, the mining task is performed either on RDF graphs or on the sets of triples. Other works focus on mining rules from assertions expressed in description logics.

In [6,7], the authors consider the RDF graph and propose the algorithm AMIE+, which mainly focuses on relations, without considering the domain and the range of the applications. The algorithm AMIE+ searches for association rules between relations of the form $r_1(x, y)$ and $r_2(x, z) \longrightarrow r_1(y, z)$ where x, y and z are random resources. For example, *people married to a person who lives in some place P also live in P* is the kind of rule that can be extracted by AMIE+.

In [4], an extension of FCA to conceptual graphs, called G-FCA, is proposed. Compared to RDF graphs, conceptual graphs are oriented bipartite graphs. The two kinds of nodes are classes in one hand and relations in the other hand. Contrasting RDF graphs which consider only binary relations, CGs handle n-ary relations. In this setting, concepts (called *Projected graph patterns*) have the following form: the intent corresponds to a graph pattern whereas the extent corresponds to the candidate solutions, i.e. the set of subgraphs matching the graph pattern. In [22], the authors use FCA in order to summarise RDF graphs. From an RDF graph, they build a formal context where objects are resources and attributes are classes and relations. The extracted concepts are used to produce a new RDF graph which summarise the original one.

Contrasting the other approaches, the authors in [20] rely on FCA and association rules to mine sets of triples. They build different formal contexts in order to discover specific relations, such as subsumption between two classes or transitivity of a relation for example. To this end, they build multiple contexts and mine association rules. The rules extracted can be interpreted as relations and they can be expressed in description logics. By contrast, in [15], the authors rely on rule mining and search for *obligatory class attributes*. Given a class, an obligatory attribute denotes a relation that every individual of the class should be involved in, e.g. every person has a birth date, and then `hasBirthdate` is an obligatory attribute of class `Person`. To this end, the authors focus on relations and are not interested in the range of relations. In [2], the authors take into account both relations and their range. They rely on pattern structures to build a context where objects are resources and attributes are pairs $\texttt{A}_\texttt{i} = (\texttt{predicate}_\texttt{i}, \texttt{object}_\texttt{i})$. Then, implications $A_i \Longrightarrow A_j$ are searched. Only implications whose converse has a high support are kept as candidate definitions.

We position ourselves in the continuity of these works. However, while most of the approaches search for implications and are based on association rules, we search for definitions and we use redescription mining. Redescription Mining (RM) [9] is predominantly used in application in the field of ecology such as for finding bioclimatic niches or properties of animal teeth. In [19] we propose a preliminary study in which we compare redescription mining to association rules mining and translation rule mining [16] applied to LOD. In this work, we extend our previous work and experiments and we focus on redescription mining and association rule mining.

6 Conclusion and Future Work

In this paper, we compared the use of redescription mining and association rule mining for discovering definitions of categories in DBpedia.

The experimental results show that the approach is well-founded, allowing to retrieve a subset of definitions. Compared to association rules, the definitions discovered by redescriptions are shorter. The Jaccard coefficient is well-suited for mining definitions in LOD. Other metrics used for association rules, such as stability and lift, could be used and compared to confidence and Jaccard coefficient.

Most of the time, the definition of a *DBpedia* category depends on only one attribute. Even if a large part of the observed results can be explained by the characteristics of the datasets, some artifacts remain, due to the data available in *DBpedia*. The dataset `Hospital_films` is one example of the results obtained when there are such artifacts.

The relation between predicates and complements have not been discussed here, and may be investigated in a future work.

References

1. Agrawal, R., Imieliński, T., Swami, A.: Mining association rules between sets of items in large databases. ACM SIGMOD Rec. **22**(2), 207–216 (1993)
2. Alam, M., Buzmakov, A., Codocedo, V., Napoli, A.: Mining definitions from RDF annotations using formal concept analysis. In: IJCAI, pp. 823–829 (2015)
3. Baader, F., Calvanese, D., McGuinness, D., Nardi, D., Patel-Schneider, P. (eds.): The Description Logic Handbook. Cambridge University Press, Cambridge (2003)
4. Ferré, S., Cellier, P.: Graph-FCA in practice. In: Proceedings of 22nd ICCS, pp. 107–121 (2016)
5. Ferré, S., Ridoux, O.: A logical generalizationof formal concept analysis. In: Ganter, B., Mineau, G.W. (eds.) Conceptual Structures: Logical, Linguistic, and Computational Issues, pp. 371–384. Springer, Heidelberg (2000). https://doi.org/10.1007/10722280_26
6. Galárraga, L.A., Teflioudi, C., Hose, K., Suchanek, F.M.: AMIE: association rule mining under incomplete evidence in ontological knowledge bases. In: WWW 2013, pp. 413–422 (2013)
7. Galárraga, L.A., Teflioudi, C., Hose, K., Suchanek, F.M.: Fast rule mining in ontological knowledge bases with AMIE+. VLDB J. **24**(6), 707–730 (2015)

8. Galbrun, E., Miettinen, P.: Siren: an interactive tool for mining and visualizing geospatial redescriptions. In: Proceedings of the 18th ACM SIGKDD International Conference on Knowledge Discovery and Data Mining. KDD 2012, pp. 1544–1547. ACM (2012)
9. Galbrun, E., Miettinen, P.: Redescription Mining. SCS. Springer, Cham (2017). https://doi.org/10.1007/978-3-319-72889-6
10. Ganter, B., Kuznetsov, S.O.: Pattern structures and their projections. In: Delugach, H.S., Stumme, G. (eds.) ICCS-ConceptStruct 2001. LNCS (LNAI), vol. 2120, pp. 129–142. Springer, Heidelberg (2001). https://doi.org/10.1007/3-540-44583-8_10
11. Ganter, B., Wille, R.: Formal Concept Analysis. Springer, Heidelberg (1999). https://doi.org/10.1007/978-3-642-59830-2
12. Kaytoue, M., Marcuola, F., Napoli, A., Szathmary, L., Villerd, J.: The coron system. In: Boumedjout, L., Valtchev, P., Kwuida, L., Sertkaya, B. (eds.) 8th International Conference on Formal Concept Analsis (ICFCA) - Supplementary Proceedings, pp. 55–58 (2010). (demo paper). http://www.loria.fr/~kaytouem/publi/ICFCA10-demo-coron.pdf
13. Kirchberg, M., Leonardi, E., Tan, Y.S., Link, S., Ko, R.K.L., Lee, B.S.: Formal concept discovery in semantic web data. In: Domenach, F., Ignatov, D.I., Poelmans, J. (eds.) ICFCA 2012. LNCS (LNAI), vol. 7278, pp. 164–179. Springer, Heidelberg (2012). https://doi.org/10.1007/978-3-642-29892-9_18
14. Klemettinen, M., Mannila, H., Ronkainen, P., Toivonen, H., Verkamo, A.I.: Finding interesting rules from large sets of discovered association rules. In: CIKM 1994, pp. 401–407 (1994)
15. Lajus, J., Suchanek, F.M.: Are all people married? Determining obligatory attributes in knowledge bases. In: International Conference WWW, Lyon, France (2018)
16. van Leeuwen, M., Galbrun, E.: Association discovery in two-view data. TKDE **27**(12), 3190–3202 (2015)
17. Lehmann, J., et al.: DBpedia-a large-scale, multilingual knowledge base extracted from Wikipedia. Semant. Web **6**(2), 167–195 (2015)
18. Ramakrishnan, N., Kumar, D., Mishra, B., Potts, M., Helm, R.F.: Turning CARTwheels: an alternating algorithm for mining redescriptions. In: KDD 2004, pp. 266–275 (2004)
19. Reynaud, J., Toussaint, Y., Napoli, A.: Three approaches for mining definitions from relational data in the web of data. In: Proceedings of the 6th International Workshop FC4AI (co-located with IJCAI/ECAI). CEUR Proceedings vol. 2149, pp. 21–32 (2018)
20. Völker, J., Niepert, M.: Statistical schema induction. In: Antoniou, G., et al. (eds.) ESWC 2011. LNCS, vol. 6643, pp. 124–138. Springer, Heidelberg (2011). https://doi.org/10.1007/978-3-642-21034-1_9
21. Zaki, M.J.: Scalable algorithms for association mining. TKDE **12**(3), 372–390 (2000)
22. Zneika, M., Lucchese, C., Vodislav, D., Kotzinos, D.: RDF graph summarization based on approximate patterns. In: Grant, E., Kotzinos, D., Laurent, D., Spyratos, N., Tanaka, Y. (eds.) ISIP 2015. CCIS, vol. 622, pp. 69–87. Springer, Cham (2016). https://doi.org/10.1007/978-3-319-43862-7_4

Enhanced FCA

A Formal Context for Closures of Acyclic Hypergraphs

Jaume Baixeries(✉)

Departament de Ciències de la Computació, Universitat Politècnica de Catalunya, 08034 Barcelona, Catalonia, Spain
jbaixer@cs.upc.edu

Abstract. Database constraints in the relational database model (RDBM) can be viewed as a set of rules that apply to a dataset, or as a set of axioms that can generate a (closed) set of those constraints. In this paper, we use Formal Concept Analysis to characterize the axioms of Acyclic Hypergraphs (in the RDBM they are called Acyclic Join Dependencies). This present paper complements and generalizes previous work on FCA and databases constraints.

Keywords: FCA · RDBM dependencies · Acyclic Join Dependencies

1 Introduction

In the Relational Database Model (RDBM), different constraints play important roles, mostly in the normalization of a database schema. Normalization is the process of splitting an original dataset into smaller units in order to prevent redundancies and anomalies in the update process. However, some of the most well-known constraints have also been used for other unrelated purposes. For instance, **Functional Dependencies** ([31]) are not only used for normalization, database design and database validation, but also for data cleaning [18].

Multivalued Dependencies (MVDs) ([31]) have a more elaborate semantics, which is capable to express how to split a table into two different tables such that their join is exactly the original table. As we have previously mentioned, this is of key importance for database normalization and design.

Acyclic Join Dependencies (AJDs) ([32]) are a generalization of MVD's. They are of key importance in the decomposition method [21] because they specify how to decompose a dataset in the so-called 4th-normal form (4NF), which prevents redundancy and update errors.

There are more constraints in the RDBM, but FD's, MVD's and AJD's are among the most widely well-known and studied. These constraints (or *dependencies*, as they are also called in the RDBM) can be characterized from two different perspectives: according to their *semantical* definitions, this is, the description of the constraints that they impose on a dataset, or from their axiomatization, this is, the set of axioms that define what dependencies will hold in a dataset, if a

D. Cristea et al. (Eds.): ICFCA 2019, LNAI 11511, pp. 259–273, 2019.
https://doi.org/10.1007/978-3-030-21462-3_17

given set of those dependencies holds. This latter concept of axiomatization is equivalent to that of *entailment* in logics.

Formal Concept Analysis (FCA) is a lattice-oriented mathematical framework strongly connected to lattice theory with many different uses, for instance, knowledge discovery and machine learning [29], among many others.

The relationship between FCA (and lattice-based approaches) and dependencies in the RDBM is a fruitful line of research that goes a long way back. The list of papers that relate both disciplines is long, and unfortunatelly, no review paper exists yet, but, just a few highlights may provide a general overview of the amount of work devoted to it. Table 1 tries to synthesize and present some of the results that are relevant to this paper. This table contains references to some papers that deal with the characterization or computation of database constraints from both semantical and syntactical perspectives. Needless to say that this table is far from being exhaustive or systematic.

Table 1. Table of references (non exhaustive) of FCA/lattice-based characterization of database constraints.

Semantical		Syntactical	
Horn clauses/Implications	[7,27]	[20,22,24,26]	Armstrong Dependencies
Functional Dependencies	[2,11,13,30]		
Similarity Dependencies	[12,16]		
MVD Clauses	[3]	[4,5,23,25]	Symmetric Dependencies
Degenerated MVD	[8]		
Multivalued Dependencies	[9]		
Acyclic Join Dependencies	[6]	**This paper**	Acyclic Hypergraphs

In this paper, we present the characterization of the axioms for AJD's in terms of FCA. This work complements that in [6], where a characterization of the AJD's that hold in a dataset (also in terms of FCA) was provided, and generalizes that in [4] and in [5], that presents a characterization of the axioms for MVD's, which are, as we previously mentioned, a special case of AJD's. Table 1 also shows the blank that this paper tries to fill. As we will see in the Sect. 6, the goal of this paper is not only to fill a gap, but to carry on the solving of a more general lattice-oriented theory for RDBM-dependencies. This is why this paper does not present any experimental evaluation of its results. However, this implies as well that the computational advantages of this proposal will need to be accounted for. So far, evaluations of FCA-based characterizations of different dependencies in the RDBM have been presented in [10,12,13].

We present first the basic notation in Sect. 2, and then, we introduce the relevant details on the dependencies that will be discussed in Sect. 3. We present also previous results that are related to this paper in Sect. 4, and finally, we present our results in Sect. 5. Finally, we present the conclusions and future work.

2 Notation

The basic elements that are used in this paper are a set of **attributes** (commonly known as column names), defined on a **dataset**, which is a set of tuples: $T = \{t_1, \ldots, t_N\}$ (we use indistinctively dataset and set of tuples as equivalent terms). Each tuple has a value associated to each attribute.

We use non capital letters for single elements of the set of attributes $\mathcal{U}$, starting with $a, b, c, \ldots, x, y, z, \ldots$, and capital letters $A, B, C, \ldots, X, Y, Z, \ldots$ for subsets of $\mathcal{U}$. We drop the union operator and use juxtaposition to indicate set union. For instance, instead of $X \cup Y$ we write XY. If the context allows, we drop the set notation, and write abc instead of $\{a, b, c\}$. We use the notation $t(X)$ to indicate the **restriction** of a tuple t to the set of attributes X.

We use the notation $\Pi_X(S)$, where $X \subseteq \mathcal{U}$ and $S \subseteq T$ as the set $\Pi_X(S) = \{t(X) \mid t \in S\}$, this is, the set of restrictions of all tuples in S to the set of attributes X. Finally, we define the **join** operator $\bowtie$ on the datasets R and S as follows: $R \bowtie S = \{r \cup s \mid r \in R \wedge s \in S \wedge r(X) = s(X)\}$, where X is the set of attributes that are common to both R and S and $r \cup s$ is the tuple that results from the *combination* of the two tuples r and s. If they have no common attributes, then, this operation becomes a cartesian product.

3 Constraints in the RDBM

Constraints in the Relational Database Model are relations between sets of attributes, which can be viewed from two different perspectives:

1. As relations between values that attributes can take in a given dataset.
2. As a specific set of axioms.

The main difference between both points of view is that in the former case, the dataset contains all the information needed to characterize the set of constraints. In the latter case, the constraints are defined with respect to a set of axioms, and do not need a dataset in order to be defined nor characterized.

As examples of what we mean by *semantical* characterization, let us take the well-known definition of (Acyclic) Join Dependencies and Multivalued Dependencies

Definition 1 ([1]). *Let T be a set of tuples and let $\mathcal{U}$ be its attribute set. A* ***join dependency*** *$R = [R_1, \ldots, R_N]$ is a set of sets of attributes such that:*

1. $R_i \subseteq \mathcal{U}, \forall i : 1 \leq i \leq N$.
2. $\mathcal{U} = \bigcup_{1 \leq i \leq N} R_i$*, this is, all attributes are present in R.*

A join dependency R holds in T if and only if:

$$T = \Pi_{R_1}(T) \bowtie \ldots \bowtie \Pi_{R_N}(T)$$

The intuition behind a join dependency is that the set of tuples T can be decomposed into different smaller (with less attributes and, maybe, tuples) sets of tuples, such that their composition according to R is lossless, this is, no information is lost [15].

Acyclic join dependencies are join dependencies that hold in a set of tuples according to the condition in Definition 1. However, they have some syntactical restrictions that make them more tractable than join dependencies, both in terms of axiomatization and computational complexity [28].

On the other hand, **Multivalued Dependencies** (MVD's) are AJD's of size two. The notation of a MVD $[X, Y]$ is $X \cap Y \Rightarrow Y \setminus X$. However, MVD's have a definition of their own, not related to the fact that they are a specific case of AJD's:

Definition 2 ([1]). *Let T be a set of tuples and let $\mathcal{U}$ be its attribute set. A **multivalued dependency** $X \cap Y \Rightarrow Y$ holds in T if and only if:*

$$\begin{aligned} \forall t_i, t_j \in T : t_i[X] = t_j[X] \Rightarrow & < t_i[X], t_i[Y], t_j[Z] > \in T \\ & < t_i[X], t_j[Y], t_i[Z] > \in T \end{aligned}$$

where $Z = \mathcal{U} \setminus X \cup Y$.

In these two previous examples, we have defined both AJD's and MVD's in terms of their relation w.r.t. a dataset. The presence of a dataset is necessary for their definition. We say that these definitions are **semantical**.

However, these two sets of dependencies also admit a different definition, which is not given in terms of their relation with data, but as a set of **axioms** for each of them. Those axioms provide a **syntactical** definition of those dependencies, *regardless of a dataset.* The purpose of those axioms is to define what dependencies can be inferred from a given set of dependencies. Given a set of dependencies Σ, we say that the **closure** of Σ is Σ^+, and consists of Σ plus the set of all the dependencies that can be derived from Σ applying recursively the axioms for that kind of dependencies. If all the dependencies in Σ hold in a dataset, then all the dependencies in Σ^+ hold as well, but we need to note that this is irrespective of the dataset.

Apart from the (explicit) absence of a dataset, the difference between both definitions is that each kind of dependency has a different semantical definition. However, a syntactical definition can be common to dependencies that have different semantical definitions. For instance, both the well-known Functional Dependencies and the implications (Horn clauses) that hold in a formal context share the same set of axioms: the Armstrong axioms (reflexivity, augmentation and transitivity). This is why when we discuss properties based on those axioms, we use the term **Armstrong Dependencies** to refer to both FD's and implications. In fact, these are not the only kind of dependencies that can also be axiomatized with the Armstrong axioms: Similarity Dependencies [12,16] (a relaxation of FD's) also follow those axioms.

A similar case is that of MVD's and two different types kinds of dependencies: Multivalued Dependency Clauses (MVDC's) and Degenerate Multivalued Dependencies (DMVD's). Although all three types of dependencies are defined differently [3], all of them share the same set of axioms. This is why we refer generically to all of them (from an axiomatic point of view) as **Symmetric Dependencies**.

Although Acyclic Join Dependencies do not share their axioms with any other known type of dependencies, we will also generalize and call them **Acyclic Hypergraphs** when discussing w.r.t. their axiomatization. The choice of this name will appear clear in the forthcoming sections. We now will focus on the two different types of dependencies that will be treated in this paper: Multivaled Dependencies and Acyclic Join Dependencies; from an axiomatic point of view: Symmetric Dependencies and Acyclic Hypergraphs.

3.1 Symmetric Dependencies

A **Symmetric Dependency** (SD) is a relation between two sets of attributes, and it is stated as $X \Rightarrow Y$. Given a set of attributes $\mathcal{U}$, we define $SD_{\mathcal{U}}$ as the set of all symmetric dependencies that can be formed with $\mathcal{U}$. The axioms for SD's are:

Definition 3 (Axioms for SD's)

1. *Reflexivity: If* $Y \subseteq X$, *then,* $X \Rightarrow Y$ *holds.*
2. *Complementation: If* $X \Rightarrow Y$ *holds, then,* $X \Rightarrow \overline{XY}$ *holds.*
3. *Augmentation: If* $X \Rightarrow Y$ *holds and* $W' \subseteq W \subseteq \mathcal{U}$, *then,* $XW \Rightarrow YW'$ *holds.*
4. *Transitivity: If* $X \Rightarrow Y$ *and* $Y \Rightarrow Z$ *hold, then,* $X \Rightarrow Z \setminus Y$ *holds.*

Because of complementation, we give a symmetric dependency as $X \Rightarrow Y \mid Z$, where $Z = \overline{XY}$. In fact, this notation condenses two different SD'd: $X \Rightarrow Y$ and $X \Rightarrow Z$ (where $Z = \overline{XY}$). We always assume that the rightest set in the right-hand side of a symmetric dependency is the complementary of the union of the other two. The set $SD_{\mathcal{U}}$ is the set of all non-trivial symmetric dependencies that can be formed using all the attributes in $\mathcal{U}$. By non-trivial we mean those SD's $X \Rightarrow Y \mid Z$ such that:

Definition 4. *A symmetric dependency* $X \Rightarrow Y \mid Z$ *is non-trivial if:*

1. $X \cup Y \cup Z = \mathcal{U}$.
2. $X \cap Y = X \cap Z = Y \cap Z = \emptyset$.
3. $X \neq \emptyset$, $Y \neq \emptyset$, $Z \neq \emptyset$.

As it can be seen, according to the axioms for symmetric dependencies, this limitation incurs in no loss of information, since the remaining symmetric dependencies can easily be derived from $SD_{\mathcal{U}}$ ([33]).

3.2 Acyclic Join Dependencies and Acyclic Hypergraphs

The notion of a join dependency (see Definition 1) is closely related to that of a hypergraph. Let S be a set of vertices, a **hypergraph** [17] extends the notion of a graph in the sense that an edge in a hypergraph is not limited to two vertices, as in a graph, but to any number of vertices. If we have a join dependency $R = [R_1, \ldots, R_N]$, the homologous hypergraph H would have $\mathcal{U}$ as the set of vertices, and R would as the set of its edges: $H = \langle \mathcal{U}, R \rangle$. The **reduction** of a JD R (reduction(R)) are the maximal sets of R inclusion-wise.

The notion of **acyclicity** also exists for hypergraphs, but it is not as intuitive as in the case of graphs. In fact, there are different definitions of acyclicity, some more restrictive than others. For instance, we have α, β and Γ acyclicity ([19]), but in this paper we use exclusively the definition of α-acyclicity ([14]).

However, α-acyclicity can also be characterized in different ways, and in this paper we use one that is based on a join-tree:

Definition 5 ([14]). *Let $H = [H_1, \ldots, H_n]$ be a hypergraph. A **join-tree** JT for H is a tree such that the set of nodes is the same as H and:*

1. *Each edge (H_i, H_j) is labeled with $H_i \cap H_j$.*
2. *For any pair $H_i, H_j \in H$, where $i \neq j$, we take the only path $P = \{H_i, H_{i+1} \ldots H_j\}$ between H_i and H_j. For all edges (H_{i+k}, H_{i+k+1}) we have that: $H_i \cap H_j \subseteq H_{i+k} \cap H_{i+k+1})$.*

Example 1. For instance, if we have that $\mathcal{U} = \{ a, b, c, d, e, f, g \}$ and the hypergraph $H = \langle \mathcal{U}, [abc, ace, abfg, abd] \rangle$, a join tree for H is that in Fig. 1. We can see, for instance, that in the path from node ace to node abd, the intersection $ace \cap abd = a$ appears in all the edges of the path.

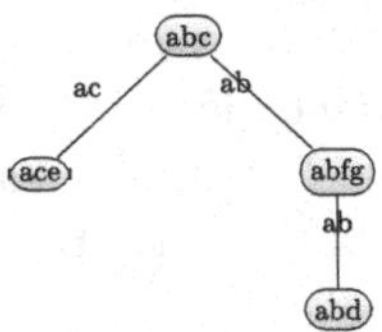

Fig. 1. Join tree for H.

For a given acyclic hypergraph, one or more join trees are possible. That is because, for instance, the choice of a root is completly arbitrary. However, switching the root is not the only possible way to have different join trees. We can now define an α-Acyclic Hypergraph.

Definition 6 ([19]). *A Hypergraph is an α-Acyclic Hypergraph (AHG) if and only if it has a **join-tree**.*

We have seen that there is a correspondence between Join Dependencies and hypergraphs. Homologously, there is a correspondence between Acyclic Join Dependencies and Acyclic Hypergraphs: an AJD is a JD that has the same syntactical restrictions of an AHG. Unlike JD's, that admit no axiomatization, Acyclic Join Dependencies have been axiomatized in [32], but we will present this axiomatization in the Sect. 3.3, once we discuss the relationship between AHG's and SD's.

3.3 Relation Between Acyclic Hypergraphs and Symmetric Dependencies

In this section, we will discuss the relationship that exists between MVD's and AJD's. However, and since we will discuss both kinds of dependencies from an axiomatic point of view, and following the same idea of generality that has been discussed in Sect. 3, we will refer to MVD's as Symmetric Dependencies, and to AJD's as Acyclic Hypergraphs, and we assume that the acyclicity which is being considered is α-acyclicity.

One important and useful property of AHGs is that a single AHG is equivalent to a set of SD'd. We remind the reader that a SD is a special case of an AHG with cardinality 2. Definition 7 is a technicality that helps to understand how to compute the set of SDs that are equivalent to an AHG in Proposition 8.

Definition 7. *Let $JT = \langle H, E \rangle$ be the join tree of an AHG H, and let $e \in E$ be and edge of that join tree. We define the function* Removal(JT, e) *that returns a SD according to the edge e as follows:*

Let $C_1 = \langle V_1, E_1 \rangle, C_2 = \langle V_2, E_2 \rangle$ be the two connected components that appear after the removal of the edge $e \in E$ in JT. Removal(JT, e) *returns a Symmetric Dependency $[\bigcup_{X \in V_1} X, \bigcup_{Y \in V_2} Y]$, such that the attributes that all the vertices of each connected component have been joined into one single set.*

Example 2. Let JT be the join tree in Example 1.

$$\text{Removal}(JT, (abc, abfg)) := [abce, abdfg]$$

We can now define the set of SDs that can be computed from a single AHG.

Definition 8 ([32]). *Let H be an AHG, and let $JT = \langle H, E \rangle$ be its join tree.*

$$\text{decomposition}(H) := \{\, \text{Removal}(JT, e) \mid \forall e \in E \,\}$$

This is, the decomposition of an AHG consists in all the SD'd that can be computed by calling the function Removal *with each of the edges of its join tree.*

Example 3. Let $H = [abc, ace, abfg, abd]$. Its join tree JT is the same as in Example 1.

$$\text{decomposition}(H) := \{\, [ace, abcdfg], [abce, abdfg], [abcefg, abd] \,\}$$

Using the usual notation for SD's, we have that:

$$\text{decomposition}(H) := \{\, ac \Rightarrow bdfg \mid e, ab \Rightarrow ce \mid dfg, ab \Rightarrow cefg \mid d \,\}$$

The equivalence between AHG's and SD's is shown in the next proposition:

Proposition 1 ([32]). *Let H be and AHG.*

$$H \equiv \text{decomposition}(H)$$

This proof is also in Theorem 7.1 in [14].

We are now ready to present and discuss the axiomatization of AHG's:

Definition 9 ([32]). *The axioms for AHG's are:*

(B1) Let $[A, B]$ be an AHG, $X \subseteq \mathcal{U}$ and $H = \text{reduction}([A \cup X, B \cup X])$.

$$[A, B] \Rightarrow H$$

(B2) Let $[A, E]$ and $[X, Y]$ be AHG's, such that $Y = A \cap E$.

$$[A, E], [X, Y] \Rightarrow [A, E \cap X]$$

(A1) $[\mathcal{U}]$ always holds.
(A2) Let $[H]$ be an AHG and let $[K] \in \text{decomposition}([H])$.

$$[H] \Rightarrow [K]$$

(A3) Let $[E, H_1, \ldots, H_n]$ and $[X, Y]$ be AHG such that $X \cap Y \subseteq E$ and $\forall H_i$: $H_i \cap E \subseteq X$ $H_i \cap E \subseteq Y$. Let $H' = \text{reduction}([E \cap X, E \cap Y, H_1, \ldots, H_n])$.

$$[E, H_1, \ldots, H_n], [X, Y] \Rightarrow [H']$$

We need to note that axiom (B1) is equivalent to the augmentation axiom for SD's and that (B2) is equivalent to the transitivity axiom for SD's. In this case, the axioms have been rephrased using the notation for AHG's. Axiom (A1) is trivial, since it states an AHG that always holds, and axiom (A2) is a consequence of Proposition 1. Therefore, the *only* non-trivial **AHG-specific axiom** is (A3).

3.4 Formal Concept Analysis

In this brief account of Formal Concept Analysis (FCA), we use standard definitions from [27]. Let G and M be arbitrary sets and $I \subseteq G \times M$ be a binary relation between G and M. The triple (G, M, I) is called a formal context. Each $g \in G$ is interpreted as an object, each $m \in M$ is interpreted as an attribute. The statement $(g, m) \in I$ is interpreted as "g has attribute m". The two following derivation operators $(\cdot)'$:

$$A' = \{m \in M \mid \forall g \in A : gIm\} \qquad \textit{for } A \subseteq G,$$
$$B' = \{g \in G \mid \forall m \in B : gIm\} \qquad \textit{for } B \subseteq M$$

define a Galois connection between the powersets of G and M. The derivation operators $\{(\cdot)', (\cdot)'\}$ put in relation elements of the lattices $(\wp(G), \subseteq)$ of objects and $(\wp(M), \subseteq)$ of attributes and reciprocally.

4 Previous Work

As we have already mentioned in Sect. 3, the characterization of RDBM dependencies with FCA can be divided in two different (yet complementary) possibilities: a *semantical* and a *syntactical* characterization of a set of dependencies. On the other hand, different kinds of RDBM dependencies have been characterized with FCA, and in this paper, we only take into account two of them: Symmetric Dependencies and Acyclic Hypergraphs. We now describe previous results that are relevant to this paper.

The **semantical** characterization of different Symmetric Dependencies has been presented in [8] for Degenerated Multivalued Dependencies, in [9] for MVD's and in [3] Multivalued Dependency Clauses. Finally, [6] presents a characterization of AJD's.

All these characterizations construct a formal context based on a dataset T and its set of attributes $\mathcal{U}$. Usually, the set of attributes is based on the set of splits of $\mathcal{U}$ (partitions of size 2) and the set of objects are based on the set of pairs of tuples of T. The relation between both sets somehow follows the definition (w.r.t. a dataset) of each kind of dependency. Since this is a semantical based characterization, there is a different definition of a formal context for each kind of dependencies that depends on a dataset. Those formal contexts characterize the set of dependencies that hold in that given dataset, and offer a (usually more compact) lattice representation of that set of dependencies.

From a **syntactical** point of view, the characterization of SD's has been presented in [4] and in [5], and also, from a lattice oriented point of view in [25]. This characterization is based on a set of attributes $\mathcal{U}$ and constructs a formal context whose set of attributes is $\wp(\mathcal{U})$, and whose set of objects is the set of all (non trivial) SD's that can be formed with $\mathcal{U}$. For instance, in [4] is defined a formal context:

$$\mathbb{K}_{SD} = (SD_{\mathcal{U}}, \wp(\mathcal{U}), \mathrm{I}\,)$$

where $SD_{\mathcal{U}}$ is the set of all the SD's that can be formed with the set of attributes $\mathcal{U}$. It is interesting to review the definition of the relation between both sets:

Definition 10. *A set $A \subseteq \mathcal{U}$ respects a Symmetric Dependency $X \Rightarrow Y \mid Z$ if and only if:*

$$X \nsubseteq A \text{ or } XY \subseteq A \text{ or } XZ \subseteq A$$

This definition will be generalized and used later to present our new results for AHG's. The main result that can be achieved from this formal context is the characterization of Σ^+ w.r.t. this context:

$$\Sigma^+ = \Sigma''$$

this is, the **closure** of a set Σ of SD's is Σ''. Again, by this closure, we mean Σ plus the set of all SD's that can be inferred recursively by the application of the axioms for SD's (3). This is why we say that this context is equivalent to the axiomatization of SD's.

Here we continue this line of work, and present a characterization of AHG's that is build upon and generalizes previous results for SD's in [5], and also, complements the semantical characterization for AJD's presented in [6].

5 Results

We now present the definition of a formal context that characterizes the closure of a set of AHG's, that is, $\Sigma^+ = \Sigma''$ in that formal context. This result is the equivalent result that has already been presented in [5] for SD's, or that in [3] for AD's. We will define a formal context between the set of all AHG's that can be formed with a set of attributes $\mathcal{U}$ and the powerset of the attributes $\wp(\mathcal{U})$, as well as a relation between both sets, as it is customary in order to define a formal context. We start by presenting this relation:

Definition 11. *Let $\mathcal{U}$ be a set of attributes, H be an AHG, and let $A \subseteq \mathcal{U}$. We say that A* **respects** *H (and we use the notation $H \quad I \quad A$) if and only if* $\forall X \Rightarrow Y \mid Z \in \text{decomposition}(H)$:

$$X \not\subseteq A \;\; or \;\; XY \subseteq A \;\; or \;\; XZ \subseteq A$$

In fact, this definition states that the set A respects an AHG H if and only if it respects all the SD's that are in decomposition(H). This is a generalization of Definition 10 for SD's, and simply takes into account the fact that an AHG is equivalent to a set of SD's. This also means that, in order to check if a set of attributes respects an AHG, we need first to compute its join tree and then, its decomposition. We now define the following formal context:

Definition 12. *Let $\mathcal{U}$ be a set of attributes, and let $\boldsymbol{AHG}(\mathcal{U})$ the set of all AHG's that can be formed with the set of attributes $\mathcal{U}$. We define the following formal context:*

$$\mathbb{K}_{\mathcal{U}} = (\boldsymbol{AHG}(\mathcal{U}), \wp(\mathcal{U}), I)$$

where $I = \{(A, H) \mid H \in \boldsymbol{AHG}(\mathcal{U}), A \in \wp(\mathcal{U}), H \quad I \quad A\}$.

The next step is to prove that this context characterizes correctly the closure of a set of AHG's. In order to do so, we need to prove that all axioms for AHG's in Definition 9 are sound and complete in the formal context $\mathbb{K}_{\mathcal{U}} = (\mathbf{AHG}(\mathcal{U}), \wp(\mathcal{U}), I)$.

As for the first two axioms for AHG's ((B1) and (B2), this is, augmentation and transitivity for SD's) we need to see, since they apply to SD's, and since decomposition$(H) = \{ H \}$ when H is a SD, then, the relation in Definition 12 boils down to the relation in Definition 10, and these two cases have already been proved in [5]. Axioms (A1) and (A2) are trivial, in the sense that they always hold. Therefore, we only have left to prove axiom (A3) of Definition 9 (the one that is particular to AHG's).

In order to prove axiom (A3), we need to characterize the SD's that are in the resulting AHG. We use the proof of Theorem 4 in [32], that states that the axioms for AHG's are sound. As a reminder, we have one AHG H and one SD $[X, Y]$ that imply the AHG S (those two dependencies fulfill the conditions in axiom [A3] of Definition 9). The part of this proof that is of interest to our case is how the join tree for S is constructed.

Definition 13 Computation of a join tree in Axiom (A3)) *(We elaborate and adapt from Theorem 4 in [32]). We take the join tree for* $H = [E, H_1, \ldots, H_n]$ *and split the edge* E *into two different edges:* $E \cap X$ *and* $E \cap Y$*. This new join tree contains, then, one more edge:* $(E \cap X, E \cap Y)$.

The first edge $E \cap X$ *will be connected to all* H_i *such that* $H_i \cap E \subseteq X$*, and the edge* $E \cap Y$ *will be connected to all* H_i *such that* $H_i \cap E \subseteq Y$*. In case that one edge* H_i *fulfills both conditions, it will be arbitrarily connected to either* $E \cap X$ *or* $E \cap Y$*. We also connect* $E \cap X$ *and* $E \cap Y$ *with one edge.*

We see that this tree contains the same edges plus one (that connects $E \cap X$ *and* $E \cap Y$*) and two new vertices, but all the rest remains the same.*

We now see that decomposition(S) = {decomposition(H), $[X, Y]$}.

Theorem 1. *Let* $H = [E, H_1, \ldots, H_n]$ *and* $[X, Y]$ *be two AHG's such that the conditions in axiom [A3] of Definition 9 hold, and let* $S = \text{reduction}([E \cap X, E \cap Y, H_1, \ldots, H_n])$.

$$\text{decomposition}(S) = \{\text{decomposition}(H), [X, Y]\}$$

Proof. It is obvious that since the join-tree adds only one more edge, then, the number of SD's that are in decomposition(S) is that of decomposition(H) plus one. We also have that all the edges in this join tree for S are the same as in the join tree for H with only one exception: the edge $(E \cap X, E \cap Y)$, which did not exist in the join tree of R. That means that all the SD'd in decomposition(H) will be also in decomposition(S).

We now see that this new SD, is, in fact, $[X, Y]$: we take the join tree and split it into two parts by removing the edge $(E \cap X, E \cap Y)$. Suppose now that there is a vertex A that is connected (even if it is by transitivity) to the vertex $E \cap X$ and that $A \not\subseteq X$. This means that, at least, there is one attribute $a \in A$ such that $a \in Y \setminus X$. Therefore, we also have that there is a vertex in the set of vertices connected to the $E \cap Y$ that will contain this attribute. Let us call Z this vertex, which is connected to $E \cap Y$. But since this tree is a join tree for S, we have that a needs to be in any path that connects the vertex A to the vertex Z. Since Z is connected to $E \cap Y$, then, by the construction of the join tree for S, this path necessarily goes through the edge $(E \cap X, E \cap Y)$, which implies that $a \in E \cap X \subseteq X$, which is a contradiction. A symmetric proof shows that all the vertices connected (also by transitivity) to $E \cap Y$, are included in Y. This proves that one part of the split will contain attributes that are all in

X and the other part will contain attributes in Y. Since all the attributes are in the join tree, then, we have that the removal of the edge $(E \cap X, E \cap Y)$ yields X and Y. ■

We are now ready to prove that the formal context in Definition 12 characterizes correctly the closure of a set of AHG's. We prove that axiom (A3) is sound in the formal context of Definition 12

Theorem 2. *Let $H = [E, H_1, \ldots, H_n]$ and $[X,Y]$ be two AHG's such that the conditions in axiom [A3] of Definition 9 hold, and let $S = \text{reduction}([E \cap X, E \cap Y, H_1, \ldots, H_n])$.*

$$S \in \{H, [X,Y]\}''$$

Proof.

$$\begin{aligned}
&\{H, [X,Y]\} \subseteq \{H, [X,Y]\}'' \\
&\qquad (\textit{by Theorem1}: \quad \text{decomposition}(S) = \{H, [X,Y]\}) \\
&\text{decomposition}(S) \subseteq \{H, [X,Y]\}'' \\
&\qquad (\textit{by Proposition1}: \quad S \equiv \text{decomposition}(S)) \\
&S \in \{H, [X,Y]\}''
\end{aligned}$$

■

As a corollary, we have that the formal context $\mathbb{K}_{\mathcal{U}} = (\mathbf{AHG}(\mathcal{U}), \wp(\mathcal{U}), I)$ is **sound** w.r.t. axioms in Definition 9 for AHG's.

Corollary 1. *Let Σ be a set of AHG's.*

$$\Sigma^+ \subseteq \Sigma''$$

in the formal context $\mathbb{K}_{\mathcal{U}} = (\boldsymbol{AHG}(\mathcal{U}), \wp(\mathcal{U}), I)$.

We now need to prove that this context is **complete**, this is, that all the AHG's that are in Σ'' are derived by the axioms in Definition 9.

Proposition 2. *Let Σ be a set of AHG's.*

$$\Sigma'' \subseteq \Sigma^+$$

in the formal context $\mathbb{K}_{\mathcal{U}} = (\boldsymbol{AHG}(\mathcal{U}), \wp(\mathcal{U}), I)$.

Proof. We prove that if $H \notin \Sigma^+$, then, $H \notin \Sigma''$. By Proposition 1, we have that if $H \notin \Sigma^+$ this is because there is a $X \Rightarrow Y \mid Z \in \text{decomposition}(H)$ such that $X \Rightarrow Y \mid Z \notin \Sigma^+$. We now reason about all the SD'd that are in Σ^+ such that their left-hand side is X, this is, all $X \Rightarrow U \mid W \in \Sigma^+$. Necessarily, there are two attributes that are one in Y and one in Z in the SD $X \Rightarrow Y \mid Z$ and that appear together in one of the dependencies $X \Rightarrow U \mid W \in \Sigma^+$, otherwise, we would have that $X \Rightarrow Y \mid Z$ could be derived by augmentativity, and we would have that $X \Rightarrow Y \mid Z \in \Sigma^+$, which is, by assumption, false.

Let those two attributes a and b, and let $X \Rightarrow U \mid W \in \Sigma^+$ such that $a, b \in U$ and U is the smallest possible set of attributes that fulfill this condition (there is only one of this sets because if $X \Rightarrow U_1 \mid W_1, X \Rightarrow U_2 \mid W_2 \in \Sigma^+$), then, $X \Rightarrow U_1 \cap U_2 \mid \mathcal{U} \setminus XU_1U_2 \in \Sigma^+$.

We take $A = XU$. We see that A respects all $X \Rightarrow Y' \mid Z' \in \Sigma^+$. We now claim that $A \in \Sigma'$. Suppose that this is not the case. This means that there is a $R \Rightarrow S \mid T \in \Sigma$ that does not respect A. That necessarily means that A contains at least one attribute from S and another form T, which are together in U. Since $R \subseteq A$, by augmentativity and assuming that S and R are disjoint, we have that $A \Rightarrow S \mid \mathcal{U} \setminus AS\Sigma^+$. We now have $X \Rightarrow P \mid \mathcal{U} \setminus XP$ and $A \Rightarrow S \mid \mathcal{U} \setminus AS$, with $X \subseteq A$. By transitivity, we have $X \Rightarrow S \setminus A \mid \mathcal{U} \setminus X(S \setminus A)$, but this means that the two attributes that are always together in a SD with X in its left-hand side now appear separated, which is contradiction. Therefore, $A \in \Sigma'$.

But A does not respect $X \Rightarrow Y \mid Z$, and, hence, $X \Rightarrow Y \mid Z \notin \Sigma''$. ∎

The core of this proof is essencially the same as in Theorem 1 in [5], but we have readopted it for the case of AHG's.

We finally can conclude from Corollary 1 and Proposition 2 that

Corollary 2. *In the formal context* $\mathbb{K}_{\mathcal{U}} = (\boldsymbol{AHG}(\mathcal{U}), \wp(\mathcal{U}), I)$*:*

$$\Sigma^+ = \Sigma''$$

In fact, in this section, we have extensivily used the property that an AHG is equivalent to a set of SD's, in order to reuse some results that were already present in [5]. We have only left the last axiom for AHG's left to proof, since this is the only axiom that is specific for AHG's.

6 Conclusions and Future Work

We have presented a new formal context for Acyclic Hypergraphs, which is a generalization of a previous result for Symmetric Dependencies (a particular case of AHG's). We have also complemented previous work for Acyclic Join Dependencies. The results in this paper have been based on the fact that an AHG is equivalent to a set of SD's, and shows the modularity of FCA to characterize RDBM dependencies.

This paper is one step towards solving many different related issues: (1) the relationship between the FCA characterization of Armstrong Dependencies and Acyclic Hypergraphs, (2) the possibility of characterizing Hypergraphs that are β of Γ acyclic, (3) the computation of minimal bases for AHGs, among many others, and (4) generalize these results to more *relaxed* versions of AHG's.

Acknowledgments. This research was supported by the recognition of 2017SGR-856 (MACDA) from AGAUR (Generalitat de Catalunya), and the grant TIN2017-89244-R from MINECO (Ministerio de Economía y Competitividad).

References

1. Abiteboul, S., Hull, R., Vianu, V.: Foundations of Databases. Addison-Wesley, Reading (1995)
2. Baixeries, J.: A formal concept analysis framework to model functional dependencies. In: Mathematical Methods for Learning (2004)
3. Baixeries, J.: Lattice characterization of Armstrong and symmetric dependencies (Ph.D. thesis). Universitat Politècnica de Catalunya (2007)
4. Baixeries, J.: A formal context for symmetric dependencies. In: Medina, R., Obiedkov, S. (eds.) ICFCA 2008. LNCS (LNAI), vol. 4933, pp. 90–105. Springer, Heidelberg (2008). https://doi.org/10.1007/978-3-540-78137-0_7
5. Baixeries, J.: A new formal context for symmetric dependencies (2011)
6. Baixeries, J.: A formal context for acyclic join dependencies. In: Kryszkiewicz, M., Appice, A., Ślęzak, D., Rybinski, H., Skowron, A., Raś, Z.W. (eds.) ISMIS 2017. LNCS (LNAI), vol. 10352, pp. 563–572. Springer, Cham (2017). https://doi.org/10.1007/978-3-319-60438-1_55
7. Baixeries, J., Balcázar, J.L.: Discrete deterministic data mining as knowledge compilation. In: Proceedings of Workshop on Discrete Mathematics and Data Mining - SIAM (2003)
8. Baixeries, J., Balcázar, J.L.: Characterization and Armstrong relations for degenerate multivalued dependencies using formal concept analysis. In: Ganter, B., Godin, R. (eds.) ICFCA 2005. LNCS (LNAI), vol. 3403, pp. 162–175. Springer, Heidelberg (2005). https://doi.org/10.1007/978-3-540-32262-7_11
9. Baixeries, J., Balcázar, J.L.: A lattice representation of relations, multivalued dependencies and Armstrong relations. In: ICCS, pp. 13–26 (2005)
10. Baixeries, J., Codocedo, V., Kaytoue, M., Napoli, A.: Characterizing approximate-matching dependencies in formal concept analysis with pattern structures. Discrete Appl. Math. **249**, 18–27 (2018). Concept Lattices and Applications: Recent Advances and New Opportunities
11. Baixeries, J., Kaytoue, M., Napoli, A.: Computing functional dependencies with pattern structures. In: Szathmary, L., Priss, U. (eds.) CLA, volume 972 of CEUR Workshop Proceedings, pp. 175–186. CEUR-WS.org (2012)
12. Baixeries, J., Kaytoue, M., Napoli, A.: Computing similarity dependencies with pattern structures. In: Ojeda-Aciego, M., Outrata, J. (eds.) CLA, volume 1062 of CEUR Workshop Proceedings, pp. 33–44. CEUR-WS.org (2013)
13. Baixeries, J., Kaytoue, M., Napoli, A.: Characterizing functional dependencies in formal concept analysis with pattern structures. Ann. Math. Artif. Intell. **72**(1–2), 129–149 (2014)
14. Beeri, C., Fagin, R., Maier, D., Yannakakis, M.: On the desirability of acyclic database schemes. J. ACM **30**(3), 479–513 (1983)
15. Beeri, C., Vardi, M.Y.: Formal systems for join dependencies. Theor. Comput. Sci. **38**, 99–116 (1985)
16. Bělohlávek, R., Vychodil, V.: Data tables with similarity relations: functional dependencies, complete rules and non-redundant bases. In: Li Lee, M., Tan, K.-L., Wuwongse, V. (eds.) DASFAA 2006. LNCS, vol. 3882, pp. 644–658. Springer, Heidelberg (2006). https://doi.org/10.1007/11733836_45
17. Berge, C.: Hypergraphs: Combinatorics of Finite Sets, volume 45 of North-Holland Mathematical Library. North-Holland, Amsterdam (1989)
18. Bohannon, P., Fan, W., Geerts, F., Jia, X., Kementsietsidis, A.: Conditional functional dependencies for data cleaning. In: ICDE, pp. 746–755 (2007)

19. Brault-Baron, J.: Hypergraph acyclicity revisited. ACM Comput. Surv. **49**, 03 (2014)
20. Caspard, N., Monjardet, B.: The lattices of closure systems, closure operators, and implicational systems on a finite set: a survey. Discrete Appl. Math. **127**(2), 241–269 (2003)
21. Codd, E.F.: Further normalization of the data base relational model. IBM Research Report, San Jose, California, RJ909 (1971)
22. Day, A.: The lattice theory of functionnal dependencies and normal decompositions. Int. J. Algebr. Comput. **02**(04), 409–431 (1992)
23. Day, A.: A lattice interpretation of database dependencies. In: Semantics of Programming Languages and Model Theory, pp. 305–325. Gordon and Breach Science Publishers Inc, Newark (1993)
24. Demetrovics, J., Hencsey, G., Libkin, L., Muchnik, I.B.: Normal form relation schemes: a new characterization. Acta Cybern. **10**(3), 141–153 (1992)
25. Demetrovics, J., Huy, N.: Representation of closure for functional, multivalued and join dependencies. Comput. Artif. Intell. **11**(2), 143–154 (1992)
26. Demetrovics, J., Libkin, L., Muchnik, I.B.: Functional dependencies in relational databases: a lattice point of view. Discrete Appl. Math. **40**(2), 155–185 (1992)
27. Ganter, B., Wille, R.: Formal Concept Analysis. Springer, Berlin (1999). https://doi.org/10.1007/978-3-642-59830-2
28. Gyssens, M.: On the complexity of join dependencies. ACM Trans. Database Syst. **11**(1), 81–108 (1986)
29. Kuznetsov, S.O.: Machine learning on the basis of formal concept analysis. Autom. Remote Control **62**(10), 1543–1564 (2001)
30. Lopes, S., Petit, J.-M., Lakhal, L.: Functional and approximate dependency mining: database and fca points of view. J. Exp. Theor. Artif. Intell. **14**(2–3), 93–114 (2002)
31. Maier, D.: The Theory of Relational Databases. Computer Science Press, Rockville (1983)
32. Malvestuto, F.: A complete axiomatization of full acyclic join dependencies. Inf. Process. Lett. **68**(3), 133–139 (1998)
33. Ullman, J.: Principles of Database Systems and Knowledge-Based Systems, vol. 1–2. Computer Science Press, Rockville (1989)

Concept Lattices as a Search Space for Graph Compression

Lucas Bourneuf(✉) and Jacques Nicolas(✉)

Univ Rennes 1, Campus de Beaulieu, 35042 Rennes cedex, France
{lucas.bourneuf,jacques.nicolas}@inria.fr

Abstract. Because of the increasing size and complexity of available graph structures in experimental sciences like molecular biology, techniques of graph visualization tend to reach their limit. To assist experimental scientists into the understanding of the underlying phenomena, most visualization methods are based on the organization of edges and nodes in clusters. Among recent ones, Power Graph Analysis is a lossless compression of the graph based on the search of cliques and bicliques, improving the readability of the overall structure. Royer et al. introduced a heuristic approach providing approximate solutions to this NP-complete problem. Later, Bourneuf et al. formalized the heuristic using Formal Concept Analysis. This paper proposes to extend this work by a formalization of the graph compression search space. It shows that (1) the heuristic cannot always achieve an optimal compression, and (2) the concept lattice associated to a graph enables a more complete exploration of the search space. Our conclusion is that the search for graph compression can be usefully associated with the search for patterns in the concept lattice and that, conversely, confusing sets of objects and attributes brings new interesting problems for FCA.

Keywords: Concept lattice · Search space formalization · Graph · Visualization

1 Introduction: Graph Compression for Graph Visualization

Visual representation of graphs remains a difficult problem either because of their size or because of the complexity of their topological analysis. Colors and spatial organization of nodes and edges are usually used to simplify the visualization [9], but such approaches remain limited, especially in experimental domains where the aim is to discover relevant graph structures.

Before displaying smartly or mining the interesting subgraphs in the graph, one can simplify the overall structure by replacing them by easier-to-read objects, for instance replacing multiple edges by a single one, which is known to increase readability [4]. This is the core idea of Power Graph Analysis: identifying bicliques and cliques in the graph in order to abstract them into simpler objects.

D. Cristea et al. (Eds.): ICFCA 2019, LNAI 11511, pp. 274–289, 2019.
https://doi.org/10.1007/978-3-030-21462-3_18

The choice of bicliques and cliques is motivated by their prevalence in organized data, such as biological [1,10] or social networks [15]. The resulting compressed graph is equivalent to the input graph, but provides high compression rate over the edges, hence increasing readability and simplifying the interpretation work for application such as protein-protein interactions [13], graph comparison [11] or community detection in social networks [15]. Figure 3 features an example of Power Graph Analysis applied to the graph in Fig. 2.

The best known effort to find a high level canonical structure in graphs that helps its analysis and visualization is Modular Decomposition [7]. However, it has been designed with different objectives in mind than Power Graph Analysis.

Modular Decomposition is a canonical representation enabling a complexity reduction of standard problems, such as MCE [8], graph drawing [12] or clustering [14]. It defines modules as sets of nodes having exactly the same neighbors outside the module. This restrictive relation enables a fast computation (linear with respect to the graph size) of a unique graph representation.

In contrast, Power Graph Analysis introduces powernodes, sets of nodes sharing common neighbors (thus a superset of modules), created together with poweredges, set of edges linking two powernodes. Thus edges and nodes are clustered at the same time. Because of the less restrictive definition of powernodes compared to modules, Power Graph Analysis is resistant to noise, allowing to extract motifs in the graph even in presence of interfering edges. This contrasts with the behaviour of Modular Decomposition, where a biclique is not guaranteed to appear simply as two parallel modules in a series module if the biclique partially interacts with other elements. The richness of structures in Power Graph Analysis leads to a NP-complete problem, but allows a more suitable formalism, to capture some relevant knowledge from the graph, and to obtain and visualize a more abstract representation of it. For instance, Power Graph Analysis provided, compared to Modular Decomposition, a more complete overview of protein-protein interaction networks, graphically extracting not only protein complexes but also hub proteins and domain induced interactions [6,13].

Because of the high complexity of Power Graph Analysis, Modular Decomposition efficiency could probably help to draw powergraphs, or as a heuristic to search for interesting clusters of nodes that could be used as seeds for more complex structures in Power Graph Analysis. To our knowledge, combining Modular Decomposition and Power Graph Analysis has never been tried.

Power Graph: Given a graph $G = (V, E)$, a *Power Graph* is defined as a special graph $PG = (PV, PE)$, with nodes $PV \subseteq 2^V$ and edges $PE \subseteq 2^E$, which fulfill the three following conditions:

subgraph condition Any pair of powernodes connected by a poweredge represents a biclique in G. As a special case, a clique in G is represented by a single powernode and a poweredge looping on this powernode.

powernode hierarchy condition Any two powernodes are either disjoint, or one is included in the other. From the point of view of classification, the sets of vertices clustered in powernodes form a hierarchy.

poweredge decomposition condition Poweredges form a partition of the set of edges.

It has been shown that finding an optimally compressed Power Graph (having the minimal number of poweredges) is an NP-complete problem [5]. A greedy strategy for graph compression consists to look for maximal cliques and bicliques iteratively. A previous article [2] dealt with the use of FCA to formalize this heuristic iterative search for concepts of maximal surface. This method however provides no guarantee to find the optimal compression This new study also show that this approach is incomplete (Sect. 6) because some particular graphs cannot be optimally covered by maximal bicliques.

The main paper contribution is to use the FCA framework and the concept lattice, to give a formalization of the Power Graph search space. Section 3 exposes an introductory example providing the necessary prerequisites for our framework, and describes how to compress a graph from a concept lattice. Section 4 proposes a simplification of the search space in order to improve its exploration. Sections 5 and 6 explain the search for two specific motifs, cliques and concept cycles. Section 7 concludes with a discussion on the potential applications of such a formalization and the open questions that this general approach raises.

2 Graph Contexts

A *graph context* is a formal context (X, Y, I) where the set of objects X and attributes Y are subsets of vertices in an undirected graph $G = \{V, E\}$ and the binary relation $I \in X \times Y$ represents its edges E. Usually, formal contexts are defined on disjointed sets X and Y. In graph contexts objects and attributes represent the same kind of elements and can intersect. For an undirected graph, the graph context is symmetric.

Chiaselotti and Gentile [3] have proposed to take $X = Y = V$ for a graph context. This corresponds to a standard representation of a graph by its (symmetrical) adjacency matrix. Note however that for a same graph, there may exist several interesting smaller representations trying to minimize the intersection of X and Y. For instance, a bipartite graph can be defined by a bipartition of disjointed sets X and Y. This can be important from a practical perspective since a number of treatments depend on the size of the matrix.

The authors shown that for simple undirected graphs (no loops, no multiple edges), the standard application of FCA leads to a coincident derivation operator for object and attributes: the derivative of a subset of vertices X in terms of graph is the neighborhood intersection of all its elements. It is equivalently the set of vertices whose neighborhood includes X. Since graphs are simple, the derivative of a subset of vertices X has no intersection with X. Thus concepts are pairs of disjointed sets. The concept lattice is by construction self-dual.

For graphs with loops or even for simple graphs if we not do not worry about reflexive edges, it is interesting to accept concepts that consider pairs of possibly intersecting sets. We study in this paper the case of simple graphs. The idea is to look at maximal rectangles in the adjacency matrix, as for standard concepts analysis, with the small but important difference that the elements on the main diagonal are don't care positions. Due to the symmetry of a formal concept, several representations are possible for a same concept and the difficulty is to properly handle the inclusion relation between concepts and retain the maximal rectangles.

	1	2	3	4
1		×	×	×
2	×		×	×
3	×	×		
4	×	×		

Fig. 1. Formal context of a small graph

Take for instance the small graph context in Fig. 1. With the classical definition proposed by Chiaselotti et al. and apart from the top and bottom concepts, concepts are $C_1 = \{1\} \times \{2,3,4\}$, $C_2 = \{2\} \times \{1,3,4\}$, and $C_3 = \{1,2\} \times \{3,4\}$, duplicated with their dual concepts $C'_1 = \{2,3,4\} \times \{1\}$, $C'_2 = \{1,3,4\} \times \{2\}$, and $C'_3 = \{3,4\} \times \{1,2\}$. C_1 and C_2 are linked to C_3 in the lattice an dually C'_3 to C'_1 and C'_2.

With the new definition allowing don't care values in the main diagonal, a tempting approach by its simplicity would be to put the diagonal at 1, then to apply standard FCA on the resulting graph context. Symmetry aside, this leads to three concepts:

$C_4 = \{1,2,3\} \times \{1,2,3\}$, $C_5 = \{1,2,4\} \times \{1,2,4\}$ and $C_6 = \{1,2\} \times \{1,2,3,4\}$. However, since the reflexive edges are don't care links these concepts can also be represented by the following equations that avoid duplicated edges: $C_4 = \{1,2\} \times \{2,3\}$, $C_5 = \{1,2\} \times \{2,4\}$ and $C_6 = \{1,2\} \times \{2,3,4\}$. Clearly, C_4 and C_5 are not maximal and C_6 is the sole admissible concept. Therefore, applying FCA on such contexts produce undesirable results. Power Graph Analysis requires in fact a more conservative approach, introducing "1" values only on demand on the diagonal. In the next section, we introduce first this graph compression issue as a search in the concept lattice of graph contexts.

3 Compression from the Lattice

A simple graph compression can be achieved by picking one concept and compressing the edges it describes in poweredges. It is possible to proceed incrementally until complete compression (remaining edges cannot be clustered in larger bicliques).

Let us consider the graph defined by the formal context in Fig. 2. It can be compressed by choosing first concept number 11 $\{d,i,j,k\} \times \{e,f,g,h\}$ in the concept lattice, then concept number 5 $\{d,e\} \times \{a,b,c,g\}$ (see Fig. 3). This compression order is noted $(11,5)$. Since these concepts overlap (they share nodes

	a	b	c	d	e	f	g	h	i	j	k
a				x	x						
b				x	x						
c				x	x						
d	x	x	x		x	x	x	x			
e	x	x	x	x			x		x	x	x
f				x					x	x	x
g				x	x				x	x	x
h				x					x	x	x
i					x	x	x	x			
j					x	x	x	x			
k					x	x	x	x			

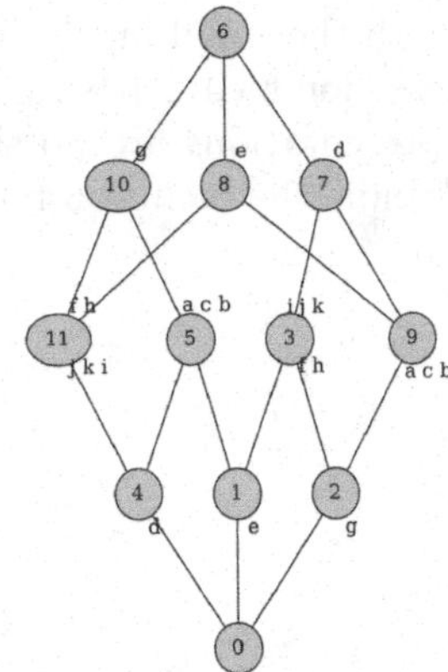

Fig. 2. Formal context describing the graph shown in Fig. 3 and its associated concept lattice.

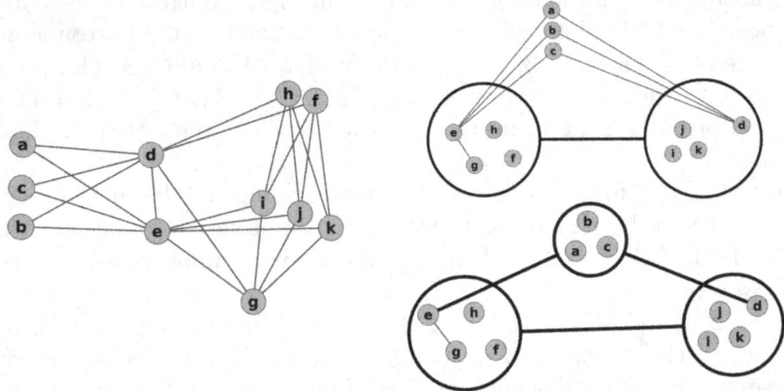

Fig. 3. The graph associated to the formal context in Fig. 2 (left) and its Power Graph compression (lower-right), where the concept/biclique $\{d, i, j, k\} \times \{e, f, g, h\}$ (number 3 or 11 in Fig. 2) was compressed first (leading to the partially compressed graph in upper-right), followed by concept/biclique $\{a, b, c, g\} \times \{d, e\}$ (number 5 or 9 in Fig. 2), that is split in 3 poweredges.

g, e and d, and edge $d \times g$), meeting the decomposition and hierarchy conditions require to split the second concept. It will be split in this case into three poweredges: $\{a, b, c\} \times \{d\}$, $\{a, b, c\} \times \{e\}$ and $\{g\} \times \{e\}$.

Choosing these concepts leads us to an optimal compression of the graph, but other concept lists may lead to different compressions with different number of poweredges: for instance, order $(1, 2, 3, 4)$ will give a 5 poweredges compression. The chosen concepts and their ordering determine the final compressed graph. The standard heuristic is to always compress first the concept, or in case of overlap the poweredge, covering the highest number of edges. This will however not necessarily yield the best compression.

In the following sections we study three features of this search space. First, due to the symmetry of the formal context, the lattice has itself a symmetry exchanging the extent and intent. In our example, 5 and 9 are symmetric as well as 3 and 11. Therefore, $(11, 9)$, $(3, 9)$, $(3, 5)$ and $(11, 5)$ will give the same compressions. The treatment of concept lattice symmetry and other properties is detailed in Sect. 4.

Second, bicliques are not sufficient. Our example does provides some cliques of 3 elements, but they cannot be compressed as such, since concepts encode only bicliques. Section 5 show how to introduce both cliques and bicliques motifs into the search space.

Finally, there is no guarantee that an optimal compression can always be represented by an optimal ordering of concepts. Section 6 gives an example of such case, and discusses how it may be handled.

4 Reducing the Search Space

4.1 Reducing the Concept Lattice

Given a set of n concepts, the set of all possible orderings is the number of permutations of size $1..n$, suggesting a factorial complexity for a complete search. It is therefore important to try to minimize the number of concepts needed to contain exponential growth.

The formal context is symmetric by construction: all nodes are present both in objects and in attributes. This property is visible in the concept lattice, where a concept (A, B) has a symmetric (B, A). One of the two can be discarded from the set of concepts to consider during the search for an order, because they cover exactly the same edges, i.e. they are redundant with respect to compression.

A naive solution to handle symmetry would be to fix an arbitrary ordering $>$ on the object/attribute sets and remove duplicates (A, B) if $A > B$. This is correct but many inclusion links between concepts may be lost in the remaining structure since the choices are independent from the lattice structure.

A structure-preserving procedure has therefore been designed, described in Algorithm 1. It uses an arbitrary fixed total ordering on concepts (numbering) to choose a direct child of the supremum to keep, and consequently a direct parent of the infimum to discard. Each direct child of the supremum is called a *root*. Their symmetrical are the direct parents of the infimum. By marking a root as *kept* (i.e. deciding to keep it in the reduced lattice), we also set its symmetrical to be *discarded* (absent in the reduced lattice). All childs of the root are also kept, and therefore all parents of the root symmetrical (which includes at least one root) are discarded. This is repeated until no more root is marked either as kept or discarded.

Because a node cannot be linked to itself in the formal context, it is not possible to have a concept marked both as kept and discarded. The resulting structure is not itself a concept lattice, it is just used to choose the list of concepts for the compression.

A reduced concept lattice of the example graph in Fig. 3 can be found in Fig. 4.

Algorithm 1. Compute a reduction of a given symmetric lattice by deciding nodes/concepts to keep. There are multiple reductions possible (depending of the root order), but all are equivalent in term of edge cover.

Require: Symmetric Concept Lattice L, symmetries between concepts
Ensure: Compute the set of concepts to keep as *taken*

$taken \leftarrow \emptyset$
$discarded \leftarrow \emptyset$
$roots \leftarrow direct\_childs(supremum(L))$
for all $root \in roots$ **do**
  if $root \notin taken \cup discarded$ **then**
    $taken \leftarrow taken \cup subconcepts(root)$
    $discarded \leftarrow discarded \cup supconcepts(symmetric(root))$
  end if
end for

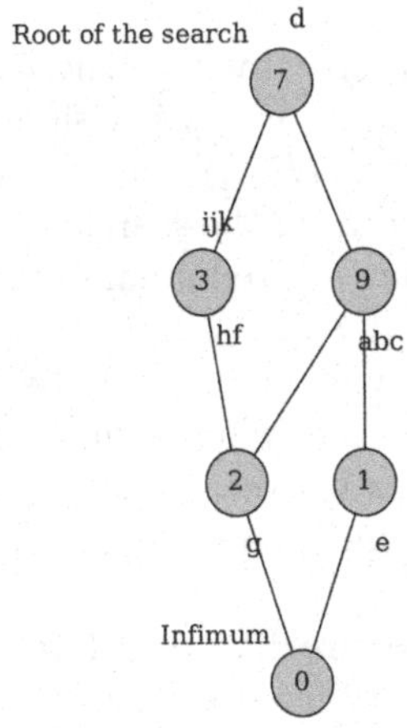

Fig. 4. A reduced concept lattice representation of the lattice in Fig. 2. Concept 7 is the only root since it is the only child of the lattice supremum.

4.2 Reducing the Number of Concept Permutations

Listing all permutations of concept sets ranging in size from 1 to n is only feasible for toy examples where there are no more than half a dozen concepts. Reducing the space of concept orderings is therefore necessary to solve real instances where lattices would typically contain thousands of concepts.

The problem can be formulated as follow: let C be a set of n concepts $\{c_1, c_2, ..., c_n\}$. Each permutation of k concepts $\in C$, $1 \leq k \leq n$, is associated with a score, so that:

- adding a concept to a permutation cannot increase its score (maximal score is given by the uncompressed graph)
- adding a concept (A, B) to a permutation will decrease its score only if there are at least two edges in $A \times B$ that are not covered, nor put in a different block by the treatment of a previous concept in the ordering.

The goal is to find a permutation associated with the smallest score without exploring the complete set of permutations.

Compression Order Cleaning: Because adding a concept covering no new edges cannot decrease the score, it is possible to add a concept to a permutation of concepts if it is not fully covered by these concepts.

Necessary Concepts (Kernel): A concept that appear in all the optimal permutations (regardless of symmetry) is necessary. An obvious optimization would be to discard any permutation that does not contains all the necessary concepts. Determining the necessary concepts is out of the scope of this article, but the following may give an insight of their properties.

Let a concept C be both an object-concept and an attribute-concept. C is the only concept (with its symmetric, removed by reduction) to cover edges $(x, y) \ \ \forall x \in X, y \in Y$. Therefore the only way to compress them is to use C. If X and Y are singletons, only a single edge is concerned, thus there are no compressible edges specific to that concept. If $|X \cup Y| > 2$, there are at least two edges to compress together.

In the example of Sect. 3, concept 11 and its symmetric 3 are the only ones to cover edges $\{f, h\} \times \{i, j, k\}$. They must be, one or the other, compressed at some point.

Note that these concept are not guaranteed to be necessary for concept cycles that will be introduced in Sect. 6.

5 Handling Cliques: Graph Contexts and Graph Concepts

Cliques are with bicliques the main structures managed by Power Graph Analysis. In particular, cliques are frequent in dense graphs. In formal contexts of simple graphs, they generate many concepts, none of them being associated directly with a clique. As a consequence, a clique will not be compressed simply (i.e. as a powernode with a reflexive poweredge), but as many bicliques. Therefore, one cannot rely only on standard concepts for the management of cliques. We are thus proposing an extension of concepts, called triplet concepts, that is tailored for the treatment of simple graphs.

Let us consider the new extended concepts formed by possibly overlapping extent and intent. These concepts can be written as pairs $(A \cup C, B \cup C)$, where A, B and C are disjoint and C is the set of vertices common to the extent and the intent of the concept. Reflexive edges on C are considered as don't care positions. To be concepts, we have further to check that it is maximal with the standard meaning in FCA: it is not possible to add an element to A, B or C that adds a non-reflexive edge, which leads to a subgraph of the context graph. We will denote such concepts by a triplet (A, B, C). Note that from a graph perspective, C is a clique and (A, B) is a biclique. A triplet concept is thus a biclique between a clique and a biclique.

Definition 1. *(**Triplet concept**) Given a graph concept, a triplet concept is a triplet of disjointed sets (A, B, C) such that $A \times B$ is a biclique, C is a clique, and $(A \cup B) \times C$ is a biclique. Moreover (A, B, C) is maximal with respect to inclusion, i.e., it is not possible to add an element to either A, B or C adding a non-reflexive edge, which leads to a subgraph of the context graph.*

Note that triplet concepts (A, B, C) and (B, A, C) are two representations of the same concept and for practical computation one can impose A to be a set of size larger than or equal to B and use a lexicographic order on sets of same cardinality to compute one of the two forms.

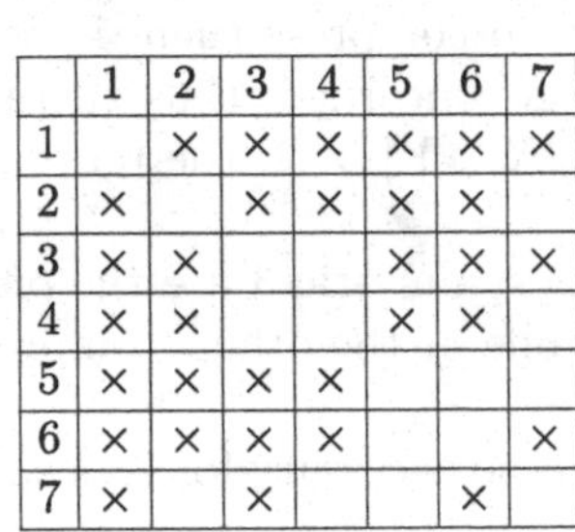

	1	2	3	4	5	6	7
1		×	×	×	×	×	×
2	×		×	×	×	×	
3	×	×			×	×	×
4	×	×			×	×	
5	×	×	×	×			
6	×	×	×	×			×
7	×		×			×	

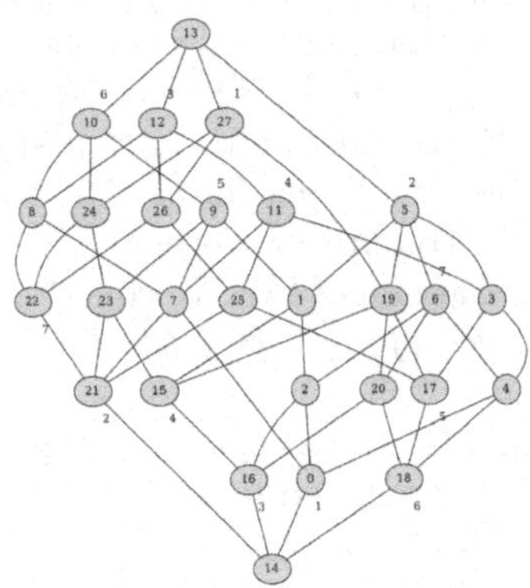

Fig. 5. A formal context and its concept lattice with reduced labelling.

Example

Consider the graph context in Fig. 5. It admits 25 concepts when applying the approach proposed by Chiaselotti et al.

If one looks for triplet concepts instead, 5 are present in the graph:

- $TC_1 = (\{3,4\}, \{5,6\}, \{1,2\})$, corresponding to the biclique $\{1,2,3,4\} \times \{1,2,5,6\}$.

- $TC_2 = (\{1, 2, 3, 4\}, \{5, 6\}, \emptyset)$, corresponding to the biclique $\{1, 2, 3, 4, 6, 7\} \times \{1, 6\}$.
- $TC_3 = (\{2, 7\}, \emptyset, \{1, 3, 6\})$, corresponding to the biclique $\{1, 2, 3, 6, 7\} \times \{1, 3, 6\}$.
- $TC_4 = (\{2, 3, 4, 7\}, \emptyset, \{1, 6\})$, corresponding to the biclique $\{1, 2, 3, 4, 6, 7\} \times \{1, 6\}$.
- $TC_5 = (\{2, 5, 6, 7\}, \emptyset, \{1, 3\})$, corresponding to the biclique $\{1, 2, 3, 5, 6, 7\} \times \{1, 3\}$.

Note that as a special case, one of the set C or B may be empty. If C is empty, one gets the standard concepts. This is the case of TC_2. Three of the triplet concepts have a B set empty: TC_3, TC_4, and TC_5.

Triplet concepts may greatly reduce the needed number of concepts in case of dense context matrices: in this example the five triplets are equivalent to the 25 concepts, they completely cover the graph's edges. From the point of view of graph visualization, compression based on triplet concepts will be able to render the standard representation of cliques in Power Graph Analysis: a powernode with a reflexive poweredge.

Proposition 1. *Let A, B, C be three disjoint set of vertices from a context graph G. Using the standard derivation operation of FCA on this context graph, the following property holds:*

The triplet of disjointed non empty sets (A, B, C) is a triplet concept if and only if C is a maximal clique in the induced subgraph $G(A \cup B \cup C)$ completed by edges $(c, c), c \in C$, and there exist three sets D, E and F disjoint of sets A, B and C and three concepts (A, A'), (B, B'), and (C, C') such that:

$$A' = B \cup C \cup D \tag{1}$$
$$B' = A \cup C \cup E \tag{2}$$
$$C' = A \cup B \cup F \tag{3}$$
$$D \cap F = E \cap F = \emptyset \tag{4}$$

Proof. Assume first (A, B, C) is a triplet concept, then $A \times (B \cup C)$, $B \times A \cup C$, and $C \times B \cup C$ are bicliques of the graph. This implies that $B \cup C \subseteq A'$, $A \cup C \subseteq B'$, and $A \cup B \subseteq C'$ and the first three equations are valid. Assume that there exists an element $x \in D \cap F$. In such a case $x \in A'$ and $x \in C'$. This implies that $(A \cup C \cup \{x\}, B \cup C)$ is a valid biclique. Since the biclique $(A \cup C, B \cup C)$ is maximal, x should belong to A or C, a contradiction.

For the reciprocal, assume concepts (A, A'), (B, B'), and (C, C') exist with the properties in equations. Then $(A \cup C, B \cup C)$ is a biclique of the graph. One has to check that this biclique is maximal. If $(A \cup C \cup x, B \cup C)$ is a biclique, $x \notin A \cup C$, then $x \in E$ from the equation of B' in concept (B, B'). Now if $x \in B$, then x appears on both sides of $(A \cup C \cup \{x\}, B \cup C)$ and C is not a maximal clique on $G(A \cup B \cup C)$. Thus $x \notin B$. One can deduce $x \in F$ from the equation of C' in concept (C, C'). The conclusion is that $x \in E \cap F$, a contradiction. The same reasoning applies if one adds an element to the right of the biclique. □

	a	b	c	d	e	f	g	h
a			×	×	×	×	×	×
b			×	×	×	×	×	×
c	×	×			×	×	×	×
d	×	×			×	×	×	×
e	×	×	×	×		×		
f	×	×	×	×	×			
g	×	×	×	×				×
h	×	×	×	×			×	

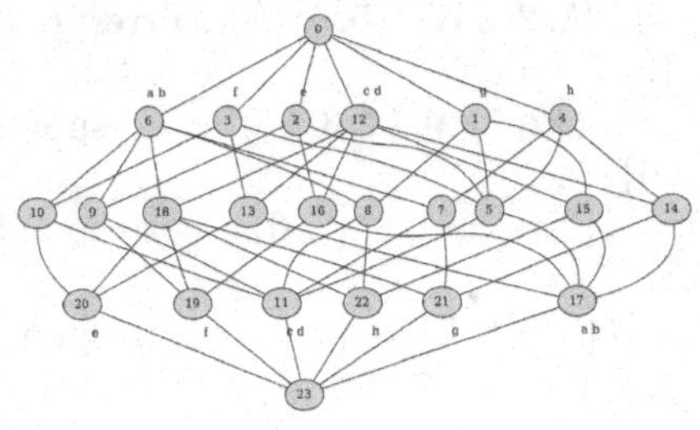

Fig. 6. A formal context and its concept lattice with reduced labelling, exhibiting the non-unicity of cliques associated with bicliques.

Note that for a given pair (A, B), the associated clique is not necessarily unique. Consider for instance the graph context in Fig. 6. Two triplet concepts exist: $(\{a, b, c, d\}, \emptyset, \{e, f\})$ and $(\{a, b, c, d\}, \emptyset, \{g, h\})$.

Computing Triplet Concepts: The generation of triplet concepts can be achieved by a direct implementation of the previous proposition, shown in Algorithm 2: first all standard concepts are produced, then triplets are formed by a combination of three concepts that have all the desired properties. Finally, concepts that are covered by a triplet concept are removed.

For example we consider the graph context in Fig. 7, containing 93 formal concepts from which 7 triplet concepts can be inferred:

$(\{a, b, c, d, e, f, g, j, k\}, \emptyset, \{h, i\})$
$(\{d, e, f, g, j, k\}, \emptyset, \{c, h, i\})$
$(\{a, b, c, g, j, k\}, \{d, e\}, \{h, i\})$
$(\{g, j, k\}, \{d, e\}, \{c, h, i\})$
$(\{d, e, f\}, \emptyset, \{c, h, i, j, k\})$
$(\{c, h, i, j, l\}, \{d, e, f\}, \{k\})$
$(\{a, b, c\}, \{d, e\}, \{g, h, i\})$

92 of the standard concepts are covered by the 7 triplet concepts. The only concept that is not covered is $(\{a, b, c, g, h, i, j, k, l\}, \{d, e\})$. As a consequence, the compression search space for this graph is constituted of 8 triplet concepts, a neat reduction with respect to the initial concept lattice. These 8 elements E can be partially ordered, so that $(A_1, B_1, C_1) \leq (A_2, B_2, C_2)$ if $A_1 \subseteq A_2$, $B_1 \supseteq B_2$ and $C_1 \supseteq C_2$. For the symmetric triplets, $\overline{E}$, the partial order reverse the direction of inclusions for the cliques: $(A_1, B_1, C_1) \leq (A_2, B_2, C_2)$ if $A_1 \subseteq A_2$, $B_1 \supseteq B_2$ and $C_1 \subseteq C_2$. For edges linking an element in E and an element in $\overline{E}$ (bold edges in the Figure), the relation of inclusion between cliques can appear in both directions. The complete line diagram is given in Fig. 8.

	a	b	c	d	e	f	g	h	i	j	k	l
a				×	×		×	×	×			
b				×	×		×	×	×			
c				×	×	×	×	×	×	×	×	
d	×	×	×				×	×	×	×	×	×
e	×	×	×				×	×	×	×	×	×
f			×					×	×	×	×	×
g	×	×	×	×	×			×	×			
h	×	×	×	×	×	×	×		×	×	×	
i	×	×	×	×	×	×	×	×		×	×	
j			×	×	×	×		×	×		×	
k			×	×	×	×		×	×	×		×
l				×	×	×					×	

Fig. 7. Formal context of a dense graph, yielding a total of 93 formal concepts. Two of the maximal cliques are intersecting: $\{d, g, h, i\}$ and $\{c, h, i, j, k\}$.

Algorithm 2. A brute-force algorithm to compute triplet concepts.

Require: Nodes $\mathcal{N}$, concepts $\mathcal{C}$
Ensure: Computation of triplet concepts

for all $A \subset \mathcal{N}$ **do**
  for all $B \subset \mathcal{N} \mid A \subseteq B'$ **do**
    for all $C \subset \mathcal{N} \mid A \cup B \subseteq C'$ *and clique*(C) **do**
      $D \leftarrow A' \setminus (B \cup C)$
      $E \leftarrow B' \setminus (A \cup C)$
      $F \leftarrow C' \setminus (A \cup B)$
      if $D \cap E = E \cap F = \emptyset$ *and* $B \cup C \subseteq A'$ *and* $A \cup C \subseteq B'$ **then**
        yield (A, B, C)
      end if
    end for
  end for
end for

6 Triplet Concepts Are Not Sufficient: The Concept-Cycle Motif

One specific class of concepts layout is resisting to the previous concept approach. We call them cycles of concepts, an example being displayed in Fig. 9 for a 4-cycle. A cycle of concepts is a series of concepts that form a circular chain by inclusion of the intent of one concept in the extent of the next one. Any cycle from 3 to any number of concepts will in fact never be optimally compressed by the procedure we have used so far iterating on the choice of concepts.

The previous section has shown the interest of clique motifs in addition to bicliques to compress graphs and we highlight here a new cycle pattern that underlines the richness of this pattern recognition approach.

As already sketched, the peculiar status of this cycling motif is due to a special organization of concepts, that the concept lattice helps to unravel. In a concept cycle motif, all involved formal concepts of the form $\{(A_1, B_1), ..., (A_n, B_n)\}$ are

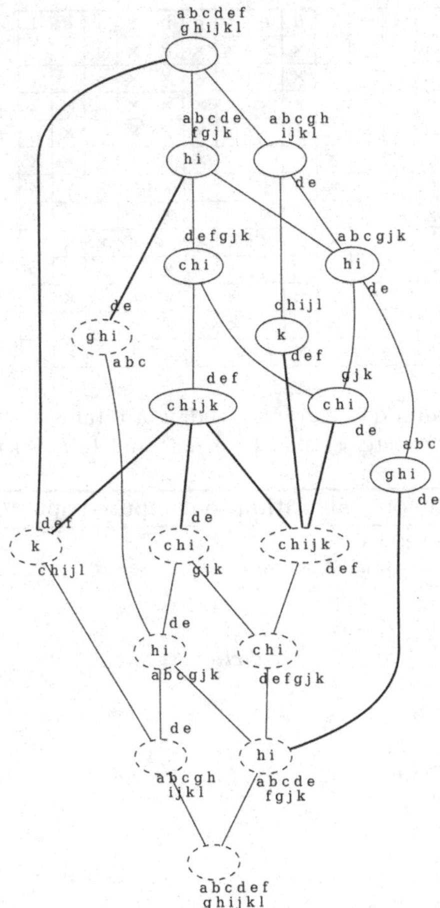

Fig. 8. A line diagram representation of the partial orders over computed triplet concepts (plain nodes) and their symmetric (dotted nodes) in graph context of Fig. 2. Bold edges symbolize the "interface" between concepts and their symmetric. For a concept (A, B, C), A appears above the node, B below and C inside.

ordered so that $A_k \subset B_{k+1} \forall k \in 1..n-1$ and $A_n \subset B_1$. In fact, the basic building block for a cycle is a pair of overlapping concepts $A_1 \times (A_2 \cup B_1)$ and $A_2 \times (A_1 \cup B_2)$, where all sets are disjoint. The two concepts could be represented by a quadruplet (B_1, A_1, A_2, B_2), where all contiguous elements form a biclique. The biclique (A_1, A_2), also not maximal, is a consequence of the fact that a set may appear either as an extend or an intend.

It appears that cycle contexts often lead to a concept lattice that is made, apart from the top and bottom concept, of a graph cycle (see Fig. 10 left) or two symmetric graph cycles (see Fig. 10 right). Note however that the 4-cycle (see Fig. 9) has a special shape where the cycle has been interrupted by intermediary concepts.

It is thus possible to find globally optimal bicliques organizations that are not based solely on the use of maximal bicliques but also use bicliques corresponding to concept overlapping. An enumeration of concepts matching these organizations can lead to a systematic detection of the cyclic pattern, and the incorporation of better motif-concepts that would extend standard or triplet formal concepts.

Further work on the search space might also point to other specific motifs helping to better compress the graph through meaningful recurrent patterns.

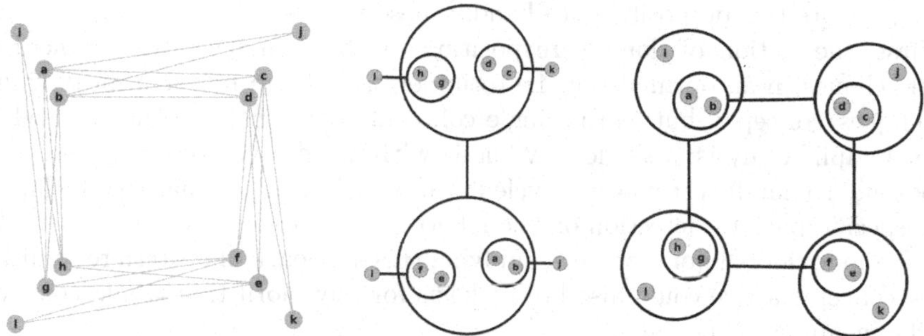

Fig. 9. A cycle of 4 concepts, non-compressed (left), compressed by a greedy method (middle), and optimally compressed (right). The greedy approach compresses first the largest concept (here $\{a, b, e, f\} \times \{c, d, g, h\}$), then 4 small bicliques, ending with 5 poweredges. The optimal compression is reached by using and splitting the four concepts $\{a, b\} \times \{c, d, e, f, j\}$, $\{c, d\} \times \{a, b, e, f, k\}$, $\{e, f\} \times \{c, d, g, h, l\}$, $\{g, h\} \times \{a, b, e, f, i\}$, leading to only 4 poweredges.

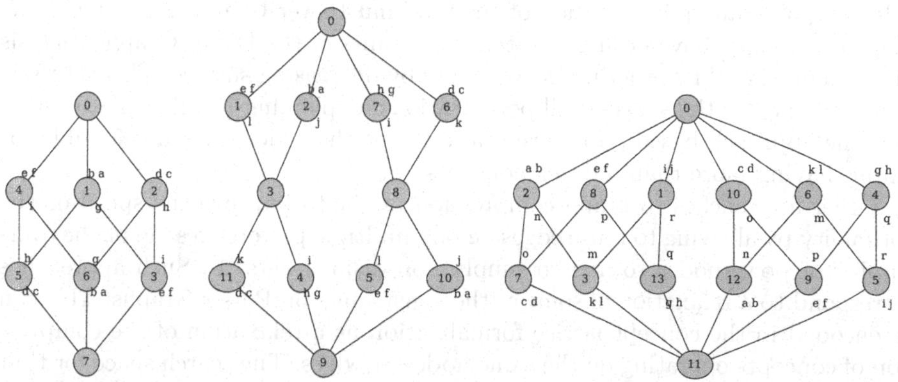

Fig. 10. The lattices of the 3-cycle (left), 4-cycle (middle) and 6-cycle (right) motifs. The latter presents two symmetric cycles, which is the consequence of a bipartite graph (an odd cycle). On the other hand, the 3-cycle is not bipartite, so the lattice symmetry is not perfectly separated.

7 Discussion and Conclusion

A general graph compression search space formalization in the framework of FCA was formulated, highlighting the main sources of difficulty of the problem. It shows that considering objects and attributes as a single set successfully renews FCA's issues.

Limits of the Concept-Only Methods. A method seeking only for concept-based compression in the graph context representation cannot reach the optimal compression, because of at least two graph motifs: cliques and concept-cycles. This paper presents two propositions to handle these motifs.

First, the notion of don't care position in the formal context allowed to handle cliques in an optimal way, i.e to understand them not as a intertwining of bicliques/concepts, but as the single coherent representation that is used by Power Graph Analysis: a single powernode with a reflexive poweredge.

Second, to handle the concept-cycle motif, which uses non maximal bicliques, some specific pattern detection on the lattice has to be designed.

For a more global point of view, applying this approach of pattern recognition in the concept lattice could also be the basis for any motif that would convey a specific meaning in the data.

A General Compression Method. Once the final (triplet and motif) concepts covering the graph have been generated, the compression process can be expressed as the choice of an ordering of a subset of all concepts. We have shown that object and attribute-concepts are useful to focus on particular subsets and the selection of subsets remain an interesting track for further researches.

The standard heuristic for the generation of graph compression is fast but only computes an approximation of the minimal Power Graph [2,13]. Once the (triplet) concepts have been generated, the results of the Power Graph Analysis can be reproduced by ordering the concepts by decreasing surface. This heuristic avoids to explore the space of all permutations, explaining its efficiency, despite that the approach based on a permutation over the concepts is not feasible for graphs having more than a dozen concepts.

Other approaches to graph compression ought to improve the speed or the optimality by allowing to reuse edges among multiple poweredges [5], or the overlapping of powernodes to handle simply non-disjoint sets [1]. Such approaches correspond to a relaxation of some of the constraints on Power Graphs. This can be encoded in the concept lattice formalization as a relaxation of the compression of concepts operating on the same nodes or edges. The search space for that matter is not different, despite the constraint relaxations.

Open Problems. We end with a series of open problems that our new framework has raised.

- From a (triplet) concept lattice find a minimal subset of concepts that are *sufficient* to cover the whole graph. Find a maximal set of concepts that are *necessary* in all optimal compression.

- Can one predict, without processing the compression itself, the number of poweredges that would result from compressing a given list of concepts in a given order?
- What is the structure of the search space as defined by extended concepts?

References

1. Ahnert, S.E.: Generalised power graph compression reveals dominant relationship patterns in complex networks. Sci. Rep. **4**, 4385 (2014)
2. Bourneuf, L., Nicolas, J.: FCA in a logical programming setting for visualization-oriented graph compression. In: Bertet, K., Borchmann, D., Cellier, P., Ferré, S. (eds.) ICFCA 2017. LNCS (LNAI), vol. 10308, pp. 89–105. Springer, Cham (2017). https://doi.org/10.1007/978-3-319-59271-8_6
3. Chiaselotti, G., Ciucci, D., Gentile, T.: Simple undirected graphs as formal contexts. In: Baixeries, J., Sacarea, C., Ojeda-Aciego, M. (eds.) ICFCA 2015. LNCS (LNAI), vol. 9113, pp. 287–302. Springer, Cham (2015). https://doi.org/10.1007/978-3-319-19545-2_18
4. Dwyer, T., Riche, N.H., Marriott, K., Mears, C.: Edge compression techniques for visualization of dense directed graphs. IEEE Trans. Vis. Comput. Graph. **19**(12), 2596–2605 (2013)
5. Dwyer, T., Mears, C., Morgan, K., Niven, T., Marriott, K., Wallace, M.: Improved optimal and approximate power graph compression for clearer visualisation of dense graphs. CoRR, abs/1311.6996 (2013)
6. Gagneur, J., Krause, R., Bouwmeester, T., Casari, G.: Modular decomposition of protein-protein interaction networks. Genome Biol. **5**(8), R57 (2004)
7. Habib, M., Paul, C.: A survey of the algorithmic aspects of modular decomposition. Comput. Sci. Rev. **4**(1), 41–59 (2010)
8. Bernard, J., Seba, H.: Solving the maximal clique problem on compressed graphs. In: Ceci, M., Japkowicz, N., Liu, J., Papadopoulos, G.A., Raś, Z.W. (eds.) ISMIS 2018. LNCS (LNAI), vol. 11177, pp. 45–55. Springer, Cham (2018). https://doi.org/10.1007/978-3-030-01851-1_5
9. King, A.D., Pržulj, N., Jurisica, I.: Protein complex prediction via cost-based clustering. Bioinformatics **20**(17), 3013–3020 (2004)
10. Navlakha, S., Schatz, M.C., Kingsford, C.: Revealing biological modules via graph summarization. J. Comput. Biol. **16**(2), 253–264 (2009)
11. Ogata, H., Fujibuchi, W., Goto, S., Kanehisa, M.: A heuristic graph comparison algorithm and its application to detect functionally related enzyme clusters. Nucleic Acids Res. **28**(20), 4021–4028 (2000)
12. Papadopoulos, C., Voglis, C.: Drawing graphs using modular decomposition. In: Healy, P., Nikolov, N.S. (eds.) GD 2005. LNCS, vol. 3843, pp. 343–354. Springer, Heidelberg (2006). https://doi.org/10.1007/11618058_31
13. Royer, L., Reimann, M., Andreopoulos, B., Schroeder, M.: Unraveling protein networks with power graph analysis. PLoS Comput. Biol. **4**(7), e1000108 (2008)
14. Serafino, P.: Speeding up graph clustering via modular decomposition based compression. In: Proceedings of the 28th Annual ACM Symposium on Applied Computing, SAC 2013, pp. 156–163. ACM, New York (2013)
15. Tsatsaronis, G., Reimann, M., Varlamis, I., Gkorgkas, O., Nørvåg, K.: Efficient community detection using power graph analysis. In: Proceedings of the 9th Workshop on Large-scale and Distributed Informational Retrieval, LSDS-IR 2011, pp. 21–26. ACM, New York (2011)

A Relational Extension of Galois Connections

Inma P. Cabrera, Pablo Cordero, Emilio Muñoz-Velasco, and Manuel Ojeda-Aciego(✉)

Dept. Matemática Aplicada, Universidad de Málaga, Málaga, Spain
{ipcabrera,pcordero,ejmunoz,aciego}@uma.es

Abstract. In this paper, we focus on a twofold relational generalization of the notion of Galois connection. It is twofold because it is defined between sets endowed with arbitrary transitive relations and, moreover, both components of the connection are relations as well. Specifically, we introduce the notion of relational Galois connection between two transitive digraphs, study some of its properties and its relationship with other existing approaches in the literature.

1 Introduction

Since its introduction in mid last century [17], Galois connections have proved to be a useful tool both for theoretical and practical purposes. In this respect, it is worth to note that the mathematical construction underlying the whole theory of FCA [8] is that of Galois connections; and one can find a number of recent publications on either its abstract generalization or its applications [2,6,10,14].

In this paper, we deal with Galois connections, particularly, we focus on its possible generalization to a relational setting. Our interest in this particular problem arises from a discussion with Bernhard Ganter about our research line initiated in [12] in which we attempted to characterize the existence of the residual (or right part of a Galois connection) of a given mapping between sets with different structure (it is precisely this condition of different structure which makes this problem to be outside the scope of Freyd's adjoint functor theorem). It is worth to remark that this problem easily adapts to any different "versions" of Galois connection obtained by considering the dual ordering either in the domain or codomain of the Galois connection.

Since then, we have obtained initial results in several frameworks: for instance, in [13], given a mapping from a (pre-)ordered set $(A, \leq_A)$ into an unstructured set B, we characterized the problem of completing the structure of B, namely, defining a suitable (pre-)ordering relation $\leq_B$ on B, such that there exists a mapping such that the pair of mappings forms an isotone Galois connection (or adjunction) between the (pre-)ordered sets $(A, \leq_A)$ and $(B, \leq_B)$.

Partially supported by the Spanish research projects TIN15-70266-C2-P-1, PGC2018-095869-B-I00 and TIN2017-89023-P of the Science and Innovation Ministry of Spain and the European Social Fund.

D. Cristea et al. (Eds.): ICFCA 2019, LNAI 11511, pp. 290–303, 2019.
https://doi.org/10.1007/978-3-030-21462-3_19

The initial steps for the extension to the fuzzy framework were done in [11], in which the construction of the right adjoint was done in terms of closure operators associated to Galois connection; this approach was later, completed in [3] by considering the corresponding problem for a function between a fuzzy preposet (A, ρ_A) and an unstructured B; moreover, this work was recently extended in [4], by considering that equality is expressed by a fuzzy equivalence relation, so that the problem considers a mapping between a fuzzy preordered structure $(A, \approx_A, \rho_A)$ and a fuzzy structure $(B, \approx_B)$. These two papers satisfactorily extend the problem to the fuzzy case in both the domain and range of the Galois connection but, in both cases, the components of the Galois connection are (crisp) functions. Hence, the next logical extension is to consider the possibility that those components are actually fuzzy functions.

In this respect, the first attempt should be to go back to the crisp case and consider a suitable generalization of Galois connection, one in which the domain and range are just sets endowed with arbitrary relations and whose components are (proper) relations, and this is the content of the present work. It is worth to note that, in order to preserve the existing construction via closures, we need to provide a relational definition of Galois connection which allows the composition of the two components of the connection, and this is something that is not guaranteed by some of the relational extensions of Galois connection that can be found in the literature.

The structure of this paper is the following: in Sect. 2, the necessary preliminaries from the theory of relations and standard Galois connections are introduced; then, in Sect. 3, we discuss the convenience of using the Smyth powerset in the definition of relational Galois connection; later, in Sect. 4 we show the relationship between the proposed definition and the standard Galois connection on the corresponding powerset and other relational generalizations in the literature; then, in Sect. 5, we study some alternative characterizations of our proposed notion of relational Galois connection. Section 6 contains some pointers to possible extensions of the definition in order to be considered not between T-digraphs, but between formal contexts. Finally, in Sect. 7 we obtain some conclusions and present prospects for future work.

2 Preliminary Definitions

We consider the usual framework of (crisp) relations. Namely, a binary relation $\mathcal{R}$ between two sets A and B is a subset of the Cartesian product $A \times B$ and it can be also seen as a (multivalued) function $\mathcal{R}$ from the set A to the powerset 2^B. For an element $(a, b) \in R$, it is said that a is related to b and denoted $a\mathcal{R}b$.

Given a binary relation $\mathcal{R} \subseteq A \times B$, the *afterset* $a^{\mathcal{R}}$ of an element $a \in A$ is defined as $\{b \in B \mid a\mathcal{R}b\}$.

In this paper, we will work with the usual notion of relational composition. Let $\mathcal{R}$ be a binary relation between A and B and $\mathcal{S}$ be a binary relation between B and C. The composition of $\mathcal{R}$ and $\mathcal{S}$ is defined as follows

$$\mathcal{R} \circ \mathcal{S} = \{(x, z) \in A \times C \mid \text{ there exists } b \in B \text{ such that } x\mathcal{R}b \text{ and } b\mathcal{S}z\}.$$

Observe that for an element $a \in A$, the afterset $a^{\mathcal{R}\circ\mathcal{S}}$ can be written as $\bigcup_{b \in a^{\mathcal{R}}} b^{\mathcal{S}}$.

As our Galois connections are intended to be defined between preordered structures, it is convenient to recall that there exist several forms to lift a preorder to the powersets. Two standard ones are given below:

Given A an arbitrary set and a preorder relation $\leq$ defined over A, it is possible to lift $\leq$ to the powerset 2^A by defining

$$X \ll Y \iff \text{for all } x \in X \text{ there exists } y \in Y \text{ such that } x \leq y\,,$$

$$X \Subset Y \iff \text{for all } y \in Y \text{ there exists } x \in X \text{ such that } x \leq y\,.$$

We will use the term *powering* to refer to the lifting of a preorder to the powerset; thus, both $\Subset$ and $\ll$ above are powerings of $\leq$.

Note that the two relations defined above are actually preorder relations, specifically those used in the construction of the, respectively, Hoare and Smyth powerdomains [20].

Naturally, each of the extensions above induces a particular notion of isotony, inflation, etc. For instance, given two preordered sets $(A, \leq)$ and $(B, \leq)$,[1] a binary relation $\mathcal{R} \subseteq A \times B$ is said to be:

- $\Subset$-isotone if $a_1 \leq a_2$ implies $a_1^{\mathcal{R}} \Subset a_2^{\mathcal{R}}$, for all $a_1, a_2 \in \mathrm{dom}(\mathcal{R})$;
- $\Subset$-antitone if $a_1 \leq a_2$ implies $a_2^{\mathcal{R}} \Subset a_1^{\mathcal{R}}$, for all $a_1, a_2 \in \mathrm{dom}(\mathcal{R})$.

A binary relation $\mathcal{R} \subseteq A \times A$ is said to be:

- $\Subset$-inflationary if $\{a\} \Subset a^{\mathcal{R}}$, for all $a \in \mathrm{dom}(\mathcal{R})$;
- $\Subset$-deflationary if $a^{\mathcal{R}} \Subset \{a\}$, for all $a \in \mathrm{dom}(\mathcal{R})$.
- $\Subset$-idempotent if $a^{\mathcal{R}\circ\mathcal{R}} \Subset a^{\mathcal{R}}$ and $a^{\mathcal{R}} \Subset a^{\mathcal{R}\circ\mathcal{R}}$, for all $a \in \mathrm{dom}(\mathcal{R})$.

We use the prefix to distinguish the powering used in the different definitions.

Traditionally, a Galois connection is understood as a pair of antitone mappings whose compositions are both inflationary, and has a number of different alternative characterizations. In our generalized relational setting, we have a wide choice for characterization used to give the formal definition, and also to the different notions of antitonicity and inflation (depending on the powering), or even the relational composition to be used.

3 A Relational Extension of Galois Connections

Our goal in this work is to define the notion of Galois connection as a pair of relations between sets *with the least possible structure*. As all of the results make use in some way of the transitive property, although we could work with arbitrary relations and use its transitive closure, in order to improve the readability of the results we will assume that the relations are transitive from the beginning.

[1] Notice that, as usual, we use the same symbol to denote both binary relations which need not be equal.

Hereinafter, we refer to a couple (A, τ), where $\tau \subseteq A \times A$, as a *digraph*,[2] and when τ is transitive we call *T-digraph.*

A T-digraph (A, τ) will be often represented as $\mathbb{A}$, to refer to the underlying set A, the accompanying relation will be written τ *whenever no ambiguities arise*, similarly to what happens with ordered sets in which the ordering relation are usually called $\leq$.

It is worth to remark that the powerings $\Subset$ and $\ll$ can be defined for any relation τ not necessarily being a preorder relation.

A well-known characterization of a Galois connection (f, g) between two posets is the so-called *Galois condition*

$$a \leq g(b) \iff b \leq f(a).$$

As stated above, in our general framework there are several possible choices, which we will distinguish by using the corresponding prefix. For instance, given two relations $\mathcal{R}$ and $\mathcal{S}$, the $\ll$-Galois condition is

$$\{a\} \ll b^{\mathcal{S}} \quad \iff \quad \{b\} \ll a^{\mathcal{R}}.$$

In [5], we studied the properties of the different extensions obtained in terms of the powerings $\ll$ and $\Subset$ used in the corresponding Galois condition. In this work we focus our attention on another desirable characterization, the definition of Galois connection in terms of closures. To begin with, we introduce below the corresponding relational extension of the notion of closure operator.

Definition 1. *Given a preordered set $(A, \leq)$, a powering $*$ of $\leq$, and $\mathcal{C} \subseteq A \times A$, we say that $\mathcal{C}$ is a* $*$-closure relation, *if $\mathcal{C}$ is $*$-isotone, $*$-inflationary, and $*$-idempotent.*

The following example shows that the definition based on the Hoare powering $\ll$ does not behave as one would expect.

Example 1. Consider the set of natural numbers together with the discrete ordering given by the equality relation $(\mathbb{N}, =)$, and consider the relation $\mathcal{R}$ given by $n^{\mathcal{R}} = \{0, \ldots, n+1\}$. The relation $\mathcal{R}$ is trivially $\ll$-antitone, and $\mathcal{R} \circ \mathcal{R}$ is obviously $\ll$-inflationary; however, it does not make sense to consider $(\mathcal{R}, \mathcal{R})$ as an extended Galois connection, since it is not difficult to check that $\mathcal{R} \circ \mathcal{R}$ is not a $\ll$-closure relation (it fails to be $\ll$-idempotent) and, furthermore, the $\ll$-Galois condition does not hold either. □

As a result, we cannot rely on the $\ll$ powering in order to obtain closure relations. This justifies the choice of the $\Subset$ ordering to provide the notion of relational Galois connection as follows:

Definition 2. *A* relational Galois connection *between two T-digraphs $\mathbb{A}$ and $\mathbb{B}$ is a pair of relations $(\mathcal{R}, \mathcal{S})$ where $\mathcal{R} \subseteq A \times B$ and $\mathcal{S} \subseteq B \times A$ such that the following properties hold:*

[2] A digraph is often called a *relational system.*

i. $\mathcal{R}$ *and* $\mathcal{S}$ *are* $\Subset$*-antitone.*
ii. $\mathcal{R} \circ \mathcal{S}$ *and* $\mathcal{S} \circ \mathcal{R}$ *are* $\Subset$*-inflationary.*

We can see below an example in which both $\mathcal{R}$ and $\mathcal{S}$ are proper (non-functional) relations.

Example 2. Consider $\mathbb{A} = (A, \tau)$ where $A = \{1, 2, 3\}$ and τ is the transitive relation $\{(1,2), (1,3), (2,2), (2,3), (3,2), (3,3)\}$. The pair of relations $(\mathcal{R}, \mathcal{S})$ given by the tables below constitutes a relational Galois connection between $\mathbb{A}$ and $\mathbb{A}$.

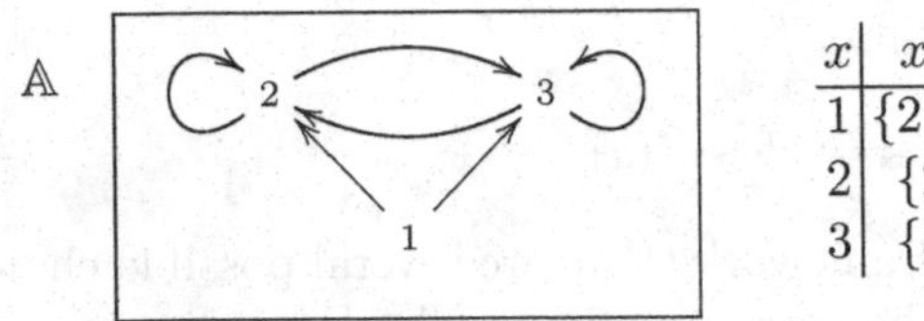

x	$x^{\mathcal{R}}$
1	$\{2,3\}$
2	$\{2\}$
3	$\{3\}$

x	$x^{\mathcal{S}}$
1	$\{2,3\}$
2	$\{2\}$
3	$\{2,3\}$

□

The interesting part is that the $\Subset$ powering guarantees that both compositions in a relational Galois connection lead to $\Subset$-closure relation. Formally, we have the following result:

Theorem 1. *Given a relational Galois connection* $(\mathcal{R}, \mathcal{S})$ *between* $\mathbb{A}$ *and* $\mathbb{B}$*, we have that* $\mathcal{R} \circ \mathcal{S}$ *and* $\mathcal{S} \circ \mathcal{R}$ *are* $\Subset$*-closure relations.*

Recall that Example 1 shows that the extension based on $\ll$ does not generate a closure relation. It is not difficult to check that conditions in Definition 2 are not satisfied either.

4 Comparison with Other Approaches

4.1 Relationship with Galois Connections Between Powersets

Given a relation $\mathcal{R} \subseteq A \times B$, we can define two mappings between powersets, from 2^A to 2^B, namely, the *direct* extension of $\mathcal{R}$, denoted by $\mathcal{R}(\cdot)$, and the *subdirect* extension of $\mathcal{R}$, denoted by $(\cdot)^{\mathcal{R}}$, which are given by

$$\mathcal{R}(X) = \bigcup_{x \in X} x^{\mathcal{R}} \qquad\qquad X^{\mathcal{R}} = \bigcap_{x \in X} x^{\mathcal{R}}.$$

Taking into account that a relation $\mathcal{R} \subseteq A \times B$ can be interpreted as a function from 2^A to 2^B, it will be worth studying the possible relationship between our notion of relational Galois connection introduced in Definition 2 and the classical Galois connections between powersets.

On the one hand, one can consider a relational Galois connection $(\mathcal{R}, \mathcal{S})$ between T-digraphs $\mathbb{A}$ and $\mathbb{B}$ and study the properties of the corresponding extension to the powersets 2^A and 2^B; we will show examples that the powerset extensions of the relations of a relational Galois connection need not form a classical Galois connection. On the other hand, the other implication does not hold either, that is, given a classical Galois connection between powersets, its relational restriction to the underlying sets by considering the images of singletons need not be a relational Galois connection. As a result, the standard definition neither implies nor is implied by our notion of relational Galois connection.

Example 3. We show here a relational Galois connection whose *direct extension* to the powerset (with the $\Subset$ powering) is not a classical Galois connection.

Let $\mathbb{A}$ and $\mathbb{B}$ be the T-digraphs shown below, and $\mathcal{R} \subseteq A \times B$ and $\mathcal{S} \subseteq B \times A$ the relations defined as follows:

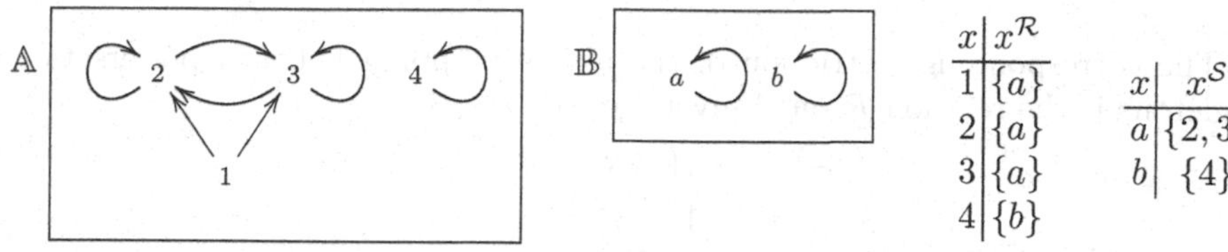

x	$x^{\mathcal{R}}$
1	$\{a\}$
2	$\{a\}$
3	$\{a\}$
4	$\{b\}$

x	$x^{\mathcal{S}}$
a	$\{2,3\}$
b	$\{4\}$

It is straightforward to see that $(\mathcal{R}, \mathcal{S})$ is a relational Galois connection, but its direct extension to the powersets is not a Galois connection. Observe that $\{1,4\} \Subset \mathcal{S}(\{a\}) = \{2,3\}$, however $\{a\} \not\Subset \mathcal{R}(\{1,4\}) = \{a,b\}$. □

The following two examples are based on preordered structures as particular cases of T-digraphs; in both cases, the depicted graphs induce the preorders via the reflexive and transitive closures.

Example 4. We show here a relational Galois connection whose *subdirect extension* to the powerset (with the $\Subset$ powering) is not a classical Galois connection.

Let $\mathbb{A}$ be the preordered set induced by the graph below, and $\mathcal{R} \subseteq A \times A$ be the relation defined as follows:

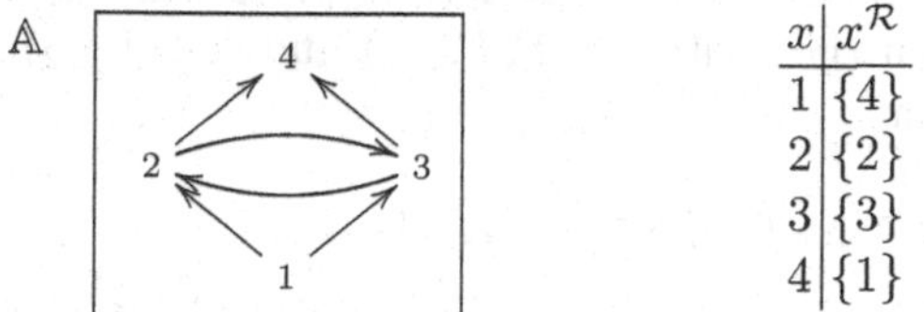

x	$x^{\mathcal{R}}$
1	$\{4\}$
2	$\{2\}$
3	$\{3\}$
4	$\{1\}$

It is straightforward to check that $(\mathcal{R}, \mathcal{R})$ is a relational Galois connection between $\mathbb{A}$ and $\mathbb{A}$, but its subdirect extension to the powersets is not a Galois connection. Observe that $\{4\} \Subset \{2,3\}^{\mathcal{R}} = \{2\}^{\mathcal{R}} \cap \{3\}^{\mathcal{R}} = \varnothing$, however $\{2,3\} \not\Subset \{4\}^{\mathcal{S}} = \{1\}$.

□

The last example shows a classical Galois connection between powersets whose restriction to singletons is not a relational Galois connection.

Example 5. Given the preordered set $\mathbb{A} = (A, \leq)$, its Smyth extension and the mapping $f\colon 2^A \to 2^A$ depicted below, the pair (f, f) is a Galois connection between $2^{\mathbb{A}}_{\Subset} = (2^A, \Subset)$ and itself.

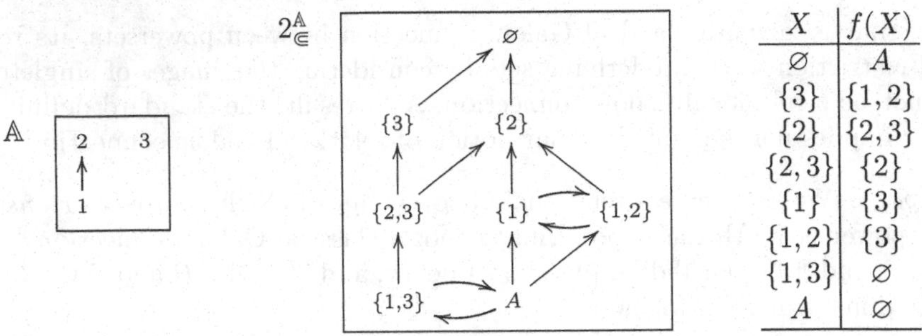

X	$f(X)$
$\varnothing$	A
$\{3\}$	$\{1,2\}$
$\{2\}$	$\{2,3\}$
$\{2,3\}$	$\{2\}$
$\{1\}$	$\{3\}$
$\{1,2\}$	$\{3\}$
$\{1,3\}$	$\varnothing$
A	$\varnothing$

The corresponding restriction of the above mapping between powersets to singletons is the relation $\mathcal{R}$ on A given by

x	$x^{\mathcal{R}}$
1	$\{3\}$
2	$\{2,3\}$
3	$\{1,2\}$

The pair $(\mathcal{R}, \mathcal{R})$ is not a relational Galois connection because $\{2\} \notin 2^{\mathcal{R}\circ\mathcal{R}}$. □

4.2 Relationship with Essential Galois Bonds

Essential Galois bonds between contexts were introduced by Xia [19], and are important for our work in that its components are relations. This definition was later renamed as relational Galois connection in [9], where a unifying language was provided in order to cope with similar attempts, previously given by Domenach and Leclerc [7] and by Wille [18].

Recall that for ordered sets $(P, \leq_1)$ and $(Q, \leq_2)$ to state that (φ, ψ) is a Galois connection between the contexts $(P, P, \leq_1)$ and $(Q, Q, \leq_2)$ translates to the usual Galois condition

$$x \leq_1 \psi(y) \iff y \leq_2 \varphi(x)\,.$$

The first approach to the natural notion of Galois connection between arbitrary contexts (G, M, I) and (H, N, J) is given in [9] as a pair of mappings (φ, ψ) where $\varphi\colon G \to N$ and $\psi\colon H \to M$, satisfying the corresponding Galois condition, namely,

$$g\ I\ \psi(h) \iff h\ J\ \varphi(g)\,.$$

Then, a further (and natural) generalisation in which the pair of mappings are replaced by a pair of arbitrary relations $\Phi \subseteq G \times N$ and $\Psi \subseteq H \times M$ should include the corresponding relational version of the Galois condition, which is called the *relational Galois condition*:

$$g\ I\ h^{\Psi} \iff h\ J\ g^{\Phi}\,.$$

This condition by itself is not strong enough in the relational framework and, hence, Xia introduced a certain optimality of Φ and Ψ. This condition was simplified in [9] in terms of intents of the corresponding contexts as follows:

Definition 3. *A* relational Galois connection *between two contexts* (G, M, I) *and* (H, N, J) *is a pair of relations* (Φ, Ψ) *where* $\Phi \subseteq G \times N$ *and* $\Psi \subseteq H \times M$ *satisfying*

- $g\ I\ h^{\Psi} \iff h\ J\ g^{\Phi}$ *for all* $g \in G, h \in H$.
- g^{Φ} *is an intent of* (H, N, J) *and* h^{Ψ} *is an intent of* (G, M, I).

This definition, as stated in [9, Lemma 1], leads back to the original definition of Galois connection between lattices, in that every classical Galois connection between the concept lattices $\underline{\mathfrak{B}}(G, M, I)$ and $\underline{\mathfrak{B}}(H, N, J)$ of two contexts (G, M, I) and (H, N, J) respectively, defines a relational Galois connection between the contexts (G, M, I) and (H, N, J) and vice versa, being this correspondence one-to-one.

The following two examples show that Definition 3 of relational Galois connection does not imply nor is implied by our definition.

Example 6. Let $\mathbb{A}$ and $\mathbb{B}$ be the T-digraphs shown below, and $\mathcal{R} \subseteq A \times B$ and $\mathcal{S} \subseteq B \times A$ the relations defined as follows:

x	$x^{\mathcal{R}}$
1	$\{a, b, c\}$
2	$\{b\}$
3	$\{c\}$

x	$x^{\mathcal{S}}$
a	$\{1, 2, 3\}$
b	$\{2\}$
c	$\{3\}$

It is easy to check that $(\mathcal{R}, \mathcal{S})$ verifies Definition 3 but does not satisfy our definition, because $\mathcal{R} \circ \mathcal{S}$ fails to be inflationary in element 1. As a consequence, $\mathcal{R} \circ \mathcal{S}$ is not a closure relation.

Example 7. Let $\mathbb{A}$ and $\mathbb{B}$ be the T-digraphs shown below, and $\mathcal{R} \subseteq A \times B$ and $\mathcal{S} \subseteq B \times A$ the relations defined as follows:

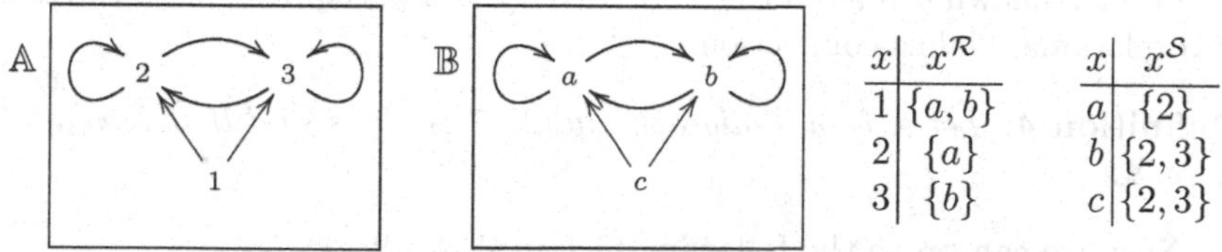

x	$x^{\mathcal{R}}$
1	$\{a, b\}$
2	$\{a\}$
3	$\{b\}$

x	$x^{\mathcal{S}}$
a	$\{2\}$
b	$\{2, 3\}$
c	$\{2, 3\}$

The pair $(\mathcal{R}, \mathcal{S})$ constitutes a relational Galois connection between (A, τ_A) and (B, τ_B) by our definition. However, it does not verify Definition 3. For instance, $2^{\mathcal{R}} = \{a\}$ is not an intent since $2^{\mathcal{R}\tau^{-1}\tau} = \{a, b\} \neq 2^{\mathcal{R}}$.

5 Characterization of Relational Galois Connections

Having in mind the different characterizations of classical Galois connections between posets in terms of the Galois condition, the definition of a relational Galois connection given above *might* also be equivalent to the corresponding Galois condition, namely:

$$\{a\} \Subset b^{\mathcal{S}} \iff \{b\} \Subset a^{\mathcal{R}}, \text{ for all } a \in A, \text{ and } b \in B. \tag{1}$$

However, the following example shows that condition (1) does not imply $(\mathcal{R}, \mathcal{S})$ being a relational Galois connection.

Example 8. Consider the T-digraph $\mathbb{A}$ and the relations $\mathcal{R}$ and $\mathcal{S}$ depicted below (note that $\mathcal{R}$ and $\mathcal{S}$ coincide in this example).

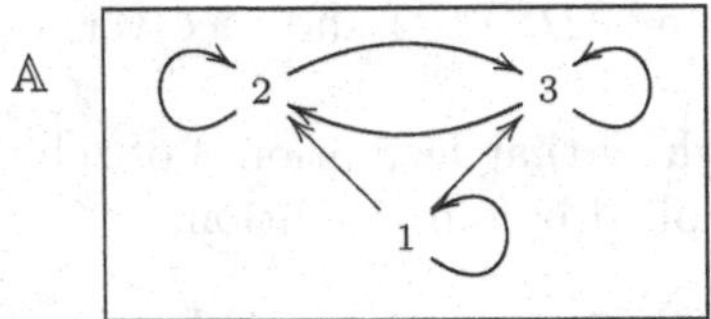

x	$x^{\mathcal{R}} = x^{\mathcal{S}}$
1	$\{3\}$
2	$\{1,2\}$
3	$\{1,3\}$

It is routine to prove that $(\mathcal{R}, \mathcal{S})$ verifies condition (1), but it does not verify Definition 2, because, for instance, $\{2\} \not\Subset_A \{2\}^{\mathcal{R}\circ\mathcal{S}}$. □

In order to characterize the notion of relational Galois connection, we will introduce an alternative powering of a relation τ to the powersets:

$$X \propto Y \iff \text{for all } x \in X \text{ and for all } y \in Y \text{ we have that } x \;\tau\; y.$$

Remark 1. Note that $\propto$ need not be either reflexive nor transitive. Nevertheless, for a T-digraph it satisfies the following weakened version of transitivity:

$$\text{For any } Y \neq \varnothing, \text{if } X \propto Y \text{ and } Y \propto Z, \text{ then } X \propto Z. \tag{2}$$

We will prove that the Galois condition (1), together with a certain technical condition somewhat related to the reflexivity of $\propto$, is equivalent to the definition of a relational Galois connection.

Definition 4. *Let $\mathbb{A}$ be a T-digraph and $X \subseteq A$. It is said that X is a clique if $X \propto X$.*

Now, we can prove the following technical result.

Lemma 1. *Let $\mathbb{A}$ be a T-digraph and $x \in X \subseteq A$. If X is a clique then, for all $Y \subseteq A$, the following statements hold:*

i. $Y \propto \{x\}$ *implies* $Y \propto X$.
ii. $\{x\} \propto Y$ *implies* $X \propto Y$.
iii. $X \Subset Y$ *if and only if* $X \propto Y$.

Proof. We just prove item (i), since the other results follow easily from this one.

Since $X \propto X$, it holds that $\{x\} \propto X$ which, together with $Y \propto \{x\}$ and (2), implies $Y \propto X$. □

Remark 2. Note that there exists a tight relationship between $\Subset$ and $\propto$, since for all $x \in A$ and all $Y \subseteq A$ we have that

$$\{x\} \Subset Y \quad \iff \quad \{x\} \propto Y\,,$$

particularly, the notions of $\Subset$-inflation and $\propto$-inflation are equivalent and, moreover, the corresponding versions of the Galois condition are also equivalent.

Our first characterization of relational Galois connections is based on the fact that the direct images of singletons should be cliques for both components of the relational Galois connection. The formal result is as follows:

Proposition 1. *$(\mathcal{R}, \mathcal{S})$ is a relational Galois connection between T-digraphs (A, τ) and (B, τ) iff the following properties hold:*

i. $\{a\} \Subset b^{\mathcal{S}}$ iff $\{b\} \Subset a^{\mathcal{R}}$ for all $a \in A$, and $b \in B$.
ii. $a^{\mathcal{R}}$ and $b^{\mathcal{S}}$ are cliques for all $a \in A$ and $b \in B$.

Proof. Assume that $(\mathcal{R}, \mathcal{S})$ is a relational Galois connection. To prove (i), suppose $\{a\} \Subset_A b^{\mathcal{S}}$, that is, $a \,\tau\, x$, for all $x \in b^{\mathcal{S}}$. Since $\mathcal{R}$ is $\Subset$-antitone, we obtain $x^{\mathcal{R}} \Subset_B a^{\mathcal{R}}$. Thus, for all $y \in a^{\mathcal{R}}$ there exists $\hat{y} \in x^{\mathcal{R}}$ such that $\hat{y} \,\tau\, y$. Furthermore, as $\mathcal{S} \circ \mathcal{R}$ is $\Subset$-inflationary, we have $b \,\tau\, \hat{y} \in b^{\mathcal{S} \circ \mathcal{R}}$. Hence, by transitivity, $b \,\tau\, y$, for all $y \in a^{\mathcal{R}}$, that is, $\{b\} \Subset_B a^{\mathcal{R}}$, proving one implication of the Galois condition. The converse implication is similar.

Let us prove now that $a^{\mathcal{R}}$ is a clique, that is, $a^{\mathcal{R}} \propto a^{\mathcal{R}}$, for all $a \in A$. As $\mathcal{R} \circ \mathcal{S}$ is $\Subset$-inflationary, it holds $\{a\} \Subset_A a^{\mathcal{R} \circ \mathcal{S}}$, which means $\{a\} \Subset_A b^{\mathcal{S}}$, for all $b \in a^{\mathcal{R}}$. This is also equivalent, by item (i) already proven, to $\{b\} \Subset_B a^{\mathcal{R}}$, that is $b \,\tau\, a^{\mathcal{R}}$; hence $a^{\mathcal{R}} \propto a^{\mathcal{R}}$. The proof of $b^{\mathcal{S}} \propto b^{\mathcal{S}}$, for all $b \in B$, is similar.

Conversely, assume that items (i), (ii) hold. Let us prove first that $\mathcal{R} \circ \mathcal{S}$ is $\Subset$-inflationary. For all $a \in A$, given $x \in a^{\mathcal{R} \circ \mathcal{S}} = \bigcup_{b \in a^{\mathcal{R}}} b^{\mathcal{S}}$, there exists $b \in a^{\mathcal{R}}$ such that $x \in b^{\mathcal{S}}$. Since, $a^{\mathcal{R}}$ is a clique by hypothesis, we have that $\{b\} \Subset_B a^{\mathcal{R}}$, which is equivalent to $a \Subset_A b^{\mathcal{S}}$. Particularly, $a \,\tau\, x$, which proves that $a \Subset_A a^{\mathcal{R} \circ \mathcal{S}}$, for all $a \in A$. The proof that $\mathcal{S} \circ \mathcal{R}$ is $\Subset$-inflationary is similar.

Finally, instead of proving that $\mathcal{R}$ is $\Subset$-antitone we will prove that it is $\propto$-antitone. Assume that $a_1 \,\tau\, a_2$, by the previous paragraph and Remark 2, we have that $\{a_2\} \propto a_2^{\mathcal{R} \circ \mathcal{S}}$, therefore we obtain that $\{a_1\} \propto a_2^{\mathcal{R} \circ \mathcal{S}}$. Now, given an arbitrary $b \in a_2^{\mathcal{R}}$ and $a_3 \in b^{\mathcal{S}}$ (so $a_3 \in a_2^{\mathcal{R} \circ \mathcal{S}}$), we have $a_1 \,\tau\, a_3$ which implies $\{a_1\} \propto b^{\mathcal{S}}$ by Lemma 1. By hypothesis, this is equivalent to $\{b\} \propto a_1^{\mathcal{R}}$. Therefore, $a_2^{\mathcal{R}} \propto a_1^{\mathcal{R}}$. □

The next result shows that our definition of $\Subset$-based relational Galois connection coincides exactly with the corresponding $\propto$-version.

Proposition 2. *$(\mathcal{R}, \mathcal{S})$ is a relational Galois connection between T-digraphs (A, τ) and (B, τ) iff the following properties hold:*

i. $\mathcal{R}$ and $\mathcal{S}$ are $\propto$-antitone.
ii. $\mathcal{R} \circ \mathcal{S}$ and $\mathcal{S} \circ \mathcal{R}$ are $\propto$-inflationary.

Proof. Assume that $(\mathcal{R}, \mathcal{S})$ is a relational Galois connection, clearly $\mathcal{R} \circ \mathcal{S}$ and $\mathcal{S} \circ \mathcal{R}$ are $\propto$-inflationary by Remark 2. As $\mathcal{R}$ is $\Subset$-antitone and, by Proposition 1, $a^{\mathcal{R}}$ is a clique for all $a \in A$, Lemma 1 (iii) states that $\mathcal{R}$ is $\propto$-antitone as well (similarly for $\mathcal{S}$).

The other implication is straightforward. □

6 Relational Galois Connections Between Formal Contexts

Given a T-digraph (A, τ), we can build as usual a context which contains all the information in the digraph, namely, (A, A, τ). Conversely, given a context $\mathbb{K} = (G, M, I)$, there are several ways to generate a T-digraph, therefore our notion of relational Galois connection could also be applied to contexts simply by considering the relational Galois connection on the underlying T-digraphs. We discuss four possibilities to build a T-digraph starting from a formal context:

i. A first one would be to consider the incidence relation I in the disjoint union $G \sqcup M$, this relation is (vacuosly) transitive and anti-reflexive (i.e. there is no $a \in G \sqcup M$ satisfying $a\ I\ a$). But, with this construction, it is not possible to define relational Galois connections $(\mathcal{R}, \mathcal{S})$ since $a^{\mathcal{R}}$ would not be a clique. We can get rid of anti-reflexivity by considering the reflexive closure of I in $G \sqcup M$, then any relational Galois connection will have functions (not proper relations) as its components since all the cliques would be singletons. Hence, this case collapses to the usual notion of Galois connection between preorders.
ii. A second possibility would be to consider the transitive closure of the incidence relation in $G \cup M$ (not necessarily disjoint). For instance, given context $\mathbb{K}$ we would obtain the T-digraph $\mathbb{A}$ below.

$\mathbb{K}$	1	2	4
1		×	
2	×		×
3		×	×

$\mathbb{A}$

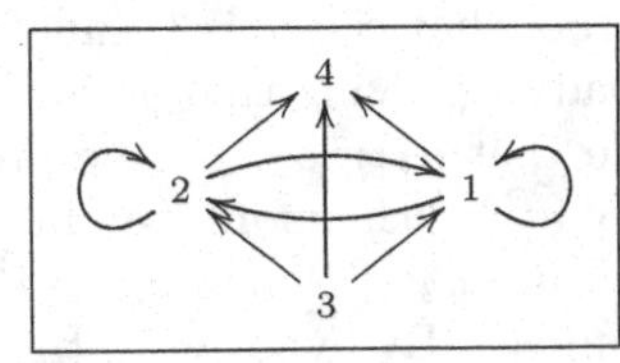

In this case, a relational Galois connection $(\mathcal{R}, \mathcal{S})$ between contexts (G, M, I) and (H, N, J), for $\mathcal{R}$ and $\mathcal{S}$ to be composable, should satisfy $a^{\mathcal{R}} \subseteq H \cap N$ for all $a \in G \cup M$ and $b^{\mathcal{S}} \subseteq G \cap M$ for all $b \in H \cup N$; specifically, we should have $\mathcal{R} \subseteq (G \cup M) \times (H \cap N)$ and also $\mathcal{S} \subseteq (H \cup N) \times (G \cap M)$.

iii. A third alternative is suggested by the mappings μ and γ of the basic theorem of FCA which embed objects and attributes in a lattice. In this case we consider the relation τ_I defined in $G \sqcup M$ which embeds objects and attributes in a T-digraph. The definition is given by
   - $g\ \tau_I\ m$ iff $g\ I\ m$, for all $g \in G$ and $m \in M$.
   - $g_1\ \tau_I\ g_2$ iff $g_2^{\uparrow} \subseteq g_1^{\uparrow}$, for all $g_1, g_2 \in G$.
   - $m_1\ \tau_I\ m_2$ iff $m_1^{\downarrow} \subseteq m_2^{\downarrow}$, for all $m_1, m_2 \in M$.

   It is easy to check that relation τ_I is transitive, moreover, it is always a preorder relation. For instance, given context $\mathbb{K}$ below we would obtain the preorder corresponding to the Hasse diagram in the right

$\mathbb{K}$	a	b	c	d
1	×			
2		×		×
3	×	×		×
4		×	×	×

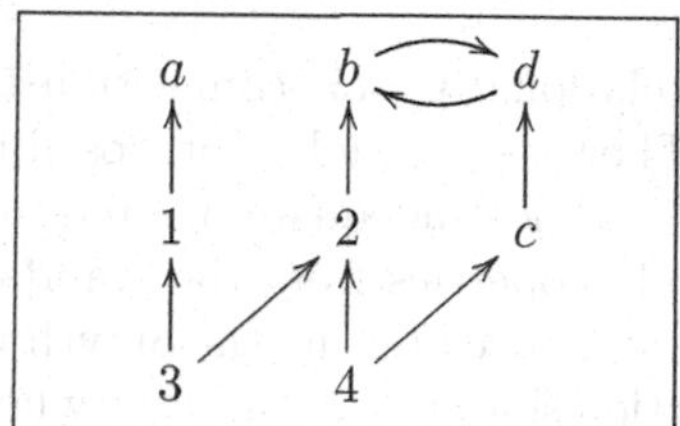

   In this case, if $(\mathcal{R}, \mathcal{S})$ is a relational Galois connection between contexts (G, M, I) and (H, N, J), for all $a \in G \sqcup M$ we have that either $a^{\mathcal{R}} \subseteq H$ or $a^{\mathcal{R}} \subseteq N$, and similarly for $b^{\mathcal{S}}$.

   Note that when the formal contexts are clarified (i.e. do not have either repeated columns or rows), the resulting relational construction again collapses to the usual Galois connection between preorders.

iv. We can also reproduce a similar construction in the non-necessarily disjoint union $G \cup M$. Specifically, for all $a, b \in G \cup M$, consider $a\ \tau_I\ b$ if and only if one of the following conditions holds:
   - $a \in G$, $b \in M$ and $a\ I\ b$.
   - $a, b \in G$ and $b^{\uparrow} \subseteq a^{\uparrow}$.
   - $a, b \in M$ and $a^{\downarrow} \subseteq b^{\downarrow}$.

   In this case, we obtain a reflexive relation which need not be transitive. For instance, for the following context

$\mathbb{K}$	1	2	3
1		×	×
2	×		
4			×

   We have that $2\ \tau_I\ 1$, since $2\ I\ 1$, and $1\ \tau_I\ 4$ since $4^{\uparrow} \subseteq 1^{\uparrow}$; however, $2 \not\tau_I\ 4$. Therefore, in this construction we have to consider the transitive closure again.

For options (ii) and (iv) above, it is remarkable that if we start from a T-digraph (A, τ) which is a preordered set, then the obtained T-digraph from this context is the initial preorder relation. For construction ii this is obvious;

for construction (iv) we have just to take into account that in the associated context (A, A, τ) the following holds

$$a \,\tau\, b \quad \text{iff} \quad b^{\uparrow} \subseteq a^{\uparrow} \quad \text{iff} \quad a^{\downarrow} \subseteq b^{\downarrow}\,.$$

It is also worth mentioning that, in any of the approaches above, apart from the "global" reflexive closure, we can also consider"partial" reflexive closures (just on the objects or just on the attributes), and study its potential relationship with the object- or attributed-oriented versions of FCA.

7 Conclusions and Future Work

A notion of relational Galois connection between transitive digraphs has been introduced. There is a number of possible extensions to a relational setting, and choosing the most adequate one requires a trade-off between generality and preservation of properties from the standard framework. In this work, we have shown the convenience of using the Smyth powerset in the definition of relational Galois connection since it is in this case where both compositions generate closure relations, and we have also shown that the provided generalization is a proper one, in that it cannot be reproduced from the standard notion of Galois connection between powersets. Moreover, some alternative characterisations have been obtained and the new notion has been related to similar approaches existing in the literature.

Concerning future work, on the one hand, this generalization enables a further step towards our aim of obtaining a fully operative notion of relational Galois connection and, hence, initiates the search for a characterization of existence of residual to a given relation between T-digraphs (a problem which should be studied later in a fuzzy setting given a fuzzy relation between unbalanced fuzzy structures); on the other hand, it also enables a new approach to FCA, especially after further development of the constructions given in Sect. 6. Furthermore, this approach could also have implications in further advances in the study of generalized Chu correspondences, which will pave the way to use the approach given in [15] to analyze more structures related to quantum logics, such as those in [1,16].

References

1. Abramsky, S.: Big toy models: representing physical systems as Chu spaces. Synthese **186**(3), 697–718 (2012)
2. Antoni, L., Krajči, S., Krídlo, O.: Representation of fuzzy subsets by Galois connections. Fuzzy Sets Syst. **326**, 52–68 (2017)
3. Cabrera, I.P., Cordero, P., García-Pardo, F., Ojeda-Aciego, M., De Baets, B.: On the construction of adjunctions between a fuzzy preposet and an unstructured set. Fuzzy Sets Syst. **320**, 81–92 (2017)
4. Cabrera, I.P., Cordero, P., García-Pardo, F., Ojeda-Aciego, M., De Baets, B.: Adjunctions between a fuzzy preposet and an unstructured set with underlying fuzzy equivalence relations. IEEE Trans. Fuzzy Syst. **26**(3), 1274–1287 (2018)

5. Cabrera, I.P., Cordero, P., Ojeda-Aciego, M.: Relation-based Galois connections: towards the residual of a relation. In: Proceedings of Computational and Mathematical Methods in Science and Engineering (CMMSE 2017) (2017)
6. Denniston, J.T., Melton, A., Rodabaugh, S.E.: Formal contexts, formal concept analysis, and Galois connections. Electr. Proc. Theor. Comput. Sci. **129**, 105–120 (2013)
7. Domenach, F., Leclerc, B.: Biclosed binary relations and Galois connections. Order **18**(1), 89–104 (2001)
8. Ganter, B., Wille, R.: Formal Concept Analysis. Springer, Heidelberg (1999). https://doi.org/10.1007/978-3-642-59830-2
9. Ganter, B.: Relational Galois connections. In: Kuznetsov, S.O., Schmidt, S. (eds.) ICFCA 2007. LNCS (LNAI), vol. 4390, pp. 1–17. Springer, Heidelberg (2007). https://doi.org/10.1007/978-3-540-70901-5_1
10. García-Pardo, F., Cabrera, I.P., Cordero, P., Ojeda-Aciego, M.: On Galois connections and soft computing. In: Rojas, I., Joya, G., Cabestany, J. (eds.) IWANN 2013. LNCS, vol. 7903, pp. 224–235. Springer, Heidelberg (2013). https://doi.org/10.1007/978-3-642-38682-4_26
11. García-Pardo, F., Cabrera, I.P., Cordero, P., Ojeda-Aciego, M.: On closure systems and adjunctions between fuzzy preordered sets. In: Baixeries, J., Sacarea, C., Ojeda-Aciego, M. (eds.) ICFCA 2015. LNCS (LNAI), vol. 9113, pp. 114–127. Springer, Cham (2015). https://doi.org/10.1007/978-3-319-19545-2_7
12. García-Pardo, F., Cabrera, I.P., Cordero, P., Ojeda-Aciego, M., Rodríguez-Sanchez, F.J.: On the existence of isotone Galois connections between preorders. In: Glodeanu, C.V., Kaytoue, M., Sacarea, C. (eds.) ICFCA 2014. LNCS (LNAI), vol. 8478, pp. 67–79. Springer, Cham (2014). https://doi.org/10.1007/978-3-319-07248-7_6
13. García, F., Cabrera, I.P., Cordero, P., Ojeda-Aciego, M., Rodriguez, F.: On the definition of suitable orderings to generate adjunctions over an unstructured codomain. Inf. Sci. **286**, 173–187 (2014)
14. Jeřábek, E.: Galois connection for multiple-output operations. Algebra Univers. **79**, 17 (2018)
15. Krídlo, O., Ojeda-Aciego, M.: Formal concept analysis and structures underlying quantum logics. In: Medina, J., et al. (eds.) IPMU 2018. CCIS, vol. 853, pp. 574–584. Springer, Cham (2018). https://doi.org/10.1007/978-3-319-91473-2_49
16. Krídlo, O., Ojeda-Aciego, M.: Relating Hilbert-Chu correspondences and big toy models for quantum mechanics. In: Kóczy, L., Medina-Moreno, J., Ramírez-Poussa, E., Šostak, A. (eds.) Computational Intelligence and Mathematics for Tackling Complex Problems. SCI, vol. 819, pp. 75–80. Springer, Cham (2019). https://doi.org/10.1007/978-3-030-16024-1_10
17. Ore, O.: Galois connexions. Trans. Am. Math. Soc. **55**(3), 493–513 (1944)
18. Wille, R.: Subdirect product constructions of concept lattices. Discrete Math. **63**, 305–313 (1987)
19. Xia, W.: Morphismen als formale Begriffe-Darstellung und Erzeugung. Verlag Shaker, Aachen (1993)
20. Zhang, G.-Q.: Logic of Domains. Springer, Boston (1991). https://doi.org/10.1007/978-1-4612-0445-9

Short Papers

Sampling Representation Contexts with Attribute Exploration

Victor Codocedo[1(✉)], Jaume Baixeries[3], Mehdi Kaytoue[2], and Amedeo Napoli[4]

[1] Departamento de Informática, Universidad Técnica Federico Santa María, Campus San Joaquín, Santiago, Chile
victor.codocedo@inf.utfsm.cl
[2] Université de Lyon, CNRS, INSA-Lyon, LIRIS, UMR5205, 69621 Lyon, France
[3] Computer Science Department, Universitat Politècnica de Catalunya, 08032 Barcelona, Catalonia, Spain
[4] Université de Lorraine, CNRS, Inria, LORIA, 54000 Nancy, France

Abstract. We tackle the problem of constructing the representation context of a pattern structure. First, we present a naive algorithm that computes the representation context a pattern structure. Then, we add a sampling technique in order to reduce the size of the output. We show that these techniques reduce significantly the size of the representation context for interval pattern structures.

1 Introduction

Formal Concept Analysis (FCA) has dealt with non binary data mainly in two different ways: either by *scaling* [6] the original dataset into a formal context or through the use **pattern structures** [7,10]. The former is known to induce large formal contexts which are difficult to handle. The latter is well preferred for a wide variety of applications [8].

Any pattern structure can be represented by an equivalent formal context, which is consequently called a **representation context** (RC) [3,7]. An RC and a pattern structure contain the same set of extents. In this short paper we propose a general approach to build an RC for a given pattern structure using a sampling strategy based on attribute exploration. Our proposition aims at calculating a reduced RC as well as reducing the computational complexity of calculating pattern structures with high dimensional object representations.

2 Notation and Theoretical Background

In the following we introduce some definitions needed for the development of this article. The notations used are based on [6]. A formal context $(\mathtt{G}, \mathtt{M}, \mathtt{I})$ is a triple where $\mathtt{G}$ is a set of objects, $\mathtt{M}$ is a set of attributes and $\mathtt{I} \subseteq \mathtt{G} \times \mathtt{M}$ is an incidence relation with $(g, m) \in \mathtt{I}$ denoting "*object g has the attribute m*". The derivation operators are denoted as $' : \wp(\mathtt{G}) \rightarrow \wp(\mathtt{M})$ and $' : \wp(\mathtt{M}) \rightarrow \wp(\mathtt{G})$. A many-valued

D. Cristea et al. (Eds.): ICFCA 2019, LNAI 11511, pp. 307–314, 2019.
https://doi.org/10.1007/978-3-030-21462-3_20

context $\mathcal{M} = (\mathtt{G}, \mathtt{N}, \mathtt{W}, \mathtt{J})$ is a data table where, in addition to G and M, we define a set of values $\mathtt{W}^m$ for each attribute $m \in \mathtt{M}$ (where $\mathtt{W} = \bigcup \mathtt{W}^m$ for all $m \in \mathtt{M}$) such that $m(g) = w$ denotes that *"the value of the attribute m for the object g is w"* (with $w \in \mathtt{W}^m$). Additionally, in this document we will consider each $\mathtt{W}^m$ as an ordered set where $\mathtt{W}^m_i$ denotes the i-th element in the set.

The pattern structure framework is a generalization of FCA [7] where objects are described by *complex* representations. A pattern structure is a triple $(\mathtt{G}, (\mathtt{D}, \sqcap), \delta)$ where G is a set of objects, $(\mathtt{D}, \sqcap)$ is a semi-lattice of complex object representations, δ is a function that assigns to each object in G a representation in D, and with $\mathtt{D}_\delta = \{d \in \mathtt{D} \mid \exists (X \subseteq \mathtt{G}) \sqcap_{g \in X} \delta(g) = d\}$, $(\mathtt{D}_\delta, \sqcap)$ is a complete subsemilattice of $(\mathtt{D}, \sqcap)$. The derivation operators for a pattern structure will be denoted as $(\cdot)^\square : \wp(\mathtt{G}) \rightarrow \mathtt{D}$ and $(\cdot)^\diamond : \mathtt{D} \rightarrow \wp(\mathtt{G})$. A pattern structure can be represented with a formal context, as the next proposition shows:

Proposition 1 ([7]). *Let* $(\mathtt{G}, (\mathtt{D}, \sqcap), \delta)$ *be a pattern structure, and let* $(\mathtt{G}, \mathtt{M}, \mathtt{I})$ *be a formal context such that* $\mathtt{M} \subseteq \mathtt{D}$ *and* $(g, m) \in \mathtt{I} \iff m \sqsubseteq \delta(g)$.

If every element of $(\mathtt{D}_\delta, \sqcap)$ *is of the form*

$$\bigsqcup X = \bigsqcap \{d \in \mathtt{D}_\delta \mid (\forall x \in X) x \sqsubseteq d\}$$

for some $X \subseteq \mathtt{M}$ *(i.e.* M *is* $\bigsqcup$*-dense in* $(\mathtt{D}_\delta, \sqcap)$*), then* $(\mathtt{G}, \mathtt{M}, \mathtt{I})$ *is a* ***representation context*** *(RC) of* $(\mathtt{G}, (\mathtt{D}, \sqcap), \delta)$ *and* $(\mathtt{D}_\delta, \sqcap, \sqcup)$ *is a complete lattice.*

The condition that M is $\bigsqcup$-dense in $(\mathtt{D}_\delta, \sqcap)$ means that any element in $\mathtt{D}_\delta$ can be represented (uniquely) as the *meet* of the filters of a set of descriptions $X \subseteq \mathtt{M}$ in $(\mathtt{D}_\delta, \sqcap)$. RCs yield concept lattices that are isomorphic to the pattern concept lattices of their corresponding pattern structures. For any pattern-intent $d \in \mathtt{D}_\delta$ in the pattern structure we have an intent $X \subseteq \mathtt{M}$ in the representation context such that $d = \bigsqcup X$ and $X = \downarrow d \cap \mathtt{M}$ where $\downarrow d$ is the ideal of d in $(\mathtt{D}, \sqcap)$.

The details on interval pattern structures can be found in [9], but we recall the following definitions. Given a many-valued context $(\mathtt{G}, \mathtt{M}, \mathtt{W}, \mathtt{I})$, we have:

$$\begin{aligned}
\delta(g) &= \langle [\ m_i(g), m_i(g)\ ] \rangle_{i \in [1..|\mathtt{M}|]} \\
\delta(g_1) \sqcap \delta(g_2) &= \langle [\ min\{m_i(g_1), m_i(g_2)\}, max\{m_i(g_1), m_i(g_2)\}\ ] \rangle_{i \in [1..|\mathtt{M}|]} \\
A \subseteq \mathtt{G}\ ;\ A^\square &= \langle [\ min\{m_i(g) \mid g \in A\}, max\{m_i(g) \mid g \in A\}\ ] \rangle_{i \in [1..|\mathtt{M}|]} \\
d \in \mathtt{D}\ ;\ d^\diamond &= \{g \in \mathtt{G} \mid d \sqsubseteq \delta(g)\}
\end{aligned}$$

3 Building a Simple Representation Context

Algorithm 1 shows a first approach to build an RC for a pattern structure $(\mathtt{G}, (\mathtt{D}, \sqcap), \delta)$ with implementations for the derivation operators $(\cdot)^\square$ and $(\cdot)^\diamond$, given a many-valued context $\mathcal{M} = (\mathtt{G}, \mathtt{N}, \mathtt{W}, \mathtt{J})$. To distinguish the attributes in $\mathcal{M}$ from those in the RC created by Algorithm 1, we will refer to N as a set of *columns* in $\mathcal{M}$. Algorithm 1 is based on the NextClosure algorithm [5] for calculating intents. Actually, it only differs in lines 12, 13, 14 (marked with an asterisk).

Algorithm 1 starts by building the RC $\mathcal{K}$: the set of objects is the same as in the pattern structure, and the set of attributes and the incidence relation are initially empty. Line 12 checks whether a set of objects in the RC is an extent in the pattern structure. If this is the case, the algorithm continues executing NextClosure. Otherwise, the algorithm adds to the RC a new attribute corresponding to the pattern associated to the mismatching closure. It also adds to the incidence relation the pairs *object-attribute* as defined in line 14. Line 24 outputs the calculated RC. In what follows, we show that Algorithm 1 calculates a proper RC for the pattern structure defined over $\mathcal{M}$.

Proposition 2.
Algorithm 1 computes an RC $(\mathtt{G}, \mathtt{M}, \mathtt{I})$ *of* $(\mathtt{G}, (\mathtt{D}, \sqcap), \delta)$.

Proof. We show that $(\mathtt{G}, \mathtt{M}, \mathtt{I})$ meets the conditions in Proposition 1. Similarly to NextClosure, Algorithm 1 enumerates all closures (given an arbitrary closure operator) in lectical order. However, Algorithm 1 uses two different closure operators, namely the standard closure operator of FCA defined over the RC under construction $(\cdot)''$, and the one defined by both derivation operators over the pattern structure, $(\cdot)^{\Box\diamond}$. Both closure operators are made to coincide by the new instructions in the algorithm. When this is not the case (this is $B'' \neq B^{\Box\diamond}$ for a given $B \subseteq \mathtt{G}$) a new attribute is added to the RC in the shape of $B^{\Box}$. Additionally, the pair $(h, B^{\Box})$ is added to the incidence relation set of the RC for all objects $h \in B$. This in turn ensures that $B'' = B^{\Box\diamond}$ in the modified RC.

A consequence of $B'' = B^{\Box\diamond}$ is that the set of extents in the RC is the same as the set of extents in the pattern structure. This also means that there is a one-to-one correspondence between the intents in the RC and the patterns in the pattern structure, i.e. for any extent $B'' = B^{\Box\diamond}$, the intent $B''' = B'$ corresponds to the pattern $B^{\Box\diamond\Box} = B^{\Box}$ ($B''' = B'$ and $B^{\Box\diamond\Box} = B^{\Box}$ are properties of the derivation operators [6]). Thus, we have that any element in $\mathtt{D}_\delta$ (which can be represented as $B^{\Box}$ for an arbitrary $B \subseteq \mathtt{G}$) is of the form $\bigsqcup B'$ or $\bigsqcap\{d \in \mathtt{D}_\delta \mid (\forall m \in B')m \sqsubseteq d\}$. Consequently, $\mathtt{M}$ is $\bigsqcup$-dense in $(\mathtt{D}_\delta, \sqcap)$ and $(\mathtt{G}, \mathtt{M}, \mathtt{I})$ is an RC of the pattern structure $(\mathtt{G}, (\mathtt{D}, \sqcap), \delta)$. $\Box$

We can consider the extreme case when the generated RC contains one attribute per pattern in the pattern structure to estimate the complexity of Algorithm 1. Taking from [5], we know that the polynomial delay of the base algorithm NextClosure is $\mathcal{O}(|\mathtt{G}|^2|\mathtt{M}|)$ (in here $\mathtt{G}$ and $\mathtt{M}$ are inverted as we are enumerating extents) which decomposes in the main loop of the algorithm (at most $|\mathtt{G}|$ repetitions), and the number of calculations taken by the closure operation in the formal context ($|\mathtt{G}||\mathtt{M}|$). Additionally, we need to consider the calculation of the closure operation over the pattern structure in line 12 of the algorithm which strongly depends on the nature of the pattern structure and its implementation. For example, for interval pattern structures, the closure takes $|\mathtt{G}||\mathtt{N}|$ operations. For such a case, the overall complexity is $\mathcal{O}(|\mathtt{G}|^2 \operatorname{argmax}(|\mathtt{N}|, |\mathtt{M}|))$.

Since we know $|\mathtt{N}|$, we need to characterize the size of $|\mathtt{M}|$. As previously stated, $|\mathtt{M}|$ may be as large as the size of patterns in $\mathtt{D}_\delta$, which would imply

Algorithm 1. Naive Representation Context Calculation

```
procedure RepresentationContextNaive(𝓜, (·)□, (·)◇)            ▷ 𝓜 = (G, N, W, J)
    M ← ∅
    I ← ∅
    𝒦 ← (G, M, I)
    A ← ∅
    while A ≠ G do
        for g ∈ G in reverse order do
            if g ∈ A then
                A ← A \ {g}
            else
                B ← A ∪ {g}
                if B'' ≠ B□◇ then                                ▷ (*)
                    M ← M ∪ {B□}                                 ▷ (*)
                    I ← I ∪ {(h, B□) | h ∈ B□◇}                  ▷ (*)
                if B''\A contains no element < g then
                    A ← B''
                    Exit For
        Output A
    return 𝒦
end procedure
```

a non-polynomial delay of the algorithm (as $|\mathtt{D}_\delta|$ may grow up to $2^{|\mathtt{G}|}$). Differently, there may be cases when $|\mathtt{M}|$ is much smaller than $|\mathtt{D}_\delta|$. If by some happy accident the first patterns included in M correspond exactly to the set of join-irreducible patterns (JIPs) in $|\mathtt{D}_\delta|$, then we can expect that this would be the case. Particularly, when $|\mathtt{D}_\delta| = 2^{|\mathtt{G}|}$, we have that the number of JIPs in $\mathtt{D}_\delta$ is $|\mathtt{G}|$. Unfortunately, as stated in [6], determining the maximal number of JIPs for a given number of objects is difficult. An asymptotic upper bound is described in [4] as a factor of $\binom{|\mathtt{G}|}{[|\mathtt{G}|/2]}$ which is still better than $2^{|\mathtt{G}|}$. Experience shows that for real-life datasets this number is much smaller, however to the best of our knowledge an actual study on this subject has not been performed yet.

Regardless, let us consider that by some clever mechanism Algorithm 1 is always able to add JIPs of the pattern structure to the RC. If this would be the case, any other pattern (this is, join-reducible pattern) could be represented as the join of a set of attributes in the RC (using Proposition 1), provided that these JIPs have been previously included in M. The problem with this is that we cannot be certain when this last condition is met. This is, given a set of objects $B \subseteq \mathtt{M}$ and its closure in the RC B'', we cannot be sure whether B'' is an extent in the pattern structure until the algorithm has finished. The single exception is when $B = B''$ as we show in Proposition 3.

Proposition 3.
Let $(\mathtt{G}, \mathtt{M}, \mathtt{I})$ *be the partial RC calculated by Algorithm 1 for the pattern structure* $(\mathtt{G}, (\mathtt{D}, \sqcap), \delta)$. *Given a set of objects* $B \subseteq \mathtt{G}$, *at any point during the execution of Algorithm 1 we have that*

$$B = B'' \implies B = B^{\square\diamond}$$

Proof. Since a closure operator is extensive we have that $B \subseteq B''$ and $B \subseteq B^{\square\diamond}$. It suffices to show that $B \subseteq B^{\square\diamond} \subseteq B''$ at any point during the execution of

Algorithm 2. A General Representation Context Calculation

```
procedure RepresentationContext(𝓜, (·)□, (·)◇)          ▷ 𝓜 = (G, N, W, J)
    M ← ∅
    I ← ∅
    𝒦 ← (G, M, I)
    A ← ∅
    while A ≠ G do
        for g ∈ G in reverse order do
            if g ∈ A then
                A ← A \ {g}
            else
                B ← A ∪ {g}
                if B ≠ B'' then
                    while B'' ≠ B□◇ do                              ▷ (*)
                        X ← Sample(B, B'', 𝓜)                        ▷ (*)
                        M ← M ∪ {m_X}              ▷ (m_X is a new attribute)
                        I ← I ∪ {(h, m_X) | h ∈ X}                   ▷ (*)
                if B''\A contains no element < g then
                    A ← B''
                    Exit For
        Output A
    return 𝒦
end procedure
```

Algorithm 1. This is a consequence of the lectical enumeration performed by the base algorithm NextClosure which ensures that whenever we calculate B'', we have already calculated and verified all closures $C'' = C^{\Box\diamond}$ with $C'' \subseteq B$. Then, B'' can be characterized as $C'' \cap g''$ for some $C'' \subseteq B$ and $g \in \mathtt{G} \backslash C''$. Simultaneously, $B^{\Box\diamond}$ can be characterized as $C^{\Box\diamond} \cap \delta(g)^{\diamond}$ for the same set C and object g. Because $C'' = C^{\Box\diamond}$, we need to show that $\delta(g)^{\diamond} \subseteq g''$. If we consider that $g' = \{X^{\Box} \in \mathtt{M} \mid g \in X^{\Box\diamond}\}$ (line 14 of Algorithm 1), it follows that $(\forall X^{\Box} \in g')X^{\Box} \sqsubseteq \delta(g) \implies \delta(g)^{\diamond} \subseteq X^{\Box\diamond}$. Additionally, g'' can be also characterized as $\bigcap_{X^{\Box} \in g'} X^{\Box\diamond}$, then we have that $\delta(g)^{\diamond} \subseteq g''$. □

Proposition 3 tells us that for a given set $B \subseteq \mathtt{G}$ in the lectical enumeration, if we have that $B = B''$, it follows that this is true in the pattern structure without need of verification of $B^{\Box\diamond}$ This is useful if the calculation of B'' in the RC is computationally cheaper than the calculation of $B^{\Box\diamond}$ in the pattern structure. A corollary of Proposition 3 is that $g \notin B'' \implies g \notin B^{\Box\diamond}$, meaning that B'' actually behaves as an estimation of $B^{\Box\diamond}$ which gets refined the more attributes we add to the RC.

Our proposition provides a reduction in the complexity of calculating extents in a pattern structure whenever the size of $\mathtt{M}$ is smaller than the number of *columns* in the many-valued formal context (i.e. $|\mathtt{M}| < |\mathtt{N}|$). This is true when the algorithm begins execution. Moreover, the algorithm can keep track on this relation allowing for an adaptive strategy, i.e. falling back to NextClosure whenever the use of the proposed strategy becomes pointless.

4 Computing Smaller Representation Contexts

Algorithm 2 shows a better implementation of Algorithm 1 considering the results discussed in the previous section. It differs slightly from Algorithm 1

Algorithm 3. An interval pattern extent sampler algorithm

```
procedure SAMPLE(B, B'', M)                                   ▷ M = (G, N, W, J)
    found ← False
    states ← list of size |N|
    for n ∈ N do
        side ← pick randomly from {0, 1}
        states[n] ← (side, 0, |W_n|)
    while not found do
        n ← pick randomly from N
        side, i, j ← states[n]
        a = i + side
        b = j + (side − 1)
        if a < b then
            X ← {g ∈ G | W^n_a ≤ n(g) ≤ W^n_b}
            side ← not side
            if B ⊆ X then
                i, j ← a, b
                if B'' ⊈ X then
                    found ← True
            states[n] = (side, i, j)
    return X
end procedure
```

in those lines marked with an asterisk. Instead of adding a single attribute per set of objects whenever $B'' \neq B^{\Box\diamond}$, it keeps on asking for *samples* until both closures coincide. Each *sample* corresponds to a set of objects $X \subseteq \mathtt{G}$ s.t. $X^{\Box} \sqsubseteq B^{\Box}$ and $(\exists g \in B''\backslash B^{\Box\diamond}) X^{\Box} \not\sqsubseteq \delta(g)$. Notice that if there is no such $g \in B''$, then necessarily $B'' = B^{\Box\diamond}$. Furthermore, when $X^{\Box} = B^{\Box}$, we fall back to Algorithm 1.

Ideally, $X^{\Box}$ would be a join-irreducible pattern. However, we cannot be certain this is the case until we have calculated the entire set of patterns in the pattern structure which is exactly what we would like to avoid. For this reason, we simply require that the set X is as large as possible. This increases it chances that it may be a join-irreducible pattern.

Sampling strongly depends to the nature of the pattern structure. Algorithm 3 presents a simple sampling technique designed for interval pattern structures (IPS). We have chosen IPS to illustrate our approach for two reasons. Firstly, since they work directly on numerical data, they are prone to suffer from the curse of dimensionality. Secondly, because of their definition, IPS lattices are very large and their corresponding RCs usually contain contranominal scales.

For the sake of brevity we do not provide a full explanation on the inner working of Algorithm 3. We simply mention that, given a set of objects B and its *estimated* closure B'', it works by using the largest possible convex region in the description space and reducing it in one of its dimensions picked randomly. Objects within the reduced convex region are assigned to a set X. If $B \subseteq X$ and $B'' \not\subseteq X$, then X is retrieved as a sample (notice that a pre-condition of Algorithm 3 given by Algorithm 2 is that $B \neq B''$). Algorithm 3 provides an answer in $\mathcal{O}(|\mathtt{G}|^2|\mathtt{N}|)$.

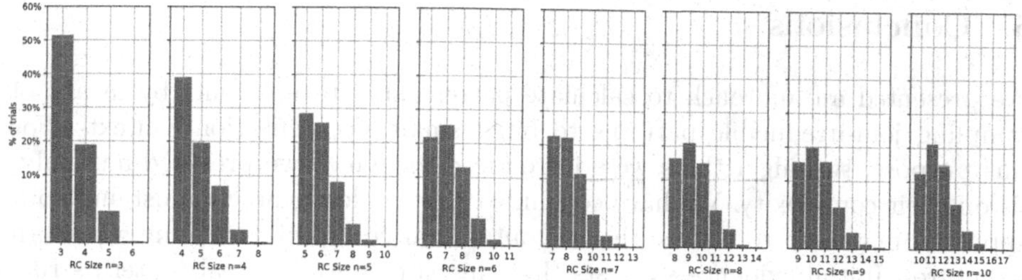

Fig. 1. Distribution on the number of attributes sampled for one thousand RCs. The x-axis has the number of attributes sampled. The y-axis represents the proportion of RCs in the total number of trials.

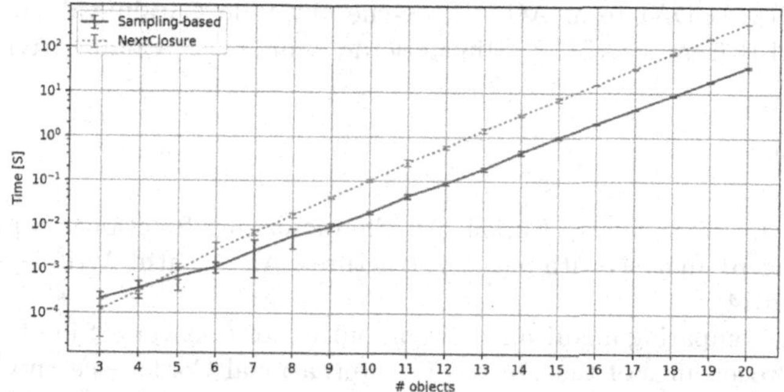

Fig. 2. Execution times for Algorithms 2 and 3. The x-axis is the number of objects generated. The y-axis (time) is in logarithmic scale.

4.1 Experiments in Synthetic Data

To illustrate how our approach samples small RCs, we performed a simple experiment. We created several synthetic many-valued contexts with objects ranging from 2 to 10 such that an IPS defined over them would yield a Boolean lattice. An RC defined naively over these IPS would contain $2^{|\mathtt{G}|}$ attributes, however an RC containing only JIPs (a contranominal scale) would contain only $|\mathtt{G}|$ attributes. Figure 1 shows the distribution on the number of attributes sampled over 1000 executions of Algorithms 2 and 3 over these datasets. From the figure, we can observe that very small RCs are sampled with very high probability. Moreover, for some many-valued context sizes, the RCs containing only JIPs are sampled with the highest probability. Figure 2 shows a comparison on running times between standard NextClosure and our approach averaged over 100 runs over many-valued contexts with the same characteristics mentioned above, with a number of objects ranging from 3 to 20. While both techniques show an exponential growth in the running time w.r.t. the number of columns of the many-valued context (as expected), our approach shows a reduction in an order of magnitude over the NextClosure baseline.

5 Conclusions

We presented an approach to calculate pattern structure extents by means of sampling join-irreducible patterns to build small representation contexts. For this purpose, two algorithms were introduced for which we provided an analysis on their complexity. We have concluded that under some circumstances our approach may help to reduce the computational complexity of mining pattern structures. Initial evidence suggests that this is true for interval pattern structures, although more research is necessary in order to claim this is true also in real-world data.

Acknowledgments. This research work has been supported by the recognition of 2017SGR-856 (MACDA) from AGAUR (Generalitat de Catalunya), and the grant TIN2017-89244-R from MINECO (Ministerio de Economía y Competitividad).

References

1. Baixeries, J., Kaytoue, M., Napoli, A.: Characterizing functional dependencies in formal concept analysis with pattern structures. Ann. Math. Artif. Intell. **72**(1–2), 129–149 (2014)
2. Bazin, A.: Comparing algorithms for computing lower covers of implication-closed sets. In: Proceedings of the Thirteenth International Conference on Concept Lattices and Their Applications, Moscow, Russia, 18–22 July 2016, pp. 21–31 (2016)
3. Buzmakov, A.: Formal concept analysis and pattern structures for mining structured data. Ph.D. thesis, University of Lorraine, Nancy, France (2015)
4. Erdös, P., Kleitman, D.J.: Extremal problems among subsets of a set. Discrete Math. **306**(10–11), 923–931 (2006)
5. Ganter, B., Obiedkov, S.: Conceptual Exploration. Springer, Berlin (2016). https://doi.org/10.1007/978-3-662-49291-8
6. Ganter, B., Wille, R.: Formal Concept Analysis. Springer, Berlin (1999). https://doi.org/10.1007/978-3-642-59830-2
7. Ganter, B., Kuznetsov, S.O.: Pattern structures and their projections. In: Delugach, H.S., Stumme, G. (eds.) ICCS-ConceptStruct 2001. LNCS (LNAI), vol. 2120, pp. 129–142. Springer, Heidelberg (2001). https://doi.org/10.1007/3-540-44583-8_10
8. Kaytoue, M., Codocedo, V., Buzmakov, A., Baixeries, J., Kuznetsov, S.O., Napoli, A.: Pattern structures and concept lattices for data mining and knowledge processing. In: Bifet, A., et al. (eds.) ECML PKDD 2015. LNCS (LNAI), vol. 9286, pp. 227–231. Springer, Cham (2015). https://doi.org/10.1007/978-3-319-23461-8_19
9. Kaytoue, M., Kuznetsov, S.O., Napoli, A.: Revisiting numerical pattern mining with formal concept analysis. In: Walsh, T. (ed.) Proceedings of the 22nd International Joint Conference on Artificial Intelligence (IJCAI 2011), pp. 1342–1347. IJCAI/AAAI (2011)
10. Kuznetsov, S.O.: Pattern structures for analyzing complex data. In: Sakai, H., Chakraborty, M.K., Hassanien, A.E., Ślęzak, D., Zhu, W. (eds.) RSFDGrC 2009. LNCS (LNAI), vol. 5908, pp. 33–44. Springer, Heidelberg (2009). https://doi.org/10.1007/978-3-642-10646-0_4

Discovering Implicational Knowledge in Wikidata

Tom Hanika[1,2], Maximilian Marx[3(✉)], and Gerd Stumme[1,2]

[1] Knowledge and Data Engineering Group, University of Kassel, Kassel, Germany
{tom.hanika,stumme}@cs.uni-kassel.de
[2] ITeG, University of Kassel, Kassel, Germany
[3] Knowledge-Based Systems Group, TU Dresden, Dresden, Germany
maximilian.marx@tu-dresden.de

Abstract. Knowledge graphs have recently become the state-of-the-art tool for representing the diverse and complex knowledge of the world. Among the freely available knowledge graphs, Wikidata stands out by being collaboratively edited and curated. Among the vast numbers of facts, complex knowledge is just waiting to be discovered, but the sheer size of Wikidata makes this infeasible for human editors. We apply Formal Concept Analysis to efficiently identify and succinctly represent comprehensible implications that are implicitly present in the data. As a first step, we describe a systematic process to extract conceptual knowledge from Wikidata's complex data model, thus providing a method for obtaining large real-world data sets for FCA. We conduct experiments that show the principal feasibility of the approach, yet also illuminate some of the limitations, and give examples of interesting knowledge discovered.

Keywords: Wikidata · FCA · Property dependencies · Implications

1 Introduction

The quest for the best digital structure to collect and curate knowledge has been going on since the first appearances of knowledge stores in the form of semantic networks and databases. The most recent, and arguably so far most powerful, incarnation is the *knowledge graph*, as used by corporations like Facebook, Google, Microsoft, IBM, and eBay. Among the freely available knowledge graphs, Wikidata [19,20] stands out due to its free and collaborative character: like Wikipedia, it is maintained by a community of volunteers, adding *items*, relating them using *properties* and *values*, and backing up claims with *references*. As of 2019-02-01, Wikidata has 52,373,284 items and 676,854,559 statements using a total 5,592 properties. Altogether this constitutes a gargantuan collection of factual data accessible to and freely usable by everyone.

Authors are given in alphabetical order. No priority in authorship is implied. We kindly remind the reader that a more detailed version of this paper [9] is available.

D. Cristea et al. (Eds.): ICFCA 2019, LNAI 11511, pp. 315–323, 2019.
https://doi.org/10.1007/978-3-030-21462-3_21

But Wikidata is more than just the collection of all factual knowledge stored within: Wikidata also contains *implicit* knowledge that is not explicitly stated, but rather holds merely due to certain statements being present (or absent, respectively). While such implicit knowledge can take many shapes, we focus on "rules" (propositional implications) stating that whenever an entity has statements for some given properties, it should also have statements for certain other properties. We believe that such rules can guide editors in enriching Wikidata and discuss potential uses in the full version of the paper [9].

Previous approaches have studied extracting rules in the form of implications of first-order logic (FOL) as a feasible approach to obtain interesting and relevant rules from Wikidata [5,10], but the expressive power of FOL comes with a steep price: to understand such rules, one needs to understand not only the syntax, but also advanced concepts such as quantification over variables, and it seems far-fetched to assume that the average Wikidata editor possesses such understanding. We thus propose to use rules that are conceptually and structurally simpler, and focus on extracting Horn implications of *propositional logic* (PL) from Wikidata, trading expressive power for ease of understanding and simplicity of presentation.

While Formal Concept Analysis (FCA) [7] provides techniques to extract a sound and complete basis of PL implications (from which all other implications can be inferred), applying these techniques to Wikidata is not straightforward: A first hurdle is the sheer size of Wikidata, necessitating the selection of subsets from which to extract rules. Secondly, the intricate data model of Wikidata, while providing much flexibility for expressing wildly different kinds of statements, is not particularly amenable to a uniform approach to extracting relevant information.

In this work, we tackle both issues by describing procedures (i) for extracting, in a structured fashion, implicational knowledge for arbitrary subsets of properties, and (ii) for deriving suitable sets of attributes from Wikidata statements, depending on the type of property. We provide an implementation of these procedures[1], and while incorporating the extracted rules into the editing process is out of scope for this paper, we nevertheless demonstrate that we are able to obtain meaningful and interesting rules using our approach.

2 Related Work

FCA has been applied to Wikidata in [8] to model and predict the dynamic behaviour of knowledge graphs using lattice structures, and in [12] to determine obligatory attributes for classes. Another related topic is rule mining, and several successful approaches to generating lists of FOL rules, e.g., in [5,10] have been proposed. This task is often connected to ranked lists of rules, like in [21], or to completeness investigations for knowledge graphs, as in [6,18]. Rule mining using FCA has been proposed for RDF graphs [1], but the extensive use of reification in the Wikidata RDF exports prohibits such an approach. Rudolph [16]

[1] https://github.com/mmarx/wikidata-fca.

describes a general method for deriving attributes from properties of relational structures, where the property can be expressed by a concept description in the description logic $\mathcal{FLE}$. We note that there is no such concept description capturing Problem 2.

3 Wikidata

Data Model. Wikidata [20] is the free and open Knowledge Graph of the Wikimedia foundation. In Wikidata, *statements* representing knowledge are made using *properties* that connect *entities* (either *items* or other properties) to *values*, which, depending on the property, can be either items, properties, *data values* of one of a few data types, e.g., URIs, time points, globe coordinates, or textual data, or either of the two special values *unknown value* (i.e., *some* value exists, but it is not known) and *no value* (i.e., it is known that there is no such value).

Example 1. Liz Taylor was married to Richard Burton. This fact is represented by a connection from item Q34851 ("Elizabeth Taylor") to item Q151973 ("Richard Burton") using property P26 ("spouse"). But Taylor and Burton were married twice: once from 1964 to 1974, and then from 1983 to 1984.

To represent these facts, Wikidata enriches statements by adding *qualifiers*, pairs of properties and values, opting for two "spouse" statements from Taylor to Burton with different P580 ("start time") and P582 ("end time") qualifiers.

Metadata and Implicit Structure. Each statement carries metadata: *references* track provenance of statements, and the statement *rank* can be used to deal with conflicting or changing information. Besides *normal* rank, there are also *preferred* and *deprecated* statements. When information changes, the most relevant statement is marked preferred, e.g., there are numerous statements for P1082 ("population") of Q1794 ("Frankfurt"), giving the population count at different times using the P585 ("point in time") qualifier, with the most recent estimate being preferred. Deprecated statements are used for information that is no longer valid (as opposed to simply being outdated), e.g., when the formal definition of a planet was changed by the International Astronomical Union on 2006-09-13, the statement that Q339 ("Pluto") is a Q634 ("Planet") was marked deprecated, and an P582 ("end time") qualifier with that date was added.

Example 2. We may write down these two statements in *fact notation* as follows, where qualifiers and metadata such as the statement rank are written as an *annotation* on the statement:

$$\begin{aligned}&\texttt{population}_{P1082}(\texttt{Frankfurt}_{Q1794}, 736414)@[\texttt{determination method}_{P459}\text{:}\\&\quad\texttt{estimation}_{Q965330}, \texttt{point in time}_{P585}\text{: } 2016\text{-}12\text{-}31, \texttt{rank:preferred}]\end{aligned} \tag{1}$$

$$\texttt{instance of}_{P31}(\texttt{Pluto}_{Q339}, \texttt{Planet}_{Q634})@[\texttt{end time}_{P582}\text{: 2006-09-13}, \texttt{rank: deprecated}] \quad (2)$$

Further structure is given to the knowledge in Wikidata using statements themselves: Wikidata contains a class hierarchy comprising over 100,000 *classes*, realised by the properties P31 ("instance of") (stating that an item is an *instance* of a certain class) and P279 ("subclass of"), which states some item q is a *subclass* of some other class q', i.e., that all instances of q are also instances of q'.

Formalisation. Most models of graph-like structures do not fully capture the peculiarities of Wikidata's data model. The generalised Property Graphs [15], however, have been proposed specifically to capture Wikidata, and we thus phrase our formalisation in terms of a *multi-attributed relational structure*.[2]

Definition 1. *Let $\mathcal{Q}$ be the set of Wikidata items, $\mathcal{P}$ be the set of Wikidata properties, and let $\mathcal{V}$ be the set of all possible data values. We denote by $\mathcal{E} := \mathcal{Q} \cup \mathcal{P}$ the set of all entities, and define $\Delta := \mathcal{E} \cup \mathcal{V}$. The Wikidata knowledge graph is a map $\mathcal{W}\colon \mathcal{P} \to (\mathcal{E} \times \Delta \times (\mathcal{P} \times \Delta))$ assigning to each property p a ternary relation $\mathcal{W}(p)$, where a tuple $\langle s, v, a\rangle \in \mathcal{W}(p)$ corresponds to a p-statement on s with value v and annotation a.*

Thus, $\langle \Delta, (\mathcal{W}(p))p \in \mathcal{P}\rangle$ is a *multi-attributed relational structure*, i.e., a relational structure in which every tuple is annotated with a set of pairs of attributes and annotation values. While technically stored separately on Wikidata, we will simply treat references and statement ranks as annotations on the statements. In the following, we refer to the Wikidata knowledge graph simply by $\mathcal{W}$. Furthermore, we assume that deprecated statements and the special values *unknown value* and *no value* do not occur in $\mathcal{W}$. This is done merely to avoid cluttering formulas by excluding these cases, and comes without loss of generality.

Example 3. Property P26 ("spouse") is used to model marriages in Wikidata. Among others, $\mathcal{W}(\texttt{spouse}_{P26})$ contains the two statements corresponding to the two marriages between Liz Taylor and Richard Burton from Example 1:

$$\langle \texttt{Elizabeth Taylor}_{Q34851}, \texttt{Richard Burton}_{Q151973}, \{\langle \texttt{start time}_{P580}, 1964\rangle, \langle \texttt{end time}_{P582}, 1974\rangle\}\rangle \quad (3)$$

$$\langle \texttt{Elizabeth Taylor}_{Q34851}, \texttt{Richard Burton}_{Q151973}, \{\langle \texttt{start time}_{P580}, 1983\rangle, \langle \texttt{end time}_{P582}, 1984\rangle\}\rangle \quad (4)$$

Next, we introduce some abbreviations for when we are not interested in the whole structure of the knowledge graph.

[2] This is merely a formalisation of Wikidata's actual data model (cf. https://mediawiki.org/wiki/Wikibase/DataModel), not a new model for conceptual data.

Definition 2. *Let $R \subseteq S^3$ be a ternary relation over S. For $t = \langle s, o, a \rangle \in S^3$, we denote by* $\operatorname{subj} t := s$ *the* subject *of t, by* $\operatorname{obj} t := o$ *the* object *of t, and by* $\operatorname{ann} t := a$ *the* annotation *of t, respectively. These extend to R in the natural fashion:* $\operatorname{subj} R := \{\operatorname{subj} t \mid t \in R\}$, $\operatorname{obj} R := \{\operatorname{obj} t \mid t \in R\}$, *and* $\operatorname{ann} R := \{\operatorname{ann} t \mid t \in R\}$, *respectively. We indicate with ˆ that a property is incident with an item as object: $\mathcal{W}(\hat{}\,\mathtt{spouse}_{P26})$ contains $\langle \mathtt{Richard\ Burton}_{Q151973}, \mathtt{Elizabeth\ Taylor}_{Q34851}, \{\langle \mathtt{start\ time}_{P580}, 1964\rangle, \langle \mathtt{end\ time}_{P582}, 1974\rangle\}\rangle$.*

4 Formal Concept Analysis

We assume familiarity with the basic notions from FCA and kindly refer the reader to the full version of the paper [9], and to [7] for an introduction to FCA.

5 Property Theory

We now describe how to harness FCA to obtain a more accessible view of the Wikidata knowledge graph and how the properties therein depend on each other. Krötzsch [11] argues that knowledge graphs are primarily characterised by three properties: (i) normalised storage of information in small units, (ii) representation of knowledge through the connections between these units, and (iii) enrichment of the data with contextual knowledge. In Wikidata, properties serve both as a mechanism to relate entities to one another, as well as to provide contextual information on statements through their use as qualifiers. Taking the structure and usage of properties into account is thus crucial to any attempt of extracting structured information from Wikidata. We now introduce the two most basic problem scenarios for selecting sets of properties from Wikidata. We outline several other approaches to exploiting different aspects of Wikidata's rich data model, but defer a detailed discussion to the full version of the paper [9].

5.1 Plain Incidence

We start by constructing the formal context that has a chosen set $\hat{\mathcal{P}} \subseteq \mathcal{P}$ as its attribute set and the entity set $\mathcal{E}$ as the object set.

Problem 1. Given the Wikidata knowledge graph $\mathcal{W}$ and some subset $\hat{\mathcal{P}} \subseteq \mathcal{P}$, compute the canonical base for the implicational theory $Th_{\hat{\mathcal{P}}}(\mathcal{E}, \hat{\mathcal{P}}, I^{\mathtt{plain}})$, where

$$\langle e, \hat{p} \rangle \in I^{\mathtt{plain}} :\Longleftrightarrow e \in \operatorname{subj} \mathcal{W}(\hat{p}), \text{ i.e.,} \tag{5}$$

an entity e coincides with property $\hat{p}$ iff it occurs as a subject in some $\hat{p}$-statement.

Although this is the most basic problem we present, with growing $\hat{\mathcal{P}}$ it may quickly become computationally infeasible, cf. Sect. 6. More importantly, however, entities occurring as objects are not taken into account: almost half of the data in the knowledge graph is ignored, motivating the next definition.

5.2 Directed Incidence

We endue the set of properties $\mathcal{P}$ with two colours {subj, obj} signifying whether an entity coincides with the property as subject or as object in some statement.

Problem 2. Given $\mathcal{W}$ and some set $\hat{\mathcal{P}} \subseteq \mathcal{P} \times \{\text{subj}, \text{obj}\}$ of directed properties, compute the canonical base for $Th_{\hat{\mathcal{P}}}(\mathcal{E}, \hat{\mathcal{P}}, I^{\mathtt{dir}})$, where an entity e coincides with $\hat{p}$ iff it occurs as subject or object (depending on the colour) of some p-statement:

$$\langle e, \hat{p} \rangle \in I^{\mathtt{dir}} :\Longleftrightarrow \big(\hat{p} = \langle p, \text{subj} \rangle \wedge e \in \text{subj}\,\mathcal{W}(p)\big) \\ \vee \big(\hat{p} = \langle p, \text{obj} \rangle \wedge e \in \text{obj}\,\mathcal{W}(p)\big). \tag{6}$$

Example 4. Let $\hat{\mathcal{P}} = \{\mathtt{\hat{}mother}_{\hat{}P25}, \mathtt{godparent}_{P1290}, \mathtt{mother}_{P25}\}$ be the set of attributes and let $\mathcal{E} = \{\mathtt{Miley\ Cyrus}_{Q4235}, \mathtt{Victoria}_{Q9439}, \mathtt{Naomi\ Watts}_{Q132616}, \mathtt{Angelina\ Jolie}_{Q13909}\}$ be the set of objects. The corresponding formal context $\langle \mathcal{E}, \hat{\mathcal{P}}, I^{\mathtt{dir}} \rangle$ (as extracted from Wikidata) is given by the following cross table:

Example	^P25 (“^mother”)	P1290 (“godparent”)	P25 (“mother”)
Q13909 (“Angelina Jolie”)	×	×	×
Q4235 (“Miley Cyrus”)		×	×
Q132616 (“Naomi Watts”)	×		×
Q9439 (“Victoria”)	×	×	×

Observe that the only valid (non-trivial) implication (and hence sole constituent of the canonical base) is $\{\} \to \{\mathtt{mother}_{P25}\}$: every entity has a mother.

Generalised Incidence. Even though Problem 2 can cope with Example 4, it still does not capture the subtleties of Example 3, as, e.g., two statements differing only in their annotations are indistinguishable, even though the meaning of, e.g., statements for P1038 (“relative”) can vary wildly among statements with different values for the P1039 (“type of kinship”) qualifier. Similarly, we might distinguish properties by the classes that objects of the property are instances of: having a P25 (“mother”) that is a Q22989102 (“Greek deity”) is significantly different from one that is merely a Q5 (“human”). In practice, we likely want to use different incidences for different properties, and we thus introduce generalised incidences. Again, we refer to the full paper [9] for a discussion of these approaches.

6 Experimental Results

We have computed canonical bases for Problems 1 and 2 on selected subsets of Wikidata, obtained by fixing a set of properties $\hat{\mathcal{P}}$ and restricting to statements involving them. We have selected the sets of properties by picking thematically related properties (as specified in Wikidata), see the full paper [9] for descriptions of methodology, data sets, and discovered rules. Already this preliminary evaluation shows several limitations: For the `family` datasets, we find the valid rule $\{\texttt{godparent}_{P1290}, \texttt{partner}_{P451}\} \to \{\texttt{sibling}_{P3373}\}$, stating that an entity with a godparent and a partner must also have a sibling, even though this need not be true in the real world; Wikidata is lacking a counterexample. Contrarily, the rule $\{\texttt{\^{}father\^{}}_{P22}, \texttt{\^{}relative\^{}}_{P1038}, \texttt{spouse}_{P26}\} \to \{\texttt{child}_{P40}\}$ witnesses that the more general $\{\texttt{\^{}father\^{}}_{P22}\} \to \{\texttt{child}_{P40}\}$ has counterexamples: indeed there are 1,634 non-fictional humans contradicting it. Besides such (unavoidable) noise and lack of completeness in the underlying data, another problem is that of computational infeasibility: computing canonical bases on state-of-the-art hardware takes several hours for subsets of one hundred properties; for larger sets (much less all of Wikidata), computation time is prohibitively long. Still, we obtain a plenitude of interesting and meaningful rules about the subject domains [9].

7 Conclusion and Outlook

We have shown how to extract, in a structured fashion, subsets of Wikidata and represent them as formal contexts. This provides a practically limitless source of real-world data to the FCA community, to which the full range of tools and techniques from FCA may be applied. Most importantly, we can obtain relevant, meaningful, and perspicuous implications valid for Wikidata, which may be useful in different ways: valid rules can highlight other entities that must be edited to avoid introducing counterexamples, and, conversely, absence of an expected rule may indicate the presence of unwanted counterexamples in the data.

Our approach is feasible for subsets of Wikidata that capture interesting domains of knowledge. For larger subsets, we propose to compute Luxenburger bases of association rules [13,17] or PAC bases [2], both of which are feasible for Wikidata as a whole. Another approach would be to employ conceptual exploration to compute canonical bases for generalised incidences over Wikidata. The missing ingredient here is a method to query Wikidata for possible counterexamples to a proposed implication, e.g., via the SPARQL endpoint, enabling Wikidata to be used as an *expert* for the exploration. Ultimately, Wikidata could be employed alongside human experts in a collaborative exploration setting to stretch the boundaries of human knowledge. Possible future work includes: (i) implementing the exploration approach, (ii) devising further incidence relations that capture aspects of the data model (e.g., aggregation for quantitative values), (iii) integrating completeness [3] tools such as COOL-WD [4] to ensure that incomplete data does not induce counterexamples, and incorporating background knowledge in the form of the MARPL rules [15] proposed for ontological reasoning on Wikidata [14].

Acknowledgements. This work is partly supported by the German Research Foundation (DFG) in CRC 248 (Perspicuous Systems), CRC 912 (HAEC), and Emmy Noether grant KR 4381/1-1 (DIAMOND).

References

1. Alam, M., et al.: Mining definitions from RDF annotations using formal concept analysis. In: Yang, Q., Wooldridge, M. (eds.) Proceedings of the 24th International Joint Conference on Artificial Intelligence (IJCAI 2015). AAAI Press (2015)
2. Borchmann, D., Hanika, T., Obiedkov, S.: On the usability of probably approximately correct implication bases. In: Bertet, K., Borchmann, D., Cellier, P., Ferré, S. (eds.) ICFCA 2017. LNCS (LNAI), vol. 10308, pp. 72–88. Springer, Cham (2017). https://doi.org/10.1007/978-3-319-59271-8_5
3. Darari, F., et al.: Completeness management for RDF data sources. In: TWEB 12.3, pp. 18:1–18:53 (2018)
4. Darari, F., et al.: COOL-WD: a completeness tool for Wikidata. In: Nikitina, N., et al. (eds.) Proceedings of the 16th International Semantic Web Conference (ISWC 2017): Posters & Demonstrations and Industry Tracks, vol. 1963. CEUR WS Proceedings (2017). CEUR-WS.org
5. Galárraga, L., et al.: Fast rule mining in ontological knowledge bases with AMIE++. VLDB J. **24**(6), 707–730 (2015)
6. Galárraga, L., et al.: Predicting completeness in knowledge bases. In: Proceedings of the 10th International Conference on Web Search and Data Mining (WSDM 2017). ACM (2017)
7. Ganter, B., Wille, R.: Formal Concept Analysis. Springer, Heidelberg (1999). https://doi.org/10.1007/978-3-642-59830-2
8. González, L., Hogan, A.: Modelling dynamics in semantic web knowledge graphs with formal concept analysis. In: Champin, P., et al. (eds.) Proceedings of the 2018 World Wide Web Conference (WWW 2018). ACM (2018)
9. Hanika, T., Marx, M., Stumme, G.: Discovering implicational knowledge in Wikidata. In: CoRR abs/1902.00916 (2019)
10. Ho, V.T., Stepanova, D., Gad-Elrab, M.H., Kharlamov, E., Weikum, G.: Rule learning from knowledge graphs guided by embedding models. In: Vrandečić, D., et al. (eds.) ISWC 2018. LNCS, vol. 11136, pp. 72–90. Springer, Cham (2018). https://doi.org/10.1007/978-3-030-00671-6_5
11. Krötzsch, M.: Ontologies for knowledge graphs? In: Artale, A., Glimm, B., Kontchakov, R. (eds.) Proceedings of the 30th International Workshop on Description Logics (DL 2017), vol. 1879. CEUR WS Proceedings (2017). CEUR-WS.org
12. Lajus, J., Suchanek, F.M.: Are all people married?: determining obligatory attributes in knowledge bases. In: Champin, P., et al. (eds.) Proceedings of 2018 World Wide Web Conference (WWW 2018). ACM (2018)
13. Luxenburger, M.: Implications partielles dans un contexte. Math. Inform. Sci. Humaines **113**, 35–55 (1991)
14. Marx, M., Krötzsch, M.: SQID: towards ontological reasoning for Wikidata. In: Nikitina, N., et al. (eds.) Proceedings of the 16th International Semantic Web Conference (ISWC 2017): Posters & Demonstrations and Industry Tracks, vol. 1963. CEUR WS Proceedings (2017). CEUR-WS.org
15. Marx, M., Krötzsch, M., Thost, V.: Logic on MARS: ontologies for generalised property graphs. In: Sierra, C. (ed.) Proceedings of the 26th International Joint Conference on Artificial Intelligence (IJCAI 2017), pp. 1188–1194 (2017). ijcai.org

16. Rudolph, S.: Exploring relational structures via $\mathcal{FLE}$. In: Wolff, K.E., Pfeiffer, H.D., Delugach, H.S. (eds.) ICCS-ConceptStruct 2004. LNCS (LNAI), vol. 3127, pp. 196–212. Springer, Heidelberg (2004). https://doi.org/10.1007/978-3-540-27769-9_13
17. Stumme, G., Taouil, R., Bastide, Y., Pasquier, N., Lakhal, L.: Intelligent structuring and reducing of association rules with formal concept analysis. In: Baader, F., Brewka, G., Eiter, T. (eds.) KI 2001. LNCS (LNAI), vol. 2174, pp. 335–350. Springer, Heidelberg (2001). https://doi.org/10.1007/3-540-45422-5_24
18. Pellissier Tanon, T., Stepanova, D., Razniewski, S., Mirza, P., Weikum, G.: Completeness-aware rule learning from knowledge graphs. In: d'Amato, C., et al. (eds.) ISWC 2017. LNCS, vol. 10587, pp. 507–525. Springer, Cham (2017). https://doi.org/10.1007/978-3-319-68288-4_30
19. Vrandečić, D.: Wikidata: a new platform for collaborative data collection. In: Mille, A., et al. (eds.) Companion of the 21st World Wide Web Conference (WWW 2012), pp. 1063–1064. ACM (2012)
20. Vrandečić, D., Krötzsch, M.: Wikidata: a free collaborative knowledge base. Commun. ACM **57**(10), 78–85 (2014)
21. Zangerle, E., et al.: an empirical evaluation of property recommender systems for Wikidata and collaborative knowledge bases. In: Wasserman, A.I. (ed.) Proceedings of the 12th International Symposium on Open Collaboration (OpenSym 2016), pp. 18:1–18:8. ACM (2016)

A Characterization Theorem for Continuous Lattices by Closure Spaces

Guozhi Ma[1], Lankun Guo[2(✉)], and Cheng Yang[2]

[1] College of Engineering and Design, Hunan Normal University, Changsha, Hunan 410012, People's Republic of China
guozhi.ma721@gmail.com

[2] College of Mathematics and Statistics, Key Laboratory of High Performance, Computing and Stochastic Information Processing, Hunan Normal University, Changsha, Hunan 410012, People's Republic of China
lankun.guo@gmail.com, 2543902060@qq.com

Abstract. The notion of closure spaces plays an important role in formal concept analysis, and there exists a close connection between formal concept analysis and lattice theory. In order to restructure continuous lattices, a special kind of complete lattices in Domain theory, this paper proposes a novel notion named relationally consistent F-augmented closure spaces. Then, the concept of F-approximable mappings between relationally consistent F-augmented closure spaces is introduced, which provides a representation of Scott continuous maps between continuous lattices. The final result is: the categories of relationally consistent F-augmented closure spaces and continuous lattices are equivalent.

Keywords: Closure space · Approximable mapping · Continuous lattice

1 Introduction

There exist close connection between topological spaces and partially ordered sets (posets). For instance, in Domain theory continuous lattices and bounded complete domains can be characterized via Scott topology by injective spaces and densely injective spaces respectively [5,6,8,12]. As a generalization of topological spaces, closure systems (or operators) serve as core notions of other research fields such as formal concept analysis [7] and have played an important role in restructuring various special posets [3,4,10,11,13]. A well-known fact is that

The first author thanks the support from the National Natural Science Foundation of China (No. 51801062), the Hunan Provincial Natural Science Foundation of China (No. 2017JJ3198) and the Educational Commission of Hunan Province of China (No. 17C0944). The second author thanks the support from the National Natural Science Foundation of China (No. 11401195), the Educational Commission of Hunan Province of China (No. 16B153), Young talents Program in Hunan Province and the Construct Program of the Key Discipline in Hunan Province.

D. Cristea et al. (Eds.): ICFCA 2019, LNAI 11511, pp. 324–331, 2019.
https://doi.org/10.1007/978-3-030-21462-3_22

the closed sets of an (algebraic) closure system ordered by set inclusion is an (algebraic) complete lattice, and conversely, every (algebraic) complete lattice is order isomorphic to the family of all closed sets of an appropriate (algebraic) closure system (see Chap. 7 of [2]). Recently, the notion of F-augmented closure spaces was introduced, which provides a new approach to representing algebraic domains [9].

In this paper, we introduce the notions of relationally consistent F-augmented closure spaces and F-approximable mappings. As you will see, the technique is to employ an appropriate binary relation onto the finite-subsets-family $\mathscr{F}$ of an F-augmented closure space $(X, \mathscr{C}, \mathscr{F})$. Of course, such binary relations are required to be consistent with the closure spaces. This is why we use the words "relationally" and "consistent" for the introduced notion. The main result is: continuous lattices and relationally consistent F-augmented closure spaces are equivalent from the categorical point of view.

2 Continuous Lattices and Scott Continuous Maps

Throughout this paper, given a set X, we use $F \sqsubseteq X$ to mean that F is a finite subset of X. For a closure space $(X, \mathscr{C})$, elements of $\mathscr{C}$ are called closed sets and $\overline{A}$ denotes the closure of $A \subseteq X$, i.e., the least closed set containing A.

A subset of a poset $(L, \leq)$ is said to be directed if it is non-empty and each of its non-empty finite subsets has an upper bound in it. A poset $(L, \leq)$ is called a *dcpo* if the least upper bound $\bigvee D$ exists for every directed subset $D \subseteq L$. For a poset $(L, \leq)$ being a complete lattice, it is sufficient to prove that $(L, \leq)$ is a dcpo and any finite subset of L has the least upper bound. Given $x, y \in L$, x is said to *way below* y, written as $x \ll y$, if for every directed subset $D \subseteq L$ for which $\bigvee D$ exists, $y \leq \bigvee D$ always implies the existence of some $d \in D$ such that $x \leq d$. For convenience, we always use the notation $\Downarrow x$ to denote the set $\{y \in L \mid y \ll x\}$. A dcpo $(L, \leq)$ is called a *continuous domain* if for any $x \in L$, $\Downarrow x$ is a directed subset and $x = \bigvee \Downarrow x$. Equivalently, $(L, \leq)$ is a continuous domain if and only if for any $x \in L$, there exists a directed subset $D \subseteq L$ such that $d \ll x$ for every $d \in D$ and $x = \bigvee D$. A continuous domain $(L, \leq)$ is called a *continuous lattice* if $(L, \leq)$ is a complete lattice. The following properties of a continuous lattice $(L, \leq)$ will be used: (i) for any $a, b, c, d \in L$, if $a \ll b$ and $c \ll d$, then $(a \vee c) \ll (b \vee d)$; (ii) for any $x, z \in L$, $x \ll z$ if and only if there exists $y \in L$ such that $x \ll y \ll z$; (iii) for any $x \in L$ and directed subset $D \subseteq L$, $x \ll \bigvee D$ if and only if there exists $d \in D$ such that $x \ll d$.

Let $(L_1, \leq_1)$ and $(L_2, \leq_2)$ be dcpos. A map $\psi : L_1 \to L_2$ is said to be *Scott continuous* if $\psi(\bigvee D) = \bigvee \psi(D)$ holds for any directed subset $D \subseteq L$. Continuous lattices and Scott continuous maps do form a category which is denoted by **CLA**. More facts about continuous lattices can be referred to [8].

3 Relationally Consistent F-augmented Closure Spaces and Continuous Lattices

We first recall the notion of consistent F-augmented closure spaces proposed in [9]: Let $(X, \mathscr{C})$ be a closure space, $\mathscr{F}$ a non-empty family of non-empty finite subsets of X. The triplet $(X, \mathscr{C}, \mathscr{F})$ is called a *consistent finite-subset-selection augmented closure space* (for short, *consistent F-augmented closure space*) if for any $F \in \mathscr{F}$ and $B \sqsubseteq \overline{F}$ there exists $F' \in \mathscr{F}$ such that $B \subseteq F' \subseteq \overline{F}$;

Definition 1. *Let $(X, \mathscr{C}, \mathscr{F})$ be a consistent F-augmented closure space and $\Subset$ a binary relation on $\mathscr{F}$. The quartette $\mathbb{X} = (X, \mathscr{C}, \mathscr{F}, \Subset)$ is called a relationally consistent F-augmented closure space (for short, RF-closure space) if*
(i) there exists $F_0 \in \mathscr{F}$ such that $F_0 \Subset F$ holds for any $F \in \mathscr{F}$;
(ii) the family of F's for which the associated $\overline{F}$'s are the same has a least one with respect to $\subseteq$;
(iii) for any $F_1, F_2 \in \mathscr{F}$, there exists $F_{(F_1,F_2)} \in \mathscr{F}$ such that $\overline{F_{(F_1,F_2)}} = \overline{F_1 \cup F_2}$;
(iv) for any $F_i \in \mathscr{F}$ $(i = 1, 2, 3, 4)$, the following axioms hold:

(C1) $F_1 \Subset F_2 \Rightarrow F_1 \subseteq \overline{F_2}$;
(C2) $F_1 \Subset F_2$ & $F_3 \Subset F_4 \Rightarrow F_{(F_1,F_3)} \Subset F_{(F_2,F_4)}$;
(C3) $F_1 \subseteq \overline{F_2}$ & $F_2 \Subset F_3 \subseteq \overline{F_4} \Rightarrow F_1 \Subset F_4$;
(C4) $F_1 \Subset F_2 \Rightarrow (\exists F \in \mathscr{F})\ F_1 \Subset F \Subset F_2$;
(C5) $F_1 \Subset F_3$ & $F_2 \Subset F_3 \Rightarrow (\exists F \in \mathscr{F})\ F_1 \cup F_2 \subseteq F \Subset F_3$;
(C6) $((\forall F \in \mathscr{F})\ (F \Subset F_1 \Rightarrow F \Subset F_2)) \Rightarrow F_1 \subseteq \overline{F_2}$.

Example 1. Consider the quartette $(X, \mathscr{C}, \mathscr{F}, \Subset)$ where X is the real interval $[0, 1]$, $\mathscr{C}$ is the family of all down-sets of X with respect to "less or equal relation" $\leq$, $\mathscr{F}$ is the family of all non-empty finite subsets of X for which we denote by q_F the greatest element of $F \in \mathscr{F}$, and $\Subset \subseteq \mathscr{F} \times \mathscr{F}$ is defined by $F_1 \Subset F_2$ if and only if $q_{F_1} = q_{F_2} = 0$ or $q_{F_1} < q_{F_2}$. Trivial check shows that $(X, \mathscr{C}, \mathscr{F}, \Subset)$ is an RF-closure space.

Definition 2. *Let $\mathbb{X}$ be an RF-closure space. A subset $\omega \subseteq X$ is called a relationally approximable closed set (for short, R-closed set) of $\mathbb{X}$ if for any finite subset $K \subseteq \omega$, there exist $F_1, F_2 \in \mathscr{F}$ such that $K \subseteq F_1 \Subset F_2$ and $\overline{F_2} \subseteq \omega$.*

Remark 1. (1) For any R-closed set ω of an RF-closure space $\mathbb{X}$, since $\emptyset$ is a finite subset of ω, by Definition 2 and Definition 1(C1), there exists $F \in \mathscr{F}$ such that $F \subseteq \overline{F} \subseteq \omega$. Since every element $F \in \mathscr{F}$ is non-empty, this implies that every R-closed set is non-empty. Moreover, it is easy to see that $\{F \in \mathscr{F} \mid F \subseteq \omega\}$ is directed with respect to $\subseteq$ and $\omega = \bigcup\{F \in \mathscr{F} \mid F \subseteq \omega\}$.

(2) It follows from Definition 1(C3) that an R-closed set can be equivalently characterized as: ω is an R-closed set of $\mathbb{X}$ if and only if for any finite subset $K \subseteq \omega$, there exists $F_1, F_2 \in \mathscr{F}$ such that $K \subseteq \overline{F_1}$, $F_1 \Subset F_2$ and $\overline{F_2} \subseteq \omega$. Moreover, to verify that a non-empty subset $\alpha \subseteq X$ is an R-closed set, it is sufficient to check that α satisfies the condition of Definition 2 for any non-empty finite subset $K \subseteq \alpha$.

(3) For any $F \in \mathscr{F}$, by Definition 1(i) $\{F' \in \mathscr{F} \mid F' \Subset F\}$ is non-empty. Furthermore, by Definition 1(C5), the set $\{F' \in \mathscr{F} \mid F' \Subset F\}$ is directed with respect to $\subseteq$. In the following, we always use $\widehat{F}$ to denote the set $\bigcup\{F' \in \mathscr{F} \mid F' \Subset F\}$. It is obvious that $\{\overline{F'} \mid F' \in \mathscr{F} \;\&\; F' \Subset F\}$ is directed with respect to $\subseteq$. In addition, $\bigcup\{F' \in \mathscr{F} \mid F' \Subset F\} = \bigcup\{\overline{F'} \mid F' \in \mathscr{F} \;\&\; F' \Subset F\}$.

In the sequel, we use $\mathfrak{C}(\mathbb{X})$ to denote the set of all R-closed sets of $\mathbb{X}$.

Example 2. Following Example 1, we have that $\omega \subseteq X$ is an R-closed set of $(X, \mathscr{C}, \mathscr{F}, \Subset)$ if and only if $\omega = \{0\}$ or there exists $x \in X \setminus \{0\}$ such that $\omega = \{y \in X \mid y < x\}$.

Proposition 1. *Let $\mathbb{X}$ be an RF-closure space. Then for any $F_1, F_2 \in \mathscr{F}$,*

$$F_1 \Subset F_2 \;\Leftrightarrow\; F_1 \subseteq \widehat{F_2} \;\Leftrightarrow\; \overline{F_1} \subseteq \widehat{F_2}.$$

Proposition 2. *Let $\mathbb{X}$ be an RF-closure space. Then*

(1) *for any $F \in \mathscr{F}$, $\widehat{F}$ is an R-closed set of $\mathbb{X}$;*
(2) *for any $F_1, F_2 \in \mathscr{F}$, $F_1 \subseteq \overline{F_2}$ if and only if $\widehat{F_1} \subseteq \widehat{F_2}$;*
(3) *for any $F_1, F_2 \in \mathscr{F}$, if $F_1 \Subset F_2$, then $\widehat{F_1} \subseteq \widehat{F_2}$;*
(4) *$\{\widehat{F} \mid F \in \mathscr{F}\}$ is a sup-semilattice with respect to $\subseteq$.*

Proposition 3. *Let ω and $\{\omega_i\}_{i\in I}$ be R-closed sets of $\mathbb{X}$. Then*

(1) *$\overline{K} \subseteq \omega$ holds for any finite subset $K \subseteq \omega$;*
(2) *$\widehat{F} \subseteq \omega$ holds for any $F \in \mathscr{F}$ with $F \subseteq \omega$;*
(3) *if $\{\omega_i\}_{i\in I}$ is directed with respect to $\subseteq$, then $\bigcup_{i\in I} \omega_i$ is an R-closed set.*

Proposition 4. *Let ω be an R-closed set of $\mathbb{X}$. Then*

(1) *$\{\widehat{F} \mid F \in \mathscr{F} \;\&\; F \subseteq \omega\}$ is directed with respect to $\subseteq$;*
(2) *if $F \in \mathscr{F}$ satisfies $F \subseteq \omega$, then $\widehat{F} \ll \omega$ in $(\mathfrak{C}(\mathbb{X}), \subseteq)$;*
(3) *$\omega = \bigcup\{\widehat{F} \mid F \in \mathscr{F} \;\&\; F \subseteq \omega\}$.*

Proof. (1) By Remark 1(1), $\{\widehat{F} \mid F \in \mathscr{F} \;\&\; F \subseteq \omega\}$ is non-empty. Suppose $F_1, F_2 \in \mathscr{F}$ satisfy $F_1 \subseteq \omega$ and $F_2 \subseteq \omega$. As $F_1 \cup F_2$ is a finite subset of ω, by Remark 1(2), there exist $F_3, F_4 \in \mathscr{F}$ such that $F_1 \cup F_2 \subseteq \overline{F_3}$, $F_3 \Subset F_4$ and $\overline{F_4} \subseteq \omega$. By Definition 1(C1), we have $F_1 \cup F_2 \subseteq \overline{F_3} \subseteq \overline{F_4}$. It follows from Proposition 2(2) that $\widehat{F_1} \subseteq \widehat{F_4}$ and $\widehat{F_2} \subseteq \widehat{F_4}$. It is clear that $F_4 \subseteq \omega$. Therefore, $\{\widehat{F} \mid F \in \mathscr{F} \;\&\; F \subseteq \omega\}$ is directed with respect to $\subseteq$.

(2) Suppose $\{\omega_i\}_{i\in I}$ is a directed family of R-closed sets of $\mathbb{X}$ such that $\omega \subseteq \bigcup_{i\in I} \omega_i$. As F is a non-empty finite subset of ω, there exists $i_0 \in I$ such that $F \subseteq \omega_{i_0}$. By Proposition 3(2), $\widehat{F} \subseteq \omega_{i_0}$. This means $\widehat{F} \ll \omega$ in $(\mathfrak{C}(\mathbb{X}), \subseteq)$.

(3) It follows from Proposition 3(2) that $\bigcup\{\widehat{F} \mid F \in \mathscr{F} \;\&\; F \subseteq \omega\} \subseteq \omega$. On the other hand, for any $x \in \omega$, by Definition 2, there exist $F_1, F_2 \in \mathscr{F}$ such that $x \in F_1 \Subset F_2$ and $\overline{F_2} \subseteq \omega$. This implies $x \in \widehat{F_2}$ and $F_2 \subseteq \omega$. Thus, $\omega \subseteq \bigcup\{\widehat{F} \mid F \in \mathscr{F} \;\&\; F \subseteq \omega\}$.

Proposition 5. *Let $\mathbb{X}$ be an RF-closure space. Then $(\mathfrak{C}(\mathbb{X}), \subseteq)$ is a continuous lattice.*

Proof. Suppose ω_1 and ω_2 are R-closed sets of $\mathbb{X}$. Consider the family $\mathfrak{U} = \{F \in \mathscr{F} \mid (\exists F_1, F_2 \in \mathscr{F})\ F_1 \subseteq \omega_1\ \&\ F_2 \subseteq \omega_2\ \&\ \widehat{F} = \widehat{F_1} \vee \widehat{F_2}\}$. By Proposition 2(4) and Proposition 4(1), $\{\widehat{F} \mid F \in \mathfrak{U}\}$ is directed with respect to $\subseteq$. By Proposition 3(3), $\bigcup_{F\in\mathfrak{U}} \widehat{F}$ is an R-closed set of $\mathbb{X}$. It is clear that $\omega_1 \cup \omega_2 \subseteq \bigcup_{F\in\mathfrak{U}} \widehat{F}$, which means that $\bigcup_{F\in\mathfrak{U}} \widehat{F}$ is an upper bound of ω_1 and ω_2 in $(\mathfrak{C}(\mathbb{X}), \subseteq)$. Suppose ω_3 is another upper bound of ω_1 and ω_2 in $(\mathfrak{C}(\mathbb{X}), \subseteq)$. Then for any $F \in \mathscr{F}$ with $F \subseteq \omega_1$ (or $F \subseteq \omega_2$), we have $F \subseteq \omega_3$ which implies $\widehat{F} \subseteq \omega_3$. By Proposition 4(1), $\bigcup_{F\in\mathfrak{U}} \widehat{F} \subseteq \omega_3$. This means that $\bigcup_{F\in\mathfrak{U}} \widehat{F}$ is the least upper bound of ω_1 and ω_2 in $(\mathfrak{C}(\mathbb{X}), \subseteq)$.

From Proposition 2(1) and Definition 1(i) it follows that $\widehat{F_0}$ is the bottom element of $(\mathfrak{C}(\mathbb{X}), \subseteq)$. Since $(\mathfrak{C}(\mathbb{X}), \subseteq)$ is a dcpo by Proposition 3(3), it is a complete lattice. Proposition 4 shows that $(\mathfrak{C}(\mathbb{X}), \subseteq)$ is a continuous domain.

Now we consider the inverse direction. Suppose $(L, \leq)$ is a continuous lattice. Let $\mathscr{C}_L = \{\downarrow x \mid x \in X\}$ and $\mathscr{F}_L$ the family of all finite subsets of L having a greatest element with respect to $\leq$. For convenience, we use c_F to denote such greatest element for $F \in \mathscr{F}_L$. Define a relation $\Subset_L \subseteq \mathscr{F}_L \times \mathscr{F}_L$ by

$$F_1 \Subset_L F_2 \Leftrightarrow c_{F_1} \ll c_{F_2}.$$

For any $F \in \mathscr{F}_L$, it is easy to check $\overline{F} = \downarrow c_F$. Trivial check shows that $\mathbb{X}_L = (L, \mathscr{C}_L, \mathscr{F}_L, \Subset_L)$ is an RF-closure space and $\widehat{F} = \Downarrow c_F$ holds for any $F \in \mathscr{F}_L$.

Proposition 6. *A subset $D \subseteq L$ is an R-closed set of $\mathbb{X}_L$ if and only if there exists $x \in L$ such that $D = \Downarrow x$.*

Proof. ($\Rightarrow$): Suppose $D \subseteq L$ is an R-closed set of $\mathbb{X}_L$. By Remark 1(1), D is non-empty. Given $u, w \in D$, by Remark 1(2), there exist $c_1, c_2 \in L$ such that $\{u, w\} \subseteq \downarrow c_1$, $c_1 \ll c_2$ and $\downarrow c_2 \subseteq D$. It is easy to see that $c_2 \in D$ and $u, w \leq c_2$. This implies that D is directed with respect to $\leq$ and thus $\bigvee D$ exists.

Now we show that $D = \Downarrow (\bigvee D)$. On the one hand, for any $x \in D$, by Remark 1(2), there exist $c_1, c_2 \in L$ such that $x \leq c_1 \ll c_2$ and $\downarrow c_2 \subseteq D$, which implies $x \ll c_2 \in D$ and thus $x \in \Downarrow (\bigvee D)$. This means $D \subseteq \Downarrow (\bigvee D)$. On the other hand, suppose $x \in \Downarrow (\bigvee D)$, i.e., $x \ll \bigvee D$. Then there exists $x' \in D$ sucht that $x \ll x'$. Set $F = \{x'\}$. Then $x \in \widehat{F}$. By Proposition 3(2), $x \in D$. This means $\Downarrow (\bigvee D) \subseteq D$.

($\Leftarrow$): Suppose there exists $x \in L$ such that $D = \Downarrow x$. Given any non-empty subset $K \sqsubseteq D$. Because $\Downarrow x$ is directed with respect to $\leq$, there exists $c \in L$ such that $K \subseteq \downarrow c$ and $c \ll x$. Then there exists $c' \in L$ such that $c \ll c' \ll x$. Set $F_1 = \{c\}$ and $F_2 = \{c'\}$. It is easy to see that $K \subseteq \overline{F_1}$, $F_1 \Subset_L F_2$ and $\overline{F_2} \subseteq D$. This means that D is an R-closed set of $\mathbb{X}_L$.

From Proposition 6 and the definition of continuous lattices, we obtain:

Theorem 1. *A continuous lattice $(L, \leq)$ is order isomorphic to $(\mathfrak{C}(\mathbb{X}_L), \subseteq)$.*

4 F-approximable Mappings Between RF-closure Spaces

In this section, we introduce the notion of F-approximable mappings and establish the one-to-one correspondence between F-approximable mappings and Scott continuous maps. This result immediately induces the equivalence between the category of continuous lattices and that of RF-closure spaces. The used notions about category theory are from [1].

Definition 3. *Let* $\mathbb{X} = (X, \mathscr{C}_X, \mathscr{F}_X, \Subset_X)$ *and* $\mathbb{Y} = (Y, \mathscr{C}_Y, \mathscr{F}_Y, \Subset_Y)$ *be RF-closure spaces. A binary relation* $\Theta \subseteq X \times \mathscr{F}_Y$ *is called an F-approximable mapping from* $\mathbb{X}$ *to* $\mathbb{Y}$ *if, for any finite subset* $K \subseteq X$*,* $F \in \mathscr{F}_X$ *and* $G, G_1, G_2 \in \mathscr{F}_Y$*,*

(FM1) $K\Theta G \Rightarrow (\exists F \in \mathscr{F}_X)\ K \subseteq F\Theta G;$
(FM2) $K\Theta G_1 \subseteq \overline{G_2} \Rightarrow K\Theta G_2;$
(FM3) $F\Theta G \Leftrightarrow\ (\exists F' \in \mathscr{F}_X, G' \in \mathscr{F}_Y)\ F \Subset_X F'\Theta G' \Subset_Y G,$

where $K\Theta G$ *means* $(x, G) \in \Theta$ *for any* $x \in K$*.*

We use **RCON** to denote the category of RF-closure spaces and F-approximable mappings.

Remark 2. Given an F-approximable mapping Θ from $\mathbb{X}$ to $\mathbb{Y}$, $\emptyset\Theta G$ holds for any $G \in \mathscr{F}_Y$. It follows from Definition 3(FM1) that: for any $G \in \mathscr{F}_Y$ there exists $F \in \mathscr{F}_X$ such that $F\Theta G$.

Proposition 7. *Let* Θ *be an F-approximable mapping from* $\mathbb{X}$ *to* $\mathbb{Y}$*. Then for any finite subset* $K \subseteq X$ *and* $G \in \mathscr{F}_Y$*,*

$$K\Theta G \Leftrightarrow\ (\exists F \in \mathscr{F}_X)\ K \subseteq \overline{F}\ \&\ F\Theta G. \qquad \text{(FM1}')$$

Proposition 8. *Let* Θ *be an F-approximable mapping from* $\mathbb{X}$ *to* $\mathbb{Y}$*. Then for any* $F \in \mathscr{F}_X$ *and* $G \in \mathscr{F}_Y$*,*

$$F\Theta G \Leftrightarrow\ (\exists F' \in \mathscr{F}_X)\ F \Subset_X F'\Theta G.$$

Proposition 9. *Let* Θ *be an F-approximable mapping from* $\mathbb{X}$ *to* $\mathbb{Y}$*. For any* $G \in \mathscr{F}_Y$*, the family* $\{F \in \mathscr{F}_X \mid F\Theta G\}$ *is directed with respect to* $\subseteq$ *and* $\bigcup\{F \in \mathscr{F}_X \mid F\Theta G\}$ *is an R-closed set of* $\mathbb{X}$*.*

In the sequel, given $G \in \mathscr{F}_Y$, we use the notation G^Θ to denote the subset $\bigcup\{F \in \mathscr{F}_X \mid F\Theta G\}$, i.e.,

$$G^\Theta = \bigcup\{F \in \mathscr{F}_X \mid F\Theta G\}\ (= \bigcup\{\widehat{F} \mid F \in \mathscr{F}_X\ \&\ F\Theta G\}).$$

Proposition 10. *Let* Θ *be an F-approximable mapping from* $\mathbb{X}$ *to* $\mathbb{Y}$*. Then for any* $F \in \mathscr{F}_X$ *and* $G \in \mathscr{F}_Y$*,*

$$F \subseteq G^\Theta \Leftrightarrow F\Theta G.$$

Proposition 11. *Let Θ be an F-approximable mapping from $\mathbb{X}$ to $\mathbb{Y}$. For any R-closed set ω of $\mathbb{Y}$, the family $\{G^\Theta \mid G \in \mathscr{F}_Y \ \&\ G \subseteq \omega\}$ is directed with respect to $\subseteq$ and thus $\bigcup\{G^\Theta \mid G \in \mathscr{F}_Y \ \&\ G \subseteq \omega\}$ is an R-closed set of $\mathbb{X}$.*

In the sequel, given an R-closed set ω of $\mathbb{X}$, we always use ω^Θ to denote the subset $\bigcup\{G^\Theta \mid G \in \mathscr{F}_Y \ \&\ G \subseteq \omega\}$, i.e.,

$$\omega^\Theta = \bigcup\{G^\Theta \mid G \in \mathscr{F}_Y \ \&\ G \subseteq \omega\}.$$

Proposition 12. *Let Θ be an F-approximable mapping from $\mathbb{X}$ to $\mathbb{Y}$. Then for any $F \in \mathscr{F}_X$ and R-closed set ω of $\mathbb{Y}$,*

$$F \subseteq \omega^\Theta \Leftrightarrow (\exists G \in \mathscr{F}_Y)\ F\Theta G \subseteq \omega.$$

Proposition 13. *Let $\mathbb{X}$ and $\mathbb{Y}$ be RF-closure spaces and ϕ a Scott continuous map from $\underline{\mathfrak{C}}(\mathbb{Y})$ to $\underline{\mathfrak{C}}(\mathbb{X})$. Then for any $F \in \mathscr{F}_X$ and $G \in \mathscr{F}_Y$, $F \subseteq \phi(\widehat{G})$ if and only if there exist $F' \in \mathscr{F}_X$ and $G' \in \mathscr{F}_Y$ such that $F \subseteq \widehat{F'}$, $F' \subseteq \phi(\widehat{G'})$ and $G' \subseteq \widehat{G}$.*

Proof. Suppose $F \subseteq \phi(\widehat{G})$. By Proposition 4, there exists $F' \in \mathscr{F}_X$ such that $F' \subseteq \phi(\widehat{G})$ and $F \subseteq \widehat{F'}$. By the continuity of ϕ and Proposition 4, we have $\phi(\widehat{G}) = \bigcup\{\phi(\widehat{G'}) \mid G' \in \mathscr{F}_Y \ \&\ G' \subseteq \widehat{G}\}$. As $\{\phi(\widehat{G'}) \mid G' \in \mathscr{F}_Y \ \&\ G' \subseteq \widehat{G}\}$ is directed with respect to $\subseteq$, there exists $G' \in \mathscr{F}_Y$ such that $F' \subseteq \phi(\widehat{G'})$ and $G' \subseteq \widehat{G}$.

Conversely, suppose there exist $F' \in \mathscr{F}_X$ and $G' \in \mathscr{F}_Y$ such that $F \subseteq \widehat{F'}$, $F' \subseteq \phi(\widehat{G'})$ and $G' \subseteq \widehat{G}$. Based on Proposition 3(2), because ϕ is order-preserving, we have $F \subseteq \widehat{F'} \subseteq \phi(\widehat{G'}) \subseteq \phi(\widehat{G})$.

Now we show that there is a one-to-one correspondence between Scott continuous maps from $\underline{\mathfrak{C}}(\mathbb{Y})$ to $\underline{\mathfrak{C}}(\mathbb{X})$ and F-approximable mappings from $\mathbb{X}$ to $\mathbb{Y}$.

Theorem 2. *Let $\mathbb{X}$ and $\mathbb{Y}$ be RF-closure spaces.*

(1) *For any F-approximable mapping Θ from $\mathbb{X}$ to $\mathbb{Y}$, define a map $\phi_\Theta : \underline{\mathfrak{C}}(\mathbb{Y}) \to \underline{\mathfrak{C}}(\mathbb{X})$ by $\phi_\Theta(\omega) = \omega^\Theta$. Then ϕ_Θ is a Scott continuous map.*

(2) *Conversely, for any Scott continuous map ϕ from $\underline{\mathfrak{C}}(\mathbb{Y})$ to $\underline{\mathfrak{C}}(\mathbb{X})$, define a relation $\Theta_\phi \subseteq X \times \mathscr{F}_Y$ by $(x, G) \in \Theta_\phi \Leftrightarrow x \in \phi(\widehat{G})$. Then Θ_ϕ is an F-approximable mapping from $\mathbb{X}$ to $\mathbb{Y}$.*

(3) *Moreover, $\Theta = \Theta_{\phi_\Theta}$ and $\phi = \phi_{\Theta_\phi}$.*

Based on Proposition 5, define a map $\mathfrak{F}_o : \mathbf{RCON}_o^{\mathrm{op}} \to \mathbf{CLA}_o$ as: for any RF-closure space $\mathbb{X}$, $\mathfrak{F}_o(\mathbb{X}) = \underline{\mathfrak{C}}(\mathbb{X})$. Based on Theorem 2(1), define a map $\mathfrak{F}_a : \mathbf{RCON}_a^{\mathrm{op}} \to \mathbf{CLA}_a$ as: for any F-approximable mapping Θ from $\mathbb{X}$ to $\mathbb{Y}$, $\mathfrak{F}_a(\Theta) = \phi_\Theta$. That is, for any R-closed set ω of $\mathbb{Y}$, $\mathfrak{F}_a(\Theta)(\omega) = \omega^\Theta$. Theorem 1 tells $\mathfrak{F}$ is representable. Theorem 2 demonstrates that $\mathfrak{F}$ is both full and faithful. As a conclusion, we obtain the main result of this paper.

Theorem 3. *The categories* $\mathbf{CLA}$ *and* $\mathbf{RCON}^{\mathrm{op}}$ *are equivalent.*

5 Conclusions

In this paper, the notions of relationally consistent F-augmented closure spaces and F-approximable mappings are proposed. It is proved that relationally consistent F-augmented closure spaces and continuous lattices are equivalent from the categorical point of view. Since closure spaces are the core mathematical structures in formal concept analysis, this work also provides a new approach to representing continuous lattices by means of formal context analysis.

References

1. Barr, M., Wells, C.: Category Theory for Computing Science, 3rd edn. Prentice Hall, New York (1990)
2. Davey, B.A., Priestley, H.A.: Introduction to Lattices and Order, 2nd edn. Cambridge University Press, Cambridge (2002)
3. Erné, M.: Lattice representations for categories of closure spaces, Categorical topology (Toledo, OH, 1983). Bentley, H.L. (ed.) Sigma Series in Pure Mathematics, vol. 5, pp. 197–222. Heldermann, Berlin (1984)
4. Erné, M.: General stone duality. Topology Appl. **137**(1–3), 125–158 (2004)
5. Eršov, Y.L.: The theory of A-spaces. Algebra Logic **12**(4), 209–232 (1973)
6. Escardó, M.H.: Properly injective spaces and function spaces. Topology Appl. **89**(1), 75–120 (1998)
7. Ganter, B., Wille, R.: Formal Concept Analysis: Mathematical Foundations. Springer, Berlin (1999). https://doi.org/10.1007/978-3-642-59830-2
8. Gierz, G., Hofmann, K.H., Keimel, K., Lawson, J.D., Mislove, M., Scott, D.S.: Continuous Lattices and Domains. Cambridge University Press, Cambridge (2003)
9. Guo, L., Li, Q.: The categorical equivalence between algebraic domains and F-augmented closure spaces. Order **32**, 101–116 (2015)
10. Hofmann, K.H., Keimel, K.: A General Character Theory for Partially Osets and Lattices, vol. 122, no. 122. Memoirs of the American Mathematical Society (1972)
11. Priestley, H.A.: Representation of distributive lattices by means of ordered Stone spaces. Bull. London Math. Soc. **2**, 186–190 (1970)
12. Scott, D.: Continuous lattices. In: Lawvere, F.W. (ed.) Toposes, Algebraic Geometry and Logic. LNM, vol. 274, pp. 97–136. Springer, Heidelberg (1972). https://doi.org/10.1007/BFb0073967
13. Stone, M.H.: The theory of representations of Boolean algebras. Trans. Am. Math. Soc. **40**, 37–111 (1936)

On Coupling FCA and MDL in Pattern Mining

Tatiana Makhalova[1,2](✉), Sergei O. Kuznetsov[1], and Amedeo Napoli[2]

[1] National Research University Higher School of Economics, Moscow, Russia
{tpmakhalova,skuznetsov}@hse.ru

[2] Université de Lorraine, CNRS, Inria, LORIA, 54000 Nancy, France
amedeo.napoli@loria.fr

Abstract. Pattern Mining is a well-studied field in Data Mining and Machine Learning. The modern methods are based on dynamically updating models, among which MDL-based ones ensure high-quality pattern sets. Formal concepts also characterize patterns in a condensed form. In this paper we study MDL-based algorithm called Krimp in FCA settings and propose a modified version that benefits from FCA and relies on probabilistic assumptions that underlie MDL. We provide an experimental proof that the proposed approach improves quality of pattern sets generated by Krimp.

1 Introduction

Pattern Mining (PM) has an important place in Data Mining (DM) and Knowledge Discovery (KD). One main concern of PM is to discover something interesting in a pattern space which can be exponentially large w.r.t. the size of the dataset. "Interestingness" can be defined as generalities or specificities underlying the data, (dis)accordance to user hypotheses, etc. One distinguishes *static* and *dynamic* approaches in PM [1].

The static approaches envelop a large number of interestingness measures [9]. The patterns are mined under non-changeable assumptions about interestingness. For example, in frequent PM, one assumes that all the patterns having support greater than a user-specified threshold are interesting. Other popular measures work under independence [2], randomization models [8] or other constraints [4,13]. The main drawbacks of the static approaches are (i) pattern redundancy, i.e., the discovered pattern set contains a lot of similar patterns, and (ii) subjectiveness, i.e., it is often not easy to provide explanation or justification about using one measure rather than some others.

Mining with interestingness measures addresses a traditional question of PM, i.e., finding *all the patterns that satisfy some given constraints*. By contrast, in [1] (Chap. 8), it is argued that one should ask for a small (easily interpretable) and non-redundant (with high diversity) set of interesting patterns. This is precisely what dynamic approaches to PM are aimed at.

D. Cristea et al. (Eds.): ICFCA 2019, LNAI 11511, pp. 332–340, 2019.
https://doi.org/10.1007/978-3-030-21462-3_23

A dynamic approach implies taking into account initial assumptions, e.g., background knowledge, and building progressively a suitable "model", i.e., a pattern set. Following this way, the mining protocol starts with a general model, possibly very simple, which is then iteratively enriched with new extracted information. A lot of existing dynamic approaches are based on Minimum Description Length (MDL) principle which allows one to select patterns compressing the dataset at the most [11,12,14]. MDL has important properties: (i) it ensures lossless compression, i.e., the encoded and decoded data are guaranteed to be the same; (ii) it does not concern itself with materialized codes, the compression is a means for model selection rather than compressors for reducing the amount of bits. In PM, a model is a set of patterns. The best pattern set H among sets of patterns $\mathcal{H}$ is one that minimizes the description length $L(H, D) = L(H) + L(D|H)$, where $L(H)$ is the length, in bits, of the model $H \in \mathcal{H}$ and $L(D|H)$ is the description length, in bits, of the data D encoded by the model. This version is called two-part MDL and is considered to be most appropriate for model selection [6].

In this paper we consider transactional (binary) datasets and propose an MDL-based approach for mining a small set of closed itemsets. This approach relies on Krimp [14], where both the pattern candidate set and code length function are revised to allow for a more natural reasoning on the description length from a probabilistic perspective. In contrast to the existing MDL-based approaches to PM [10,11,15], our method does not involve any sampling techniques and iterative adjustments of probability estimates. It relies on "likeliness" of co-occurrence of attributes under the independence model. We give an experimental proof that the proposed method returns a small set of non-redundant patterns that describe a big portion of data.

The paper is organized as follows. Section 2 introduces Krimp, the seminal algorithm for PM. In Sect. 3 we discuss Krimp from a probabilistic perspective and propose its modified version where patterns are formal concepts. Section 4 provides an experimental justification that the proposed approach improves quality of pattern sets. In Sect. 5 we conclude and give directions of future work.

2 Krimp: An Example of Pattern Mining with MDL

In Krimp [14], model H is a two-column *code table* CT, itemsets and their associated prefix codes are arranged in the left- and right-hand columns, respectively. The order of itemsets is important. The initial code table contains only singleton patterns and is called *standard code table.*

A cover function $cover(CT, I_g)$ encodes a set of items I_g of object g and returns a set of mutually disjoint itemsets $\mathcal{S} \subseteq CT$, such that $\cup_{I \in \mathcal{S}} I = I_g$. In Krimp *cover* is implemented under a greedy strategy. Items I_g are initialized as uncovered and denoted by I_u. Starting from the top-itemset I_i of CT the cover function tries to cover uncovered items I_u. If $I_i \subseteq I_u$ then I_i is appended to $\mathcal{S}$ and $I_u = I_u \setminus I_i$. Once I_u is empty, *cover* returns $\mathcal{S}$. The probability distribution for $I \in CT$ is given by

$$P(I) = \frac{usage(I)}{\sum_{J \in CT} usage(J)}, \tag{1}$$

where $usage(I) = |\{g \in G \mid I \in cover(CT, I_g)\}|$. We call the probability estimates given in Formula 1 *usage-based* estimates. The length of $I \in CT$ is computed with the Shannon codes, i.e., $length(I) = -\log P(I)$.

The patterns being added to CT are chosen from a *candidate set*. The candidate set is a subset of frequent (closed) itemsets that are ordered w.r.t. their frequencies, lengths and lexicographically (this order is called *standard*). At each iteration a non-singleton itemset is considered as a candidate to CT. It is added to CT if it allows for a smaller encoding length, otherwise it is removed from the code table and candidate set.

The two-part description length of dataset D encoded with CT is given by

$$L(D, CT) = L(CT|D) + L(D|CT), \tag{2}$$

where $L(CT|D)$ is the length of the code table CT computed on dataset D and $L(D \mid CT)$ is the length of D encoded with CT. These values are given by

$$L(D|CT) = \sum_{g \in D} \sum_{I \in cover(CT, I_g)} length(I) = -\sum_{I \in CT} usage(I) \log P(I) \tag{3}$$

$$L(CT|D) = \sum_{I \in CT} length(I) = -\sum_{I \in CT} \log P(I).$$

$L(D|CT)$ is the negative log-likelihood of data and $L(CT|D)$ is the negative log-likelihood of a model CT. The code table is filled up in a greedy manner. Initially, all items in D are set as uncovered.

Example. An example of Krimp-based PM is given in Fig. 1. The candidate set (CS) is closed itemsets with frequency threshold 0.25 in the standard order. At *Step 1* top-ranked pattern ac is used to cover objects g_1, g_4 and g_5, the usage of single attributes a and c decreases by 3 (to 0 and 1, respectively), ac is accepted for CT, since adding it to CT allows for a shorter description length (Formula 2). At *Step 2* de is also added to CT. At *Step 3* the last candidate bc is examined. It can cover only object g_2, which does not minimize $L(D, CT)$, thus, bc is discarded. The subset of MDL-optimal patterns is $\{ac, de\}$.

3 Krimp in FCA

3.1 Motivation

In [14], it was noticed that *"for datasets with little noise the closed frequent pattern set can be much smaller and faster to mine and process"*, i.e., closed itemsets in Krimp provide better compression than arbitrary ones. However in practice, datasets are usually noisy and the number of formal concepts increases exponentially with the size of the dataset and noise rate. In this study propose to

	Dataset	CS	CT	P(X)	Covering	CT	P(X)	Covering	CT	P(X)	Covering
g_1	*abc*	*ac*	*c*	4/15	$(a)(b)(c)$	*ac*	3/12	**(ac)**(b)	*ac*	3/9	**(ac)**(b)
g_2	*bcde*	*de*	*a*	3/15	$(b)(c)(d)(e)$	*d*, *e*	3/12	$(b)(c)(d)(e)$	*de*	3/9	$(b)(c)$**(de)**
g_3	*de*	*bc*	*d*	3/15	$(d)(e)$	*b*	2/12	$(d)(e)$	*b*	2/9	**(de)**
g_4	*acde*		*e*	3/15	$(a)(c)(d)(e)$	*c*	1/12	**(ac)**$(d)(e)$	*c*	1/9	**(ac)(de)**
g_5	*ac*		*b*	2/15	$(a)(c)$	*a*	0	**(ac)**	*a*, *d*, *e*	0	**(ac)**
			Initial state.			*Step 1.* *ac* is added.			*Step 3.* *bc* is discarded.		

Fig. 1. Some stages of Krimp algorithm. "Covering" tables show the dataset with covering by itemsets from the corresponding code table. For *CT*s we show left-hand columns. Probability $P(X)$ is used to compute code lengths.

use Krimp in FCA settings to benefit better compression by concepts (see basic notions of FCA in [5]). We consider Krimp from a probabilistic perspective to make it more suitable for compression with concepts even for real-world noisy data. We illustrate weaknesses of the original Krimp in FCA settings by means of a small example.

Example. The best pattern set covers a big portion of data by a small number of patterns having a small overlapping area. A dataset composed of two patterns is given in Fig. 2. Top-ranked frequent itemsets with the threshold 0.4 in the standard order are **cd**, **acd**, **cdf**, *ac*, *ad*, *cf*, *df*, **abcd**, *abc*, *abd*, *ab*, *bc*, *bd*, **cdef**, *cde*, etc. The Kripm-based covering with the given list of candidates contains an extra pattern *cd*, see Fig. 2, (b). The set of closed candidates (they are highlighted in bold) is much smaller, but since only disjoint patterns are permitted the major portion of data remains uncovered, see Fig. 2, (c). For closed itemsets with frequency in range [0.2, 0.4] we get "true" patterns, see Fig. 2, (d). However, according to the model, pattern *abcd* is more important (interesting) than *cdef*, since it has higher usage. It is a side effect of the greedy strategy rather than capture of the regularity underlying the data.

The example shows that closed itemsets provide more condensed data representation, however, PM with Krimp is affected by heuristics. The latter hampers revealing true patterns underlying the data. In this section we address the following questions: (i) Can we make Krimp less affected by heuristics? (ii) Can we further reduce the number of closed patterns and what is the most suitable way to do that? We study these questions in Sects. 3.2 and 3.3, respectively.

3.2 Probability Estimates

Replacing frequency by usage in Formula 1 allows for cutting the sum in the denominator and increases the probability of the top-ranked patterns in *CT*. Put differently, usage-based estimates favor the frequent patterns that bring "new information" to *CT*. However, the estimates are affected by the greedy cover strategy and do not always capture true data structure, e.g., an "artificial"

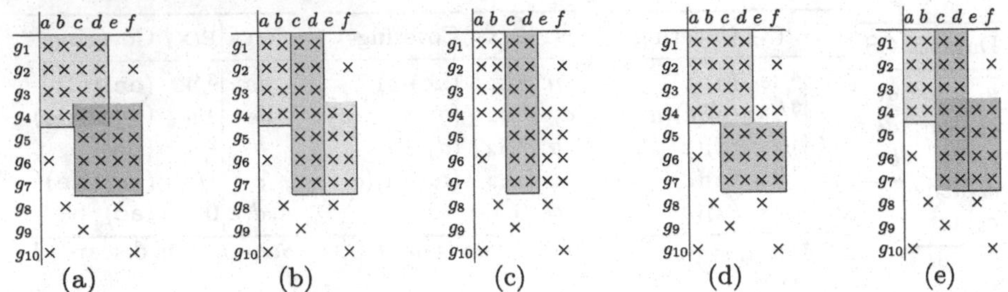

Fig. 2. Formal context and its coverings. The best data covering w.r.t. the number of patterns and covering rate is given in (a). The Krimp-based coverings are shown in (b)–(d): covering by frequent itemsets, $fr \geq 0.4$ (b), by closed frequent itemsets $fr \geq 0.4$ (c), by closed frequent itemsets, $fr \in [0.4, 0.6]$ (d). Subfigure (e) shows the covering by frequent itemsets that overlap.

splitting of an itemset into smaller patterns is given in Fig. 2, (b) and assignment of different lengths to itemsets caused by heuristics and the lexicographical order is given in Fig. 2, (d).

Replacing the usage in Formula 1 by frequency allows us to get rid of the bias introduced by the greedy strategy, but at the same time that increases the total length of the model and makes frequent patterns less probable. Nevertheless, this estimate is more intuitive and resistant to side effects caused by heuristics. We call such estimates *frequency-based* estimates. Using these estimates is equivalent to covering with overlaps. Permitting patterns to overlap has a positive impact since it does not entail an artificial splitting of patterns.

Example. Let us demonstrate the benefits of frequency-based estimates by means of the running example. We use the same approach as before, but now we permit patterns to overlap and replace $usage$ with $frequency$ in Formula 1. Covering with closed itemsets of frequency between $[0.3, 0.6]$ gives us a pattern set shown in Fig. 2 (a) with correct lengths (compare with (d)). Covering by frequent closed itemsets with frequency threshold $fr \geq 0.4$ is given in Fig. 2, (e). Being overlapped with *cd*, "true" itemsets *abcd* and *cdef* are in the code table, but the pattern set is redundant (it contains an extra pattern *cd*). Thus, the proposed estimates are less affected by heuristics and are able to identify "true" patterns.

3.3 New Approach for Computing Candidates to Code Table

In this section we propose a new approach for candidate computing that complies with the independent pattern model (analogously to the independent attribute model).

The cover function $cover(CT, I_g)$ returns a set of patterns $I_{i_1} \cup \ldots \cup I_{i_k} = I_g$. The length of the object g is given by $length(g) = \sum_{j=i_1,\ldots,i_k} length(I_j) = -\sum_{j=i_1,\ldots,i_k} \log P(I_j)$. It means that under the given model object g is observed with probability $2^{-length(g)} = \prod_{j=i_1,\ldots,i_k} P(I_j)$. The probability is computed under the independent pattern model. We call an itemset I *likely-occurring* (LO) if $\exists I_r \subset I$, such that $P(I \setminus I_r)P(I_r) < P(I)$, i.e., itemsets I_r and $I \setminus I_r$ more likely occur together in I than separately. Consistent with this reasoning we introduce a tree-based depth-first algorithm that sequentially grows closed itemsets (concepts). It is closely related to FP-trees [7], however it grows trees in an attribute-wise manner (while in the FP-Growth algorithm transactions are added sequentially to grow a tree). Another important difference is an additional "likely-occurring" constraint on adding/updating nodes instead of a frequency threshold.

The basic idea is the following. The Input of the algorithm is a dataset, the output is a subset of closed likely-occurring itemsets. At the beginning, the attributes are ordered in the frequency-descending order. At each iteration an attribute is added to a tree. An attribute may extend the label of the current node, be added as a new node to the tree, or be ignored. An attribute m extends label I of node n, if each object that shares attributes of the node n has attribute m. Otherwise we check if itemset I and m are likely to appear together, i.e., $P(I)P(\{m\}) < P(I \cup \{m\})$, and we add a node labeled by $I \cup \{m\}$ if it is the case. The computed tree consists of likely-occurring closed itemsets.

Example. Let us consider how a tree is growing using the dataset from Fig. 2. Attributes are ordered in the frequency-descending order and lexicographically, i.e., c, d, a, f, b, e. The iteration when a is added to the tree is given in Fig. 3. We start from the root and add a new child node labeled by $I \cup \{m\}$ to the node labeled by I, if the new node satisfies the LO-condition. We stop traversal towards to leafs if the parent node labeled I and attribute m does not satisfy the LO-condition.

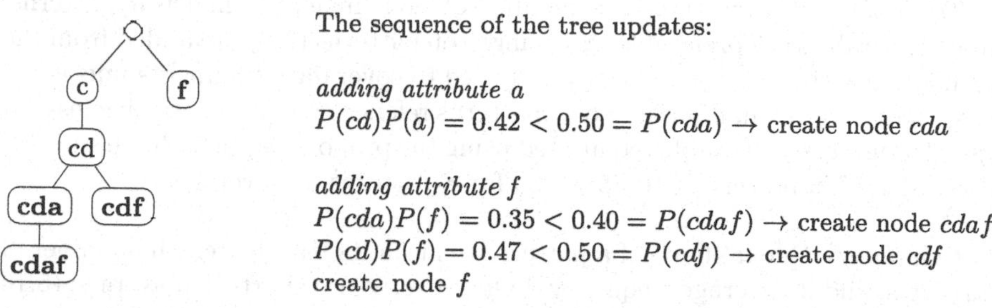

Fig. 3. An intermediate step of tree building. Attributes a and f are added to a tree built on attributes c and d.

4 Experiments

We use the discretized datasets from LUCS-KDD repository [3], their parameters are given in Table 1. We compare pattern sets computed by Krimp, where candidate sets are frequent closed patterns with usage-based estimates, and the

Table 1. The average values of characteristics of pattern sets for the overlapping covering strategy.

Dataset	Size	den.	#FC	#LO	CT size.#NS/#S		Overlapping		Uncovered		Avg NS freq.		Avg pat.len.	
					Krimp	LOF	Krimp	LOF	Krimp	LOF	Krimp	LOF	Krimp	LOF
anneal	898 × 71	0.20	9 611	1 204	83/58	52/38	2.39	2.90	0.12	0.05	0.03	0.14	11.83	6.67
breast	699 × 16	0.64	641	74	24/10	7/9	1.56	2.14	0.03	0.10	0.07	0.48	9.04	5.71
carEv	1728 × 25	0.29	12 638	1 030	94/5	36/4	1.47	1.04	0.01	0.04	0.04	0.08	3.47	2.00
ecoli	327 × 29	0.29	694	77	25/24	16/15	2.88	1.66	0.10	0.09	0.14	0.19	6.08	3.75
heartDis	303 × 50	0.29	36 738	1 683	54/45	48/37	3.27	2.29	0.13	0.11	0.2	0.13	5.09	5.38
hepat	155 × 52	0.36	199 981	7 716	44/48	71/35	2.40	3.43	0.13	0.10	0.20	0.12	5.59	8.03
horseCol	368 × 83	0.21	228 477	30 626	101/78	140/72	2.58	2.41	0.19	0.12	0.11	0.07	3.92	5.46
iris	150 × 19	0.25	162	43	13/9	11/15	1.39	1.10	0.17	0.26	0.11	0.11	3.92	2.82
led7	3200 × 24	0.50	7 037	150	152/16	11/14	2.33	1.48	0.01	0.17	0.03	0.32	6.80	2.55
mush	8124 × 90	0.25	186 331	8 046	211/47	103/55	1.00	3.89	0.14	0.05	0.00	0.10	19.53	10.84
pageBl	5473 × 44	0.26	994	68	45/30	28/15	2.49	1.38	0.00	0.11	0.10	0.06	10.27	6.79
pima	768 × 38	0.22	3 203	202	50/34	32/26	5.53	1.76	0.04	0.13	0.20	0.08	5.86	4.28
ticTacToe	958 × 29	0.33	59 503	1 298	160/21	47/27	2.89	1.43	0.05	0.13	0.07	0.11	4.02	2.55
wine	178 × 68	0.20	14 554	4 757	52/65	115/55	1.91	2.21	0.29	0.12	0.10	0.05	3.90	5.43
avg	**1 666 × 46**	**0.31**	**54 326**	**4 070**	**79/35**	**51/30**	**2.43**	**2.08**	**0.10**	**0.11**	**0.10**	**0.15**	**7.09**	**5.16**

proposed approach, where the candidate set is likely-occurring concepts with frequency-based estimates (LOF). Pattern sets are compared following the overlapping strategy, i.e., we take the patterns and cover objects enabling the patterns to overlap. For pattern sets we study the following characteristics.

The size of a candidate set, i.e., the number of closed ($\#FC$) and LO ($\#LO$) closed itemsets for Krimp and the proposed approach, respectively. The candidate set size for the proposed approach is much smaller. More than that, it does not require any thresholds.

The code table size, i.e., the number of non-singleton/singleton patterns. Small code tables are preferable. Non-singleton patterns (NS) are added from the candidate set, singleton patterns (S) are used to cover data fragments uncovered by NS-patterns. A small number of S-patterns refers to good "descriptiveness" of NS-patterns. The code tables computed using the proposed approach are smaller, 79 vs. 51 for NS-patterns and 35 vs. 30 for S-patterns, on average.

Pattern Shape. Under shape of a pattern we mean its length, i.e., the number of attributes, and its average frequency. LOF patterns are shorter on average (5,16 vs 7,09) and more frequent (0,15 vs 0,10) than Krimp-selected ones.

Overlapping Rate and Rate of Uncovered Cells. Overlapping rate is the average number of NS-patterns that cover an item (cell) in a dataset. Non-redundant pattern sets have overlapping rate close to 1. The rate of uncovered cells, i.e., the rate of cells that are not covered by NS-patterns characterizes how well patterns describe the dataset, the uncovered cell rate close to 0 is preferable. It is important to check these values in a pair, since overlapping rate about 1 does not justify high quality of a pattern set. Krimp- and LOF-generated pattern sets on average have almost the same rate of uncovered cells, however, the overlapping rate of LOF is smaller, which allows %Таня, that используется B restrictive clause, a здесь non-restrictive clause us to conclude that LOF-pattern sets have better quality in terms of "descriptiveness". More than that, LOF pattern sets cover the same portion of data with a smaller number of patterns (see $\#NS$).

5 Conclusion

In this paper we have studied how MDL principle can be applied in FCA settings. We have considered Krimp algorithm, that is an MDL-based approach for Pattern Mining, and we have studied it from the probabilistic point of view. Relying on the probabilistic interpretation of MDL we have proposed an algorithm for generating a subset of closed itemsets (that compose a pattern search space) and new probability estimates for patterns that are used to compute pattern code length. The experiments have shown that the proposed approach allows for a smaller set of patterns that are less redundant than Krimp-based generated ones. The modified estimates of probability permit patterns to overlap and are less affected by the greedy cover strategy used to build a pattern set. The encoding function can be further improved by introducing an error code function for singleton patterns with low usage and more complex constraints on generating candidates.

Acknowledgment. The work of Tatyana Makhalova and Sergei O. Kuznetsov was supported by the Russian Science Foundation under grant 17-11-01294 and performed at National Research University Higher School of Economics, Moscow, Russia.

References

1. Aggarwal, C.C., Han, J.: Frequent Pattern Mining. Springer, Heidelberg (2014). https://doi.org/10.1007/978-3-319-07821-2
2. Brin, S., Motwani, R., Silverstein, C.: Beyond market baskets: generalizing association rules to correlations. ACM SIGMOD Rec. **26**, 265–276 (1997)
3. Coenen, F.: The LUCS-KDD discretised/normalised ARM and CARM data library. Department of Computer Science, The University of Liverpool, UK (2003). http://www.csc.liv.ac.uk/~frans/KDD/Software/LUCS_KDD_DN
4. Gallo, A., De Bie, T., Cristianini, N.: MINI: mining informative non-redundant itemsets. In: Kok, J.N., Koronacki, J., Lopez de Mantaras, R., Matwin, S., Mladenič, D., Skowron, A. (eds.) PKDD 2007. LNCS (LNAI), vol. 4702, pp. 438–445. Springer, Heidelberg (2007). https://doi.org/10.1007/978-3-540-74976-9_44
5. Ganter, B., Wille, R.: Formal Concept Analysis: Logical Foundations. Springer, Berlin (1999). https://doi.org/10.1007/978-3-642-59830-2
6. Grünwald, P.D.: The Minimum Description Length Principle. MIT Press, Cambridge (2007)
7. Han, J., Pei, J., Yin, Y.: Mining frequent patterns without candidate generation. ACM SIGMOD Rec. **29**, 1–12 (2000)
8. Hanhijärvi, S., Ojala, M., Vuokko, N., Puolamäki, K., Tatti, N., Mannila, H.: Tell me something I don't know: randomization strategies for iterative data mining. In: Proceedings of the 15th ACM SIGKDD, pp. 379–388. ACM (2009)

9. Kuznetsov, S.O., Makhalova, T.: On interestingness measures of formal concepts. Inf. Sci. **442–443**, 202–219 (2018)
10. Mampaey, M., Vreeken, J., Tatti, N.: Summarizing data succinctly with the most informative itemsets. TKDD **6**(4), 16 (2012)
11. Siebes, A., Kersten, R.: A structure function for transaction data. In: Proceedings of SDM, pp. 558–569. SIAM (2011)
12. Smets, K., Vreeken, J.: SLIM: directly mining descriptive patterns. In: Proceedings of SDM, pp. 236–247. SIAM (2012)
13. Tatti, N.: Maximum entropy based significance of itemsets. Knowl. Inf. Syst. **17**(1), 57–77 (2008)
14. Vreeken, J., Van Leeuwen, M., Siebes, A.: KRIMP: mining itemsets that compress. Data Min. Knowl. Disc. **23**(1), 169–214 (2011)
15. Wang, C., Parthasarathy, S.: Summarizing itemset patterns using probabilistic models. In: Proceedings of the 12th ACM SIGKDD, pp. 730–735. ACM (2006)

A Study of Boolean Matrix Factorization Under Supervised Settings

Tatiana Makhalova[1,2(✉)] and Martin Trnecka[3]

[1] National Research University Higher School of Economics, Moscow, Russia
tpmakhalova@hse.ru
[2] LORIA, (CNRS – Inria – University of Lorraine), Vandœuvre-lès-Nancy, France
[3] Department of Computer Science, Palacký University Olomouc, Olomouc, Czech Republic
martin.trnecka@gmail.com

Abstract. Boolean matrix factorization is a generally accepted approach used in data analysis to explain data or for data preprocessing in the supervised settings. In this paper we study factors in the supervised settings. We provide an experimental proof that factors are able to explain not only data as a whole but also classes in the data.

Keywords: Boolean matrix factorization · Supervised settings · Classification · Quality of factors

1 Introduction

Boolean matrix factorization (BMF) is a powerful tool that is widely used in data mining to describe data. It allows for data explanation by means of factors, i.e. hidden variables that rely on a solid algebraic foundation.

In general, BMF is used in the unsupervised settings, where the input data are not labeled, classified or categorized. However, evaluation of quality of generating factors did not received appropriate attention in the scientific literature on BMF. An exception is a pioneer work [4] that provides basic ideas of how the quality of BMF algorithms can be assessed in the unsupervised settings. In this paper we evaluate BMF algorithms in the supervised settings. To the best of our knowledge, the quality of factors in this settings has not been studied yet.

It was shown that BMF algorithms used as a preprocessing stage [2,3,18] or as neurons in a simple (one layer) artificial neural network [13] can improve classification quality. Other relevant works come from the Formal Concept Analysis [11] (FCA), since factors are often formal concepts [6]. In [1,12,14] closed sets of attributes, i.e. intents of formal concepts, were studied as basic classifiers (hypothesis) in different voting and inference schemes. In the mentioned studies the whole set of (frequent) factors was used to build classifiers. One may consider factors as a result of the selection of only relevant concepts (hypotheses) w.r.t. to coverage or MDL principle, e.g. in [17] MDL principle is used to select concepts

D. Cristea et al. (Eds.): ICFCA 2019, LNAI 11511, pp. 341–348, 2019.
https://doi.org/10.1007/978-3-030-21462-3_24

that then were evaluated under supervised settings. From the FCA perspective, our study can be considered as evaluation of BMF-optimal concepts (intents or their generators) in the supervised settings. Under BMF-optimal concepts we mean those that are generated by a BMF algorithm.

Our contribution is twofold. First, we evaluate the ability of factors to explain classes of objects rather than the data as a whole. Second, we propose different models of factor-based classifiers and study their quality.

The paper is organized as follows. Section 2 introduces the used notation and the basic notions of BMF. In Sect. 3 we discuss how factors can be used and evaluated in supervised settings. Section 4 provides the results of a comparative study of factor sets generated by different BMF algorithms as well as evaluation of different models of factor-based ensembles of classifiers. In Sect. 5 we conclude and discuss direction of future work.

2 Preliminaries

In this section we recall the main notions used in this paper. Matrices are denoted by upper-case bold letters. $\mathbf{I}_{ij}$ denotes the entry of matrix $\mathbf{I}$ corresponding to the row i and column j. $\mathbf{I}_{i\_}$ and $\mathbf{I}_{\_j}$ denotes the ith row and jth column of matrix $\mathbf{I}$, respectively. The set of all $m \times n$ Boolean matrices is denoted by $\{0,1\}^{m\times n}$. The number of 1s in Boolean matrix $\mathbf{I}$ is denoted by $||\mathbf{I}||$, i.e $||\mathbf{I}|| = \sum_{i,j} \mathbf{I}_{ij}$.

For matrices $\mathbf{A} \in \{0,1\}^{m\times n}$ and $\mathbf{B} \in \{0,1\}^{m\times n}$ we define the following element-wise operations: (i) *Boolean sum* $\mathbf{A} \oplus \mathbf{B}$, i.e. the normal matrix sum where $1+1 = 1$. (ii) *Boolean subtraction* $\mathbf{A} \ominus \mathbf{B}$, i.e. the normal matrix subtraction where $0 - 1 = 0$.

The objective of BMF is the following one: for a given Boolean matrix $\mathbf{I} \in \{0,1\}^{m\times n}$ to find matrices $\mathbf{A} \in \{0,1\}^{m\times k}$ and $\mathbf{B} \in \{0,1\}^{k\times n}$ such that

$$\mathbf{I} \approx \mathbf{A} \circ \mathbf{B}, \tag{1}$$

where $\circ$ is Boolean matrix multiplication, i.e. $(\mathbf{A} \circ \mathbf{B})_{ij} = \max_{l=1}^{k} \min(\mathbf{A}_{il}, \mathbf{B}_{lj})$, and $\approx$ represents an approximate equality assessed by $|| \cdot ||$, see [5] for details. The matrices $\mathbf{I}$, $\mathbf{A}$, and $\mathbf{B}$ describe the object-attribute, object-factor, and factor-attribute relations, correspondingly.

Under this model, the decomposition of $\mathbf{I}$ into $\mathbf{A} \circ \mathbf{B}$ may be interpreted as discovery of k factors that exactly or approximately explain the data, i.e. with $\mathbf{I}_{ij} = 1$ the object i has the attribute j, if and only if there exists factor l such that l applied to i and j is one of the particular manifestations of l.

3 Factors Under Supervised Settings

Quality of factors is most often understood as their ability to explain data [4]. However, a lot of problems is needed to be solved in the supervised settings, where class labels of objects are available.

In supervised settings, Boolean matrix $\mathbf{I} \in \{0,1\}^{m \times n}$ corresponds to m objects described by n attributes. A special target attribute refers to an object class. More formally, we define a function *class* that maps row $\mathbf{I}_{i\_}$ to its class label $c = class(\mathbf{I}_{i\_}) \in \mathcal{Y}$, the size of set $\mathcal{Y}$ is equal to the number of classes.

3.1 Key Components of Classifiers

Representation and Labeling. For the Boolean matrix factorization $\mathbf{I} = \mathbf{A} \circ \mathbf{B}$ we consider *factor-classifier* as a tuple (f_i, c, sim), where f_i is the ith Boolean factor (represented by the ith column and ith row of matrices $\mathbf{A}$ and $\mathbf{B}$, respectively), c is a class label given by *class* function, and *sim* is a classification strategy (see details below). In our study we assign to c a class label of the majority of objects from column $\mathbf{A}_{\_i}$. If the majority is not unique, we do not consider the factor as a classifier.

Strategy of Classification. We focus on two common classification strategies, namely *rule-based* and *similarity-based.*

According to the first strategy, object $g = \mathbf{I}_{j\_}$ (given by n-dimensional vector) is classified by factor-classifier (f_i, c, sim) if $\mathbf{B}_{i\_} \cdot g = \mathbf{B}_{i\_}$, i.e. the object g has all attributes of factor f_i, "$\cdot$" denotes the element-wise multiplication.

With the second strategy, the object g is classified by factor-classifier (f_i, c, sim) if $similarity(\mathbf{B}_{i\_}, g) > \varepsilon$, i.e. the attributes of factor f_i are quite similar to the attributes of object g. The similarity can be defined by means of either a distance measure or an asymmetrical operator.

It should be noted that the rule-based classification strategy is a particular case of the similarity-based one, where for $g = \mathbf{I}_{j\_}$ $similarity(\mathbf{B}_{i\_}, \mathbf{I}_{j\_}) \equiv \sum_{l=1}^{n}(\mathbf{B}_{il} \rightarrow \mathbf{I}_{jl}) \equiv \sum_{l=1}^{n}(\overline{\mathbf{B}_{il}} \,|\, \mathbf{I}_{jl}) = n$. Operations $\rightarrow$ and $|$ represent logical implication and logical OR, respectively.

For the sake of simplicity, we will use (f_i, c) to denote a classifier, because in our experiments we use only the *similarity* function.

Responses of Classifiers. We say that object g is classified by (f_i, c, sim) if $sim(\mathbf{B}_{i\_}, g) > \varepsilon$. To assign a class label to g, the responses of classifiers (f_i, c, sim) can be accounted with weights $w^g_{(f_i,c,sim)}$, e.g. precision, accuracy of f_i, or similarity between $\mathbf{B}_{i\_}$ and g. We assume that f_i does not contribute to the final decision on a class of g (the response is 0) if g is not classified by (f_i, c, sim). Again, for the sake of simplicity, we will use $w^g_{(f_i,c)}$ instead of $w^g_{(f_i,c,sim)}$.

To compute a class label of an object, the responses of classifiers (weights) are aggregated. We discuss aggregation strateges in Sect. 4.2.

4 Experimental Evaluation

To evaluate factors under supervised settings, we use 11 different real-world datasets from UCI repository [8] binarized with tools from [7]. The characteristics of the datasets are shown in Table 1. In our experiments we use 10-fold cross-validation.

Table 1. Datasets and their characteristics.

Dataset	Size	Density **I**	Class distribution
anneal	898×66	0.20	0.76/0.04/0.11/0.07/0.01
breast	699×14	0.64	0.34/0.66
hepatitis	155×50	0.36	0.79/0.21
horse colic	368×81	0.21	0.63/0.37
iris	150×16	0.25	0.33/0.33/0.33
led7	3200×14	0.50	0.11/0.09/0.10 (×8 classes)
mushroom	8124×88	0.25	0.52/0.48
nursery	1000×27	0.30	0.32/0.34/0.34
page block	5473×39	0.26	0.90/0.02/0.01/0.05/0.02
pima	768×36	0.22	0.650/0.35
wine	178×65	0.20	0.33/0.40/0.27

We compare most common BMF algorithms, namely 8M [9], GreConD [6], GreEss [5], Hyper [19], MDLGreConD [16], NaiveCol [10] and PaNDa$^+$ [15].

4.1 Factor as Classification Rule

In this section we examine factors as single classifiers. We study (i) the connection between factor ranks given by unsupervised and supervised quality measures, and which factors are best ones w.r.t. the supervised quality measures, (ii) how well the factors summarize classes.

Connection Between Supervised and Unsupervised Quality Measures. The mentioned BMF algorithms are based on a greedy strategy. The generated factors are ordered w.r.t. their importance. The importance of factors is estimated by a particular objective of an algorithm. Put it differently, the factors generated first might best explain data. Since some factor sets are very small, we cannot use correlation analysis to examine the dependence between the importance of factors (unsupervised quality measure) and their precision (supervised quality measure). To assess the connection between these measures we count how many factors we need to compute to get the best k factors w.r.t. precision. The less the number of factors we need to compute, the stronger connection between unsupervised and supervised quality measures.

The average number of factors is given on Fig. 1. We note that the PaNDa$^+$ factor sets are small, but it does not mean that most important factors provide best precision. These small values are caused by the small sizes of the PaNDa$^+$-generated factor sets. The extremely small size of factor sets produced by PaNDa$^+$ affects also the factor quality in the unsupervised settings [4,5].

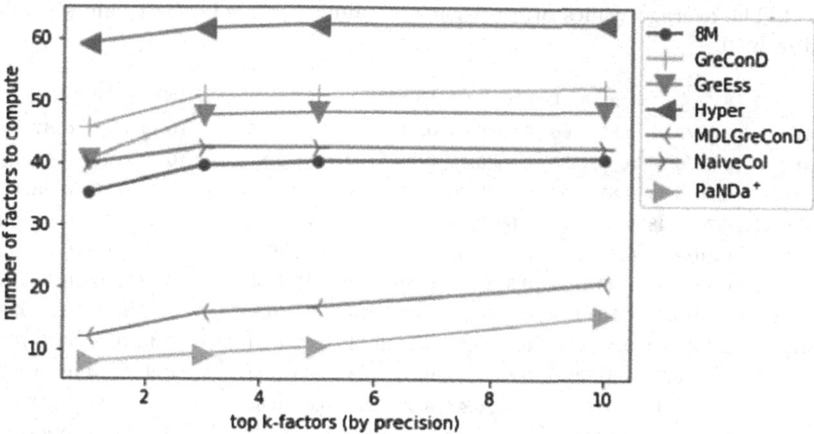

Fig. 1. The no. of factors required to be computed in order to get k best factors w.r.t. precision.

Figure 1 shows that the lowest values correspond to the MDLGRECOND factors. It means that we need to compute only few factors to get the most precise classifiers. The most important factors w.r.t. the MDLGRECOND objective have relatively higher precision than the most important factors generated by other BMF algorithms.

In the next section we discuss the ability of factors to explain classes rather than data as a whole, i.e. their ability to distinguish a single class from others.

Summary of Classes. For every factor-classifier (f_i, c) that corresponds to row $\mathbf{A}_{\_i}$ and column $\mathbf{B}_{i\_}$ we compute precision, recall and accuracy as follows:

$$prec(f_i, c) = \frac{tp}{tp + fp}, \; recall(f_i, c) = \frac{tp}{tp + fn}, \; accuracy(f_i, c) = \frac{tp + tn}{m},$$

where $tp = |\{\mathbf{A}_{ji} \mid \mathbf{A}_{ji} = 1, class(\mathbf{I}_{j\_}) = c, j = 1, \ldots, m\}|$ is the true positive rate, $fp = |\{\mathbf{A}_{ji} \mid \mathbf{A}_{ji} = 1, class(\mathbf{I}_{j\_}) \neq c, j = 1, \ldots, m\}|$ is the false positive rate, $fn = |\{\mathbf{A}_{ji} \mid \mathbf{A}_{ji} = 0, class(\mathbf{I}_{j\_}) = c, j = 1, \ldots, m\}|$ is the false negative rate.

Precision and accuracy characterize how well f_i describes class c. The factors with high values of these measures summarize better the given class c. The only difference between accuracy and precision is the following one: precision is the "local" class specificity (it shows how well objects from c are distinguished among the classified objects), while accuracy is the "global" class specificity (it shows how well objects from c are distinguished among all objects). Precision and accuracy give preference to classifiers with low values of fp and $fp + fn$, respectively.

The results of the experiments given in Table 2 show that the highest average precision is achieved for factors computed by PANDA$^+$ (0.78, on average), the MDLGRECOND factors also have quite high values of precision (0.74, on average). The MDLGRECOND factors have the most stable quality measures (precision on test sets is smaller by 0.07 than on training sets).

Table 2. The average values of precision on training/test sets. Best values are highlighted in bold.

	8M	GreConD	GreEss	Hyper	MDLGreConD	NaiveCol	PaNDa$^+$
anneal	0.86/0.67	0.85/0.66	0.86/0.64	0.84/0.62	0.85/0.75	0.84/0.63	**0.87**/**0.87**
breast	0.88/0.73	0.88/0.85	0.84/0.84	**0.93**/0.64	0.87/0.87	0.80/0.80	0.85/0.81
hepatitis	0.80/0.64	0.81/0.61	0.81/0.60	0.81/0.68	**0.83**/0.75	0.79/0.59	**0.83**/0.55
horse colic	0.70/0.48	0.69/0.60	0.69/0.60	0.72/0.61	0.70/0.63	0.69/0.59	**0.80**/0.56
iris	0.80/0.75	0.80/0.61	0.80/0.61	0.79/0.67	0.92/0.86	0.79/0.67	**0.96**/0.53
led7	0.40/0.44	0.33/0.32	0.33/0.32	**0.50**/0.19	0.37/0.36	0.23/0.22	0.43/0.42
mushroom	0.82/0.76	0.82/0.79	0.83/0.79	0.85/0.70	**0.87**/0.84	0.78/0.75	0.81/0.00
nursery	0.45/0.44	0.45/0.44	0.45/0.44	0.45/0.44	0.42/0.41	0.45/0.44	**0.58**/0.53
page blocks	0.82/0.35	0.82/0.46	0.84/0.43	0.78/0.33	0.80/0.51	**0.83**/0.51	0.80/0.74
pima	0.70/0.43	0.68/0.49	0.68/0.48	0.69/0.44	0.68/0.61	0.67/0.45	**0.77**/0.73
wine	0.66/0.40	0.69/0.57	0.68/0.56	0.67/0.49	0.84/0.77	0.64/0.50	**0.88**/0.66
Average	0.72/0.53	0.71/0.58	0.71/0.57	0.73/0.53	0.74/**0.67**	0.68/0.56	**0.78**/0.65

More than that, Table 2 provides precision of factor-classifiers on training and test data. Precision on training data for all algorithms is quite similar (the best algorithm is PaNDa), while MDLGreConD demonstrates the best precision on test sets. It should be noticed that MDLGreConD has the smallest difference in precision for training and test data. That might indicate its ability to generalize well (i.e. it is less likely to overfit). Almost the same quality of factors, but in the unsupervised settings, was described in [4].

4.2 Factors as Ensemble of Classifiers

The modern state-of-the-art classifiers, e.g. Random Forests, Multilayer Networks, Nearest Neighbour classifiers, are comprised of a set single classifiers, i.e. the single classifiers make ensembles. In this section we examine a set of factor-classifiers as an ensemble and evaluate its accuracy.

It should be noticed that some factor sets are incomplete, in other words, they do not contain factors for several classes. It is caused by unbalanced training sets, where some classes contain only few objects. Here we examine the datasets where there are enough factors for every class, namely `iris`, `mushroom`, `pima` and `wine` datasets. We study rule-based ensembles.

As it was mentioned in Sect. 3.1 the responses of classifiers can be taken into account in several ways. We focus on two strategies, where the responses of all voted classifiers or the best one are considered, we call them "all-votes" and "best-vote", respectively. For a rule-based ensemble of factor-classifiers $\mathcal{C} = \{(f_i, c) \mid j = 1, \ldots, k\}$, where k is the number of factors, the class label is assigned to object g as follows:

$$\textit{all-votes-class}(g, \mathcal{C}) = \arg\max_{c \in \mathcal{Y}} \sum_{\substack{(f_i,c) \in \mathcal{C} \\ \mathbf{B}_{i\_} \cdot g = \mathbf{B}_{i\_}}} w^g_{(f_i,c)},$$

$$best\text{-}vote\text{-}class(g, \mathcal{C}) = \arg\max_{c \in \mathcal{Y}} \max_{\substack{(f_i, c) \in \mathcal{C} \\ \mathbf{B}_{i\_} \cdot g = \mathbf{B}_{i\_}}} w^g_{(f_i, c)}.$$

Table 3. The average accuracy of classifier ensembles computed on `iris`, `mushroom`, `pima` and `wine` datasets. Best values are highlighted in bold.

	All-votes		Best-vote	
	Precision	Accuracy	Precision	Accuracy
`iris`	0.84/0.82	0.84/0.82	0.84/0.82	0.84/0.82
`mushroom`	0.93/0.93	0.89/0.89	0.99/0.99	0.88/0.88
`pima`	0.66/0.66	0.67/0.66	0.72/0.70	0.73/0.73
`wine`	0.77/0.75	0.76/0.75	0.79/0.75	0.76/0.75
Average	**0.80/0.79**	0.79/0.78	**0.83/0.81**	0.80/0.79

The results of the experiments, given in Table 3, show that the most accurate ensembles are those that are based on the precision-weighed votes. According to the examined datasets, the "best-vote" scheme (where the response of the best classifier is considered) provides best results.

5 Conclusion

In this paper we examine the factors computed on unlabeled data in the supervised settings. We provided an experimental justification that in case of factors the data explanation problem is closely related to the class explanation problem, i.e. a factor is able to explain specificity of a particular (sub)class. Based on the results of the supervised factor evaluation we propose several models of factor-based ensembles of classifiers. We show that factor-based classifiers can achieve accuracy comparable to the state-of-the-art ensembles of classifiers.

An important direction of further work is to study factors computed in supervised settings for each class separately rather than for the whole dataset. Incorporating precision or accuracy to a BMF objective might improve accuracy of the model as well as provide a deeper insight on a class structure.

Acknowledgment. The work of Tatiana Makhalova was supported by the Russian Science Foundation under grant 17- 11-01294 and performed at National Research University Higher School of Economics, Moscow, Russia.

References

1. Belohlavek, R., Baets, B.D., Outrata, J., Vychodil, V.: Inducing decision trees via concept lattices. Int. J. Gen. Syst. **38**(4), 455–467 (2009)
2. Belohlavek, R., Grissa, D., Guillaume, S., Nguifo, E.M., Outrata, J.: Boolean factors as a means of clustering of interestingness measures of association rules. Ann. Math. Artif. Intell. **70**(1–2), 151–184 (2014)

3. Belohlavek, R., Outrata, J., Trnecka, M.: Impact of Boolean factorization as preprocessing methods for classification of boolean data. Ann. Math. Artif. Intell. **72**(1–2), 3–22 (2014)
4. Belohlavek, R., Outrata, J., Trnecka, M.: Toward quality assessment of Boolean matrix factorizations. Inf. Sci. **459**, 71–85 (2018)
5. Belohlavek, R., Trnecka, M.: From-below approximations in Boolean matrix factorization: geometry and new algorithm. J. Comput. Syst. Sci. **81**(8), 1678–1697 (2015)
6. Belohlavek, R., Vychodil, V.: Discovery of optimal factors in binary data via a novel method of matrix decomposition. J. Comput. Syst. Sci. **76**(1), 3–20 (2010)
7. Coenen, F.: The LUCS-KDD discretised/normalised ARM and CARM data library (2003). http://www.csc.liv.ac.uk/~frans/KDD/Software/LUCS_KDD_DN
8. Dheeru, D., Karra Taniskidou, E.: UCI machine learning repository (2017). http://archive.ics.uci.edu/ml
9. Dixon, W.: BMDP statistical software manual to accompany the 7.0 software release, vols. 1–3 (1992)
10. Ene, A., Horne, W.G., Milosavljevic, N., Rao, P., Schreiber, R., Tarjan, R.E.: Fast exact and heuristic methods for role minimization problems. In: Ray, I., Li, N. (eds.) 13th ACM Symposium on Access Control Models and Technologies, SACMAT 2008, Estes Park, CO, USA, 11–13 June 2008, Proceedings, pp. 1–10. ACM (2008)
11. Ganter, B., Wille, R.: Formal Concept Analysis Mathematical Foundations. Springer, Heidelberg (1999). https://doi.org/10.1007/978-3-642-59830-2
12. Ganter, B., Kuznetsov, S.O.: Hypotheses and Version Spaces. In: Ganter, B., de Moor, A., Lex, W. (eds.) ICCS-ConceptStruct 2003. LNCS (LNAI), vol. 2746, pp. 83–95. Springer, Heidelberg (2003). https://doi.org/10.1007/978-3-540-45091-7_6
13. Kueti, L.T., Tsopzé, N., Mbiethieu, C., Nguifo, E.M., Fotso, L.P.: Using Boolean factors for the construction of an artificial neural networks. Int. J. Gen. Syst. **47**(8), 849–868 (2018)
14. Kuznetsov, S.O.: Machine learning and formal concept analysis. In: Eklund, P. (ed.) ICFCA 2004. LNCS (LNAI), vol. 2961, pp. 287–312. Springer, Heidelberg (2004). https://doi.org/10.1007/978-3-540-24651-0_25
15. Lucchese, C., Orlando, S., Perego, R.: A unifying framework for mining approximate top-k binary patterns. IEEE Trans. Knowl. Data Eng. **26**(12), 2900–2913 (2014)
16. Makhalova, T., Trnecka, M.: From-below boolean matrix factorization algorithm based on MDL. arXiv preprint arXiv:1901.09567 (2019)
17. Makhalova, T.P., Kuznetsov, S.O., Napoli, A.: A first study on what MDL can do for FCA. In: Ignatov, D.I., Nourine, L. (eds.) Proceedings of the Fourteenth International Conference on Concept Lattices and Their Applications. CEUR Workshop Proceedings, vol. 2123, pp. 25–36 (2018)
18. Outrata, J.: Preprocessing input data for machine learning by FCA. In: Kryszkiewicz, M., Obiedkov, S.A. (eds.) Proceedings of the 7th International Conference on Concept Lattices and Their Applications, Sevilla, Spain, 19–21 October 2010. CEUR Workshop Proceedings, vol. 672, pp. 187–198. CEUR-WS.org (2010)
19. Xiang, Y., Jin, R., Fuhry, D., Dragan, F.F.: Summarizing transactional databases with overlapped hyperrectangles. Data Min. Knowl. Discov. **23**(2), 215–251 (2011)

Author Index

Printed in the United States
By Bookmasters